Techniques for Reducing
Pesticide Use

Techniques for Reducing Pesticide Use

Economic and Environmental Benefits

Edited by
David Pimentel
Cornell University, USA

JOHN WILEY & SONS

Chichester · New York · Weinheim · Brisbane · Singapore · Toronto

Copyright © 1997 by John Wiley & Sons Ltd,
Baffins Lane, Chichester,
West Sussex PO19 1UD, England

National   01243 779777
International (+44) 1243 779777

e-mail (for orders and customer service enquiries): cs-books@wiley.co.uk
Visit our Home Page on http://www.wiley.co.uk
or http://www.wiley.com

Other Wiley Editorial Offices

John Wiley & Sons, Inc., 605 Third Avenue,
New York, NY 10158-0012, USA

VCH Verlagsgesellschaft mbH, Pappelallee 3,
D-69469 Weinheim, Germany

Jacaranda Wiley Ltd, 33 Park Road, Milton,
Queensland 4064, Australia

John Wiley & Sons (Asia) Pte Ltd, 2 Clementi Loop #02-01,
Jin Xing Distripark, Singapore 129809

John Wiley & Sons (Canada) Ltd, 22 Worcester Road,
Rexdale, Ontario M9W 1L1, Canada

Library of Congress Cataloging-in-Publication Data

Techniques for reducing pesticide use : economic and environmental
    benefits / edited by David Pimentel.
        p. cm.
    Includes bibliographical references and index.
    ISBN 0-471-96838-2
    1. Agricultural pests—Control. 2. Agricultural pests—Control—Economic aspects.
    3. Agricultural pests—Control—Environmental aspects. 4. Pesticides—Economic aspects.
    5. Pesticides—Environmental aspects. I. Pimentel, David, 1925– .
    SB950.T43 1997
    632′.95042—dc20
                                                                96-27169
                                                                CIP

British Library Cataloguing in Publication Data

A catalogue record for this is available from the British Library

ISBN 0-471-96838-2

Typeset in 10/12pt Times by Vision Typesetting, Manchester
Printed and bound in Great Britain by Bookcraft (Bath) Ltd
This book is printed on acid-free paper responsibly manufactured from sustainable forestation,
for which at least two trees are planted for each one used for paper production.

Contents

List of Contributors

M.A. Arshad Northern Agriculture Research Centre, Agriculture and Agri-Food Canada, Beaverlodge, Alberta T0H 0C0, Canada, 403-354-2212 (TEL), 403-354-8171 (FAX)

K.L. Bailey Agriculture and Agri-Food Canada, Research Centre, 107 Science Place, Saskatoon, Saskatchewan S7N 0X2, Canada, 306-956-7260 (TEL), 306-956-7247 (FAX), baileyk@em.agr.ca

Torgeir Edland Department of Entomology and Nematology, The Norwegian Crop Research Institute, Plant Protection Centre, N-1432 Ås, Norway, 47-64-94-9400 (TEL), 47-64-94-9226 (FAX), torgeir.edland@planteforsk.nlh.no

Jason Friedman College of Agriculture and Life Sciences, Cornell University, Comstock Hall, Ithaca, New York 14853-0901, USA

Cesare Gessler Institute of Plant Sciences/Phytopathology, Swiss Federal Institute of Technology, Universittstrasse 2, 8092 Zurich, Switzerland, gessler@ipw.agrl.ethz.ch

K.S. Gill Northern Agriculture Research Centre, Agriculture and Agri-Food Canada, Beaverlodge, Alberta T0H 0C0, Canada

Jennifer A. Grant Cornell University, New York State Integrated Pest Management Program, New York State Agricultural Experiment Station, Geneva, New York 14456, USA, 315-787-2342 (TEL), 315-787-2326 (FAX), jag7@cornell.edu

Anthony Greiner College of Agriculture and Life Sciences, Cornell University, Comstock Hall, Ithaca, New York 14853-0901, USA

Heikki M.T. Hokkanen Department of Applied Zoology, P.O. Box 27 (Viikki, C), FIN-00014 University of Helsinki, Finland, 358-0-70-85-371 (TEL), 358-0-70-85-463 (FAX), heikki.hokkanen@helsinki.fi

David J. Horn   Department of Entomology, The Ohio State University, 103 Botany and Zoology Building, 1735 Neil Avenue, Columbus, Ohio 43210-1220, USA

David Kahn   College of Agriculture and Life Sciences, Cornell University, Comstock Hall, Ithaca, New York 14853-0901, USA

Hugh Lehman Department of Philosophy, University of Guelph, Guelph, Ontario N1G 2W1, Canada, 519-824-4120 (TEL), 519-837-8634 (FAX)

David Marshall   Department of Plant Pathology, Texas Agricultural Experiment Station, Texas A & M University REC, 17360 Coit Road, Dallas, Texas 75252-6599, USA, 214-952-9252 (TEL), 214-231-9010 (FAX), dmarshal@dallas-ctr.tamu.edu

J.R. Moyer   Lethbridge Research Centre, Agriculture and Agri-Food Canada, Lethbridge, Alberta T1J 4B1, Canada

Maurizio G. Paoletti   Universita Degli Studi di Padova, Dipartimento di Biologia, Via Trieste 75, 35121 Padova, Italy, 39-49-827-6304 (TEL), 39-49-827-6300 (FAX), paoletti@civ.bio.unipd.it

Jan E. Parlevliet   Durable Resistance Program, Department of Plant Breeding, Agricultural University Wageningen, P.O. Box 386, 6700 AJ Wageningen, The Netherlands, 31-317-483597 (TEL), 31-317-483457 (FAX), jan.parlevliet@users.pv.wau.nl

Barbara R. Patterson   Washington State Department of Ecology, Olympia, Washington 98506, USA

John H. Perkins   Member of the Faculty in Environmental Studies, The Evergreen State College, Olympia, Washington 98505, USA, 360-866-6000 ext. 6503 (TEL), perkinsj@elwha.evergreen.edu

Olle Pettersson   The Swedish University of Agricultural Sciences, Research Information Centre, Box 7014, S-75007 Uppsala, Sweden, 46-1867-1000 (TEL), 46-1867-2795 (FAX), olle.pettersson@info.slu.se

David Pimentel   College of Agriculture and Life Sciences, Cornell University, Comstock Hall, Ithaca, New York 14853-0901, USA, 607-255-2212 (TEL), 607-255-0939 (FAX), dp18@cornell.edu

Helmut F. van Emden   Department of Horticulture and Landscape, School of Plant Sciences, The University of Reading, Reading, Berkshire RG6 2AT, UK, 44-1734-318071 (TEL), 44-1734-750630 (FAX), h.f.vanemden@reading.ac.uk

Frank G. Zalom   Director, Statewide Integrated Pest Management Project, Department of Entomology, University of California at Davis, Davis, California 95616-8621, USA, 916-752-8350 (TEL), 916-752-6004 (FAX), frank.zalom@wserver.ipm.ucdavis.edu

Preface

More than 2 billion humans are malnourished according to reports of the United Nations. Indeed, this is the largest number of hungry humans that has been recorded in history. Concurrently, the world population continues to expand, and – based on projected rates of increase – is expected to double from about 6 billion to more than 12 billion in less than 50 years. Augmented food production has, therefore, become the priority of world agriculturists as they endeavor to provide an adequate food supply for the world population. Important efforts are needed to conserve and protect the basic resources of the environment that sustain all agriculture. These include land, water, energy and biological resources.

Unfortunately, approximately 40% of world food production is lost because crops are destroyed by insects, diseases, and weeds, even though 2.5 million tons of pesticides are used on crops each year. Once the remaining food is harvested, other pests destroy an additional 20% of the food. This means approximately half of all food produced is being lost to pests worldwide, despite human efforts to protect their food. As the world population continues to increase rapidly, providing an adequate food supply will be difficult, if not impossible, unless improved pest control technologies are implemented.

Pesticides adversely affect public health and inflict considerable damage on the environment. Consider there are about 3 million human pesticide poisonings worldwide, with an estimated 220,000 deaths each year. Widespread heavy use of pesticides is responsible for bird and fish deaths, destruction of many beneficial natural enemies, loss of vital plant pollinators, the development of pesticide resistance, and other diverse environmental problems. In the US, the yearly environmental costs of pesticide use are estimated to be more than $8 billion; worldwide damage reaches $100 billion annually.

The contributors to this book recognize the value of pesticide use for pest control. However, they suggest techniques that can be employed to reduce pesticide use while maintaining crop yields and not lowering the 'cosmetic standards' of our fruits and vegetables. Although not every pest insect, plant pathogen, and weed and/or crop grown is assessed in this volume, a sufficient number is analyzed to confirm that it is possible to reduce pesticide use by more than 50% in North America and Europe while improving pest control economics, public health, and the environment. In fact, successful programs using various techniques that reduce pesticide use are already being used in Sweden, Denmark, the Netherlands, Indonesia, and the Province of Ontario in Canada.

We expect that the information presented in this volume will assist pest control specialists, policy makers, and all others interested in protecting the environment and public health. Many techniques already exist that could help reduce pesticide use, while at the same time enhancing food supplies. As entomologists, plant pathologists, and weed specialists, we are confident that world crop losses can, and indeed must, be reduced below the current level. Now, and certainly in future decades, population pressure makes it imperative that we improve not only pest control but all aspects of agricultural production.

Special thanks to Tad Bashore and M.H. Pimentel for their help with administrative and technical aspects of the book. At the same time, I am indebted to John Moseley of John Wiley for his interest in this project and for assisting us with the many details involved in publishing this volume.

David Pimentel

CHAPTER 1

Pest Management in Agriculture

David Pimentel

Cornell University, Ithaca, NY, USA

PEST MANAGEMENT IN AGRICULTURE

Throughout time a significant portion of the world's agricultural production of food and fibre crops has been damaged by pests and thereby made unavailable for human use or consumption. This loss continues at present, while the need for these crops escalates in response to the rapid growth in the human population.

HISTORY OF PEST CONTROL

Prior to 1945, farmers were able to control some pests by using such cultural methods as crop rotations, tillage and field sanitation. Only a few chemically based products such as lead arsenate, nicotine and pyrethrums were available for use on crops.

In 1945 the development and production of DDT – soon followed by BHC, dieldrin and related chlorinated pesticides – began the era of chemical pest control. Initially, DDT and the other chemicals fulfilled their promise. They were simple to apply, fast acting and killed most pests. Enthusiasm for these new chemical weapons was great, and their use spread rapidly throughout the US and the rest of the world.

However, problems soon became apparent both in the effectiveness of pest control and in unexplained changes occurring in some bird and fish populations. Within two years after the first use of DDT, resistance to the chemical was observed in houseflies and other insect pests (Pimentel et al., 1951). Over time, this resistance has meant that more insecticides had to be used to ensure control. In addition, not all pests were susceptible to the chlorinated insecticides.

The natural enemies of some pests were also killed by these pesticides, allowing many non-pest species to explode in numbers and become pests themselves. In apple orchards, for instance, pest mites increased in numbers because their natural insect predators were destroyed by DDT. As a result, apple trees turned brown from the heavy mite feeding and yields declined.

In addition to on-site pest problems caused by insecticides, their impacts extended beyond croplands and into the environment. The large kills of fish and birds which were observed in nature were found to be caused by insecticides and other pesticides. Concern

Techniques for Reducing Pesticide Use. Edited by D. Pimentel
© 1997 John Wiley & Sons Ltd.

heightened when milk and other foods were found to be heavily contaminated with pesticides.

Finally, in 1972, the use of DDT and related chlorinated insecticides were banned in the United States. Production and use of pesticides has continued, but many of the newer types are extremely potent based on current dosages per hectare. Thus, while smaller amounts are applied per hectare, their toxicity is much greater than that of the early pesticides. Their benefit is that they do not persist in the environment like DDT and some of the other early pesticides.

CROP LOSSES TO PESTS

Worldwide an estimated 67 000 different pest species damage agricultural crops. Included in these are approximately 9000 species of insects and mites, 50 000 species of plant pathogens (USDA, 1960) and 8000 weeds (Ross and Lembi, 1985). In general, less than 5% of these are considered major pests. In most instances, the pests specific to a particular region have moved from feeding on native vegetation to feeding on crops which were introduced into the region (Pimentel, 1988; Hokkanen and Pimentel, 1989).

Despite the yearly investment of $26 billion for the application of 2.5 million metric tons of pesticides (Table 1.1), in addition to the use of various biological and other non-chemical controls, between 35 and 42% of the world food and fibre production is destroyed by pests (Pimentel, 1991; Oerke et al., 1994). Worldwide, insect pests cause an estimated 13 to 16% loss, plant pathogens cause a 12–13% loss, and weeds a 10–13% loss. The value of this crop loss is estimated to be $244 billion per year, yet there is still a $3 to $4 return per dollar invested in control.

In the United States, yearly crop losses caused by pests are estimated to reach 37% (13% to insects, 12% to plant pathogens and 12% to weeds; Pimentel et al., 1991). In total, pests are destroying an estimated $50 billion per year in food and fibre crops despite all efforts to control them with pesticides and various non-chemical controls. Currently, the US invests about $5 billion in pesticidal controls, which saves about $20 billion per year in crops, providing a $4 return per dollar invested (Pimentel et al., 1991). Non-chemical pest controls also save crops valued at an estimated $20 billion per year (D. Pimentel, unpublished data, 1995).

Without pesticides and non-chemical controls, the damage inflected by pests would be

Table 1.1   Estimated annual pesticide use

Country/region	Pesticide use (10^6 metric tons)
United States	0.5
Canada	0.1
Europe	0.8
Other, developed	0.5
Asia, developing	0.3
Latin America	0.2
Africa	0.1
Total	2.5

Source: Data from Pimentel, 1995b.

more severe than it is at present. Oerke et al. (1994) estimated that world crop losses would then increase to 70%. Such an increase would cause an estimated economic loss of $400 billion each year and would obviously have a negative impact on the world's food supply. Similarly, estimates are that US crop losses would increase to about 63% and represent an economic loss of $90 billion.

Although pesticide use has increased over the past five decades, US crop losses have not shown a concurrent decline, mainly because various changes have occurred in agricultural practices. According to survey data collected from 1942 to the present, losses from weeds fluctuated, but declined slightly from 13.8% to 12% (Pimentel et al., 1991). A combination of improved chemical, mechanical and cultural weed control practices were responsible for this decline. Over that same period, losses from plant pathogens, including nematodes, have increased slightly from 10.5% to about 12%. This happened, in part, because crop rotations were abandoned, field sanitation was reduced and more stringent cosmetic standards for many crops were implemented by the government, wholesalers and retailers.

Unfortunately, the share of crops lost to insects has nearly doubled during the last 40 years (Pimentel et al., 1993a) despite a more than 10-fold increase in both the amount and the toxicity of synthetic insecticides used (Arrington, 1956; USBC, 1971; 1994). This increase in crop losses is associated with several major changes taking place in US agricultural practices. These include: the planting of some crop varieties that are more susceptible to insect pests than those used previously; the destruction of natural enemies of certain pests by insecticides, thereby creating the need for additional pesticide treatments (van den Bosch and Messenger, 1973); resistance to pesticides developing in pest populations (Roush and McKenzie, 1987; Georghiou, 1990); reduction in crop rotations which caused further increases in pest populations; increases in monoculture and reduced crop diversity (Pimentel, 1961, 1977; Cromartie, 1991); lowering of Food and Drug Administration (FDA) tolerance for insects and insect parts in foods, and the enforcement of more stringent 'cosmetic standards' by fruit and vegetable processors and retailers (Pimentel et al., 1993b); increased use of aircraft application technology; reduction in field sanitation, including less attention to the destruction of pest-infected fruit and crop residues (Pimentel, 1986); reduced tillage, with more crop residues left on the land surface; culturing of crops in climatic regions when they become susceptible to insect attack; use of herbicides that alter the physiology of some crop plants and increase their vulnerability to insect attack (Oka and Pimentel, 1976; Pimentel, 1995b).

Added to the damage pests inflict during the growing season are the substantial losses that occur during the often lengthy time many crops are stored prior to their use. Worldwide, an estimated average of 20% of food crops (ranging from 10 to 50%) is destroyed by pests during the post-harvest period (Oerke et al., 1994). In the US post-harvest food losses to pests are estimated to be less than 10% (Pimentel and Pimentel, 1979).

Thus, despite all efforts to prevent crop losses caused by pests, they are destroying between 50 and 60% of all world food production. Up to now, the losses have been offset by increased crop yields achieved by planting high-yielding varieties, in addition to the increased use of fertilizers, irrigation and other fossil-energy inputs (Pimentel and Wen, 1990). Concern and doubts that this kind of compensation for crop losses cannot be sustained are growing because fertile land is being lost to erosion, water supplies are being stressed, aquifers are being mined and fossil-energy supplies – especially in the US – are being depleted (Pimentel et al., 1995).

ENVIRONMENTAL AND PUBLIC HEALTH COSTS OF PESTICIDE USE

Some pesticide benefits are offset by the cost of the public health problems and the pollution of the environment caused by pesticide use in the US. These problems are detailed in Chapter 4. To summarize, the economic costs of human and animal health hazards, plus the costs of other environmental impacts associated with US pesticide use, total more than $8.3 billion each year. This conservative estimate does not include costs of pesticide damage to microorganisms and wild mammals. However, if the more than $8.3 billion is added to the yearly cost of $5 billion for pesticide treatments, the total cost of using pesticides in the US would rise to about $13.3 billion per year. Thus, based on the estimated savings in crops of $20 billion per year by pesticide use, the crop value per dollar invested in pesticidal control would be only about $1.50 (Pimentel et al., 1991). Nonetheless, based on a strictly cost/benefit basis, the benefits of pesticide use are financially positive.

In contrast to the United States, the negative impacts of pesticides on public health and the environment in developing nations are great, and conceivably could reach about $100 billion each year. In these regions, the number of human deaths and illnesses caused by pesticides is high because regulations on the use of pesticides, both in the field and during storage, are lax and are frequently not followed by industries, farmers or labourers (WHO/UNEP, 1989).

The differences that exist in effective regulation of pesticide use between the United States and most developing nations are illustrated by the pesticide residues found in foods. In the US about 35% of the foods sold to consumers contain measurable levels of pesticide residues (FDA, 1993); about 1.1% of all foods have pesticide residues above the tolerance level set by the FDA. In addition, approximately 35% of the imported food sold in the US market contains measurable levels of pesticide residues, while between 1% and 3% of all imported foods have residues above the FDA tolerance level (FDA, 1993). Because residue analyses are conducted after the food has been sold, this means that the public is consuming some foods which contain residues above the tolerance level (FDA, 1993).

Pesticide residues in foods are much higher in developing countries than in the United States. For example, the measurable pesticide contamination of foods sold in Indian markets ranges as high as 80% (Singh, 1993). Another troubling dimension of the pesticide problem in India is that 70% of all insecticides used in India are the chlorinated insecticides DDT and BHC, and their use is growing at the rate of 6% per year (Singh, 1993). Furthermore, DDT and BHC, long banned in the United States, are persistent pesticides that accumulate in soil, water and biota. Thus, food contamination can be expected to increase with the growing use of these chlorinated insecticides in the Indian agricultural system. These conditions are similar to those existing in other developing countries.

NEW DIRECTIONS IN PEST CONTROL

Over time, many changes have been made in US pest control, not only because of Rachel Carson's book *Silent Spring* (1962), but also because the public has become concerned about the health and environmental problems associated with pesticides. As a result, pest control options have been enlarged to include numerous non-chemical methods.

The four broad classes of pest control include pesticides and integrated pest management (IPM) in addition to cultural and biological controls. Initially, IPM was designed to emphasize the use of non-chemical controls as a first line of defence, and pesticides as the fall-back defence. IPM has evolved to rely on the judicious and reduced use of pesticides by monitoring both pest populations and natural enemies to ascertain if and when pesticides should be used. However, some pro-pesticide groups now use the term IPM to justify their continued heavy use of pesticides in pest control.

Cultural technologies which have often been ignored in recent decades are being used more frequently today. These include crop rotations, crop diversity, host-plant resistance, soil, water and nutrient management practices, use of short-season crops, altered planting time, trap crops, pest sex-attractants, and various combinations of these. Sometimes a relatively simple change in the agricultural ecosystem, such as how the soil is tilled or when the crop is planted, provide control of a troublesome pest (Pimentel, 1991).

Before selecting the most appropriate strategy for pest control, the agroecosystem and the diverse ecological factors that cause the pest to reach outbreak levels must be understood (Pimentel, 1977). Then the cultural and biological control procedures must be tailored to regional ecosystems, including soils and climate. This approach substitutes ecological knowledge for pesticides, and opens up the possibility of including diverse strategies. Although more complex than spraying, over the long term the pay-offs are significant and their benefits will continue long into the future. Generally, biological controls include the use of predators, parasites and microbial agents for pest population control, improved use of all natural enemies in the agroecosystem, introduced natural enemies for control, and the release and/or application of natural enemies, including microbial agents, for pest population suppression.

Classical biological control relies on the use of natural enemies introduced from the native home of the pest insect for control. Some successful biological controls associated with this approach include the introduction of the vedelia beetle for the control of cottony cushion scale in California (DeBach, 1974), and recently the introduction of parasites into Africa to control the mealy bug attacking cassava (Herren and Neuenschwander, 1991; Mwanza, 1993).

Even with success, there have been limitations in the use of classical biocontrol. One fact that was commonly overlooked in the implementation of such biocontrol strategies was that in any given geographic region, between 30 and 80% of the pests are native to the region and have moved from feeding on native vegetation to feeding on the introduced crop (Pimentel, 1988). For example, in Colorado the potato beetle moved from feeding on a weed to feeding on the introduced potato crop (Pimentel, 1988).

The difficulty of using classical biocontrol to control large numbers of native pests, and the fact that about 20 introductions usually have to be made to achieve one success, prompted the development of the 'new association' method in biological control (Pimentel, 1961; Hokkanen and Pimentel, 1989). This approach can be illustrated by the attempts to control the European rabbit which was introduced into Australia, where it soon became a pest. All the natural enemies associated with the rabbit in Europe that were brought to Australia failed to control the rabbit there (Pimentel, 1961).

Finally, the myxomatosis virus, associated with the South American rabbit, was discovered and introduced into Australia (Levin and Pimentel, 1981). The myxomatosis virus, it should be noted, had little or no effect on the South American rabbit. However, the new association of the South American myxomatosis virus with the European rabbit was devastating. The initial spread of the virus in the European rabbit population killed more than 90% of the rabbits (Levin and Pimentel, 1981). Gradually the remaining rabbits developed some resistance to the virus, while the virus evolved a virulence toward the rabbit until a degree of natural balance was achieved between the parasite and its new host. The rabbit population has increased, but the myxomatosis virus still is providing about 40% control. This level of control is sufficient to allow many other predators to be effective in keeping the rabbit population under satisfactory control (Levin and Pimentel, 1981).

Another successful use of the 'new association' method was the control of the native pine moth in Colombia, South America. In this case, a wasp parasite of a related moth species found in Virginia was introduced into Colombia (Drooz et al., 1977). The new association between pest pine moth and wasp parasite has provided effective control. In general, about 40% of the successes in biocontrol are due to 'new associations'. The use of this approach is growing because it is providing successful control for both native and introduced pests (Pimentel, 1961; Hokkanen and Pimentel, 1989).

For many decades, host-plant resistance has been the dominant non-chemical method for the control of plant pathogens. Between 75 and 100% of all cultured crops grown have some degree of host-plant resistance to plant pathogens as a result of plant breeding (Oldfield, 1984; Pimentel, 1991).

Scientists have also been successful in breeding plant resistance to some insect pests such as the hessian fly (Pimentel, 1991). Now with the availability of genetic engineering, the use of host-plant resistance has greater potential for the control of insects, pests and plant pathogens (Paoletti and Pimentel, 1996).

CONVENTIONAL VERSUS ECOLOGICALLY SOUND AGRICULTURE

The implementation of various cultural technologies that reduce the need for large amounts of chemical inputs, including pesticides and fertilizers, has the advantage of decreasing chemical pollution of soil, water and food. Furthermore, use of chemicals related to human illness and death are reduced and the degradation of the agroecosystem is diminished. With careful management, soil erosion and associated rapid water runoff can be controlled to preserve soil and water resources (Pimentel et al., 1995). In addition, effective care and application of livestock wastes enhances soil nutrition and decreases environmental pollution (Pimentel et al., 1987; Pimentel, 1995a).

The differences between a conventional corn production system and a modified system that includes the implementation of several environmentally sound practices are illustrated in Table 1.2 (Pimentel, 1993). The conventional system relied on chemicals for pest control and fertilizers to provide soil nutrients. In the modified system, pesticides were not used, tillage was substituted for herbicides, crop rotation was employed for insect control, and manure was substituted for a large proportion of the fertilizers.

In the conventional system, the annual yield was $7000 \, kg \, ha^{-1}$ of corn at a cost of $523 \, ha^{-1}$, and the total energy used was more than 7.8 million $kcal \, ha^{-1}$ (Table 1.2). Crop loss caused by insects associated with the conventional system was 12%, while the estimated cost of environmental damage was $230 \, ha^{-1}$. Approximately $20 \, t \, ha^{-1} \, year^{-1}$ of soil was eroded with this system.

The modified system not only yielded more corn than the conventional system (a total of $8000 \, kg \, ha^{-1}$), but did so at a lower cost ($337 \, ha^{-1}$). Crop loss to insect pests was

Table 1.2 Energy and economic inputs per hectare for conventional and modified corn production systems

	Conventional			Modified		
	Quantity	10^3 kcal	Economic ($)	Quantity	10^3 kcal	Economic ($)
Labour (h)	10	7	50	12	9	60
Machinery (kg)	55	1485	91	45	1215	75
Fuel (l)	115	1255	38	70	764	23
N (kg)	152	3192	81	27	5591	17
P (kg)	75	473	53	34	214	17
K (kg)	96	240	26	15	38	4
Limestone (kg)	426	134	64	426	134	64
Corn seed (kg)	21	520	45	21	520	45
Cover crop seed (kg)	–	–	–	10	120	10
Insecticides (kg)	1.5	150	15	0	0	0
Herbicides (kg)	2	200	20	0	0	0
Electricity (10^3 kcal)	100	100	8	100	100	8
Transport (kg)	322	89	32	140	39	14
Total		7845	523		3712	337
Yield (kg)	7000	24 746		8100	29 160	
Output/input ratio		3.15			7.86	

Source: Data from Pimentel, 1993.

3.5%, considerably below the 12% in the conventional system. Soil erosion was reduced from approximately $20\,t\,ha^{-1}$ $year^{-1}$ in the conventional system to less than $1\,t\,ha^{-1}$ $year^{-1}$ in the modified system. Note that the $1\,t\,ha^{-1}$ $year^{-1}$ erosion rate equals soil sustainability (Pimentel et al., 1995). Furthermore, in the modified system fossil energy inputs were half those of the conventional system. The total cost of production was reduced by 36% to $337\,ha^{-1}$ $year^{-1}$ (Pimentel, 1993). As fossil energy resources continue to decline and become more costly, reducing energy inputs will become critical in agricultural production.

Several additional sound management practices were employed in the modified system (Pimentel, 1993). Careful use of manure reduced the pollution of ground water and/or adjacent water ways. Also, more effective use was made of manure and its valuable nutrients. The use of manure and recycling organic matter to the soil helps reduce soil erosion (Pimentel et al., 1987; Pimentel, 1995a).

The selection of an appropriate crop such as soybeans for rotation with corn reduced corn rootworm (Pimentel et al., 1993a), corn disease (Pearson, 1967; Mora and Moreno, 1984) and many typical weed problems (NAS, 1968, 1989; Mulvaney and Paul, 1984). Furthermore, the corn and soybean rotation system was more profitable than raising either crop alone (Helmers et al., 1986). This rotation eliminated the corn rootworm problem and the need for expensive insecticides.

Cover crops – especially legume cover crops such as clover or winter vetch – not only reduce weed problems, but, more importantly, reduce soil erosion and water runoff and conserve soil nutrients. Furthermore, soil nutrients picked up and stored by the legume cover crop are subsequently released when the cover crop is tilled into the soil.

Although mulch and tillage substituted for herbicides in the modified system, this was done only to demonstrate that herbicides could be replaced in the corn system (Pimentel, 1993). In some situations, combinations of herbicides and tillage is advantageous (Pimentel, 1991).

In summary, in the modified system, pesticide and fertilizer use was reduced and soil and water resources were conserved, while a yield higher than that of the conventional corn system was achieved.

REDUCED PESTICIDE USE

Reports from several regions of the world detail that when pest control research is focused on the ecology of pests, appropriate ways to decrease pesticide use without diminishing crop yields can be developed, and pesticide use can be reduced from 33% to 75% (Pimentel and Lehman, 1993).

In Guatemala, for instance, the amount of insecticide used for pest control in cotton was reduced by more than one-third once a strategy was developed to save many natural enemies that usually controlled potential pest problems. Under this system cotton yields increased by 15%, and some large cotton farmers increased their profits by more than $1 million $year^{-1}$ (ICAITI, 1977).

In Indonesia, the investment of $1 million $year^{-1}$ in ecological research, followed by extension programmes to train farmers to conserve natural enemies, is paying large dividends. Based on an approach similar to that used in Guatemala, pesticide use in Indonesia has been reduced by 65% for rice while rice yields increased by 12% (I.N. Oka, Bogor Food Research Station, Bogor, Indonesia, personal communication, 1995). The

Indonesian government was able to eliminate the $20 million in pesticide subsidies it paid to farmers.

By implementing IPM programs in New York State, sweet corn processors saved $500 000 per year and maintained high yields while reducing pesticide treatments by 55–65% (Koplinka-Loehr, 1995). Pesticide use has been reduced on other crops in New York State. Pesticide use has been reduced in the US with an estimated savings of at least $500 million per year.

In addition to the US, Indonesia and Guatemala, other nations have adopted effective programs to reduce pesticide use by 50–75%. These nations include Sweden, Norway, the Province of Ontario, Denmark and The Netherlands. The programs in Sweden and Norway are described in detail in Chapters 5 and 10, respectively.

CONCLUSION

There is great concern that more than 40% of all potential food and fibre production is being lost to pests despite the 2.5 million tons of pesticides and the diverse non-chemical controls being applied worldwide to agricultural crops. Losses of this magnitude continue at a time when agricultural production is strained to provide for the basic needs of the escalating human population. In fact, these supplies will have to be increased at least three-fold during the next 20 years, while per capita arable land is declining because of population growth and soil erosion. In addition, shortages of fresh water are intensifying and fossil fuel supplies are declining.

The crop losses caused by pests could be reduced substantially if pest control research was focused on the entire agroecosystem. Pesticide use will continue, especially for some crops, but will be applied wisely and only when necessary. Estimates are that pesticide use can be reduced as much as 50% without reducing crop yields or substantially reducing 'cosmetic standards' of fresh fruits and vegetables (Pimentel et al., 1993b). Reducing pesticides will reduce the costs of pest control, protect public health and improve the environment in all countries.

REFERENCES

Arrington, L.G. (1956). *World Survey of Pest Control Products*. US Government Printing Office, Washington, DC.

Carson, R. (1962). *Silent Spring*. Riverside Press, Cambridge.

Cromartie, W.J. (1991). The environmental control of insect using crop diversity. In Pimentel, D. (Ed.) *Handbook of Pest Management in Agriculture*. CRC Press, Boca Raton, pp. 183–216.

DeBach, P. (1974). *Biological Control by Natural Enemies*. Cambridge University Press, New York.

Drooz, A.T., Bustillo, A.E., Fedde, F.G. and Fedde, V.H. (1977). North American egg parasite successfully controls a different host genus in South America. *Science* 197, 390–391.

FDA (1993). Food and Drug Administration monitoring program. *Journal of AOAC International* 76(5), 127A–141A.

Georghiou, G.P. (1990). Overview of insecticide resistance. In Green, M.B., LeBaron, H.M. and Moberg, W.K. (Eds) *Managing Resistance to Agrochemicals: From Fundamental Research to Practical Strategies*. American Chemical Society, Washington, DC.

Helmers, G.A., Langemeir, M.R. and Atwood, J. (1986). An economic analysis of alternative cropping systems for east-central Nebraska. *American Journal of Alternative Agriculture* 4, 253–258.

Herren, H.R. and Neuenschwander, P. (1991). Biological control of cassavam pests in Africa.

Annual Review of Entomology **36**, 257–283.

Hokkanen, H.M.T. and Pimentel, D. (1989). New association in biological control: Theory and practice. *Canadian Entomology* **121**, 828–840.

ICAITI (1977). *An Environmental and Economic Study of the Consequences of Pesticide Use in Central American Cotton Production.* Central American Research Institute for Industry, United National Environment Programme, Guatemala City.

Koplinka-Loehr, C. (1995). Integrated pest management fulfills vision of creators. *Cornell Focus* **4**(1), 18–19.

Levin, S. and Pimentel, D. (1981). Selection of intermediate rates of increase in parasite–host systems. *American Naturalist* **117**, 308–315.

Mora, L.E. and Moreno, R.A. (1984). Cropping pattern and soil management influence on plant diseases. I. *Diplodia macrospora* leaf spot of maize. *Turrialbo* **341**, 35–40.

Mulvaney, D.L. and Paul, L. (1984). Rotating crops and tillage: Both sometimes better than just one. *Crop Soils* **367**, 8–19.

Mwanza, F. (1993). South American wasp comes to the rescue of cassavam growers in Africa. *BioScience* **43**(7), 452–453.

NAS (1968). *Weed Control.* National Academy of Sciences, Washington, DC.

NAS (1989). *Alternative Agriculture.* National Academy of Sciences, Washington, DC.

Oerke, E.C., Dehne, H.W., Schonbeck, F. and Weber, A. (1994). *Crop Production and Crop Protection: Estimated Losses in Major Food and Cash Crops.* Elsevier, Amsterdam.

Oka, I.N. and Pimentel, D. (1976). Herbicide (2,4-D) increases insect and pathogen pests on corn. *Science* **193**, 239–240.

Oldfield, M.L. (1984). *The Value of Conserving Genetic Resources.* US Department of Interior, National Park Service, Washington, DC.

Paoletti, M.G. and Pimentel, D. (1996). Genetic engineering in agriculture: Limits and options for maintaining biodiversity and sustainability. In Nath, B., Hens, L. and Devuyst, D. (Eds) *Sustainable Development.* VUB University Press, Brussels pp. 233–263.

Pearson, L.C. (1967). *Principles of Agronomy.* Reinhold, New York.

Pimentel, D. (1961). Species diversity and insect population outbreaks. *Annals of the Entomology Society of America* **54**, 76–86.

Pimentel, D. (1977). Ecological basis of insect pest, pathogen, and weed problems. In Cherrett, J.M. and Sagar, G.R. (Eds) *The Origins of Pest, Parasite, Disease, and Weed Problems.* Blackwell, Oxford.

Pimentel, D. (1986). Agroecology and economics. In Kogan, M. (Ed.) *Ecological Theory and Integrated Pest Management Practice.* John Wiley & Sons, New York, pp. 299–319.

Pimentel, D. (1988). Herbivore population feeding pressure on plant hosts: Feedback evolution and host conservation. *Oikos* **53**, 289–302.

Pimentel, D. (1991). Diversification of biological control strategies in agriculture. *Crop Protection* **10**, 243–253.

Pimentel, D. (1993). Environmental and economic benefits of sustainable agriculture. In Paoletti, M.G., Napier, T., Ferro, O., Stinner, B.R. and Stinner, D. (Eds) *Socio-Economic and Policy Issues for Sustainable Farming Systems.* Cooperativa Amicizia S.r.l., Padova, Italy, pp. 5–20.

Pimentel, D. (1995a). Agriculture, technology and natural resources. In *Technology and Global Environmental Issues.* Makofske, W.J. and Karlin, E.F. (Eds) Harper Collins College Publishers, New York, pp. 38–50.

Pimentel, D. (1995b). Protecting crops. In Olsen, W.C. (Ed.) *The Literature of Crop Science.* Cornell University Press, Ithaca, NY. pp. 49–66.

Pimentel, D. and Lehman, H. (Eds) (1993). *The Pesticide Question: Environment, Economics, and Ethics.* Chapman & Hall, New York.

Pimentel, D. and Pimentel, M. (1979). *Food, Energy, and Society.* Edward Arnold, London.

Pimentel, D. and Wen, D. (1990). Ecological resource management for a productive sustainable agriculture in Northeast China. In Tso, T.C. (Ed.) *Agricultural Reform and Development in China.* IDEALS, Beltsville, MD, pp. 297–313.

Pimentel, D., Dewey, J.E. and Schwardt, H.H. (1951). An increase in the duration of the life cycle of DDT-resistant strains of the house fly. *Journal of Economic Entomology* **44**, 477–481.

Pimentel, D., Allen, J., Beers, A., Guinand, L., Linder, R., McLaughlin, P., Meer, B., Musonda, D.,

Perdue, D., Poisson, S., Siebert, S., Stoner, K., Salazar, R. and Hawkins, A. (1987). World agriculture and soil erosion. *BioScience* **37**, 277–283.

Pimentel, D., McLaughlin, L., Zepp, A., Lakitan, B., Kraus, T., Kleinman, P., Vancini, F., Roach, W.J., Graap, E., Keeton, W.S. and Selig, G. (1991). Environmental and economic impacts of reducing US agricultural pesticide use. In Pimentel, D. (Ed.) *Handbook of Pest Management in Agriculture*. Vol. I. 2nd edn. CRC Press, Boca Raton, pp. 679–718.

Pimentel, D., Acquay, H., Biltonen, M., Rice, P., Silva, M., Nelson, J., Lipner, V., Giordano, S., Horowitz, A. and D'Amore, M. (1993a). Assessment of environment and economic impacts of pesticide use. In Pimentel, D. and Lehman, H. (Eds) *The Pesticide Question: Environment, Economics and Ethics*. Chapman & Hall, New York, pp. 47–84.

Pimentel, D., Kirby, C. and Shroff, A. (1993b). The relationship between 'cosmetic standards' for foods and pesticide use. In *The Pesticide Question: Environment, Economics and Ethics*. Pimentel, D. and Lehman, H. (Eds) Chapman & Hall, New York, pp. 85–105.

Pimentel, D., Harvey, C., Resosudarmo, P., Sinclair, K., Kurtz, D., McNair, M., Crist, S., Spritz, L., Fitton, L., Saffouri, R. and Blair, R. (1995). Environmental and economic costs of soil erosion and conservation benefits. *Science* **267**, 1117–1123.

Ross, M.A. and Lembi, C.A. (1985). *Applied Weed Science*. Burgess, Minneapolis, MN.

Roush, R.T. and McKenzie, J.A. (1987). Ecological genetics of insecticide and acaricide resistence. *Annual Review of Entomology* **32**, 361–380.

Singh, B. (1993). Pesticide residues in the environment: A case study of Punjab. In Sengupta, S. (Ed.) *Green Revolution Impact on Health and Environment*. Voluntary Association of India, New Delhi, pp. 21–28.

USBC (1971). *Statistical Abstract of the United States 1971*. US Bureau of the Census, US Government Printing Office, Washington, DC.

USBC (1994). *Statistical Abstract of the United States 1994*. US Bureau of the Census, US Government Printing Office, Washington DC.

USDA (1960). *Index of Plant Diseases in the United States*. Crop Research Division, ARS, US Department of Agriculture, Washington, DC.

van den Bosch, R. and Messenger, P.S. (1973). *Biological Control*. Intext Educational, New York.

WHO/UNEP (1989). *Public Health Impact of Pesticides Used in Agriculture*. World Health Organization/United National Environment Programme, Geneva.

CHAPTER 2

Pests, Pesticides and the Environment: A Historical Perspective on the Prospects for Pesticide Reduction

John H. Perkins[1] and Barbara R. Patterson[2]

[1]*The Evergreen State College, Olympia, WA, USA.*
[2]*Washington State Department of Ecology, Olympia, WA, USA*

Pests are organisms that appear at the wrong place at the wrong time, according to human wishes. What is of interest is how humans deal with pests, especially through the use of pesticides. The uses of pesticides seldom attracted major government attention or regulation before 1900. In contrast, by the late 1900s pesticides are major sources of political and policy conflict in many countries. In fact, the proposition of pesticide reduction, the major theme of this book, is partly a response to the frustration of devising appropriate regulatory programs for pesticides. Reduction of pesticide use, however, inevitably affects the mastery of nature and land-use control which lie at the heart of cultural identity. Efforts to reduce pesticide use will benefit from a better understanding of the history of pest control and pesticides. This chapter provides a brief sketch of the central points and illustrates them with some examples from Washington State.

SHIFTING PERCEPTIONS ABOUT PESTS AND PESTICIDES

Pests as environmental threats

By the 19th century, pest problems became serious enough to warrant concerted attention. With the industrial revolution, agriculture became a part of capitalist enterprise and the subsistence farmer passed from the scene. In commercial agriculture, pest problems took on a new dimension: cash marketing of the harvest, rather than subsistence use, meant that farmers were dependent upon the cash income. If borrowed money had been used to finance the crop, an important dimension of commercial agriculture, then a certain level of cash income was imperative for debt servicing. Under these conditions, losses to pests were a threat to the continued existence of the enterprise. Control of pests thus became a solution to an environmental problem.

    Pesticides were one of several new methods introduced to control pests. Products that

Techniques for Reducing Pesticide Use. Edited by D. Pimentel
© 1997 John Wiley & Sons Ltd.

were developed in or before the 19th century included oil, sulfur, nicotine, rotenone, pyrethrum, Paris green, London purple and lead arsenate (Howard, 1930; Perkins, 1982, p. 1). Pesticides were not the only pest control technology introduced in the 19th century, but the development of alternatives was also tied to increased commercialization of agriculture. For example, one of the most successful applications of biological control ever known was the introduction of the vedalia beetle (*Rodolia cardinalis*) and a dipteran parasite (*Cryptochaetum iceryae*) to control the cottony cushion scale (*Icerya purchasi*) on California citrus (DeBach and Rosen, 1991). As historian Richard Sawyer has convincingly argued, the enthusiasm for biological control was drawn largely from the highly commercial, entrepreneurial citrus growers (Sawyer, 1996).

Regardless of the pest control method, therefore, these new practices all indicated that pest damages threatened the viability of commercial farmers. Pest control, including use of pesticides, was a way to control an environmental hazard, and pesticides became the technology of choice during the 20th century. After 1900, industries other than agriculture also began to rely on pesticides for commercial viability. A major source of change in the 1940s and 1950s was the invention of many new synthetic, organic pesticides, stimulated by the successes of DDT insecticides after 1939 (Perkins, 1978). By 1950, the chemical industry marketed a wide range of pesticides that were cheap, versatile and effective in many places. As a result, forestry, homes and gardens, commercial and service establishments, and industrial and recreation sites all began to use pesticides. For broad reviews of the introduction and spread of pesticides see Whorton (1974), Dunlap (1981) and Perkins (1982).

Despite the commercial successes of pesticides, however, several factors changed the way people saw pests and pesticides. Especially important were resistance to pesticides, destruction of natural enemies and beneficial insects by pesticides, the development of alternative control strategies, public health problems, damage to wildlife and non-target organisms, ground water contamination and, most recently, the spectre of endocrine system disruption.

Resistance

Resistance of a pest organism to a pesticide was the first new wave of altered perceptions. This phenomenon was first noticed as an important event in the apple orchards of Washington State in 1908, when San José scale (*Quadraspidiotus perniciosus*) no longer responded to treatments of lime-sulfur washes (Perkins, 1982, pp. 34–36). Although troublesome, resistance did not threaten a major commercial industry until entomologists in Louisiana concluded in 1954 that boll weevil (*Anthonomus grandis*) had evolved resistance to commonly used chlorinated hydrocarbon insecticides (Roussel and Clower, 1955). Officers of the National Cotton Council, the trade association representing all segments of the cotton industry, quickly moved to advocate research for longer-term solutions, including pest control practices less dependent upon chronic use of pesticides (US Congress, 1958). Canadian entomologist A.W.A. Brown concluded in 1960 that the 'golden age' of pesticides had passed because resistance was becoming so prevalent (Brown, 1961).

Resistance has continued its relentless rise to the present day. Moreover, its first and most serious occurrences in insects began to be matched by examples of weeds resistant to herbicides and fungal pathogens resistant to fungicides. Resistance was also accompanied by cross-resistance, which meant that pest populations showed insensitivity to chemicals that had never been used against them (Hollomon, 1993; Moss and Rubin, 1993; Brun et al., 1994; Tabashnik, 1994; US Congress, 1995). Resistance suggested that the ability of pesticides to control nature was limited and fragile.

Destruction of natural enemies and beneficial insects

Insect pest problems are only one class of pest problems, but the control of unwanted insects is hampered by a serious constraint. Insecticides aimed at troublesome insects also kill non-target insects. Some of these unintended casualties are parasitic and predatory insects that feed upon pest insects, i.e. they are *natural enemies* of pest species and, when present, provide significant pest suppression. Others with unintended mortalities also provide benefits, for example pollination and honey production from bees.

Understanding this second set of problems required a more complete understanding of pest population ecology. An increase in insect pests following insecticide application was the first symptom of a problem. Insecticides, however, were designed to kill pests, so the concept that they could increase a pest population was counter-intuitive. Experimental and theoretical work over about three decades, from the mid-1920s to the mid-1950s, established the principle that destruction of natural enemies was a likely consequence of insecticide use. (Folsom, 1927; Boyce, 1936; Clausen, 1936; Sweetman, 1936; Steiner, 1938; Thompson, 1939; DeBach, 1946; Ripper, 1956; Perkins, 1982, pp. 38–44).

Destruction of pollinating insects and honey producers was also a problem that became serious after 1945, due to the increased use of the new synthetic organic insecticides. Honey bees are an extremely important pollinating insect, leading to nearly $10 billion per year of increased crop production. Farmers expend about $40 million per year to rent bee colonies for enhanced pollination and production. As much as $4 billion per year may be lost due to inadequate pollination, much of it caused by pesticide damage to honey bees. (Kagel, 1993; see also Erickson et al., 1994). Together with destruction of natural enemies, this destruction of honey bees and other pollinators was a signal to entomologists and farmers that pesticide-based technology was potentially unreliable. New pest control practices were needed.

Development of alternatives

Advocates of organic agriculture gained increasing strength after 1945, but their enthusiasm for reduced reliance on pesticides did not have major impacts on research scientists. Instead, resistance and destruction of natural enemies and beneficials stimulated entomologists, who had been most affected by the phenomena. Starting in the 1940s, research entomologists in the state universities and in the US Department of Agriculture sought new strategies of control that did not rely entirely on pesticides. They were conscious of the need to develop new pest-control methods that would work for farmers and for other users of pesticides, i.e. they wanted their innovative efforts to fit the social and economic conditions of the country. Two new strategies emerged from this work: integrated pest management (IPM) and total population management (TPM) (Perkins, 1982; pp. 61–126).

IPM had its origins in the biological control research done in California and the necessity of Californian agricultural research to respond to a highly commercialized agricultural industry. IPM relied on the use of host-plant resistance and the suppression provided by natural enemies. All other means of pest suppression, including pesticides, were subordinated to the foundation of host-plant resistance and natural enemies. IPM was designed to work by site-specific scouting of individual fields for sizes of pest populations and natural enemy populations. Eradication was considered generally unattainable and unnecessary. All the pest control operator had to do was suppress the pest below levels that were economically worth controlling. After 1959, IPM became the major research strategy in American entomology, even though it has taken many years to get only limited successful implementation of IPM on farms.

TPM had its origins in the US Department of Agriculture, especially around work to respond to the screwworm fly (*Cochliomyia hominovorax*) and the boll weevil. In 1952, researchers demonstrated the ability of the sterile-male technique to suppress certain populations of screwworm fly to extremely low levels, very likely levels that truly were 'eradications'. From this work came the TPM strategy, which was based on attack against the pest population over a very large area of land, not just individual farmers' fields. Pesticides or other techniques were used first to reduce the population to low levels, after which the sterile-male technique was used to suppress it even further, possibly even to the point of eradication. TPM guided much research during the 1960s and 1970s, but it was a cumbersome technology to introduce. All land owners in a given area had to cooperate with the program, a social achievement that was hard to obtain.

Both IPM and TPM were designed to free insect control from resistance and the destruction of natural enemies. Each was successful within limits, but neither has yet supplanted pesticide-based practices. Ease and economy of pesticide use compared to IPM, and especially TPM, have continued to drive farmers and other pest controllers to use the chemicals. Only when resistance and destruction of natural enemies became severe did IPM and TPM win much allegiance. Nevertheless, research agendas are driven by the alternatives, especially IPM.

Health problems

Chemicals are dangerous to human health, wildlife health and plant life. In time, health problems became the major driving force of reform of pesticide regulation. Of primary importance were (a) problems to consumers of pesticide residues on and in food, (b) acute and chronic toxicity of pesticides to farmworkers and to others exposed occupationally, and (c) acute and chronic toxicity to wildlife.

Pesticide residues on food

Even before the advent of the synthetic organic pesticides, residues generated heated controversy. Especially troublesome were lead, arsenic and fluorine (Whorton, 1974). For the most part, compounds containing these materials were used only on fruits and vegetables, the only crops with a high enough value to warrant the expenses of protecting them. Unfortunately, efforts to regulate these residues before 1938 proved contentious and impractical.

The Federal Food, Drug and Cosmetic Act 1938 opened the way to what the Food and Drug Administration (FDA) hoped would be a workable resolution to the disputes about residues (Whorton, 1974; pp. 243–247). The outbreak of the Second World War, however, derailed implementation of the new law. In addition, invention of DDT in 1939 stimulated an enormous innovative effort to find other new pesticides.

Continued Congressional concern about the toxicological hazards posed by the new pesticides stimulated the formation of the Select Committee to Investigate the Use of Chemicals in Foods and Cosmetics in 1950. In well-publicized hearings between 1950 and 1952, this committee, under the chairmanship of James J. Delaney (Democrat, New York), thoroughly explored the situation of residues. In 1954, Representative A.L. Miller (Republican, Nebraska), who served on the Delaney Committee, successfully led the passage of an amendment to the Federal Food, Drug and Cosmetic Act (Perkins, 1982, pp. 29–34). Under the new provisions, manufacturers who wanted to market a pesticide that left residues on the crop were required to provide data from toxicological studies to the FDA before the product could marketed. The FDA used the data, plus other information on the benefits of the chemical, to establish a tolerance for the residues.

Miller's amendment, however, did not fully resolve the increased concern felt about pesticide residues in food. Congressman Delaney responded to concern about the potential for food additives to cause cancer by proposing another amendment to the Federal Food, Drug and Cosmetic Act. Passed in 1958, the Delaney amendment held that no additive could be considered safe if it caused cancer when ingested by people or animals, and thus must be prohibited. Pesticides that concentrated when food was processed were considered food additives subject to the Delaney amendment. In the early 1980s, EPA moved to regulate pesticides that concentrated during food processing by limiting their residues to levels that caused less than one-in-a-million chance of cancer over a lifetime, a policy not strictly in keeping with the Delaney amendment. EPA practice was also to deny tolerances for such a pesticide on unprocessed food, even though the law governing unprocessed foods (the Miller Amendment) directed a risk–benefit approach (Merrill, 1994; Flamm, 1995).

Scientific controversy swirled around the growing list of food stuffs that, under the Delaney amendment, should be seized. Some argued that the best way to protect consumer health was to follow Delaney and absolutely prohibit any carcinogen. Others argued that rodent tests for carcinogenicity may not be valid. Additional arguments in 1987 by the National Academy of Sciences held that a 'Delaney Paradox' was possible: a rigorous effort to follow Delaney could lead to increased, not decreased, health problems (National Research Council, 1987).

In 1992 the Natural Resources Defense Council (NRDC) won a suit against the Environmental Protection Agency (EPA) (Cushman, 1994; Merrill, 1994). The NRDC demanded that the EPA should enforce the Delaney amendment by prohibiting carcinogens. However, an outcome quite different from the one envisioned by the NRDC is a possibility. In 1993 the Clinton Administration signalled that they wanted to repeal the strict prohibition against carcinogens in processed foods and regulate tolerances in all food on the risk–benefit basis of the Miller amendment (Schneider, 1993). When the Republicans swept to power in Congress in 1994 with a vow to reduce federal environmental regulations, it appeared that the remaining days for the Delaney amendment might be few. Environmental activist groups, however, believe the Delaney amendment is the key law to move the nation away from heavy dependency on pesticides and are likely to oppose any repeal (Feldman, 1994).

Occupational exposure to pesticides

Whoever applies a pesticide must be concerned about 'acute toxicity' or immediate poisoning. Ingestion, inhalation and dermal contact are the three major routes of exposure in workers.

Pesticides used before 1940 were potentially harmful, but most of the chemicals then in use had relatively low acute toxicities. After 1940, however, some of the new, synthetic, organic pesticides had very high acute toxicities. Generally the most dangerous materials were insecticides, but some fungicides, herbicides and other materials also required care in their use. By the 1950s, the occupational health literature began to fill with reports of pesticide-induced occupational illnesses (Maddy et al., 1990). In the 1960s, hazards to farmworkers from pesticides became one of the rallying cries for Cesar Chavez and the United Farm Workers in California (United Farm Workers Organizing Committee, 1965). In some cases, outbreaks of occupational poisoning occurred in conjunction with problems of resistance and destruction of natural enemies. For example, a physician in southern Texas noted a 400% increase in cases of poisoning in 1964 when cotton growers switched from chlorinated hydrocarbons to organophosphates because of the resistance of boll weevils to chlorinated hydrocarbons (Gallaher, 1967; Reich et al., 1968).

Acute toxicity, in which the onset of sickness is fast and unambiguously connected to the use of pesticides, is relatively easy to document. More difficult to study are chronic conditions that may develop over very long periods of time. Nevertheless, some studies have found an association between pesticides and subtle, chronic effects. For example, in the 1970s, sterility in male workers in the plant producing DBCP, a nematocide, was attributed to the pesticide (Giglio, 1993). In the 1980s, a series of studies raised questions about the ability of herbicides such as 2,4-D to cause cancer in forestry workers and grain farmers, a claim that still sparks controversy among epidemiologists (Ibrahim et al., 1991).

Ironically, the scientific, legal and political efforts to address the effects of pesticide residues consumed considerably dwarf those from occupational exposure. Acute toxicity episodes, especially among farmworkers, persist to this day, despite the accumulation of scientific evidence that clearly identifies the cause. The powerlessness of these workers compared with the land owners using the pesticides underlies this tragic and unfair situation. Below, we recount in more detail a particularly needless outbreak of poisoning in 1993 in the orchards of Washington State from the insecticide mevinphos.

Toxicity to wildlife

People are not the only species adversely affected by pesticides. Many species of wildlife were also inadvertently exposed to pesticides used in agriculture, forestry, landscape architecture and other situations. It is likely that some populations of a few species were adversely affected by the arsenical and other pesticides used before 1940. However, the vast increase in the use of pesticides that occurred after 1945 inaugurated a far more visible wave of morbidity and mortality in wildlife populations by the early 1960s. A very famous book jolted the issue into the political arena: Rachel Carson's *Silent Spring* (Carson, 1962). Carson provided the first comprehensive ecological theory about the effects of pesticides.

A close study of *Silent Spring* suggests that some spectacular problems with wildlife poisoning formed the heart of her thinking. Specifically, she made powerful use of the death of wildlife from (a) treatment of forests for gypsy moth, (b) treatment of pastures and forests for imported fire ant, (c) treatment of marshlands for mosquitoes, and (d) treatment of university campuses for Dutch elm disease. Perhaps the most important reason for *Silent Spring*'s success, however, was its strong environmental ethic. It was expertise with a soul and the imagination to envision a better way of dealing with pests.

As successful as Carson was in raising the alarm about pesticide dangers, her book was not sufficient to eliminate the use of the chemicals most dangerous to wildlife. Further research linked DDT and other chlorinated hydrocarbons to the demise of certain raptorial birds in both North America and Europe through thinning of their eggshells. Birds such as the peregrine falcon could no longer lay an egg that was strong enough to bring a new fledgling into life. As reproduction shut down, the birds disappeared.

No one, not even the harshest critic of pesticides, would ever have imagined in 1940 that small amounts of these compounds could interfere with the calcium metabolism of birds. Thus the eggshell thinning effect was probably impossible to detect before the use of the materials. Once the effect was understood, however, it demonstrated that seemingly small amounts of the chemicals could have extremely potent, wide-ranging and unanticipated consequences. Shell thinning by DDT in bald eagles eggs was probably influential in the decision to ban the material in the United States after 1972. Suspicion that DDT might be a human carcinogen also played an important role (Dunlap, 1981, pp. 231–235).

High-profile problems of the 1980s and 1990s

Concerns about resistance, destruction of natural enemies and beneficials, and health problems all remain active in the political arena. However, two specific problems have achieved particular prominence.

Ground water contamination

Rachel Carson clearly identified ground water contamination from pesticides as an issue in *Silent Spring*. Carson had direct evidence that pesticides contaminated surface water, but her case for harm to ground water was indirect (Carson, 1962, pp. 42–43). Disagreement attended her argument, because some experts thought it unlikely that pesticides would move into ground water reservoirs (Bouwer, 1990).

In the late 1970s and early 1980s, the picture began to change rapidly. The EPA and state regulatory agencies identified four pesticides as ground water contaminants: aldicarb, an insecticide, in New York and Wisconsin; DBCP, a fumigant and nematocide, in California, Arizona, South Carolina and Maryland; EDB, a fumigant and fuel additive, in Georgia, Hawaii, California and Florida; atrazine, a herbicide, in Nebraska. More recently, a report in Washington State documented the occurrence of pesticide contamination in ground water (Bouwer, 1990; Cohen, 1990; Pignatello and Cohen, 1990; Ritter, 1990; Washington Toxics Coalition and Cooperative Extension – Washington State University, 1994; Ostertag, 1995). Resolution of the matter has not yet occurred, and it is likely that concern over pesticides and other contaminants in ground water will increase in the years to come.

Endocrine disruption

Endocrine glands are organs that secrete hormones which are transported to all parts of the body. They cause an effect at a site remote from their production. Secretions from endocrine glands, combined with chemicals produced by nerve cells, play an important role in coordinating the different functions of the body and in regulating the development and function of organs from the embryo to the adult.

Although many functions are governed by endocrine and neural hormones, the development and functioning of the reproductive system in all vertebrate animals is especially dependent upon the levels and kinds of several hormones. In both females and males, disruptions or abnormalities of the endocrine system cause errors in the development and functioning of the organs and structures involved in reproduction.

Long and complex research was needed to put this story together. In the late 1970s and early 1980s, a new and more comprehensive theory began to emerge: all organisms, from conception to death, are bathed in a sea of *xenoestrogens*, chemicals that mimic the effects of natural estrogens. In the late 1980s and early 1990s, Theo Colborn and others expanded the xenoestrogen theory even more: both wildlife and human beings are now suffering acute harm from the universal bath of xenoestrogens, many of which were pesticides. They emphasized that organs especially at risk were those that responded normally to sex hormones, primarily reproductive organs. The ubiquity of endocrine disruption, as portrayed by Colborn and her colleagues, was staggering. Widely across the vertebrates, these chemicals could affect thyroid function, fertility and normal sexual differentiation in males and females (Colborn et al., 1994; Environment Impact Assessment Review Team, 1994).

Further fuel was added to this fire by two other types of report. First, epidemiological evidence in 1993 and 1994 addressed the question of whether DDT and its residues were associated with breast cancer in women. The 1993 study showed a significant association (Wolff et al., 1993), but the 1994 study produced only a few cases that were of borderline significance (Krieger et al., 1994). Considerable debate surrounded the interpretation of these two articles. Interestingly, residue levels were higher in the 1994 study than in the earlier one, a finding that may be difficult to reconcile with the significant trend found in the 1993 study.

Second, results published in 1992 and 1995 addressed the quality of semen and sperm from men. The 1992 study reviewed 61 reports published between 1940 and 1990 and argued that the quantity of semen produced and the density of sperm had both declined significantly (Carlsen et al., 1992). The 1995 study found no change in the volume of semen but a significant decline in sperm densities and motilities, and the proportion of normal sperm (Auger et al., 1995). Both of these studies elicited both criticism and support for their methodologies and findings (Brake and Krause, 1992; Irvine, 1994; Sherins, 1995). In mid-1995, a new study reported that DDT and its metabolites acted not as an estrogen but as an anti-androgen, a property that could also explain human male reproductive system abnormalities (Kelce et al., 1995).

Taken together, studies on breast cancer and male reproductive health suggested that endocrine disruption has been a growing health problem since 1940, the very time that the synthetic organic pesticides became so common. As of now, however, neither the scientific community nor the regulatory system have reached closure on endocrine disruption.

FROM THE GENERAL TO THE SPECIFIC: PESTICIDES AND ENVIRONMENTAL ISSUES IN WASHINGTON STATE

The environmental problems discussed above have affected virtually every pest control situation in every location. Much of the attention has been riveted on agricultural pest control, the largest single category of pesticide use. However, in order to gain a better understanding of the shifting perceptions of environmental problems that have developed during the history of pesticide use it is necessary to look at specifics. Here we examine the events around two crops in Washington State, the cranberry and the apple (Figure 2.1) (US Department of Commerce, 1992). Problems in these two crops in this location demonstrate that pesticide and environmental issues have many situation-specific features, which have required different solutions. Prospects for pesticide reduction depend upon the specifics of each case.

Although both cranberries and apples are classified as 'minor crops' on a national basis, their significance to Washington State agriculture is much larger than such a term implies. Approximately 1400 acres of cranberries are cultivated in Pacific and Grays Harbor Counties in Coastal western Washington. The value of cranberries in 1994 was $9.3 million (WA Department of Agriculture, 1995). There are approximately 150 000 acres in apple production in Washington, primarily in the eastern part of the state. From 1991 to 1994 apples were the number one agricultural commodity in Washington, with a value of almost one billion dollars per year. This difference in scale has at least partially influenced pest control methods and the environmental problems that have emerged on these crops.

Before DDT

Prior to the 1950s, farmers had few methods for fighting pests in either crop, and insects, diseases and weeds could quickly threaten to devastate yields. Plant and fruit diseases (such as fruit rot and false blossom disease) were the primary problem facing cranberry farmers soon after bogs were established in the 1880s (Chandler, 1956, p. 4). Insects were a secondary problem, followed by weeds. Some of the most common remedies included copper and lime to fight fungal diseases, hand-pulling weeds, and flooding bogs to fight frost and control the cranberry worm (Eck, 1990). Lead arsenate was most probably used for insect control. In 1918, 40% of the Washington crop was lost to damage by an insect pest introduced from the East, the black-headed fireworm (Chandler, 1956, p. 14). Ironically this is still one of the most troublesome pests.

As mechanization occurred in cranberry production, growers became more dependent upon stable incomes and thus more dependent upon effective pest control. In 1930, Ocean Spray Cooperative Inc. was founded in order to expand the market for cranberries. Expanded markets increased income for growers but also increased incentives for achieving effective pest control. The Washington State Coastal Agricultural Experiment Station, opened in 1923 to provide assistance to farmers, helped find pest control practices as cranberry production increased.

Apple production began in the 19th century in Washington and increased as new irrigation works brought water to orchard areas. This crop was intensely commercial from its earliest days, with markets in the eastern United States and Europe. Pests, especially coddling moth, continually threatened to undermine profitable production.

22

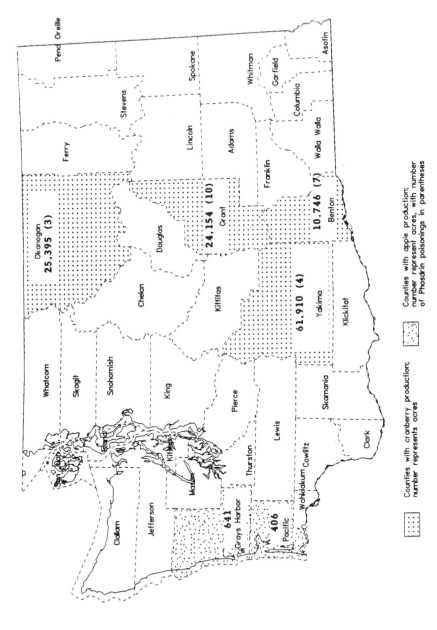

Figure 2.1 Geographic distribution of cranberry production and incidences of Phosdrin poisonings in apple orchards

Before 1950, lead arsenate was the primary insecticide used for control of coddling moth, but it was only marginally effective and as a result was applied at high rates. As early as the 1920s these high application rates led to concerns regarding health effects from exposure to residues on fruit (Dunlap, 1981, p. 43) and resistance of the coddling moth (Perkins, 1982, p. 19). Although these effects were well known (but disputed) for many years, it was only in 1927 that the FDA was able to establish tolerance limits. Even then, the International Apple Association was instrumental in keeping residue levels set well below those that had been established in Britain (Dunlap, 1981, pp. 35–55, 255–256).

DDT era

As the 1950s began, cranberry and apple growers gained a new set of tools that provided better control of pests than they had ever before experienced. In addition, federal regulations ensured that there was a steady supply of pesticides (Bosso, 1987, p. 63). The tripling of pesticide production from 1945 to 1950 (Bosso, 1987, p. 63), was a probable contributory cause of the dramatic increase in cranberry bog acreage in Washington during the 1950s. The combination of DDT and parathion effectively controlled the black-headed fireworm which had previously devastated yields (Chandler, 1956, p. 14).

In orchards, DDT had almost replaced lead arsenate by 1950 because it was dramatically more effective in its control of coddling moth. However as early as the 1940s there was evidence of chronic health effects from exposure to DDT (Dunlap, 1981, pp. 64–72) and by the early 1950s there was evidence of coddling moth resistance to DDT (Perkins, 1982, p. 36). Use of DDT soon led to the destruction of natural enemies and the emergence of secondary pests such as aphids, although in some cases aphids can be a pest in their own right. In order to control aphids, growers were instructed to use parathion, a highly toxic organophosphate insecticide (Perkins, 1982, pp. 18–20).

The aphid complex, *Aphis pomi* DeGeer and *Aphis spireacola* Patch, is one of the most serious pests in Washington's orchards (second to coddling moth). Aphid damage can stunt tree growth and damage fruit. While feeding, aphids produce a sticky excrement which provides a medium for mould growth. The damage to fruit is primarily of a cosmetic nature, since most of the mould can be washed off except for that in the stem cavity. Owing to cosmetic standards and consumer demand for good-looking fruit, apples damaged by aphids are diverted away from the fresh fruit market. Growers receive a much lower price for apples used for processing, so there is a strong economic incentive to minimize aphid damage.

As farmers began to depend on the effective new pesticides, problems with residues became more important and were particularly devastating to cranberry growers. In 1956, the herbicide amino-triazole (ATZ) was tested on Washington bogs and seemed 'to have the possibilities of a perfect cranberry weed killer.' (Chandler, 1956, p. 16). ATZ was registered in March 1956, but could only be used after harvest because ATZ was a suspected carcinogen. In 1957, cranberries were seized by the FDA, which suspected that farmers had not been following guidelines, and residues were found on the fruit. Tolerance levels had not yet been established for ATZ, however, so there were no ramifications. In 1959, after the Delaney amendment had been passed, the FDA again found cranberries tainted with ATZ. Just prior to the holiday season the FDA made an announcement that cranberries were tainted with a carcinogen. Cranberries were banned nationwide, devastating the industry. Although the issue of pesticide residues gave way to different concerns about pesticides, the cranberry industry experienced a 10-year decline in sales (Bosso, 1987, pp. 94–100). As a result, Ocean Spray Cooperative adopted its own pesticide residue monitoring program and set some tolerance levels lower than those set by the FDA (A. Broaddus, personal communication, 1994). This event was clearly a major clash between farmers and the government. The government had limited the tools that farmers could use to gain control over the pests that plagued them.

DDT's legacy – environmental impacts

While the cranberry industry worked to regain the public trust during the 1960s, widespread use of organochlorine insecticides led to concern regarding the environmental fate of the pesticides and their effects on wildlife. These concerns are still prevalent today. Given that cranberries are native to wetlands and marsh areas, the potential for pesticide impact appears to be considerable. In Washington, like Massachusetts, the peat that had accumulated between the sand dunes along the coast provided an excellent environment for cranberry culture. Because of increasing pressure to protect natural wetlands, cranberry bogs have more recently been developed in non-characteristic areas and less 'valuable' wetland areas. Although the amount of bog acreage in coastal Washington is relatively small in comparison to other cranberry growing regions, and in comparison to other agricultural crops grown in Washington, the concentration of bogs in this part of the state potentially threatens local water resources and wildlife.

Since the 1960s, several studies have been conducted on the cranberry to determine the movement of insecticides through bogs. These studies have found that the organochlorine pesticides persist in drainage water and sediment. In general, though, the organic content of cranberry bogs helps prevent vertical leaching (Miller et al., 1967; Deubert and Zuckerman, 1969; Lum, 1984). More recent monitoring, conducted as part of the Washington State Pesticides Monitoring Program, has found concentrations of organophosphate and organochlorine insecticides in a cranberry bog drainage ditch in the Grayland area (Grays Harbor County). Insecticides diazinon, chlorpyrifos and azinphos-methyl were found above EPA recommended levels. DDD and DDE (metabolites of DDT) were also above criteria (Davis and Johnson, 1995). Results of these studies indicate that aquatic biota and wildlife may be at risk from pesticide use on bogs. Effects on ground water are less well known and need additional study. The results of all these studies, however, present a threat to farmers because they could lead to loss of effective controls.

As a legacy of apple production, the environmental impacts of both lead arsenate and DDT are still evident in the Yakima River Valley, where many of the nation's apples are grown. In a survey of wells northwest of Yakima, arsenic was detected above Washington State ground water quality standards in 13 of 27 wells sampled (Erickson, 1992). DDT is still found in high concentrations in agricultural soils. Since DDT adsorbs to sediment, DDT is transported to rivers by erosion. Fish in the Yakima River have concentrations of DDT that are 13 times higher than those found in fish from other western streams (Rinella et al., 1993). As a result of these findings, the Washington State Department of Health issued several advisories. This measure alone, however, does not help those people, especially Native Americans, who rely on fish as a means of sustenance. Minimizing erosion would be the only significant measure that might lead to declining levels of DDT in fish.

A new regulatory system

DDT was banned in 1972, the same year that Congress significantly amended the Federal Insecticide, Fungicide, and Rodenticide Act (FIFRA). In important yet indirect ways, both of these actions stemmed from Rachel Carson's *Silent Spring* and signalled a new regulatory regime on pesticides.

After DDT was banned and the regulatory pressure on other organochlorine pesticides continued, the use of the more acutely toxic organophosphates increased. From 1966 to 1982 organochlorine use dropped from 70% of insecticides used to 6%, while organophosphate use rose from 20% to 67% in the same time period (Osteen, 1993). Although the organophosphate insecticides break down more quickly in the environment than the organochlorines, the acute toxicity of this class of insecticides led to new problems of aquatic and wildlife poisonings and farmworker poisonings.

The wave of cancellations of organochlorines after 1972 was due primarily to evidence of their persistence in the environment. The rate at which pesticide registrations were cancelled increased greatly after 1988, when further amendments to FIFRA required re-registration of pesticides registered before 1984. The purpose of this amendment was to protect health and the environment, but it may have had serious, unintended, deleterious consequences. Many manufacturers chose not to renew registrations, especially on minor crops (Patterson, 1994).

The cancellation of parathion in 1991 represented a significant loss to cranberry farmers. They were left to find effective alternatives for the two most prevalent cranberry bog pests in Washington, the black vine weevil and the black headed fireworm. On the environmental and worker safety front, the cancellation of this highly toxic organophosphate was a blessing, but in order to maintain acceptable pest control it is possible that farmers have been forced to use greater quantities of other less effective but still highly toxic insecticides.

Since the EPA began the re-registration process, 955 products have been voluntarily cancelled while only 613 products have been re-registered (US EPA, 1995). Given this ratio, farmers can expect to continue to lose pesticides. The process of cancellation and substitution is what ultimately led to the poisoning of 27 farmworkers in Washington orchards in the summer of 1993. Information about this incident was drawn primarily from a report prepared in the Master's Program in Environmental Studies at the Evergreen State College, directed by J.H. Perkins (Flores et al., 1994). In 1992 the manufacturer of Phosphamidon, an aphicide, voluntarily cancelled the registration for apples. This was the third aphicide to be cancelled since 1986 (Cystox and Metasystox were the other two). Some growers decided that Phosdrin (mevinphos) was the only effective substitute that would not interfere with IPM efforts and for which aphids had not developed a resistance.

Many objected to the use of Phosdrin in orchards. Washington State University scientists strongly discouraged growers from using Phosdrin because they did not feel it could be used safely in orchards. Interestingly, the manufacturer and distributors also warned orchardists of the extreme toxicity of Phosdrin, in part because they did not want to lose the use of Phosdrin in vegetable and seed crops, where it had been a successful part of an IPM program for several years. The manufacturer also requested that Washington State Department of Agriculture mandate the use of closed systems, which keep the chemical contained and away from workers until application.

Washington State Department of Agriculture felt that costs for growers to acquire closed systems would be prohibitive. Instead, they required increased training and restricted re-entry intervals and specific application requirements. The agency essentially chose to protect the rights of farmers over farm-workers. The first poisoning occurred in June 1993, shortly after Phosdrin entered the fields, and by mid-August, 26 more cases had been recorded. The agency first adopted stricter guidelines for use, and on 30 August finally banned Phosdrin from orchards. Most poisonings involved individuals who had been mixing or spraying the pesticide.

Several factors contributed to these incidents and to the problem of farm-worker safety in general. The extreme toxicity of Phosdrin is the most obvious aspect. Prior to its use in Washington orchards, there were many well-documented Phosdrin poisonings. Between 1982 and 1990 there were 1154 cases of organophosphate poisoning in California; Phosdrin was associated with 43% of these reported cases. Phosdrin was also the leading cause of hospitalization due to exposure to pesticides from 1974 to 1982 (Epidemiologic Notes and Reports, 1994).

The nature of the workforce also plays a role in the hazards of pesticide use. The apple industry is highly dependent on an adequate supply of seasonal migrant workers. In 1991, 59% of the labour for the apple harvest were inter-state farmworkers. It is estimated that 80–90% of farm labour in Washington is of Hispanic or Mexican origin. The majority of farmworkers of Mexican origin have less than an 8th-grade education and have low proficiency in the English language. The nature of this constantly changing workforce and the hazards associated with this type of work are factors which may explain the problems of pesticide exposure in orchards.

A 1991 report from the Washington State Department of Labor and Industries states that pesticides and agricultural machinery are the two most important dangers in agriculture, which is the nation's most hazardous industry (Karr et al., 1991). In 1993, 55% of all agricultural pesticide claims received by the Washington State Department of Labor and Industries were from orchards (WA State Department of Health, 1995, p. 26). Many exposure incidents occur from failure to use personal protective equipment properly. In 78% of the Phosdrin incidents, personal protective equipment guidelines had not been followed (Epidemiologic Notes and Reports, 1994). Factors such as high air temperatures and an inability to adjust equipment when wearing gloves make it difficult to wear personal protective equipment 100% of the time. It is questionable whether or not even the most highly trained individual can safely use a pesticide as toxic as Phosdrin.

Federal and state regulations fail to protect farmworkers. The substitution process created through FIFRA re-registration unintentionally exacerbated worker safety problems. It is unlikely that farmworkers are provided with specific details regarding the toxicity of pesticides, or if they are informed, they may not have the education to make sense of the information. In the case of the substitution of Phosdrin for Phosphamidon, both are Category I organophosphate insecticides, the highest level of toxicity. If workers were aware of this information they may have assumed that both could be handled similarly. EPA and Washington Department of Labor and Industries have developed standards for the protection of agricultural workers. Ironically, most of the standards were supposed to have been implemented prior to the summer of 1993 when the Phosdrin incidents occurred.

In early 1994 Washington State Department of Agriculture cancelled all uses of Phosdrin in Washington State. In July 1994, the manufacturer voluntarily cancelled all registered uses of Phosdrin (Schreiber, 1995). Essentially, this left orchardists as well as vegetable and seed-crop farmers to find a new substitute. Farmworkers are probably no safer, however, since many highly toxic materials are still used in Washington's orchards (WA State Department of Health, 1995).

IPM's role in pesticide reduction

Integrated pest management approaches have been developed for the apple and cranberry industries. Implementation of these approaches has not been as thorough as development, however. Development of IPM in apple orchards in the 1950s and 1960s was prompted by the resistance of mites to organophosphates. Scientists soon discovered that if some levels of mites were tolerated, natural predators would gain control. This approach was so successful that orchardists have not used chemical controls for mites since the 1960s. Other IPM approaches have been slower to become established (Beers et al., 1993, p. 18).

IPM is a relatively new concept in cranberry farming. The first programme was initiated in 1983 in Massachusetts. Ocean Spray Cooperative actively researches IPM methods and biological controls, and provides technical assistance to farmers. The Washington State University Cooperative Extension is also actively involved in educating farmers in IPM techniques and maintains communication through the Cranberry Vine Newsletter.

Monitoring is a fundamental aspect of IPM to determine if and when chemical control is needed. Pheromone traps are used to monitor the presence of the black headed fireworm on cranberry (Broaddus, 1991), and for several moth species in orchards (Beers et al., 1993, p. 21). Cranberry bogs are monitored at night with a sweep net to determine whether control is necessary for the black vine weevil (Broaddus, 1992).

Biological controls are an important component of an IPM programme. The use of insect parasitic nematodes for black vine weevil control on the cranberry is increasing, and appears to be a promising alternative to synthetic chemical pesticides. Biological controls of apple aphids include natural predators such as the ladybird, syrphid fly larvae, green lacewings and the predatory bug *Deraeocoris brevis* (Beers, 1993, p. 163). These controls are not being used effectively, and more research is necessary to make maximum use of these natural predators.

Bees are an important and effective method of pollination in cranberry bogs and apple orchards. Before 1950 and the introduction of synthetic pesticides, bees were plentiful enough around Washington bogs for natural pollination (Shawa, 1984). In orchards, honey bees have been found to increase fruit yields if there are two or more hives per hectare (Kagel, 1993). Unfortunately, organophosphates are extremely toxic to bees. The timing of pesticides to avoid poisoning pollinators is an essential part of an effective IPM program. This practice also encourages farmers to look for alternatives to traditional insecticides.

Prospects for reduction of pesticides in Washington State

Pesticide use in Washington has gone through all the standard traumas during the time (approximately 100 years) that the materials have been used. Despite the entrenchment of pesticides in standard production practices, however, innovative programmes and coordination among federal and state agencies, cooperative extension offices and farmers can make pesticide reduction a reality.

Nevertheless, it is important to note that farmers generally feel overwhelmed by regulation. Cranberry farmers, for example, deal not only with the regulations on pesticide use that have been discussed above, but also with regulations concerning wetland protection. Frustration with regulations appears to stem primarily from regulations affecting the availability of pesticides. However, this fosters frustration with any additional regulations that may further affect crop yields or how a farmer does business. This position was stated clearly in a newsletter from Washington State University: 'Too many regulations from too many agencies with too many fingers in the pie. Most pesticide issues are overblown and out of proportion.' (Patten, 1994). Given this frame of mind, it is clear that regulatory efforts need to be coordinated and have objectives that are clearly communicated.

Reduction of pesticide use currently seems more likely for the Washington cranberry than for the apple. The devastating brush with the Delaney amendment over 35 years ago has probably left an indelible sensitivity to pesticide issues in the minds of cranberry producers. Cranberry growers are strongly encouraged by Ocean Spray Cooperative to develop IPM plans and implement best management practices. Since all but two Washington cranberry growers are members of Ocean Spray Cooperative, and growers are dependent on Ocean Spray for their income, there is a strong incentive to follow 'company policy'.

Washington cranberry bogs are small, an average of only 12 acres, and are primarily family owned. Most farmers and their families live adjacent to their bogs and rely on well water. Given this situation, farmers are inclined to have a stewardship ethic and be concerned about how pesticides affect the health of themselves and their families. Since farms are relatively small, bog owners are also more likely to know, or become acquainted with, their workers and be concerned with their safety.

The Coastal Cranberry Initiative is a cooperative project that is further encouraging IPM and best management practice implementation. 'Best Management Practices' are those which prevent or reduce non-point pollutant discharges. These practices become best management practices when they are approved by the Washington State Department of Ecology (Washington Administrative Code 173-201A-020). Initiated by the Pacific County Conservation District, this project was funded by an EPA grant and involves the USDA Natural Resource Conservation District, Washington State University Cooperative Extension, Washington Department of Ecology and the cranberry farmers.

Half of the grant has been spent on research which has primarily been conducted by Washington State University. Research projects have included the timing of insecticides to minimize harm to pollinating bees, raptor perches to attract raptors for vole control, carbon filters to remove pesticide residues in drainage water, and temperature sensors for determining frost protection measures. The rest of the grant money was used to develop a best management practice guide for cranberry growers (M. Norman, personal communication, 1995). While cranberry farmers attract wildlife to bogs as a means of pest control (for voles), this program should include minimizing the use of highly toxic pesticides. The Coastal Cranberry Initiative has been described as allowing cranberry farmers to control their regulatory destiny (Patten, 1994). This program also is an example of interagency cooperation; something that may help to improve farmer and government relations.

Apple orchards, on the other hand, are becoming corporate owned. The family-owned orchard is becoming a thing of the past, and there are few profitable orchards of less than 40 acres in size (M. Skylstad, personal communication, 1993; E. Beers, personal communication, 1994). As corporations purchase orchards, the absentee-landlord syndrome is in place. Managers are hired to run the orchard and hire a work crew. The manager's job is to manage aggressively for the greatest yield and greatest profit. Worker safety, including training and personal protective equipment, may in some cases be seen as an inconvenience and an expense.

Apple growers, too, have had a historic brush with pesticide problems in the scare over Alar contamination in 1989 (Hathaway, 1993). It was, perhaps, only pure blind luck that kept the Phosdrin poisonings of 1993 from erupting into a full-blown public panic that could have devastated the market for Washington's apples. In contrast to the cranberry producers, however, the apple industry in Washington has not made a visible effort to move toward pesticide reduction other than for reasons of reducing problems of resistance, destruction of natural enemies and beneficials, and reduction of costs. Orchards continue to be a hazardous workplace, and occupational exposure probably remains the biggest source of environmental damage from pesticides on apples.

HISTORY'S IMPLICATIONS FOR PESTICIDE REDUCTION

Our discussion of the history of pesticides, both generally and in Washington State, leads us to three major observations that are relevant to the prospects for achieving pesticide reduction.

First, pest control practices have changed substantially over time, for a variety of reasons. Achievement of pesticide reduction for its own sake, however, has never been a factor driving change. Instead changes have occurred owing to the invention of new chemicals, problems with resistance and loss of natural enemies, and regulatory decisions that have banned certain chemicals because of their dangers to human health and wildlife. IPM and other practices that reduce pesticide use have been voluntarily adopted in some cases. All of these changes were reforms of pest control practice, but generally they were not linked.

Reformers from the inside, generally agricultural scientists, sought reform through technological innovations such as IPM and TPM. Similarly, growers and their representatives, like the Ocean Spray Cooperative, sought reform when they had suffered great economic damage due to pesticides. Reformers from the outside, especially the environmental movement, focused more on health issues, generally safeguarding the health of consumers from the effects of residues. Farm labour organizations tended to emphasize the occupational hazards of pesticides.

It is not that these different groups cannot work together, or have not done so from time to time, but the general pattern has tended to be narrowly conceived reform efforts with little serious attention being given to linking arms for a general reform programme. Advocates of pesticide reduction will probably need to develop a comprehensive program that can enlist the support of all those who desire reform in pesticide affairs. Coordination of research, extension advice, industry initiatives and regulatory policies will all be essential.

Second, health problems deserve a special comment. We have described health problems as having two foci: dangers to consumers from residues and dangers to workers from occupational exposure. Our observation is that the former has tended to be given more attention and to assume more importance in Congress. Yet the largest number of morbidities of unequivocal etiology is in occupational exposure. We find it utterly reprehensible and shocking that an epidemic of poisoning by Phosdrin could erupt as recently as 1993. In no way do we conclude that the threat to consumers from residues is not a problem worthy of attention, but if society wants to reduce sickness and anxiety from pesticide poisoning faster, then more political and policy attention needs to be directed toward occupational exposures. Pesticide reduction could play a vital role in ending one of the greatest scandals in contemporary labour relations.

Third, our review of the history of pesticide use suggests an important element: legal and cultural institutions are not geared to achieving pesticide reduction. In fact, just the opposite is true. Our economic and political institutions are geared to the promotion of efficient production in agriculture and elsewhere. Based on accounting systems currently in use (i.e. 'externalities' from pesticide damage not charged to the pesticide user), pesticides are often, but not always, cheaper than alternative practices. Furthermore, we have elaborate systems of patent and copyright laws to protect the intellectual property rights of technologies such as pesticides, but those institutions do not offer comparable protection to strategies like IPM. Additionally, in the era of GATT and global free trade, our export policies treat exports of agricultural goods and pesticides as a valuable contribution to American well being.

In short, it seems that legal and institutional leanings are towards pesticides and away from pesticide reduction. Advocates of pesticide reduction must therefore realize that they are up against some barriers that will not easily be breached. Despite the problems of achieving pesticide reduction, we believe that it is a goal worth pursuing. The project of pesticide reduction offers a chance to initiate a new pathway for pest-control reform. Pests will remain a problem, often a serious one, but reduction of pesticide use has the potential to reduce many other problems. We believe the creativity of people in agriculture, industry and the universities can create new pest-control practices less dependent upon pesticides. The challenge of pesticide reduction is one which should stimulate and then celebrate, human ingenuity.

ACKNOWLEDGEMENTS

We are indebted to Hugo Flores, Veronica Schmitt and Darius Sivin for all the work they did on *The Phosdrin Lesson*, which we have used extensively in this article. We are also indebted to unpublished work by David Giglio, Susan Kagel and Nicola Ostertag. We thank David Pimentel for helpful comments on a draft of this manuscript. All errors of commission and omission, of course, remain our responsibility.

REFERENCES

Auger, J., Kunstmann, J.M., Czyglik, F. and Jouannet, P. (1995) Decline in semen quality among fertile men in Paris during the past 20 years. *New England Journal of Medicine* **332**(5), 2 February, 281–285.

Beers, E.H., Brunner, J.F., Willett, M.J. and Warner, G.M. (eds) (1993) *Orchard Pest Management*. Good Fruit Grower Publishing, Yakima.

Bosso, C.J. (1987) *Pesticides and Politics. The Life Cycle of a Public Issue*. University of Pittsburgh Press, Pittsburgh, PA.

Bouwer, H. (1990) Agricultural chemicals and ground water quality: Issues and challenges. *Ground Water Monitoring Review* **10**(1), 71–79.

Boyce, A.M. (1936) The citrus red mite *Paratetranychus citri* McG. in California and its control. *Journal of Economic Entomology* **29**, 125–130.

Brake, A. and Krause, W. (1992) Decreasing quality of semen. *BMJ* **305**, 12 December, 1498.

Broaddus, A. (1991) *Cranberry Vine Newsletter*. WA State University Cooperative Extension, Long Beach, WA, May.

Brown, A.W.A. (1961) The challenge of insecticide resistance. *Bulletin of the Entomological Society of America* **7** 6–19.

Brun, L.O., Marcillaud, C., Gandichon, V. and Suckling, D.M. (1994) Cross resistance between insecticides in coffee berry borer, *Hypothenemus hampei* (Colleoptera: Scolytidae) from New Caledonia. *Bulletin of Entomological Research* **84**, 175–178.

Carlsen, E., Giwercman, A., Keiding, N. and Skakkebaek, N.E. (1992) Evidence for decreasing quality of semen during the past 50 years. *British Medical Journal* **305**, 12 September, 609–613.

Carson, R. (1962) *Silent Spring*. Houghton Mifflen, Boston, MA, 368 pp.

Chandler, F.B. (1956) *Survey of the Cranberry Industry in Washington State*. WA State Department of Agriculture, Olympia, WA, September.

Clausen, C.P. (1936) Insect parasitism and biological control. *Annals of the Entomological Society of America* **29**, 201–223.

Cohen, S.Z. (1990) Pesticides in ground water: An overview. In Hutson, D.H. and Roberts T.R. (Eds) *Environmental Fate of Pesticides*. John Wiley & Sons, New York, pp. 13–25.

Colborn, T., vom Saal, F.S. and Soto, A.M. (1994) Developmental effects of endocrine-disrupting chemicals in wildlife and humans. *Environmental Impact Assessment Review* **14**, 469–489.

Cushman Jr., J.H. (1994) EPA settles suit and agrees to move against 36 pesticides. *New York Times*, 13 October.

Davis, D. and Johnson, A. (1996) *Washington State Pesticide Monitoring Program, 1994 Surface Water, Fish and Sediment Sampling Report*. Washington State Department of Ecology, Olympia, WA, Publication 96–305, February.

DeBach, P. (1946) An insecticidal check method for measuring the efficacy of entomophagous insects. *Journal of Economic Entomology* **39**, 695–697.

DeBach, P. and Rosen, D. (1991) *Biological Control by Natural Enemies* 2nd edn. Cambridge University Press, Cambridge, pp. 140–148.

Deubert, K.H. and Zuckerman, B.M. (1969) Pesticides in soil: Distribution of Dieldrin and DDT in cranberry bog soil. *Pesticides Monitoring Journal* **2**(4), 172–175.

Dunlap, T.R. (1981) *DDT: Scientists, Citizens, and Public Policy*. Princeton University Press, Princeton, 209 pp.

Eck, P. (1990) *The American Cranberry* Rutgers University Press, New Brunswick. p. 23.

Environmental Impact Assessment Review Team (1994) An interview with Theo Colborn. *Environmental Impact Assessment Review* **14**, 491–497.

Epidemiologic Notes and Reports (1994) Occupational pesticide poisoning in apple orchards—Washington, 1993. *Morbidity and Mortality Weekly Report* **42**, January, 993–995.

Erickson, D. (1992) *Gleed Agricultural Chemicals Ground Water Quality Assessment* WA State Department of Ecology, Olympia, WA.

Erickson, E.H., Erickson, B.J. and Wyman, J.A. (1994) Effects on honey bees of insecticides applied to snap beans in Wisconsin: chemical and biotic Factors. *Journal of Economic Entomology* **87**, 596–600.

Feldman, J. (1994) A landmark worth keeping. *The Environmental Forum*, January/February, p. 25.

Flamm, W.G. (1995) Regulatory aspects of carcinogenicity studies and the Delaney clause. *Regulatory Toxicology and Pharmacology* **21**, 71–74.

Flores, H., Patterson, B., Perkins, J.H., Schmitt, V. and Sivin, D.D. (1994) *The Phosdrin lesson: Study in the environmental politics of an agricultural technology*. Master's Program in Environmental Studies, Working Paper 94-1, The Evergreen State College, Olympia, WA, 49 pp.

Folsom, J.W. (1927) Calcium arsenate as a cause of aphid infestation. *Journal of Economic Entomology* **20**, 840–843.

Gallaher, G.L. (1967) Recent experiences with parathion poisoning. Paper presented at Texas Medical Association Annual Meeting, Dallas, May.

Giglio, D. (1993). DBCP and the EPA, a case study in pesticide regulation. Unpublished paper prepared at The Evergreen State College, Olympia, WA.

Hathaway, J.S. (1993) Alar: The EPA's mismanagement of an agricultural chemical. In Pimentel, D. and Lehman, H. (Eds) *The Pesticide Question: Environment, Economics, and Ethics.* Chapman & Hall, New York, pp. 337–343.

Hollomon, D.W. (1993) Pesticide resistance. *Chemistry & Industry* **22**, 892–895.

Howard, L.O. (1930) *A History of Applied Entomology.* Smithsonian Institution, Washington, DC, p. 64.

Ibrahim, M.A. et al. (1991) Weight of the evidence on the human carcinogenicity of 2,4-D. *Environmental Health Perspectives* **96**, 213–222.

Irvine, D.S. (1994) Falling sperm quality. *BMJ* **309**, 13 August, 476.

Kagel, S. (1993) The Effects of Pesticide Use on Honey Bees. Unpublished paper prepared at The Evergreen State College, Olympia, WA.

Karr, C., Kalat, J., Locke, D. and Lirette, P. (1991). *Farmworker Health and Safety in Washington State.* WA State Department of Labor and Industries, Olympia, WA, p. 6.

Kelce, W.R., Stone, C.R., Laws, S.C., Earl Gray, L., Kemppalnen, J.A. and Wilson, E.M. (1995) Persistent DDT metabolite p,p'-DDE is a potent androgen receptor antagonist. *Nature* **375**, 15 June, 581–585.

Krieger, N., Wolff, M.S., Hiatt, R.A., Rivera, M., Vogelman, J. and Orentreich, N. (1994) Breast cancer and serum organochlorines: A prospective study among white, black, and Asian women. *Journal of the National Cancer Institute* **86**(8), 20 April, 589–599.

Lum, W.E. (1984) *A Reconnaissance of the Water Resources of the Shoalwater Bay Indian Reservation and Adjacent Areas, Pacific County, WA, 1978–1979.* USGS Report 83-4165, Tacoma, WA.

Maddy, K.T., Edmiston, S. and Richmond, D. (1990) Illness, injuries, and deaths from pesticide exposures in California, 1949–1988. *Reviews of Environmental Contamination and Toxicology* **114**, 57–123.

Merrill, R.A. (1994) Congress as scientists. *The Environmental Forum*, January/February, pp. 20–26.

Miller, C.W., Zuckerman, B.M. and Gunner, H.B. (1967) Pesticide occurrence, concentration, and degradation in free water systems (WR-10). In *Proceedings of the Water Resources Research Symposium 2 June 1967.* Water Resources Research Symposium, University of Massachusetts, pp. 94–99.

Moss, S.R. and Rubin, R. (1993) Herbicide resistant weeds: A worldwide perspective. *Journal of Agricultural Science, Cambridge* **120**, 141–148.

National Research Council (1987) Committee on Scientific and Regulatory Issues Underlying Pesticide Use Patterns and Agricultural Innovations. *Regulatory Pesticides in Food: The Delaney Paradox.* National Academy Press, Washington, DC, 272 pp.

Osteen, C. (1993) Pesticide use trends, and issues in the United States. In Pimentel, D. and Lehman, H. (Eds) *The Pesticide Question—Environment, Economics, and Ethics.* Chapman & Hall, New York, p. 310.

Ostertag, N. (1995) Ground water contamination by atrazine and its metabolites in Midwestern states and policy issues. Unpublished paper prepared at The Evergreen State College, Olympia, WA.

Patten, K. *The Cranberry Vine Newsletter* (1994) WA State University Cooperative Extension, Long Beach, WA, November.

Patterson, B. (1994) Pesticide use on the cranberry in Washington: An evaluation of the environmental risks and regulations. Essay, The Evergreen State College, Olympia, WA.

Perkins, J.H. (1978) Reshaping technology in wartime: The Effect of military goals on entomological research and insect-control practices. *Technology and Culture* **19**(2), 169–186.

Perkins, J.H. (1982) *Insect, Experts, and the Insecticide Crisis.* Plenum Press, New York, 304 pp. 68 Stat 511–517.

Pignatello, J.J. and Cohen, S.Z. (1990) Environmental chemistry of ethylene dibromide in soil and ground water. *Reviews of Environmental Contamination and Toxicology* **112**, 1–47.

Reich, G.A., Gallaher, G.L. and Wiseman, J.S. (1968) Characteristics of pesticide poisoning in south Texas. *Texas Medicine* **64**(9), 56–58.

Rinella, J.F., Hamilton, P.A. and McKenzie, S.W. (1993) *Persistence of the DDT Pesticide in the Yakima River Basin Washington.* US Geological Survey Circular 1090, USGPO.

Ripper, W.E. (1956) Effect of pesticides on balance of arthropod populations. *Annual Review of Entomology* **1**, 403–438.

Ritter, W.F. (1990) Pesticide contamination of ground water in the United States—A review. *Journal of Environmental Science and Health* **B25**(1), 1–29.

Roussel, J.S. and Clower, D. (1955) Resistance to the chlorinated hydrocarbon insecticides in the boll weevil (*Anthonomus grandis* Boh.). Louisiana Experiment Station Circular No. 41, Louisiana State University and Agricultural and Mechanical College, 5 pp., plus tables.

Sawyer, R.C. (1996) To make a spotless orange: Biological control in California. Iowa State University Press, Ames, 290 pp.

Schneider, K. (1993) EPA plans to ask Congress to relax rule on pesticides: Chief calls law outdated. *New York Times*, 2 February.

Schreiber, A. (1994) *Agrichemical and Environmental News.* WA State University, 108, February, p. 4.

Shawa, A.Y., Shanks, Jr., C.H., Bristow, P.R., Shearer, M.N. and Poole, A.P. (1984) *Cranberry Production in the Pacific Northwest.* WA State University, PNW 247, September, p. 34.

Sherins, R.J. (1995) Are semen quality and male fertility changing? *New England Journal of Medicine* **332**,(5), 2 February, 327–328.

Steiner, H.M. (1938) Effects of cultural practices on natural enemies of the white apple leafhopper. *Journal of Economic Entomology* **31**, 232–240.

Sweetman, H.L. (1936) *The Biological Control of Insects.* Comstock, Ithaca, NY, pp. 278–280.

Tabashnik, B.E. (1994) Evolution of resistance to *Bacillus thuringiensis. Annual Review of Entomology* **39**, 47–79.

Thompson, W.L. (1939) Cultural practices and their influence on citrus pests. *Journal of Economic Entomology* **32**, 782–789.

[United Farm Workers Organizing Committee (1965?)] *The Delano Grape Strike and Boycott.* United Farm Workers Organizing Committee, Delano, CA, 15 pp.

US Congress (1995) *Biologically Based Technologies for Pest Control.* Office of Technology Assessment, OTA-ENV-636, Government Printing Office, Washington, DC, pp. 19–21.

US Congress (1958) Committee on Appropriations, Department of Agriculture Appropriations for 1959, Hearings, Part 2, 85th Congress, 2nd session, pp. 449–466.

US Department of Commerce (1992) *1992 Census of Agriculture*: Vol. I. Economic and Statistics Administration, Bureau of the Census, Geographic Area Series, Part 47, WA State and County Data, AC92-A-47 (acreage for Grays Harbor County was based on 1987 data).

US EPA (1995) *Pesticide Reregistration Report.* Office of Prevention, Pesticides, and Toxic Substances, EPA 738-R-95-010, January, p. 19.

WA Department of Agriculture (1995) *Washington Agricultural Statistics 1994–1995.* Olympia, WA.

Washington Toxics Coalition and Cooperative Extension—Washington State University (1994) *The State of Our Groundwater: A Report on Documented Chemical Contamination in Washington.* Washington State University, Pullman, WA, EB1756.

WA State Department of Health (1995) *Annual Report 1994. Pesticide Incident Reporting and Tracking Review Panel.* Olympia, WA, March.

Whorton, J. (1974) *Before Silent Spring.* Princeton University Press, Princeton, NJ, 288 pp.

Wolff, M.S., Toniolo, P.G., Lee, E.W., Rivera, M. and Dubin, N. (1993) Blood levels of organochlorine residues and risk of breast cancer. *Journal of the National Cancer Institute* **85**(8), 21 April, 648–652.

ADDENDUM

On 3 August 1996, President Clinton signed the Food Quality Protection Act of 1996 (P.L. 104–170), which took effect immediately. This new law eliminated the Delaney amendment to the Federal Food, Drug, and Cosmetic Act. EPA will now establish tolerances on all food so that they are 'safe', which is defined to mean 'a reasonable certainty that no harm will result from aggregate exposure'. In addition, the new law requires special considerations to protect infants and children and new protocols to test for endocrine disruption. This new law, therefore, has the potential to substantially alter the regulation of pesticides in the United States.

CHAPTER 3

Environmental Ethics and Pesticide Use

Hugh Lehman

University of Guelph, Guelph, Ontario, Canada

INTRODUCTION

In a recent paper, I argued that at the present time we ought to reduce our use of synthetic pesticides in agriculture (Lehman, 1993). That does not mean that everyone must discontinue use of pesticides immediately. Defenders of pesticide use continue to stress the benefits for humans to be derived from pesticide use in agriculture, and predict that people will require them in agriculture for the foreseeable future (Furtick, 1976; Reding, 1993). Where use of synthetic pesticides is required for the production or preservation of sufficient affordable nutritious food for human beings or to enable agricultural workers or others to achieve a significantly better quality of life through relief from disease or drudgery, there are strong ethical arguments in favour of continued use of pesticides. The same considerations could even justify limited extensions of the use of pesticides in some circumstances where they are not currently used. However, where pesticides are used to produce cash crops rather than food for local populations, this justification for their continued use is undermined. Furthermore, there is ample evidence that it is possible to reduce pesticide use significantly in many instances without causing significant crop losses, or adversely affecting agricultural workers (Postel, 1988; Pettersson, 1993; Pimentel, et al., 1993b,c; Surgeoner and Roberts, 1993). Even avid defenders of pesticide use admit that DDT, and perhaps other pesticides as well, have been overused (Ray and Guzzo, 1994). Given the extensive documented harm to human beings and other creatures which has resulted from exposure to pesticides, it is clear that nations around the world should aim at reducing the opportunity for such exposure where that can be achieved without compromising human nutrition or worker well-being (Metcalf, 1993; Pimentel et al., 1993a). The arguments of my paper cited above support continued development and deployment of techniques for integrated pest management or other means of reducing dependence on synthetic pesticides. That conclusion is warranted on the basis of either utilitarian or Kantian moral principles.

To a great extent, the arguments offered by scientists and others to support the criticism of pesticide use have focussed primarily on harm to humans. For example, humans have been poisoned by exposure to pesticides. Adverse effects on human health

---

Techniques for Reducing Pesticide Use. Edited by D. Pimentel
© 1997 John Wiley & Sons Ltd.

have been documented (Pimentel et al., 1993a). Even where non-humans have been damaged or injured through exposure to pesticides, the harm is often perceived in terms of economic losses to humans. As indicated in my paper, utilitarian ethical principles imply that harm to certain animals also must be taken into account to determine what activities involving pesticide use are ethically justifiable. In this chapter I shall develop that point in greater depth. I shall do this through considering a context framed by consideration of environmental ethics. Subsequently, I shall consider a number of theories of environmental ethics which diverge in various ways from both utilitarian and Kantian principles.

For the past few decades there has been increasing interest among philosophers in the development of environmental ethics. Some theories of environmental ethics suggest criticisms of pesticide use on the basis of harm to animals and, even further, on the basis of harm to the environment itself. Such theories of environmental ethics imply that even were all harm to human beings eliminated through development of improved pesticides or improved methods of using pesticides, for example pesticides which quickly break down into benign or even beneficial substances or which are toxic only to the target pest, as well as methods of ensuring that only the absolute minimum amount necessary is used and that it is all used directly on the target pest, there would remain further strong reasons to reduce or, where possible, eliminate pesticide use. In this chapter I shall introduce and discuss some theoretical considerations underlying such reasons. To facilitate such discussion it is useful to develop a classification of theories of environmental ethics.

Theories of environmental ethics may be subdivided into three categories. In the first category are theories which imply that we must protect our environment in order to fulfill our obligations to human beings. Much discussion of our obligations concerning the use of pesticides has, as we noted above, been justified by reference to theories of this first category. Since I have explained the theoretical basis of such theories in my paper cited above, I shall not devote space here explicitly to consideration of theories of the first category.

Theories in the second category imply that we must protect our environment in order to fulfill our moral obligations to certain animals as well as humans. While I mentioned this possibility in my earlier paper, in the following section I will develop and discuss such ideas further. Theories in the third category imply that we must protect our environment in order to fulfil our moral obligations to living beings, to ecosystems, to species and ultimately to the biosphere as a whole. I shall refer to such theories as *environmental ethics proper*. I shall explain theories of this third category in the second main section of this chapter. Were such theories true, restrictions on acceptable uses of pesticides would go far beyond restrictions aimed primarily at avoiding serious harm to humans or animals. In the third section I shall consider elements of a rationale for environmental ethics proper, including consideration of some criticisms of the value theories implied by utilitarianism. I shall also suggest some objections to theories of the third category. In the final part I shall briefly explain and criticize deep ecology. Deep ecology diverges even further from utilitarian and Kantian principles than other theories of environmental ethics proper, and probably entails that even more stringent restrictions on the use of pesticides would be required.

SPECIESISM

Tom Regan, in several publications, has argued that certain sorts of animals have basic moral rights, and consequently that we have moral obligations not to harm those animals (Regan, 1982b, 1983). Although he does not discuss pesticide use, Regan's view would imply that we ought not to kill those animals that have moral rights to life through use of pesticides. For example, his view would inhibit use of rodenticides in situations in which such use was required to defend humans against attack by rodents. A utilitarian view which would also imply that we are required to restrict our use of pesticides to avoid harm to animals has been developed by Peter Singer. Singer elaborated a concept of speciesism which may be employed to criticize many uses of pesticides in agriculture. I shall discuss Singer's view here. Either view implies that when we are trying to determine whether use of a pesticide is acceptable, we are obligated to consider the well-being of animals as well as humans. I have discussed and criticized Regan's view elsewhere (Lehman, 1984, 1988). Regan's view has also been criticized by Jamieson (1990).

In his book *Animal Liberation*, Peter Singer explained a concept which he called speciesism (Singer, 1975). Speciesism is an attitude of bias against members of other species. In Singer's view, speciesism is strictly analogous to racism, sexism, ethnocentrism or religious bias. Behaviour expressive of such biases is, unless very remarkable conditions obtain, morally unacceptable.

To explain speciesism (or racism, etc., as Singer conceived them) we must explain a moral principle which is commonly called the principle of equal consideration of interests. To explain this principle let us introduce the terms 'moral standing' and 'moral agent'. A being is a moral agent if that being can have moral obligations. Normally, mature human beings are moral agents whereas most animals are not moral agents. We do not attribute moral agency to most animals because such creatures are incapable of understanding moral concepts, such as the concept of duty, and further are incapable of voluntarily regulating their behaviour in order to conform to moral rules. A creature or object is said to have moral standing if some moral agents have moral obligations to it. For example, in conventional ways of thinking, mature human beings are not only moral agents, they also have moral standing. However, in modern conventional moral thought, human children have moral standing but are not moral agents. Moral agents have obligations to children even if the children are too young to have moral obligations themselves.

In conventional moral thinking, we (moral agents), with one major exception, do not have obligations *to* other sorts of objects. In such ethical thought, if we have an obligation concerning such an object, for example an obligation not to damage a chair or a statue, that obligation derives from some obligation we have to one or more human beings. Our obligation not to damage the chair is not an obligation *to* the chair even though it involves the chair. The exception to this general rule is that an obligation to treat animals humanely is now recognized as valid. This was not always the case, as animals in some places and times have been regarded merely as property. Further, some philosophers have suggested that we do not have moral obligations to animals (Narveson, 1980; Kant, 1989).

Singer's view concerning the proper scope of the term 'moral standing' derives from his acceptance of utilitarian ethical theory. According to utilitarianism, our fundamental moral obligation is to act so as to produce minimum pain, maximum pleasure or maximum satisfaction of desires. In this view, all creatures capable of feeling pleasure or pain or of having desires or preferences have moral standing. Singer refers to such creatures as sentient. We (moral agents) have moral obligations to sentient creatures. To determine our obligations, we have to determine how our actions will affect the pleasure, pain or satisfaction of such creatures (including other human beings). To express this idea more concisely, Singer used the term 'interests' in place of more complex references to pleasure, pain or desires. In his view, to determine our obligations we have to consider the interests of all sentient creatures.

Reflection about interests suggests that interests vary in regard to how much they count in the determination of our obligations. Consider, for example, a desire to eat. This desire may be strong or forceful or it may be mild or weak. Since, according to utilitarian thinking, we are obligated to act so as to maximize satisfaction or pleasure, we must take the forcefulness of desires into account. Satisfaction of a forceful desire counts for more than does satisfaction of a mild desire. Similarly, production of a small pleasure counts for less than does production of a great amount of pleasure, and production of a great pain counts for more than does production of a small pain. To express these ideas concisely we may use the term 'weight'. A weightier interest is one which satisfies a more forceful desire or which derives from a greater amount of pleasure or pain.

The principles of equal consideration of interests implies that we are obligated to give due consideration to all interests. In other words, according to this principle we fail in our obligations to someone or some creature if we ignore or discount that creature's interests in our efforts to determine which of the actions we can perform will yield the maximum pleasure or satisfaction. For example, suppose that a wild mammal has an interest in not suffering pain, then if when we are deciding whether to use a pesticide we ignore that interest, we are not giving that creature due consideration, we are not acting in accord with the principle of equal consideration. Speciesism consists in ignoring or discounting some creature's interests merely because those creatures are members of some species other than *Homo sapiens*. If, in our agricultural practices, we decide whether to use pesticides without taking the interests of sentient creatures such as mammals or birds into account, then we are being speciesist.

To avoid speciesism in regard to our use of pesticides would not be easy. Often in trying to determine whether to pursue an activity, for example use of a particular pesticide, we try to weigh the harm and the benefits. We assume that an activity is acceptable if the difference between its benefit and harm is positive, that is produces more pleasure than pain or more satisfaction of desires than frustration, and if there is no alternative activity which yields a greater positive difference. To make these assessments we calculate the benefits and the harms in terms of prices. We consider the dollar value of the benefits and the dollar value of the harm. Of course, using prices as a way of measuring benefits and harm yields only an approximation to the real values associated with the activity. To the extent that the activity produces pleasures, pains or satisfactions that are not expressed in monetary terms, monetary assessments of the value of such activities will be inaccurate. In this context it is easy to see that by using price as a way of estimating values we will almost inevitably manifest speciesism. Since animals do not express their interests in monetary terms and, for the most part, no one else expresses animal interests in that way for the animals, using monetary estimates of utility will automatically result in ignoring the animals' interests. Consequently, the activities which result from making decisions in this way are speciesist.

There are several claims that a defender of the use of pesticides might make in an effort to rebut this criticism. First, the defender might claim that sentient animals lack moral standing. Second, the defender of pesticide use might claim that sentient animals do not suffer pain or other harm as a result of pesticide use. Third, the defender of pesticide use might claim that while sentient animals do suffer pain or other harm, use of pesticides is justified in order to protect human interests against infringement by the animals. Let us consider these attempted rebuttals in order.

Do sentient animals have moral standing? Utilitarians assume that they do. This assumption can be supported by the following argument. We are morally obligated to produce the greatest total amount of goodness or the least amount of harm. Pleasure or satisfaction of desires is good. Pain or frustration of desires is bad. Since some animals have the capacity to experience pleasure, pain, satisfaction or frustration, what we do to those animals influences the amount of goodness we produce. Thus, we are morally obligated to take the interests of those animals into account, i.e. those animals have moral standing.

In my judgment this argument is compelling, i.e. a rational person who is aware of the relevant facts should accept its conclusion. However, this argument, like any argument, can be challenged further. It rests on the assumptions that we are obligated to act so as to produce the greatest total goodness, that pleasure, satisfaction of desire, etc. are good, and that such good or bad is part of animal existence. Since these assumptions cannot be proved conclusively to be true, they are open to challenge. Libertarianism is an ethical perspective that rejects the fundamental utilitarian moral premise (Narveson, 1980). It is possible to doubt that pleasure or satisfaction of desires is good, or that pain or frustration of desires is bad. Some philosophers have challenged the assumption that animals have desires or preferences (Frey, 1980). Finally, some thinkers continue to doubt that animals have morally significant experiences of pleasure or pain (Carruthers, 1989). Whether animals are sentient has been discussed widely. I have discussed this issue at length elsewhere (Lehman, 1995).

The second effort to rebut the claim that some of our uses of pesticides in agriculture are speciesist involves denying that sentient animals suffer harm as a result of pesticide use. I do not regard such a claim as rationally credible. Among the pesticides we use to prevent damage to crops some are explicitly aimed at rodents. Some pesticides are aimed at birds. Further, other pesticides, for example insecticides or herbicides, also cause harm to mammals or birds. Where such uses of pesticides cause animals to suffer pain, etc., they harm such animals. To the extent that our use of pesticides is based on ignoring or discounting the pain that such creatures suffer, such use of pesticides is speciesist.

The most plausible ground for challenging the claim that our use of pesticides in agriculture is often speciesist rests on the assumption that we are entitled to protect our own interests and that use of synthetic pesticides is essential for this purpose. However, in my judgment, even this line of defence fails. I would agree that we are entitled to use synthetic pesticides where that is necessary to produce sufficient food to provide adequate affordable nutrition to human beings or to protect humans against serious diseases such as malaria. However, human beings are not entitled to use any means whatsoever to protect any human interests no matter how trivial. With respect to some uses of synthetic pesticides, it is arguable that the interests we try to protect are too trivial to justify inflicting pain or suffering on animals. For example, if use of a herbicide or insecticide causes sentient animals to suffer, and the objective of that pesticide use is to achieve excessively high cosmetic standards of fruits or vegetables where human nutritional needs can be amply satisfied by less 'perfect' products, then that use of the herbicide or insecticide is open to the charge of speciesism. Given that human interests in nutrition and even good taste can be well satisfied without such excessive use of pesticides, the charge of speciesism appears valid. Similarly, such a charge appears valid in such cases as the use of herbicides or other pesticides in the production of grass for human recreation, such as golf, where such uses cause suffering to sentient creatures such as mammals or birds. The only satisfactory way to rebut the charge of speciesism in such cases is to show that the pleasure or satisfaction of desires which humans experience from such pesticide use is of such magnitude that even taking the pain or suffering of animals into account, such use of pesticides yields greater overall utility than any alternative course of action.

ENVIRONMENTAL ETHICS PROPER

We may distinguish several forms of utilitarianism by reference to distinct theories concerning the basic nature of goodness or value. Hedonists maintain that all goodness is ultimately reducible to pleasure or pain. One form of utilitarianism is hedonistic conse-quentialism. This is the view that every person is morally obligated to act in ways which yields a maximum of pleasure and minimum of pain. Some philosophers, while rejecting hedonism, maintain that ultimately goodness consists in the satisfaction of desires or preferences, while that which frustrates desires is bad. Combining this view of goodness with the assumption that we are all obligated to act so as to maximize goodness yields another form of utilitarianism, commonly called preference utilitarianism. According to both forms of utilitarianism the only objects that have moral standing are sentient creatures, beings who experience pleasure, pain, or satisfaction or frustration of desires. However, one may raise the question whether any non-sentient beings have moral standing. I argued above that sentience is a sufficient condition for moral standing. However, a number of thinkers have argued that sentience, or even bare consciousness, is not a necessary condition for moral standing (Leopold, 1949; Goodpaster, 1978; Regan, 1982; Callicott, 1986; Taylor, 1986; Johnson, 1991). In defence of their views on moral standing, such thinkers advocate alternative theories concerning the basis of value.

Perhaps the most well-known exponent of such an alternative theory of value is Albert Schweitzer (Schweitzer, 1923). For example, Schweitzer makes claims which imply that life itself is a basic value. An elaborate philosophical defence of a position such as Schweitzer's was developed by Taylor and is often referred to as *biocentrism* (Taylor, 1986). Johnson's ethical theory is not only biocentrist, it is holistic. That is to say, in Johnson's view, not only do individual organisms have moral standing, but so also do entities such as species or ecosystems. Contrary to the value assumptions of biocentrist or holistic theories, utilitarian or Kantian ethical theories affirm that life has value only insofar as it is conducive to the attainment of pleasure or satisfaction or to the attainment of the legitimate objectives of rational beings. In the utilitarian or Kantian view life is said to be merely of instrumental value, value as a means to an end. This, of course, is not to assert that the value of life is small. The instrumental value of life is directly proportional to the values of the goals or ends for which it is required. However, if life is valuable merely as a means for achieving pleasure (or something else deemed to be a basic value) that would have a number of important implications. On a utilitarian theory we could not be obligated to sacrifice the interests of sentient beings merely to promote the existence of non-sentient forms of life. According to a Kantian ethical theory we are obligated never to use rational beings merely as resources. However, in a Kantian view, there is nothing morally unacceptable about using non-rational living beings merely as resources. The basic value and ethical assumptions of biocentrist or holistic views imply that it is morally unacceptable to use non-rational living beings or holistic entities merely as resources. Some thinkers have suggested that such theories of environmental ethics should be distinguished from theories which rest on utilitarian or Kantian assumptions about moral standing; thus we suggest that such theories be referred to as environmental ethics proper (Regan, 1982a). A third form of environmental ethics proper may be distinguished, namely deep ecology. I shall discuss deep ecology below.

Many people will raise an immediate objection to biocentrism on the grounds that living beings are entitled to do what is necessary to survive and reproduce and that, with the exception of plants that derive energy and nutrients directly from abiotic entities, all living beings can survive only through killing other living beings. While this objection might appear to be decisive, the biocentrist can reply to it in plausible ways. Given that creatures must kill to survive, biocentrists can argue that killing a creature, where this is necessary for one's own survival, is not incompatible with the thesis that all living creatures are equally inherently valuable. To do what is necessary to prevent one's own death is not to assume that one's own life is more valuable than the life of another creature. Such behaviour is justified if one's life is as valuable as that of the other creature. The biocentrist can maintain that living creatures find themselves in a situation in which bad consequences are inevitable. Thus, limited killing, while regrettable, is morally acceptable.

To compare the implications of a theory of environmental ethics of this sort with the implications of a utilitarian or Kantian theory it will be useful to consider a hypothetical example. For this purpose let us suppose that a new pesticide, which we shall call 'pest-1-kill, has been developed. Pest-1-kill might appear to be an almost ideal insecticide. It is 100% effective against the target insect, thereby eliminating any likelihood of developing resistant pests. It can be applied directly to the pests, thereby reducing wastage that results from applying pesticides which often miss the target pests. Further, pest-1-kill does not kill beneficial insects that normally prey on the target insect. Indeed, pest-1-kill breaks down quickly into substances which are apparently not harmful to humans, as has been shown by tests conducted over many generations on a wide range of animal models. Throughout all those tests, pest-1-kill showed not the slightest indication of being a mutagen, teratogen or carcinogen, and it also gave no indication that it could cause neurological disorders or disorders in any vital organs. Owing to the small amount of pesticide required and the rapid breakdown of pest-1-kill, it does not affect the quality of water for human consumption. Further, the LD 50s for pest-1-kill were so high that it was reasonable to conclude that consumption of this substance by sentient animals would be insufficient to cause pain or discomfort. Such an effective, apparently safe and even environmentally friendly insecticide would surely be approved for use if it were screened in accordance with criteria derived from utilitarian or Kantian considerations.

However, in spite of all of the above features, pest-1-kill could be objectionable on grounds that derive from an environmental ethic such as that advocated by Taylor or Johnson. As safe as pest-1-kill is, conceivably it would have such profound effects on other elements of the agricultural ecosystem as to modify them in extensive ways. Taylor's view could imply that use of pest-1-kill is unacceptable on the grounds that it causes unnecessary deaths of living organisms. Johnson's view could imply that use of pest-1-kill is unacceptable on the grounds that is causes unnecessary modifications of ecosystems. Conceivably some populations of insects in the ecosystem would be significantly reduced in number, or their behaviour would be modified to such a degree that the roles formerly played by those insects were vacated. Conceivably the roles would go unfulfilled for a time, perhaps leading to a build-up of some material that the insects had degraded. As a result, on the basis of his ethical theory, Johnson could claim that the original ecosystem had been significantly damaged. Even if the changes in the ecosystem were not immediately harmful to human beings, Johnson could suggest that the resulting system of organisms and materials no longer functioned as effectively. Alternatively, he could say that the integrity of the original ecosystem had been undermined, leading eventually to the disintegration of the ecosystem. Regulations based on an environmental ethic such as that of Johnson or Taylor could imply that pest-1-kill should not be approved.

CONSIDERATIONS RELEVANT TO RATIONAL JUSTIFICATION OF THEORIES OF ENVIRONMENTAL ETHICS

In order to try to establish an alternative theory of basic value, it is necessary to develop sufficiently strong reasons for thinking that non-sentient living beings or holistic entities are not merely instrumentally valuable. Taylor has attempted to do this through elaborating a biocentrist theory of value and arguing that such a perspective is acceptable on rational grounds. Johnson has attempted to support his theory of environmental ethics through arguing that basic value does not originate merely through the satisfaction of any desires or the production of any pleasure. Satisfaction of desires or production of pleasure may be valuable, in his view, providing that the satisfaction of desires contributes to fulfillment of the interests of the individuals in question. Johnson's view of interests differs from that of Peter Singer discussed above. In Johnson's view, the good of living organisms consists in the exercise of their capacities while preserving their identity and integrity over time. Something satisfies an organism's interests if it contributes to the preservation of its integrity or identity. Further, Johnson argues, living beings and holistic entities also have interests and, since the fulfillment of interests realizes value, the fulfillment of the interests of non-sentient living beings or holistic entities also has value. Consequently, such beings have moral standing also.

I do not believe it is appropriate, in book such as this which is not addressed primarily to philosophers, to undertake a thorough investigation of the rationale for theories of environmental ethics. However, I suspect that some readers of this may be tempted to dismiss such theories of environmental ethics as clearly absurd and thus false. I do not believe that such cavalier dismissal is rationally justified. These theories deserve serious consideration. To support this contention I shall review a number of considerations which might be offered on behalf of such theories. Since some people may be led to accept such a theory by appeal to common ways of speaking, I shall briefly formulate and criticize such a rationale. A stronger reason for accepting environmental ethics may be developed on the basis of criticism of the utilitarian theories of value. Thus, I shall discuss such criticisms. As part of this discussion, I will explain and criticize Johnson's criticism of the theory that value derives from the satisfaction of preferences. I shall conclude this section with a few considerations relevant to assessing Johnson's assumptions concerning the basis of value.

In defence of views such as those of Schweitzer, Johnson or Taylor, we might consider common ways of speaking. Consider a tree. Trees are neither conscious nor sentient. Yet it is not uncommon for people to speak of some condition as being *good for* or *bad for* trees. In the same vein, people speak of trees, or other plants, as thriving or being healthy, or alternatively as doing poorly. A healthy tree is one which is living well. This form of words suggests that it makes sense to speak of the good of the tree. Similarly, it is not uncommon for people to speak of some collection of entities as thriving or doing well. For example, a community may be said to be thriving and we may speak of the good of the community.

To speak of the good of a tree is to say that there is a state or condition of the tree which is good regardless of whether any person or other sentient creature values that condition. If there is a good *of* a tree or of a community, then those things which are good *for* trees (or communities) could well be things which promote or protect the good *of* the tree (the community); perhaps something is good for a tree if it promotes the growth of the tree or enhances the tree's capacity to resist parasites. These things, that is the growth or the resistance to parasites, either constitute the good of the tree or tend to yield other things which constitute the good of the tree. Things which are bad for trees would be those things which interfere with the attainment of the good of the tree. However, it is possible to speak of something as being good or bad *for* trees, and also of trees as living well or poorly even if there is no good of the tree, i.e. even if trees do not have a good of their own. Saying that something is good for a tree does not, in itself, imply that the tree has a good of its own. Someone might speak of something being good for trees in the course of saying that trees are good for people without any implication that any condition of the tree is or would be good in the absence of people who value trees. The defenders of environmental ethics cannot establish that trees, etc., have a good of their own merely by referring to such common expressions as 'Fertilizer is good for trees'. Such ways of speaking need not be taken as implying that trees, etc., have a good of their own.

Further, even if common ways of speaking implied that there is a good of trees, that would show only that common speech reflects an underlying theory of values. Common speech reflects acceptance of various theories which need not be true. For example, we still speak of a tsunami as a tidal wave, even though it is not a high tide. To defend an environmental ethic which implies that people are obligated to promote the good of trees, or communities, or other non-sentient entities, it is necessary to defend some value theories other than those on which utilitarianism rests, some value theories which imply that trees which attain this condition or state are doing well. Appeals to common ways of speaking are not strong evidence for a theory of basic or intrinsic value.

Let us turn next to consideration of hedonist or preference theories of value. Such theories might be criticized by observing that sometimes pain is good, as when it serves as a warning of an injury that needs care. For example, a pain in a person's chest can be a warning of injury to their heart. Similarly it may be argued that sometimes satisfaction of desires is bad, as when a creature desires to consume an object which would make it ill or lead to some other harm. However, advocates of hedonism or of preference theories of value have replies to these objections readily available, and so cannot be refuted in this way. Such replies appeal to the distinction between basic and instrumental value to which we have already made reference.

According to the hedonist, positive value ultimately derives from pleasure and negative value derives from pain. This might be expressed by saying that on this view pleasure is good in itself and pain is bad in itself. The phrase 'in itself' is used to take into account the fact that something may be pleasant but also conducive to pain. Philosophers since Aristotle have distinguished between what is of value in itself and what is valuable as a means to some other objective. Objects of the latter sort are said to be instrumentally valuable. In saying that value ultimately derives from pleasure (or pain) the hedonist is recognizing the fact that many unpleasant activities, e.g. studying verb endings, memorizing the multiplication tables, etc., may be of great instrumental value in that knowledge of verb endings or of how to multiply may be an essential component of activities that lead to pleasure.

The attempt to refute hedonism by pointing out that a pain may often benefit us through warning us of injury is an attempt to refute hedonism by producing some examples of pain that are not bad. The hedonist's rebuttal of this objection consists in showing that when pain occurs as a warning it is part of a larger process which is more pleasant or less painful than it would have been had the warning pain not occurred. To challenge hedonism we need to produce cases in which such a rebuttal cannot be made.

It is at least plausible to argue that hedonism is mistaken because pleasure or pain are merely feelings, and that while often a life without pleasure is less good than a life with some pleasure, it is not true that a life which consisted solely of pleasant experiences is the best or most desirable type of life. We might support this by considering hypothetical cases. Consider the life of a scientist or an artist. Each life consists of activity focussed on achievement of some objective. In the case of the scientists it may be to develop a method of curing a disease. In the case of the artist it may be to create an artistic work, for example a piece of sculpture. In either case the goodness of the life appears to consist either in the activity itself or in the successful accomplishment of its goal. If the artist or scientist succeeds, they may experience great pleasure or joy. However, the pleasure is, at most, an additional good. The scientist or artist deems life to be good regardless. Similarly, the activity may be painful. While such pain is perhaps bad in itself, even if the pain is intense and long-lasting, the activity of the life may be good on the whole. In this way, we might argue that even if pleasure is good in itself, it is not the only such thing. Since hedonists maintain that pleasure, and pleasure alone, is good in itself, hedonism is mistaken.

In my paper cited above I expressed, without argument, the theory that value derives ultimately from attitudes. Objects acquire value either through being desired or through satisfying desires or preferences. An object acquires positive value or goodness if it is an object that some creature favours, for example, if either the creature desires to experience or possess the object, or if the object satisfies a desire for that object. An object acquires negative value if it is an object which a creature shuns or wishes to destroy, or if the object frustrates fulfillment of a desire. We have referred to this view as the attitude or preference theory of value. Johnson rejects this theory for reasons that parallel the above initial effort to refute hedonism. Johnson argues that while we may often desire something which is good, it is not good solely in virtue of our desiring it. Rather, it is good regardless of whether or not we desire it. In support of this he argues that sometimes what we desire is not good. Our desiring it does not render it good. As an example, consider a person suffering from depression. Such a person, although physically healthy and capable of leading a life of fruitful activity, may desire their own destruction. Having this desire, Johnson would argue, does not make their death or destruction good.

How could the defender of the desire theory of values respond to this criticism? Conceivably he or she would make a distinction between saying that satisfying a desire has some positive value and saying that overall the value produced through satisfying the desire in question has more positive value than would be derived from frustrating the desire. If the desire of the depressed person to die were frustrated, and the person were subjected to appropriate therapy, the desire to die would disappear. The strength of other desires would increase, the desires of others for the individual's survival could be satisfied, etc. In such circumstances frustrating the desire, while of some immediate negative value, would in time have so much positive value as to more than counter-balance the basic negative value incurred by frustrating this desire. We would not say that an act is good unless we mean that its positive value overall in the circumstances is large. The positive value of satisfying the desire to die is not large. Very likely such an act would have a negative overall value. Thus, a defender of the desire theory of value would not agree that allowing the depressed person to die is good.

Is this defence of the desire theory of value satisfactory? That is a question about which the reader must use their own judgment. Many people, in a range of cultures, would agree with Johnson that goodness or value is independent of preference or attitude. Consider that there are people who would maintain that a certain sort or form of life is a good life regardless of whether it is desired. Some people would say that a life as a homemaker is a good life for a woman regardless of whether she or anyone desires her to lead such a life. Again, some people have suggested that a life in which a person manifests certain virtues, e.g. courage or renunciation of bodily appetites, is a good life regardless of any person's attitude toward such a life. As noted, I have rejected such views. I would argue that courage is a value because it tends to contribute to accomplishing what people desire to accomplish. A homemaker's life, even if it is not the life the homemaker prefers, can be a good life through its contribution to the satisfactions of the lives of others. (Clearly, saying this does not imply that women are morally required to sacrifice their own desires and become homemakers.) I would argue further that a life of renunciation, in the absence of a desire to conform to such a pattern of life, is not worthwhile.

Finally, let us briefly give further consideration to Johnson's assumptions concerning the basis of value. Rational justification of a theory of environmental ethics proper requires the development of such a theory. Johnson appears to assume that a human being has a wide range of capacities, and that a good life for a person consists in orderly or effective activity which exercises a wide range of such capacities while preserving the integrity and identity of the person. Such a life can take many forms. It can be a life of rigorous scientific activity or life of imaginative creativity.

Since, in Johnson's view, goodness consists in acting effectively so as to preserve the entity's integrity or coherence and identity in carrying out a wide range of capacities, he argues that goodness is not restricted to the good of people. He suggests that living things, both animals and plants, have a good, or in Johnson's terms a wellbeing. Further, he suggests that various holistic entities such as species or ecosystems also have a good. Their good consists in effective activity involving a wide range of their capacities, activity which both manifests and tends to preserve the coherence and identity of the ecosystem in question. The good of a dog is different from the good of a person as well as from the good of a tree or of the ecosystem in which people, dogs, trees etc. are all functioning parts.

One might doubt, however, as I do, whether a life of scientific activity would be a good life if it were neither desired nor tended to satisfy desires. However, for me to make this criticism of Johnson's view could be dismissed as question-begging. A second criticism of Johnson's view is that it is unclear what sorts of organization of activities count as effective ways of preserving the identity of things such as forests or ecosystems. A development of this criticism would involve consideration of deep philosophical issues concerning identity, and particularly of the identity of entities such as ecosystems. While such issues are of great interest to philosophers, I do not wish to pursue such matters here except to note that, in my judgment, Johnson needs to explain and justify his views concerning the identity and integrity of ecosystems such as lakes or forests. He has not done so.

There is, I believe, another criticism of Johnson's theory of goodness which it is appropriate to consider here. We can envisage a human's life which involves activity over a wide range of capacities, including the activities of reason, emotion, imagination, etc., which is effectively organized for harmful purposes even though the person's identity and integrity is preserved. Consider the ingenuity of leaders of criminal gangs. The activity of such a person may be effectively organized and so appears to qualify as good according to Johnson's theory concerning goodness. Yet, in light of the pain and suffering resulting from such activity or from the widespread frustration of desires which it causes, I would not agree that such a life achieves goodness. We might allow that the criminal's activity is good for the criminal, but not that it is good, or good in itself. Unless the activity is good in itself, reference to such activity yields no grounds for claiming that we are morally obligated to the criminal. Again, we may ask, is any effective organization of activity in accord with the capacities of a tree (or a forest), which preserves the identity of that tree (or forest), a realization of goodness? I would answer in the negative. Conceivably a forest, just like a person, can preserve its identity while acting effectively to produce either good or its opposite.

DEEP ECOLOGY

The thought of deep ecologists has attracted a lot of attention from some well-respected environmentalists in recent years and thus is worth taking into consideration. However, the formulations of deep ecology offered by its defenders are anything but logically rigorous. The basic principles are not clearly stated and the derivation of ethical rules based on these principles is *not*, for the most part, by valid deductions. Under these circumstances the possibility of misinterpretation on my part is great. Thus, the reader is urged to consult the sources. The ethical rules alleged by deep ecologists to derive from acceptance of these basic principles amount to an austere environmental ethic (Devall and Sessions, 1994).

Just as utilitarianism may be conceived as resting on two basic assumptions, a theory of value, namely that value derives from pleasure or satisfaction of desires, and an ethical principle, namely that we are obligated to act so as to produce a maximum of goodness, so also deep ecology may be conceived as resting on two basic principles. One principle is a form of biocentrism. According to this principle, every individual has an 'equal right to live and blossom' (Naess, 1994b, p. 103). As we have discussed biocentrism above we shall turn directly to the second principle.

Deep ecologists refer to the second principle as a principle of self-realization (Naess, 1994a). I shall attempt a brief explanation of this idea. The principle of self-realization can best be understood as combining an ethical principle and a metaphysical theory concerning the distinction between one's self and one's environment. Ethical theories based on the principle of self-realization may be contrasted with consequentialist theories. According to consequentialist ethical theories, one's primary moral duty is to produce a maximum of non-moral goodness, e.g. pleasure, satisfaction of desires, etc. Theories based on self-realization are derived from Kantian rather than consequentialist assumptions. According to Kantian theories, non-moral goodness is not the axiological basis of moral obligation. (For some discussion of the contrast between consequentialist and Kantian perspectives, see the discussion of ethical theory in my paper cited above. (Lehman, 1993).)

According to Kantian ethical theories, moral goodness is goodness of a person's character and is determined by reference to the person's motives. Some may ask what conditions a person must satisfy, in regard to his or her motives, to have a morally good character. In answering this question some people may assume that a person's motives should be to act in accordance with moral requirements, otherwise they are selfish motives. Since a morally good action is not an action done out of selfishness, it must be an action done because it is morally required, or a duty. This appears to have been Kant's view (Kant, 1949). However, we may wonder whether we are driven by our basic ethical principles and logic to hold that we have a morally good character if and only if we are unselfish. There are grounds for holding that acts can be morally good, i.e. done out of ethical motives, even if they are not unselfish.

A utilitarian can readily allow that some actions are both morally right and done out of selfish motives, since the utilitarian can recognize that there are actions which a person might perform to benefit him or herself but which also lead to maximum goodness. However, deep ecologists are not utilitarians, or even consequentialists of some other sort. Their rejection of the assumption that we either act in accordance with moral requirements or act selfishly is based on other grounds. On this view, our actions can be morally good if they are the actions of a magnanimous or great-souled person, i.e. a person who emphatically identifies with others. Some such idea underlies deep ecological thought.

In discussing 'self-realization', Naess suggests that we can enlarge ourselves. We can think of ourselves as the centre of consciousness and of desire of a living body, and of our goal as being the gratification of that body. Alternatively, we can think of ourselves as beings whose deepest or strongest desires include, or even require, the well-being of others. Naess's writings suggest that we become morally better beings through enlarging ourselves. To accomplish this we must come to desire the well-being of others. Moral goodness is achieved through becoming great-souled, i.e. through enlarging ourselves.

While ethical theories such as those of Singer, Taylor and Johnson challenge assumptions about what entities have moral standing, deep ecology requires more radical transformation of commonly held views about ourselves and our relations to the rest of the biosphere. These other views may all be regarded as non-utilitarian forms of consequentialism. These views pose challenges to the use of pesticides on the grounds of the harm that such use involves to humans, to animals, to all living organisms or to ecosystems. Deep ecology is more concerned with what sort of beings we are than with what acts we perform. We can be greater or lesser beings. Greater beings are beings who achieve more inclusive and more unified selves. In this view, we achieve greatness through incorporating into ourselves ever increasing parts of what lesser people regard as parts of their environment. The process of widening of self, or maturation, involves incorporating the desires of others as one's own. This process can be carried on beyond family to community, tribe, ethnic group, nation and, according to Naess, to every other living individual in the biosphere. This clearly suggests that we can incorporate the desires of animals, plants, simple life-forms and perhaps even parts of the biosphere that are not normally regarded as alive, e.g. rocks or lakes.

The principles of deep ecology suggest that extensive transformations of the Earth such as occur through widespread use of pesticides cause great harm. Such transformations violate biocentrist principles. Further, such transformations simplify our environment through eliminating entities that we cannot use for our immediate benefit; in this way such uses of pesticides prevent us from realizing our greatest potential as human beings (Naess, 1994c). Widespread use of pesticides has occurred in Western culture and is, within a deep ecologist's perspective, equivalent to behaving in a way which prevents our achieving our full maturity.

The principles of deep ecology are open to many critical questions. One question may concern the right of human beings to survive. As with other biocentrists, it appears that deep ecologists reply to this question by allowing that humans are entitled to find ways to realize their basic vital needs, for example to protect themselves from immediate harm, to feed themselves, etc. Other questions concern the clarity of the formulation of the principles of deep ecology. The principles of deep ecology are very obscure. It is not clear where deep ecologists would draw the line between activities which count as self-protection, etc., and activities which are unacceptable. Apparently, they believe that the legitimate flourishing of human beings is incompatible with the extensive transformations of ecosystems which humans have undertaken in the course of developing society along the lines found in North America and Western Europe.

Deep ecology, including the assumptions that the best life for people involves maximum self-realization and that maximum self-realization requires leaving the ecosystems of the Earth almost exactly as we find them, is open to many further questions. Even if we agree with the idea that a morally good person strives to achieve maximum self-realization, we may doubt that this requires broadening or enlarging ourselves beyond the bounds of the normal mature ego of the Western person. Further, even if we agree that such broadening is possible, we may doubt that it is possible to push it to the extent, envisaged in Indian philosophy to which Naess refers with approval, in which all living beings are merged with oneself. Can we really identify with the desires of a snail or a jellyfish? Again, even if we agree that broadening is possible, we may doubt that broadening requires leaving the Earth's ecosystems as we find them. Some people might maintain that greater scope for human experience can be achieved through works of human creation; that, for example, it is possible to enlarge ourselves to a greater extent through creating a great symphony or city than through experiencing a jungle or a desert. Nonetheless, even though there are serious questions with which we may challenge the principles of deep ecology, we can agree that there may well be limits to the extent to which transforming the Earth's ecosystems leads to greater overall well-being. There are, almost certainly, many people (and animals) whose lives would be diminished or impoverished were wild ecosystems all developed or managed in accord with human desires.

SUMMARY

In an earlier paper I argued that, subject to certain limitations, we are obligated to reduce pesticide use from current levels. In this paper I have reviewed the implications of a number of distinct theories of environmental ethics with respect to the use of pesticides.

The first of these, the claim that we ought to avoid speciesism, is compatible with the principles espoused in my earlier paper. To avoid speciesism, we must take the interests of any sentient creatures who may be harmed from exposure to pesticides into account in order to determine whether use of a pesticide in particular circumstances is acceptable. The implication of theories of environmental ethics proper, concerning which uses of pesticides are acceptable, are far more restrictive.

REFERENCES

Callicott, J.B. (1986) The search for an environmental ethic. *Matters of Life and Death: New Introductory Essays in Moral Philosophy.* 2nd edn. In Regan, T. (Ed.) Random House, New York.

Carruthers, P. (1989) Brute experience. *Journal of Philosophy* **XXXVI**(5), 258–269.

Devall, B. and Sessions, G. (1994) Deep ecology. In Van DeVeer, D. and Pierce, C. (Eds) *The Environmental Ethics and Policy Book: Philosophy, Ecology, Economics.* Wadsworth, Belmont, CA.

Frey, R.G. (1980) *Interests and Rights: The Case Against Animals.* Clarendon Press, Oxford.

Furtick, W.R. (1976) Uncontrolled pests or adequate food. In Gunn, D.L. and Stevens, J.G.R. (Eds) *Pesticides and Human Welfare.* Oxford University Press, Oxford, pp. 3–12.

Goodpaster, K.E. (1978) On being morally considerable. *Journal of Philosophy* **75**, June.

Jamieson, D. (1990) Rights, justice, and duties to provide assistance: A critique of Regan's theory of rights. *Ethics* **100**(2), 349–362.

Johnson, L.E. (1991) *A Morally Deep World: An Essay on Moral Significance and Environmental Ethics.* Cambridge University Press, Cambridge.

Kant, I. (1949) *Fundamental Principles of the Metaphysic of Morals.* Liberal Arts Press, New York.

Kant, I. (1989) Duties in regard to animals. In Regan, T. and Singer, P. (Eds) *Animal Rights and Human Obligations.* 2nd edn. Prentice Hall, Englewood Cliffs, NJ.

Lehman, H. (1984) The case for animal rights. *Dialogue* **XXIII**.

Lehman, H. (1988) On the moral acceptability of killing animals. *Journal of Agricultural Ethics* **1**, 155–162.

Lehman, H. (1993) Values, ethics and the use of synthetic pesticides in agriculture. In Pimentel, D. and Lehman, H. (Eds) *The Pesticide Question: Environment, Economics and Ethics.* Routledge, Chapman & Hall, London.

Lehman, H. (1995) *Rationality and Ethics in Agriculture.* University of Idaho Press, Moscow, ID.

Leopold, A. (1949) *A Sand County Almanac.* Oxford University Press, Oxford.

Metcalf, R.L. (1993) An increasing public concern. In Pimentel, D. and Lehman, H. (Eds) *The Pesticide Question: Environment, Economics and Ethics.* Routledge, Chapman & Hall, London.

Naess, A. (1994a) Ecosophy T: Deep versus shallow ecology. In Pojman, L.P. (Ed.) *Environmental Ethics: Readings in Theory and Application.* Jones and Bartlett, Boston, MA.

Naess, A. (1994b) The shallow and the deep, long-range ecological movement. In Pojman, L.P. (Ed.) *Environmental Ethics: Environment, Economics and Ethics. Readings in Theory and Application.* Jones and Bartlett, Boston, MA.

Naess, A. (1994c) Self realization: An ecological approach to being in the world. In Van DeVeer, and Pierce, C. (Eds) *The Environmental Ethics and Policy Book: Philosophy, Ecology, Economics.* Wadsworth, Belmont, CA.

Narveson, J. (1980) Animal rights revisited. *Animal Regulation Studies* **2**(3), 223–236.

Pettersson, O. (1993) Swedish pesticide policy in a changing environment. In Pimentel, D. and Lehman, H. (Eds) *The Pesticide Question: Environment, Economics and Ethics.* Routledge, Chapman & Hall, London.

Pimentel, D., Acquay, H., Biltonen, M., Rice, P., Silva, M., Nelson, N., Lipner, V., Giordano, S., Horowitz, A. and D'Amore, M. (1993a) Assessment of environmental and economic impacts of pesticide use. In Pimentel, D. and Lehman, H. (Eds) *The Pesticide Question: Environment, Economics and Ethics.* Routledge, Chapman & Hall, London.

Pimentel, D., Kirby, C. and Shroff, A. (1993b) The relationship between 'cosmetic standards' for foods and pesticide use. In Pimentel, D. and Lehman, H. (Eds) *The Pesticide Question: Environ-*

ment, Economics and Ethics. Routledge, Chapman & Hall, London.

Pimentel, D., McLaughlin, L., Zepp, A., Lakitan, B., Kraus, T., Kleinman, P., Vancini, F., Roach, W.J., Graap, E., Keeton, W.S. and Selig, G. (1993c) Environmental and economic impacts of reducing US agricultural pesticide use. In Pimentel, D. and Lehman, H. (Eds) *The Pesticide Question: Environment, Economics and Ethics.* Routledge, Chapman & Hall, London.

Postel, S. (1988) Controlling toxic chemicals. In Starke, L. (Ed.) *State of the World, 1988: A Worldwatch Institute Report on Progress Toward a Sustainable Society.* W.W. Norton, New York.

Ray, D.L. and Guzzo, L. (1994) The blessings of pesticides. In Pojman, L.P. (Ed.) *Environmental Ethics: Readings in Theory and Application.* Jones and Bartlett, Boston, MA.

Reding, N.L. (1993) Seeking a balanced perspective. In Pimentel, D. and Lehman, H. (Eds) *The Pesticide Question: Environment, Economics and Ethics.* Routledge, Chapman & Hall, London.

Regan, T. (1982a) The nature and possibility of an environmental ethic. In *All That Dwell Therein: Essays on Animal Rights and Environmental Ethics.* University of California Press, Berkeley, CA.

Regan, T. (1982b) The moral basis of vegetarianism. In *All That Dwell Therein: Essays on Animal Rights and Environmental Ethics.* University of California Press, Berkeley, CA.

Regan, T. (1983) *The Case for Animal Rights.* University of California Press, Berkeley, CA.

Schweitzer, A. (1923) *Philosophy of Civilization. Part II. Civilization and Ethics.* (Trans. John Naish) A. and C. Black, London.

Singer, P. (1975) *Animal Liberation: A New Ethics for our Treatment of Animals.* Jonathan Cape, London.

Surgeoner, G.A. and Roberts, W. (1993) Reducing pesticide use by 50% in the Province of Ontario: Challenge and progress. In Pimentel, D. and Lehman, H. (Eds) *The Pesticide Question: Environment, Economics and Ethics.* Routledge, Chapman & Hall, London.

Taylor, P.W. (1986) *Respect for Nature.* Princeton University Press, Princeton, NJ.

CHAPTER 4

Environmental and Socio-Economic Costs of Pesticide Use*

David Pimentel and Anthony Greiner

Cornell University, Ithaca, NY, USA

INTRODUCTION

Worldwide, over 2.5 million tons of pesticides are applied each year (PN, 1990) with a purchase price today of $21 billion (WRI/UNEP/UNDP, 1994). In the United States approximately 500 000 tons of 600 different types of pesticides are used annually (Pimentel et al., 1991) at a cost of $6.5 billion, including application costs (USBC, 1994).

Despite the widespread use of pesticides in the United States, pests (principally insects, plant pathogens and weeds) destroy 37% of all potential food and fibre crops (Pimentel et al., 1991). Estimates are that losses to pests would increase by 10% if no pesticides were used at all; specific crop losses would range from zero to nearly 100% (Pimentel et al., 1978). Thus, pesticides make a significant contribution to maintaining world food production. In general, each dollar invested in pesticide control returns about $4 in crops saved (Pimentel et al., 1991).

However, since the publication of Rachel Carson's *Silent Spring* in 1962, the attitude towards pesticides and other chemicals has changed. Although pesticides are generally profitable, their use does not always decrease crop losses. For example, even with the 10-fold increase in insecticide use in the United States from 1945 to 1989, total crop losses from insect damage have nearly doubled from 7% to 13% (Pimentel et al., 1991). This rise in crop losses to insects is, in part, caused by changes in agricultural practices. For instance, the replacement of rotating corn with other crops with the continuous production of corn on about half of the hectarage has resulted in nearly a four-fold increase in corn losses to insects despite a more than 1000-fold increase in insecticide use in corn production (Pimentel et al., 1991).

Most benefits of pesticides are based only on direct crop returns. Such assessments do not include the indirect environmental and economic costs associated with pesticides. It has been estimated that only 0.1% of applied pesticides reach the target pests, leaving the bulk of the pesticides (99.9%) to impact the environment (Pimentel, 1995). To facilitate

* This paper is a version of a paper entitled "Assessment of environmental and economic impacts of pesticide use" by D. Pimentel published in *The Pesticide Question: Environment, Economics and Ethics.* Routledge, Chapman & Hall, New York, 1993.

the development and implementation of a balanced, sound policy of pesticide use, these costs must be examined. Over 15 years ago the US Environmental Protection Agency pointed out the need for such a risk investigation (EPA, 1977). Thus far, only a few papers on this difficult subject have been published. However, recently the EPA, the FDA and USDA have jointly announced that emergency exemptions for 'cancer-causing' pesticides will no longer be given. Perhaps this is the first major step towards severely limiting hazardous pesticide use (FDA, 1993). In 1995, the EPA and the chemical manufacturer DuPont reached an agreement which will result in the elimination of the potentially carcinogenic herbicide cyanazine by the year 2002 (JAWWA, 1995).

The obvious need for an up-dated and comprehensive study prompted the investigation by Pimentel et al. (1992) on the complex of environmental and economic costs resulting from the nation's dependence on pesticides. Included in the assessment are analyses of pesticide impacts on: human health effects; domestic animal poisonings; increased control expenses resulting from pesticide-related destruction of natural enemies and from the development of pesticide resistance; crop pollination problems and honey bee losses; crop and crop product losses; groundwater and surface water contamination; fish, wildlife and microorganism losses; governmental expenditure to reduce the environmental and social costs of pesticide use. This chapter is an updated version of that work.

PUBLIC HEALTH EFFECTS

Human pesticide poisonings and illnesses are clearly the highest price paid for pesticide use. A recent World Health Organization report (WHO, 1990) estimated that there are 3 million severe human pesticide poisonings in the world each year, of which approximately 220 000 are fatal. In the United States, pesticide poisonings reported by the American Association of Poison Control Centers total about 110 000 each year (Benbrook, 1996). J. Blondell (EPA, personal communication, 1990) has indicated that because of demographic gaps, this figure represents only 73% of the total. The number of accidental (no suicide or homicide) fatalities is about 27 per year.

Developed countries, including the United States, use approximately 80% of all the pesticides produced in the world annually (WRI/UNEP/UNDP, 1994), but less than half of the pesticide-induced deaths occur in these countries (Committee, House of Commons Agriculture, 1987). Clearly, a higher proportion of pesticide poisonings and deaths occur in developing countries, where there are inadequate occupational and other safety standards, insufficient enforcement, poor labeling of pesticides, illiteracy, inadequate protective clothing and washing facilities, and a lack of knowledge concerning pesticide hazards (Forget, 1991).

Both the acute and chronic health effects of pesticides warrant concern. Unfortunately, while the acute toxicity of most pesticides is well documented (Ecobichon et al., 1990), information on chronic human illnesses is not as sound (Wilkinson, 1990). Regarding cancer, the International Agency for Research on Cancer found 'sufficient' evidence of carcinogenicity for 18 pesticides, and 'limited' evidence of carcinogenicity for an additional 16 pesticides based on animal studies (WHO/UNEP, 1989). With humans the evidence concerning cancer is mixed. For example, a recent study in Saskatchewan indicated no significant difference in non-Hodgkin's lymphoma mortality between farmers and non-farmers (Wigle et al., 1990), whereas other studies have reported an increased prevalence of some cancers in farmers (WHO/UNEP, 1989; Cantor et al., 1992).

A realistic estimate of the number of cases of cancer in humans in the US due to pesticides is given by D. Schottenfeld (University of Michigan, personal communication, 1991), who estimates that less than 1% of the nation's cancer cases are caused by exposure to pesticides. Considering that there are approximately 1.2 million cancer cases per year (USBC, 1994), Schottenfeld's assessment suggests that fewer than 12 000 cases of cancer per year are due to pesticides.

Many other acute and chronic maladies appear to be associated with pesticide use. For example, the recently banned pesticide dibromochloropropane (DBCP), which is used for plant pathogen control, caused testicular dysfunction in animal studies (Shaked et al., 1988; Shemi et al., 1988, 1989) and was linked with infertility among human workers exposed to DBCP (Potashnik and Yanai-Inbar, 1987). Also, a large body of evidence suggesting that pesticides can produce immune dysfunction has been accumulated over recent years from animal studies (Luster et al., 1987; Thomas and House, 1989). In a study of women who had chronically ingested groundwater contaminated with low levels of aldicarb (mean 16.6 p.p.b.), Fiore et al. (1986) reported evidence of significantly reduced immune response, although these women did not exhibit any overt health problems.

In addition, there is growing evidence of sterility in humans and various other animals, especially in males, due to various chemicals and pesticides in the environment (Colborn et al., 1996). Sperm counts in males in Europe have declined by about 50% and continue to decrease by 2% per year. In the lower Columbia River, young male river otters have smaller reproductive organs than males in unpolluted regions of the river. Also, in Florida's Lake Apopka, male alligators have penises that are one-quarter normal length. With 3 billion tons of 70 000 different chemicals being released into the environment in the US, it is difficult to determine the effects of individual pesticides or other chemicals.

Of particular concern are the chronic health problems associated with the effects of organophosphorus pesticides, which have largely replaced the banned organochlorines (Ecobichon et al., 1990). The malady OPIDP (organophosphate induced delayed poly-neuropathy) is well documented and includes irreversible neurological defects (Lotti, 1992). Other defects in memory, mood and abstraction have been documented by Savage et al. (1988). Well-documented cases indicate that persistent neurotoxic effects may be present even after the termination of an acute poisoning incident (Ecobichon et al., 1990).

Such chronic health problems are a public health issue, because everyone, everywhere is exposed to some pesticide residues in food, water and the atmosphere. About 35% of the foods purchased by US consumers have detectable levels of pesticide residues (FDA, 1990). Of this, from 1 to 3% of the foods have pesticide residue levels above the legal tolerance level (Hundley et al., 1988; FDA, 1990). Residue levels may be higher than have been recorded because the US analytical methods now employed detect only about one-third of the more than 600 pesticides in use (OTA, 1988; Minyard and Roberts, 1991). Certainly the contamination rate is higher for fruits and vegetables because these foods receive the highest dosages of pesticides, and one USDA study has shown that even after washing, peeling or coring, there is still some pesticide residue left in the fruit or vegetable (Wiles and Campbell, 1994). Therefore, there are many good reasons why 97% of the public is genuinely concerned about pesticide residues in their food (FDA, 1989).

Food residue levels in developing nations often average higher than those found in developed nations, either because there are no laws governing pesticide use, or because the numbers of skilled technicians available to enforce laws concerning pesticide tolerance levels in foods are inadequate, or because other resources are lacking. For instance, most milk samples assayed in a study in Egypt had high residue levels (60 to 80%) of 15 pesticides included in the investigation (Dogheim et al., 1990). In contrast, 50% of the milk samples analyzed in a US milk study had pesticide residues, and these were all in trace quantities, well below the EPA and FDA regulatory limits (Trotter and Dickerson, 1993).

In all countries, the highest levels of pesticide exposures occur in pesticide applicators, farm workers, and people who live adjacent to heavily treated agricultural land. Because farmers and farm workers directly handle 70–80% of all pesticides used, the health of these population groups is at the greatest risk of being seriously affected by pesticides (Ciesielski et al., 1994; McDuffie, 1994). The epidemiological evidence suggests significantly higher cancer incidence among farmers and farm workers in the United States and Europe than among non-farm workers in some areas (e.g. Brown et al., 1990; Cantor et al., 1992). A consistent association has been documented between lung cancer and exposure to organochlorine insecticides (Pesatori et al., 1994). Evidence is also strong for an association between lymphomas and soft-tissue sarcomas and certain herbicides (Zahm et al., 1990).

An increasingly growing concern is the effect of pesticides on children. Children can be exposed to pesticides every day through the foods that they eat (Wiles and Campbell, 1994), in the house (Davis et al., 1992), or in the community where they live (Moses et al., 1993). With the increased realization that there are distinct physiological differences between adults and children, it has become obvious that the present pesticide tolerance and regulatory system is severely lacking when it comes to children. All of the regulations to date are based on adult tolerances. Children have much higher metabolic rates than adults, and their ability to activate, detoxify and excrete xenobiotic compounds is different from that of adults. Also, because of their smaller physical size, children are exposed to higher levels of pesticides per unit of body weight. This is evidenced by a study which showed that 50% of all pesticide poisoning incidents in England and Wales involved children under 10 years of age (Casey and Vale, 1994). There has also been work linking the use of pesticides in the home with childhood cancer (Leiss and Savitz, 1995). In general, children's sensitivities to toxicants are very different from those of adults and so there has been a movement towards setting specific pesticide regulations with children in mind (NRC, 1993; Wiles and Campbell, 1994).

Medical specialists are concerned about the lack of public health data about pesticide usage in the United States. Based on an investigation of 92 pesticides used on food, GAO (1986) estimates that 62% of the data on health problems associated with registered pesticides contains little or no information on tumors and even less on birth defects.

Although no one can place a precise monetary value on a human life, the 'costs' of human pesticide poisonings have been estimated. Studies done for the insurance industry have computed monetary ranges for the value of a 'statistical life' between $1.6 and $8.5 million (Nash, 1994). For our assessment, we use the conservative estimate of $2.2 million per human life (the average value that the surviving spouse of a slain New York City policeman receives (Nash, 1994)). Based on the available data, estimates are that human pesticide poisonings and related illnesses in the United States total about $933 million each year (Table 4.1).

DOMESTIC ANIMAL POISONINGS AND CONTAMINATED PRODUCTS

In addition to pesticide problems that affect humans, several thousand domestic animals are poisoned by pesticides each year, with dogs and cats representing the largest number

Table 4.1   Estimated economic costs of human pesticide poisonings and other pesticide-related illnesses in the United States each year

Human health effects from pesticides	Total costs ($)
Cost of hospitalized poisonings	
2380[a] × 2.84 days at $1000 day^{-1}	6 759 000
Cost of outpatient treated poisonings	
27 000[b] × $630[c]	17 010 000
Lost work due to poisonings	
4680[a] workers × 4.7 days × $80 day^{-1}	1 760 000
Pesticide cancers	
<12 000[d] cases × $70 700[c] case^{-1}	848 400 000
Cost of fatalities	
27 accidental fatalities[c] × $2.2 million	59 400 000
Total	933 329 000

[a]Keefe et al., 1990.
[b]J. Blondell, EPA, Washington, DC, personal communication, 1991.
[c]Includes hospitalization, foregone earnings, and transportation.
[d]See text for details.

(Table 4.2; Murphy, 1994). For example, of 25 000 calls made to the Illinois Animal Poison Control Center in 1987, nearly 40% of all calls concerned pesticide poisonings in dogs and cats (Beasley and Trammel, 1989). Similarly, Kansas State University reported that 67% of all animal pesticide poisonings involved dogs and cats (Barton and Oehme, 1981). This is not surprising because dogs and cats usually wander freely about the home and farm and therefore have greater opportunity to come into contact with pesticides than other domesticated animals.

The best estimates indicate that about 20% of the total monetary value of animal production, or about $4.2 billion, is lost to all animal illnesses, including pesticide poisonings (Gaafar et al., 1985). Colvin (1987) reported that 0.5% of animal illnesses and 0.04% of all animal deaths reported to a veterinary diagnostic laboratory were due to pesticide toxicosis, and thus $22 and $9.5 million, respectively, are lost to pesticide poisonings (Table 4.2).

This estimate is considered low because it is based only on poisonings reported to veterinarians. Many animal pesticide poisonings that occur in the home and on farms go undiagnosed and are attributed to factors other than pesticide poisonings. In addition, when a farm animal poisoning occurs and little can be done for an animal, the farmer seldom calls a veterinarian but rather waits for the animal to recover or destroys it (G. Maylin, Cornell University, personal communication, 1977). Such cases are usually unreported.

Additional economic losses occur when meat, milk and eggs are contaminated with pesticides. The animals can become contaminated from pesticides that are applied to crops, to farm buildings for pest control, or directly to the animal for veterinary purposes. While studies to date have shown that these residues do build up in the fatty tissues of the livestock, the residues are well below federal tolerance limits (Robertson et al., 1990; Garrido et al., 1994; Herrera et al., 1994; Spiric and Raicevic, 1994).

In the United States, all animals slaughtered for human consumption, if shipped inter-state, and all imported meat and poultry must be inspected by the US Department of Agriculture. This inspection is to ensure that the meat and products are wholesome,

Table 4.2 Estimated domestic-animal pesticide poisonings in the United States

Livestock	Number × 1000	$ per head	Number 111[a]	$ Cost per poisoning[b]	$ Cost of poisonings	Number of deaths[c]	$ Cost of deaths × 1000[d]	Total $ × 1000
Cattle	100 987[e]	659[e]	101	131.80	13 312	8	5272	18 584
Dairy cattle	9528[e]	1160[e]	10	232.00	2320	1	1160	3480
Dogs	52 500[f]	125[g]	53	25.00	1325	4	500	1825
Horses	5480[i]	1000[h]	6	200.00	1200	1	1000	2200
Cats	57 000[f]	20[g]	57	4.00	228	5	100	328
Swine	57 904[e]	74.90[e]	58	14.98	869	5	375	1244
Chickens	6 689 110[e]	1.55[e]	6690	0.31	2074	535	830	2904
Turkeys	287 220[e]	8.72[f]	288	1.75	504	23	201	705
Sheep	9079[e]	70.30[e]	9	14.06	127	1	71	198
Total	7 268 808				21 959		9509	31 468

[a]Based on a 0.1% illness rate (see text).
[b]Based on each animal illness costing 20% of the total production value of that animal.
[c]Based on a 0.008% mortality rate (see text).
[d]The death of the animal equals the total value for that animal.
[e]USDA (1994).
[f]USBC (1994).
[g]Estimated.
[h]USBC (1990).
[i]FAO (1994).

properly labeled, and do not present a health hazard. One part of this inspection, which involves monitoring meat for pesticide and other chemical residues, is the responsibility of the National Residue Program (NRP). The samples taken are intended to ensure that if a chemical is present in 1% of the animals slaughtered it will be detected (NAS, 1985; Crawford and Schor, 1991). Generally, violative residues are found less frequently in chickens and cattle, and are found most frequently in swine (Cordle, 1988).

However, of more than 600 pesticides now in use, NRP tests are made for only 41 (D. Beerman, Cornell University, personal communication, 1991) which have been determined by FDA, EPA and FSIS to be of public health concern (NAS, 1985). While the monitoring program records the number and type of violations, there is no significant cost to the animal industry because the meat, including poultry, is generally *sold and consumed* before the test results are available. About 3% of the chickens with illegal pesticide residues are sold in the market (NAS, 1987).

Compliance sampling is designed to prevent meat and milk contamination with pesticides. When a producer is suspected of marketing contaminated livestock, the carcasses are detained until the residue analyses are reported. If there are illegal residues present, the carcasses or products are condemned and the producer is prohibited from marketing other animals until it is confirmed that all the livestock are safe (NAS, 1985). If carcasses are not suspected of being contaminated, then by the time the results of the residue tests are reported the carcasses have been sold to consumers. This points to a major deficiency in the surveillance program.

In addition to animal carcasses, pesticide-contaminated milk cannot be sold and must be disposed of. In certain incidents these losses are substantial. For example, in Oahu, Hawaii, in 1982, 80% of the milk supply, worth more than $8.5 million, was condemned by public health officials because it had been contaminated with the insecticide heptachlor (van Ravenswaay and Smith, 1986). This incident had immediate and far-reaching effects on the entire milk industry on the island. Initially, reduced milk sales due to the contaminated milk alone were estimated to cost each dairy farmer $39 000. Subsequently, the structure of the island milk industry has changed. Because island milk was considered by consumers to be unsafe, most of the milk supply is now imported. The $500 million lawsuit against the producers brought by consumers is still pending (van Ravenswaay and Smith, 1986).

When the costs attributable to domestic animal poisonings and contaminated meat, milk and eggs are combined, the economic value of all livestock products in the United States lost to pesticide contamination is estimated to be at least $31.5 million annually (Table 4.2).

Similarly, other nations lose significant numbers of livestock and large amounts of animal products each year due to pesticide-induced illness or death. Exact data concerning these livestock losses do not exist, and the available information comes only from reports of the incidence of mass destruction of livestock. For example, when the pesticide leptophos was used by Egyptian farmers on rice and other crops, 1300 draft animals were poisoned and lost (Sebae, 1977, in Bull, 1982). The estimated economic losses were significant, but exact figures are not available.

In addition, countries exporting meat to the United States can experience tremendous economic losses if the meat is found to be contaminated with pesticides. In a 15-year period, the beef industries in Guatemala, Honduras and Nicaragua lost more than $1.7 million due to pesticide contamination of exported meat (ICAITI, 1977). In these countries, meat which is too contaminated for export is sold in local markets. Obviously such policies contribute to public health problems.

DESTRUCTION OF BENEFICIAL NATURAL PREDATORS AND PARASITES

In both natural and agro-ecosystems, many species, especially predators and parasites, control or help to control herbivorous populations. Indeed these natural beneficial species in theory make it possible for ecosystems to remain foliated and 'green'. With the parasites and predators keeping herbivore populations at low levels, only a relatively small amount of plant biomass is removed each growing season. Natural enemies play a major role in keeping populations of many insect and mite pests under control (Pimentel, 1988).

Like pest populations, beneficial natural enemies are adversely affected by pesticides (Croft, 1990; Wills et al., 1990). For example, the following pests have reached outbreak levels in cotton and apple crops following the destruction of natural enemies by pesticides: *cotton*, bollworm, tobacco budworm, cotton aphid, spider mites and cotton loopers (OTA, 1979; Murray, 1994); *apple*, European red mite, red-banded leafroller, San José scale, oystershell scale, rosy apple aphid, wooly apple aphid, white apple leafhopper, two-spotted spider mite and apple rust mite (Croft, 1990; Kovach and Agnello, 1991). Significant pest outbreaks have also occurred in other crops (OTA, 1979; Pimentel et al., 1980; Croft, 1990; Murray, 1994). Also, because parasitic and predaceous insects often have complex search and attack behaviours, sublethal insecticide dosages may alter this behaviour and in this way disrupt effective biological controls (L.E. Ehler, University of California, personal communication, 1991).

Fungicides can also contribute to pest outbreaks when they reduce fungal pathogens that are naturally parasitic on many insects. For example, the use of benomyl, used for plant pathogen control, reduces populations of entomopathogenic fungi, resulting in increased survival of velvet bean caterpillars and cabbage loopers in soybeans. The increased number of insects eventually leads to reduced soybean yields (Johnson et al., 1976).

When outbreaks of secondary pests occur because their natural enemies are destroyed by pesticides, additional and sometimes more expensive pesticide treatments have to be made in efforts to sustain crop yields. This raises overall costs, and contributes to pesticide-related problems.

An estimated \$520 million can be attributed to costs of additional pesticide applications and increased crop losses, both of which follow the destruction of natural enemies by pesticides applied to crops (Table 4.3).

As in the United States, natural enemies are being adversely affected by pesticides worldwide. Although no reliable estimate is available concerning the impact of the loss of natural enemies in terms of increased pesticide use and/or reduced yields, general observations by entomologists indicate that the impact of loss of natural enemies is severe in many parts of the world. For example, from 1980 to 1985 insecticide use in rice production in Indonesia drastically increased (Oka, 1991). This caused the destruction of beneficial natural enemies of the brown planthopper and pest populations exploded. Rice yields dropped so much that rice had to be imported into Indonesia for the first time in many years. The estimated loss in rice in just a 2-year period was \$1.5 billion (FAO, 1988).

Following that incident, entomologist Dr. I.N. Oka and his associates, who had

Table 4.3   Losses due to the destruction of beneficial natural enemies in US crops ($millions)

Crops	Total expenditure for insect control with pesticides ($)[a]	Added control costs ($)
Cotton	320	160
Tobacco	5	1
Potatoes	31	8
Peanuts	18	2
Tomatoes	11	2
Onions	1	0.2
Apples	43	11
Cherries	2	1
Peaches	12	2
Grapes	3	1
Oranges	8	2
Grapefruit	5	1
Lemons	1	0.2
Nuts	160	16
Other	500	50
Total		$257.4 ($520)[b]

[a]Pimentel et al. (1991).
[b]Because the added pesticide treatments do not provide as effective control as the natural enemies, we estimate that at least an additional $260 million in crops are lost to pests. Thus, the total loss due to the destruction of natural enemies is estimated to be at least $520 million year^{-1}.

previously developed a successful, low insecticide program for rice pests in Indonesia, were consulted by Indonesian President Soeharto's staff to determine what should be done to rectify the situation (I.N. Oka, Bogor Food Research Institute, Indonesia, personal communication, 1990). Their advice was to reduce insecticide use substantially and return to a sound 'treat-when-necessary' program that protected the natural enemies. Following Oka's advice, President Soeharto mandated in 1986 that 57 of 64 pesticides would be withdrawn from use on rice, and that pest management practices should be improved. Also, pesticide subsidies were reduced to zero. Subsequently rice yields increased to levels well above those recorded during the period of heavy pesticide use (FAO, 1988).

D. Rosen (Hebrew University of Jerusalem, personal communication, 1991) estimates that natural enemies of pests account for up to 90% of the control of pest species achieved in agroecosystems and natural systems; we estimate that about half of the control of pest species is due to natural enemies. Pesticides give an additional control of 10% (Pimentel et al., 1991), while the remaining 40% is due to host-plant resistance and other limiting factors present in the agroecosystem (Pimentel et al., 1991).

Parasites, predators and host-plant resistance are estimated to account for about 80% of the non-chemical control of pest insects and plant pathogens in crops (Pimentel et al., 1991). Many cultural controls such as crop rotations, soil and water management, fertilizer management, planting time, crop-plant density, trap crops, polyculture and others provide additional pest control. Together these non-chemical controls can be used effectively to reduce US pesticide use by as much as one-half without any reduction in crop yields (Pimentel et al., 1991).

PESTICIDE RESISTANCE IN PESTS

In addition to destroying natural enemy populations, the extensive use of pesticides has often resulted in the development of pesticide resistance in insect pests, plant pathogens and weeds. In a report by the United Nations Environment Programme, pesticide resistance was ranked as one of the top four environmental problems in the world (UNEP, 1979). About 504 insect and mite species (Georghiou, 1994), a total of nearly 150 plant pathogen species (Georghiou, 1986; Eckert, 1988) and about 273 weed species (LeBaron and McFarland, 1990) are now resistant to pesticides. In all, over 1600 insect species have developed resistance to pesticides since the 1940s (FAO, 1990).

Increased pesticide resistance in pest populations frequently results in the need for several additional applications of the commonly used pesticides to maintain expected crop yields. These additional pesticide applications compound the problem by increasing environmental selection for resistance traits. Despite attempts to deal with it, pesticide resistance continues to develop (Murray, 1994).

The impact of pesticide resistance, which develops gradually over time, is felt in the economics of agricultural production. A striking example of this occurred in northeastern Mexico and the Lower Rio Grande of Texas (NAS, 1975). Over time, extremely high pesticide resistance had developed in the tobacco budworm population on cotton. Finally, in early 1970 approximately 285 000 ha of cotton had to be abandoned because pesticides were ineffective and there was no way to protect the crop from the budworm. the economic and social impacts on these Texan and Mexican farming communities dependent upon cotton was devastating.

The study by Carrasco-Tauber (1989) indicates the extent of costs attributed to pesticide resistance. They reported a yearly loss of $45 to $120 ha^{-1} to pesticide resistance in Californian cotton. Thus, approximately $348 million of the Californian cotton crop was lost to resistance. Since $3.6 billion of US cotton were harvested in 1984, the loss due to resistance for that year was approximately 10%. Assuming a 10% loss in other major crops that receive heavy pesticide treatments in the United States, crop losses due to pesticide resistance are estimated to be $1.4 billion per year.

A detailed study by Archibald (1984) further demonstrated the hidden costs of pesticide resistance in Californian cotton. She reported that 74% more organophosphorus insecticides were required in 1981 to achieve the same kill of pests, such as *Heliothis* spp., as in 1979. Her analysis demonstrated that the diminishing effect of pesticides plus the intensified pest control reduced the economic return per dollar of pesticide invested to only $1.14.

Furthermore, efforts to control resistant *Heliothis* spp. exact a cost on other crops when large, uncontrolled populations of *Heliothis* and other pests disperse onto them. In addition, the cotton aphid and the whitefly exploded as secondary cotton pests because of their and their natural enemies' exposure to the high concentrations of insecticides. They have also developed resistance to insecticides.

The total external cost attributed to the development of pesticide resistance is estimated to range between 10 and 25% of current pesticide treatment costs (Harper and Zilberman, 1990), or approximately $400 million each year in the United States alone. In other words, at least 10% of pesticide used in the United States is applied just to combat increased resistance that has developed in various pest species.

In addition to plant pests, a large number of insect and mite pests of both livestock and humans have become resistant to pesticides (Georghiou, 1986). Although a relatively small quantity of pesticide is applied for control of livestock and human pests, the cost of resistance has become significant. Based on available data, we estimated the yearly cost of resistance in insect and mite pests of livestock and humans to be about $30 million for the United States.

Although the costs of pesticide resistance are high in the United States, its costs in tropical developing countries are significantly greater, because pesticides are used not only to control agricultural pests, but are also vital for the control of disease vectors (Faber, 1993; Murray, 1994). One of the major costs of resistance in tropical countries is associated with malaria control. By 1961, the incidence of malaria in India after early pesticide use declined to only 41 000 cases. However, because mosquitoes developed resistance to pesticides and malarial parasites developed resistance to drugs, the incidence of malaria in India now has exploded to about 59 000 000 cases per year (Reuben, 1989). Similar problems are occurring not only in India but also in the rest of Asia, Africa and South America, with the total world incidence of malaria estimated to be 100 to 120 million cases with 1 to 2 million deaths (WHO, 1992; NAS, 1993).

HONEY BEES AND WILD BEE POISONINGS AND REDUCED POLLINATION

Honey and wild bees are absolutely vital for pollination of fruits, vegetables and other crops. It has been estimated that the production of approximately one-third of all human food is dependent on bee pollination (Williamson, 1995). Their direct and indirect benefits to agricultural production range from $10 to $33 billion each year in the United States (E.L. Atkins, University of California, personal communication, 1990). Estimates of the benefits in Canada and Australia are $1.2 billion (Winston and Scott, 1984) and A$156 million (Gill, 1991), respectively. Because most insecticides used in agriculture are toxic to bees (MacKenzie and Winston, 1989), pesticides have a major impact on both honey bee and wild bee populations. D. Mayer (Washington State University, personal communication, 1990) estimates that approximately 20% of all honey bee colonies are adversely affected by pesticides. He includes the approximately 5% of US bee colonies that are killed outright or die during winter because of pesticide exposure. Mayer calculates that the direct annual loss reaches $13.3 million (Table 4.4). Another 15% of the bee colonies either are seriously weakened by pesticides, or suffer losses when apiculturists have to move colonies to avoid pesticide damage.

According to Mayer, the yearly estimated loss from partial bee kills and reduced honey production, plus the cost of moving colonies, totals about $25.3 million. Also, as a result of heavy pesticide use on certain crops, beekeepers are excluded from 4–6 million ha of otherwise suitable apiary locations (D. Mayer, Washington State University, personal communication, 1990). He estimates that the yearly loss in potential honey production in these regions is about $27 million.

In addition to these direct losses caused by the damage to bees and honey production, many crops are lost because of the lack of pollination. In California, for example, approximately 1 million colonies of honey bees are rented annually, at $20 per colony, to augment the natural pollination of almonds, alfalfa, melons, and other fruits and vegetables (R.A. Morse, Cornell University, personal communication, 1990). Since California produces nearly 50% of our bee-pollinated crops, the total cost for bee rental for the

Table 4.4   Estimated honey bee losses and pollination losses from honey bees and wild bees

	$ million year^{-1}
Colony losses from pesticides	13.3
Honey and wax losses	25.3
Losses of potential honey production	27.0
Bee rental for pollination	4.0
Pollination losses	200.0
Total	319.6

entire country is estimated at $40 million. Of this cost, we estimate at least one-tenth, or $4 million, is attributed to the effects of pesticides (Table 4.4).

Estimates of annual agricultural losses due to the reduction in pollination by pesticides may range as high as $4 billion per year (J. Lockwood, University of Wyoming, personal communication, 1990). For most crops both crop yield and quality are enhanced by effective pollination. For example, Tanda (1984) demonstrated that for several cotton varieties, effective pollination by bees resulted in yield increases from 20 to 30%. Assuming that a conservative 10% increase in cotton yield would result from more efficient pollination and subtracting charges for bee rental, the net annual gain for cotton alone could be as high as $400 million. However, using bees to enhance cotton pollination is impossible at present because of the intensive use of insecticides on cotton (McGregor, 1976).

Several studies emphasize that pollination is required for increased crop yields and, more importantly, it will increase the quality of crops such as melons and other fruits and vegetables (Currie et al., 1992; Cribb et al., 1993). In experiments with melons, E.L. Atkins (University of California, personal communication, 1990) reported that with adequate pollination melon yields were increased 10% and quality was raised 25% as measured by the dollar value of the crop.

Based on the analysis of honey bee and related pollination losses caused by pesticides, pollination losses attributed to pesticides are estimated to represent about 10% of pollinated crops, and have a yearly cost of about $200 million. Adding the cost of reduced pollination to the other environmental costs of pesticides on honey bees and wild bees, the total annual loss is calculated to be about $320 million (Table 4.4). Clearly, the available evidence confirms that the yearly cost of direct honey bee losses, together with reduced yields resulting from poor pollination, are significant.

CROP AND CROP PRODUCT LOSSES

Basically, pesticides are applied to protect crops from pests in order to increase yields, but sometimes the crops are damaged by pesticide treatments. This occurs when: (1) the recommended dosages suppress crop growth, development and yield; (2) pesticides drift from the targeted crop to damage adjacent nearby crops; (3) residual herbicides either prevent chemical-sensitive crops from being planted in rotation or inhibit the growth of crops that are planted; (4) excessive pesticide residues accumulate on crops, necessitating the destruction of the harvest. Crop losses translate into financial losses for growers, distributors, wholesalers, transporters, retailers, food processors and others. Potential profits as well as investments are lost. The cost of crop losses increases when the related costs of investigations, regulation, insurance and litigation are added to the equation. Ultimately the consumer pays for these losses in high market-place prices.

Data on crop losses due to pesticide use are difficult to obtain. Many losses are never reported to the state and federal agencies because the parties often settle privately (B.D. Berver, Office of Agronomy Services, South Dakota, personal communication, 1990; R. Batteese, Board of Pesticide Control, Maine Department of Agriculture, personal communication, 1990; J. Peterson, Pesticide/Noxious Weed Division, Department of Agriculture, North Dakota, personal communication, 1990; E. Streams, EPA, Region VII, personal communication, 1990). For example, in the State of North Dakota, only an estimated one-third of the pesticide-induced crop losses are reported to the State Department of Agriculture (J. Peterson, personal communication, 1990). Furthermore, according to the Federal Crop Insurance Corporation, losses due to pesticide use are not insurable because of the difficulty of determining pesticide damage (E. Edgeton, Federal Crop Insurance Corp., Washington, DC, personal communication, 1990).

Damage to crops may occur even when recommended dosages of herbicides and insecticides are applied to crops under normal environmental conditions (J. Neal, Chemical Pesticides Program, Cornell University, personal communication, 1990). Heavy dosages of insecticides used on crops have been reported to suppress growth and yield in both cotton and strawberry crops (ICAITI, 1977; Trumbel et al., 1988). The increased susceptibility of some crops to insects and diseases following normal use of 2,4-D and other herbicides was demonstrated by Oka and Pimentel (1976) and Rovira and McDonald (1986). Furthermore, when weather and/or soil conditions are inappropriate for pesticide application, herbicide treatments may cause yield reductions ranging from 2 to 50% (Akins et al., 1976).

Crops are lost when pesticides drift from the target crops to non-target crops located sometimes several miles downwind (Barnes et al., 1987). Drift occurs with almost all methods of pesticide application including both ground and aerial equipment, although the potential problem is greatest when pesticides are applied by aircraft (Ware et al., 1969). With aircraft, 50–75% of pesticides applied miss the target area (ICAITI, 1977; Ware, 1983; Akesson and Yates, 1984; Mazariegos, 1985). In contrast, 10–35% of the pesticide applied with ground application equipment misses the target area (Hall, 1991). The most serious drift problems are caused by 'speed sprayers' and 'mist-blower sprayers', because with these application technologies about 35% of the pesticide drifts away from the target area, plus large amounts of pesticide are applied by these sprayers compared with aircraft (E.L. Atkins, University of California, personal communication, 1990).

Crop injury and subsequent loss due to drift is particularly common in areas planted with diverse crops. For example, in southwest Texas in 1983 and 1984, nearly $20 million of cotton was destroyed from drifting 2,4-D herbicide when adjacent wheat fields were aerially sprayed (Hanner, 1984).

Clearly, drift damage, human exposure and widespread environmental contamination are inherent in the process of pesticide application and add to the cost of using pesticides. As a result, commercial applicators are frequently sued for damage inflicted during or after treatment. Therefore, most US applicators now carry liability insurance at an estimated cost of about $245 million per year (FAA, 1988; D. Witzman, US Aviation Underwriters, Tennessee, personal communication, 1990; H. Collins, National Agricultural Aviation Association, Washington, DC, personal communication, 1990).

When residues of some herbicides persist in the soil, crops planted in rotation may be injured (Keeling et al., 1989). In 1988/1989, an estimated $25 to $30 million of Iowa's soybean crop was lost due to the persistence of the herbicide Sceptor in the soil (R.G. Hartzler, Cooperative Extension Service, Iowa State University, personal communication, 1990).

Herbicide persistence can sometimes prevent growers from rotating their crops, and this situation may force them to continue planting the same crop (Altman, 1985; T. Tomas, Nebraska Sustainable Agriculture Society, Hartington, NE, personal communication, 1990). For example, the use of Sceptor in Iowa, as mentioned, has prevented farmers from implementing their plan to plant soybeans after corn (R.G. Hartzler, personal communication, 1990). Unfortunately, the continuous planting of some crops in the same field often intensifies insect, weed and pathogen problems (PSAC, 1965; NAS, 1975; Pimentel et al., 1991). Such pest problems not only reduce crop yields, but often require added pesticide applications.

Although crop losses caused by pesticides seem to be a small percentage of total US crop production, their total value is significant. For instance, an average of 0.14% of the total crop production of San Joaquin County, California, was lost to pesticides from 1986 to 1987 (OACSJC, 1990; OACSJC, Agricultural Commissioner, San Joaquin County, personal communication, 1990). Similarly, in Yolo County, CA, approximately 0.18% of its total crop production was lost in 1989 (OACYC, Agricultural Commissioner, Yolo County, personal communication, 1990; OACYC, 1990). Estimates from Iowa indicate that less than 0.05% of its annual soybean crop is lost to pesticides (R.G. Hartzler, personal communication, 1990).

An average 0.1% loss in the annual US production of corn, soybeans, cotton and wheat, which together account for about 90% of the herbicides and insecticides used in US agriculture, was valued at $40.9 million in 1993 (USDA, 1994). Assuming that only one-third of the incidents involving crop losses due to pesticides are reported to authorities, the total value of all crops lost because of pesticides could be as high as three times this amount, or $123 million annually.

However, this $23 million does not take into account other crop losses, nor does it include major but recurrent events such as the large-scale losses that occurred in Iowa in 1988–1989 ($25–$30 million), in Texas in 1983–1984 ($20 million) and in California's aldicarb/watermelon crisis in 1985 ($8 million, see below). These recurrent losses alone represent an average of $30 million each year, raising the estimated average crop loss value from the use of pesticides to approximately $153 million.

Additional losses are incurred when food crops are disposed of because they exceed the EPA regulatory tolerances for pesticide residue levels. Assuming that all crops and crop products that exceed the EPA regulatory tolerances (reported to be at least 1%) were disposed of as required by law, then about $550 million in crops annually would be destroyed because of excessive pesticide contamination (calculation based on data from FDA (1990) and USDA (1989a)). Because most of the crops with pesticides above the tolerance levels are neither detected nor destroyed, they are consumed by the public, avoiding financial loss but creating public health risks.

A well-publicized incident in California during 1985 illustrates this problem. In general, excess pesticides in food go undetected unless a large number of people become ill after the food is consumed. Thus, when more than 1000 persons became ill from eating the contaminated watermelons, approximately $1.5 million worth of watermelons were

Table 4.5   Estimated loss of crops and trees due to the use of pesticides

Impacts	Total costs ($ millions)
Crop losses	153
Crop applicator insurance	245
Crop destroyed because of excess pesticide contamination	550
Investigations and testing	
Governmental	10
Private	1
Total	959

ordered to be destroyed (R. Magee, State of California Department of Food and Agriculture, Sacramento, personal communication, 1990). After the public became ill, it was learned that several Californian farmers treated watermelons with the insecticide aldicarb (Temik), which is not registered for use on watermelons (Kizer, 1986; Taylor, 1986). Following this crisis, the California State Assembly appropriated $6.2 million to be awarded to claimants affected by state seizure and freeze orders (Legislative Counsel's Digest, 1986). According to the California Department of Food and Agriculture an estimated $800 000 in investigative costs and litigation fees resulted from this one incident (R. Magee, CDFA, personal communication, 1990). The California Department of Health Services was assumed to have incurred similar expenses, putting the total cost of the incident at nearly $8 million.

When crop seizures, insurance and investigation costs are added to the costs of direct crop losses due to the use of pesticides in commercial crop production, the total monetary loss is estimated to be about $959 million annually in the United States (Table 4.5).

GROUND AND SURFACE WATER CONTAMINATION

Certain pesticides applied to crops eventually end up in ground and surface waters and have even been detected in rain and fog (NRDC, 1993). It has been estimated that the total runoff of all pesticides from non-irrigated farmland is 1%, from irrigated farmland 4%, and the runoff is 33% when applied by aircraft (Ananyeva et al., 1992).

Pesticide contamination of surface waters—lakes, rivers, streams—is an important concern due to its extensive use for drinking water and recreation. Particularly disturbing is the fact that conventional drinking water treatment is not designed to remove, and consequently does not remove, pesticides. One study showed that after conventional treatment, 90% of the drinking water samples contained at least one pesticide, while 58% of the samples contained at least four different pesticides (Kelley, 1986). After studying the drinking water in 29 US cities over a 4-month period, Cohen et al. (1995) observed that the herbicide atrazine was present in the tap water of 28 of the cities (97%) and cyanazine was present in 25 of the cities (86%). In addition, Federal health levels for atrazine and cyanazine were exceeded in 17% and 35% of all samples taken, respectively. At present, water treatment plants do not regularly monitor for pesticide contamination. Estimates for the cost of daily monitoring for the triazine herbicides are $1500 per city per year, or $0.10 per person (Cohen et al., 1995). There are over 11 000 surface water treatment systems, of which approximately 4000 are large systems and 7000 are small systems (Auerbach, 1994; Natarajan and Rajagopal, 1994). Assuming $1500 per year for the large systems and $3000 per year for the small systems, the total cost would be approximately $27 million per year.

The three most common pesticides found in groundwater are aldicarb (an insecticide), alachlor and atrazine (two herbicides) (Osteen and Szmedra, 1989). Estimates are that nearly one-half of the groundwater and well water in the United States is or has the potential to be contaminated (Holmes et al., 1988; NRDC, 1993). EPA (1990a) reported that 10.4% of community wells and 4.2% of rural domestic wells have detectable levels of at least one pesticide of the 127 pesticides tested in a national survey. It would cost an estimated $1.3 billion annually in the United States if well and groundwater were monitored for pesticide residues (Nielsen and Lee, 1987).

Two major concerns about groundwater contamination with pesticides are that about one-half of the population obtains its water from wells (Beitz et al., 1994), and that once groundwater is contaminated, the pesticide residues remain for long periods of time (Gustafson, 1993). Not only are there very few microorganisms that have the potential to degrade pesticides (Pye and Kelley, 1984), but the groundwater recharge rate averages less than 1% per year.

Monitoring pesticides in groundwater is only a portion of the total cost of US groundwater contamination. Also, there is the high cost of clean-up. For instance, at the Rocky Mountain Arsenal near Denver, Colorado, the removal of pesticides from the groundwater and soil was estimated to cost approximately $2 billion (NYT, 1988). If all pesticide-contaminated groundwater were cleared of pesticides before human consumption, the cost would be about $500 million (based on the costs of cleaning water (van der Leeden et al., 1990)). Note that the clean-up process requires a water survey to target the contaminated water for clean-up. Thus, adding the monitoring costs plus the cleaning costs, the total cost regarding pesticide-polluted groundwater is estimated to be about $1.8 billion annually.

FISHERY LOSSES

Pesticides are washed into aquatic ecosystems by water runoff and soil erosion. About 18 $t\,ha^{-1}\,year^{-1}$ of soil are washed and/or blown from pesticide-treated cropland into adjacent locations, including streams and lakes (USDA, 1989b). Pesticides also drift into streams and lakes and contaminate these aquatic systems (Clark, 1989). Some soluble pesticides are easily leached into streams and lakes (Nielsen and Lee, 1987) and are readily taken up by aquatic organisms (Wang et al., 1994; Chevreuil et al., 1995). A nationwide survey of fish in the United States showed pesticide residues present in almost all of the 119 fish species that were examined (Kuehl and Butterworth, 1994).

Once in aquatic systems, pesticides cause fishery losses in several ways. These include high pesticide concentrations in water that directly kill fish, low-level doses that may kill highly susceptible fish fry, the elimination of essential fish foods like insects and other invertebrates, or the reduction of dissolved oxygen levels in the water due to the decomposition of aquatic plants killed by pesticides. In addition, because government safety restrictions ban the catching or sale of fish contaminated with pesticide residues, such unmarketable fish are considered an economic loss (ME and MNR, 1990). Furthermore, health advisories from the New York State Department of Environmental Conservation restrict the amount of sportfish than can be eaten, which impacts revenues that can be gained from tourism.

Each year large numbers of fish are killed by pesticides. Based on EPA (1990b) data, we calculate that from 1977 to 1987 the number of fish kills due to all factors has been 141 million fish per year; from 6 to 14 million fish per year are killed by pesticides. These estimates of fish kills are considered to be low for the following reasons. First, in 20% of the fish kills, no estimate is made of the number of fish killed, and second, fish kills frequently cannot be investigated quickly enough to determine accurately the primary cause. In addition, fast-moving waters in rivers dilute pollutants, so that these causes of kills frequently cannot be identified. Moving waters also wash away some of the poisoned fish, while other poisoned fish sink to the bottom and cannot be counted. Perhaps most important is the fact that, unlike direct kills, few, if any, of the widespread and more frequent low-level pesticide poisonings are dramatic enough to be observed, and therefore go unrecognized and unreported (EPA, 1990b).

The average value of a fish was estimated in 1982 to be about $1.70, using the guidelines of the American Fisheries Society (AFS, 1982). However, it was reported that Coors Beer might be 'fined up to $10 per dead fish, plus other penalties' for an accidental beer spill in a creek (Barometer, 1991). Using the estimated value of $4.00, the value of the low estimate of 6–14 million fish killed per year ranges from $24 to $56 million. This calculation only takes direct impacts into account, indirect impacts, such as tourism, are more difficult to define but can be substantial (Connelly and Brown, 1991). For instance, the revenue generated in Massachusetts by marine recreational fishing has been estimated to be $637 million annually (Storey and Allen, 1993). So the actual loss is probably several times the $24–$56 million when all of the indirect impacts are taken into account.

WILD BIRDS AND MAMMALS

Wild birds and mammals are also damaged by pesticides, and these animals make excellent 'indicator species'. Deleterious effects on wildlife include death from direct exposure to pesticides or secondary poisonings from consuming contaminated prey, reduced survival, growth and reproductive rates from exposure to sublethal dosages, and habitat reduction through elimination of food sources and refuges (Fluetsch and Sparling, 1994). In the United States, approximately 3 kg of pesticide per hectare are applied on about 160 million ha year^{-1} of land (Pimentel et al., 1991). With such a large proportion of the land area being treated with heavy dosages of pesticide, it is to be expected that the impact on wildlife is significant.

The full extent of bird and mammal destruction is difficult to determine because these animals are often secretive, camouflaged, highly mobile and live in dense grass, shrubs and trees. Typical field studies of the effects of pesticides often obtain extremely low estimates of bird and mammal mortality (Mineau and Collins, 1988). This is because bird carcasses disappear quickly, well before the dead birds can be found and counted. Studies show that only 50% of birds are recovered even when the bird's location is known (Mineau, 1988). Furthermore, where known numbers of bird carcasses were placed in identified locations in the field, 62–92% disappeared overnight due to vertebrate scavengers (Balcomb, 1986). Then too, field studies seldom account for birds that die some distance from the treated areas. Finally, birds often hide and die in inconspicuous locations.

Nevertheless, many bird casualties caused by pesticides have been reported. White et al. (1982) reported that 1200 Canada geese were killed in one wheat field that was sprayed with a 2:1 mixture of parathion and methyl parathion at a rate of 0.8 kg ha^{-1}. Carbofuran applied to alfalfa killed more than 5000 ducks and geese in five incidents, while the same chemical applied to vegetable crops killed 1400 ducks in a single incident (Flickinger et al., 1980, 1991). Carbofuran is estimated to kill 1–2 million birds each year (EPA, 1989). Another pesticide, diazinon, applied on just three golf courses, killed 700 Atlantic Brant Geese of the wintering population of 2500 geese (Stone and Gradoni, 1985).

Several studies report that the use of herbicides in crop production results in the total elimination of weeds that harbour some insects (Potts, 1986; R. Beiswenger, University of Wyoming, personal communication, 1990). This has led to significant reductions in the grey partridge in the United Kingdom and common pheasant in the United States. In the case of the partridge, population levels have decreased more than 77%, because partridge chicks (also pheasant chicks) depend on insects to supply them with the protein needed for their development and survival (Potts, 1986; R. Beiswenger, University of Wyoming, personal communication, 1990).

Frequently the form of a pesticide influences its toxicity to wildlife (Hardy, 1990). For example, treated seed and insecticide granules, including carbofuran, fensulfothion, fonofos and phorate, are particularly toxic to birds when consumed. Many birds will ingest these granules either on purpose or by accident, thereby ingesting the pesticide directly. Some recent research has focussed on the spraying of the pesticide-treated area with a solution of a naturally occurring plant substance which is unpalatable to many types of birds (Chen, 1995). Another approach includes treating the granules with taste-repellents before field application (Mastrota and Mench, 1995). However, estimates are that from 0.23 to 1.5 birds ha^{-1} year^{-1} are killed in Canada, while in the United States the estimates ranged from 0.25 to 8.9 birds ha^{-1} killed per year by the pesticides (Mineau, 1988).

Pesticide toxicity can also be species-specific. For instance, in the United Kingdom, about 1500 greylag and pink-footed geese died over a 4-year period due to exposure to carbophenothion. This occurred despite pre-approval toxicity tests using Canada Geese. Further tests revealed that these geese were much more sensitive to carbophenothion than were the Canada geese (Greig-Smith, 1990).

Pesticides also adversely affect the reproductive potential of many birds and mammals. Exposure of birds, especially predatory birds, to chlorinated insecticides has caused reproductive failure, sometimes attributed to eggshell thinning (Mineau et al., 1994). Most of the affected populations have recovered after the ban on DDT in the United States (Bednarz et al., 1990). However, DDT and its metabolite DDE remain a concern, because DDT continues to be used in some South American countries, which are the wintering areas for numerous bird species (Stickel et al., 1984).

Several pesticides, especially DBCP, dimethoate and deltamethrinare, are reported to reduce sperm production in certain mammals (Foote et al., 1986, Salem et al., 1988). Clearly, when this occurs the capacity of certain wild mammals to survive is reduced.

Habitat alteration and destruction can be expected to reduce mammal populations. For example, when glyphosphate was applied to forest clearcuts to eliminate low-growing vegetation, the southern red-backed vole population was greatly reduced because its food source and cover were practically eliminated (D'Anieri et al., 1987). In another vivid example, the decline of the alligator population in Florida's Lake Apopka, was linked to a large spill of DDT. Not only were there problems with eggs in every nest, but it was observed that the male alligators had abnormally low testosterone levels and that 25% of them had shrunken penises which would never allow them to reproduce (Begley and Glick, 1994). Similar effects have also been observed in 16 predator species in the Great Lakes region (Fantle, 1994). Similar effects from herbicides have been reported on other mammals (Pimentel, 1971). These findings have raised the concern over how much humans are being affected by these chemicals, especially since Carlsen et al. (1992) reported that worldwide, the average male sperm count has decreased by more than 40% since 1938. Overall, however, the impacts of pesticides on mammals have been inadequately investigated.

Although the gross values for wildlife are not available, expenditure involving wildlife made by humans are one measure of the monetary value. Non-consumptive users of wildlife spent an estimated $14.3 billion on their sport in 1985 (USFWS, 1988). US bird watchers spend an estimated $600 million annually on their sport and an additional $500 million on birdseed, or a total of $1.1 billion (USFWS, 1988). The money spent by bird hunters to harvest 5 million game birds was $1.1 billion, or approximately $216 per bird (USFWS, 1988). In addition, estimates of the value of all types of birds ranged from $0.40 to more than $800 per bird. To $0.40 per bird was based on the costs of birdwatching and the $800 per bird was based on the replacement costs of the affected species (Walgenbach, 1979; Tinney, 1982; Dobbins, 1986; James, 1995).

If it is assumed that the damage which pesticides inflict on birds occurs primarily on the 160 million hectares of cropland that receive most of the pesticide, and the bird population is estimated to be 4.2 birds per hectare of cropland (Blew, 1990), then 672 million birds are directly exposed to pesticides. If it is conservatively estimated that only 10% of the bird population is killed, then the total number killed is 67 million birds. Note that this estimate is at the lower end of the range of 0.25–8.9 birds ha^{-1} killed per year by pesticides mentioned earlier in this section. Also, this is considered to be a conservative estimate because secondary losses to pesticide reductions in invertebrate prey poisonings were not included in the assessment. Assuming the average value of a bird is $30, then an estimated $2 billion in birds are destroyed annually.

Also, a total of $102 million is spent yearly by the US Fish and Wildlife Service on their Endangered Species Program, which aims to re-establish species such as the bald eagle, peregrine falcon, osprey and brown pelican that in some cases were reduced by pesticides (USFWS, 1991).

When all the above costs are combined, we estimate that the US bird losses associated with pesticide use represent a cost of about $2.1 billion year^{-1}.

GOVERNMENT FUNDS FOR PESTICIDE POLLUTION CONTROL

A major environmental cost associated with all pesticide use is the cost of carrying out state and federal regulatory actions, as well as pesticide monitoring programs needed to control pesticide pollution. Specifically, these funds are spent to reduce the hazards of pesticides and to protect the integrity of the environment and public health.

At least $1 million is spent each year by the state and federal government to train and register pesticide applicators (D. Rutz, Cornell University, personal communication, 1991). Also, more than $40 million is spent each year by EPA for just registering and re-registering pesticides (GAO, 1986). Based on these known expenditures, estimates are

Table 4.6   Total estimated environmental and social costs from pesticides in the United States

Costs	$ million year^{-1}
Public health impacts	933
Domestic animal deaths and contamination	31
Loss of natural enemies	520
Cost of pesticide resistance	1400
Honey bee and pollination losses	320
Crop losses	959
Surface water monitoring	27
Groundwater contamination	1800
Fishery losses	56
Bird losses	2100
Government regulations to prevent damage	200
Total	8346

that the federal and state governments spend approximately \$200 million year^{-1} for pesticide pollution control (USBC, 1994) (Table 4.6).

Although enormous amounts of government money are being spent to reduce pesticide pollution, many costs of pesticides are not taken into account. Also, many serious environmental and social problems remain to be corrected by improved government policies.

CONCLUSION

An investment of about \$6.5 billion dollars in pesticide control saves approximately \$26 billion in US crops, based on direct costs and benefits (Pimentel et al., 1991; USBC, 1994). However, the indirect costs of pesticide use to the environment and public health need to be balanced against these benefits. Based on the available data, the environmental and social costs of pesticide use total approximately \$8.3 billion each year (Table 4.6). Users of pesticides in agriculture pay for only about \$3.2 billion of this cost, which includes problems arising from pesticide resistance and destruction of natural enemies. Society eventually pays this \$3.2 billion, plus the remaining \$5.1 billion in environmental and public health costs (Table 4.6).

Out assessment of the environmental and health problems associated with pesticides faced problems of scarce data that made this assessment of the complex pesticide situation incomplete. For example, what is an acceptable monetary value for a human life lost or a cancer illness due to pesticides? Equally difficult is placing a monetary value on killed wild birds and other wildlife, invertebrates, microbes or contaminated food and groundwater.

In addition to the costs that cannot be measured accurately, there are additional costs that have not been included in the \$8.3 billion year^{-1} figure. A complete accounting of the indirect costs should include accidental poisonings like the 'aldicarb/watermelon' crisis; domestic animal poisonings; unrecorded losses of fish, wildlife, crops, trees and other plants; losses resulting from the destruction of soil invertebrates, microflora and micro-fauna; true costs of human pesticide poisonings; water and soil pollution; human health effects such as cancer and sterility. If the full environmental and social costs could be measured as a whole, the total cost would be significantly greater than the estimate of $8.3 billion year^{-1}. Such a complete long-term cost/benefit analysis of pesticide use would reduce the perceived profitability of pesticides.

Human pesticide poisonings, reduced natural enemy populations, increased pesticide resistance and honey bee poisonings account for a substantial portion of the calculated environmental and social costs of pesticide use in the United States. Fortunately, some losses of natural enemies and pesticide resistance problems are being alleviated through carefully planned use of integrated pest management (IPM) practices, but a great deal remains to be done to reduce these important environmental costs (Pimentel et al., 1991).

This investigation not only underscores the serious nature of the environmental and social costs of pesticides, but emphasizes the great need for more detailed investigation of the environmental and economic impacts of pesticides. Pesticides are, and will continue to be, a valuable pest control tool. Meanwhile, with more accurate, realistic cost/benefit analyses, we will be able to work to minimize the risks and develop and increase the use of non-chemical pest controls to maximize the benefits of pest control strategies for all of society.

REFERENCES

AFS (1982) *Monetary Values of Freshwater Fish and Fish-kill Counting Guidelines*. American Fisheries Society Special Publication No. 13, Bethesda, MD.

Akesson, N.B. and Yates, W.E. (1984). Physical parameters affecting aircraft spray application. In Garner, W.Y. and Harvey, J. (Eds) *Chemical and Biological Controls in Forestry*. American Chemical Society Series 238, Washington, DC, pp. 95–111.

Akins, M.B., Jeffery, L.S., Overton, J.R. and Morgan, T.H. (1976) Soybean response to preemergence herbicides. *Proceedings of the Southern Weed Science Society* 29, 50.

Altman, J. (1985) Impact of herbicides on plant diseases. In Parker, C.A., Rovia, A.D., Moore, K.J. and Wong, P.T.W. (Eds) *Ecology and Management of Soilborne Plant Pathogens*. Phytopathological Society, St. Paul, MN, pp. 227–231.

Ananyeva, N.D., Naumova, N.N., Rogers, J. and Steen, W.C. (1992) Microbial transformation of selected organic chemicals in natural aquatic systems. In Schnoor, J.L. (Ed.) *Fate of Pesticides and Chemicals in the Environment*. John Wiley & Sons, New York, pp. 275–294.

Archibald, S.O. (1984) A dynamic analysis of production externalities: Pesticide resistance in California. Ph.D. Thesis, University of California, Davis.

Auerbach, J. (1994). Costs and benefits of current SDWA regulations. *Journal of the American Water Works Association* 86, 69–78.

Balcomb, R. (1986) Songbird carcasses disappear rapidly from agricultural fields *Auk* 103, 817–821.

Barnes, C.J., Lavy, T.L. and Mattice, J.D. (1987). Exposure of non-applicator personnel and adjacent areas to aerially applied propanil. *Bulletin of Environmental Contamination and Toxicology* 39, 126–133.

Barometer (1991) Too much beer kills thousands. Oregon State University Barometer, 14 May.

Barton, J. and Oehme, F. (1981) Incidence and characteristics of animal poisonings seen at Kansas State University from 1975 to 1980. *Veterinary and Human Toxicology* 23, 101–102.

Beasley, V.R. and Trammel, H. (1989). Insecticide. In Kirk, R.W. (Ed.) *Current Veterinary Therapy: Small Animal Practice*. W.B. Saunders, Philadelphia, PA, pp. 97–107.

Bednarz, J.C., Klem, D., Goodrich, L.J. and Senner, S.E. (1990) Migration counts of raptors at Hawk Mountain, Pennsylvania, as indicators of population trends, 1934–1986. *Auk* 107, 96–109.

Begley, S. and Glick, D. (1994) The estrogen complex. *Newsweek*, 21 March, pp. 76–77.

Beitz, H., Schmidt, H. and Herzel, F. (1994) Occurrence, toxicological and ecotoxicological significance of pesticides in groundwater and surface water. In Borner, H. (Ed.) *Pesticides in Ground and Surface Water*. Springer, Berlin, pp. 1–56.

Benbrook, C.M. (1996) *Pest Management at the Crossroads*. Consumer Union, Yonkers, New York.

Blew, J.H. (1990) Breeding bird census 92 conventional cash crop farm. *Journal of Field Ornithology* **61** (Suppl.): 80–81.

Brown, L.M., Blair, A., Gibson, R., Everett, G.D., Cantor, K.P., Schuman, L.M., Burmeister, L.F., Van Lier, S.F. and Dick, F. (1990) Pesticide exposures and other agricultural risk factors for leukemia among men in Iowa and Minnesota. *Cancer Research* **50**, 6585–6591.

Bull, D. (1982) *A Growing Problem: Pesticides and the Third World Poor.* Oxfam, Oxford.

Cantor, K.P., Blair, A., Everett, G., Gibson, R., Burmeister, L.F., Brown, L.M., Schuman, L. and Dick, F.R. (1992). Pesticides and other agricultural risk factors for non-Hodgkin's lymphoma among men in Iowa and Minnesota. *Cancer Research* **52**, 2447–2455.

Carlsen, E., Giwercman, A., Keiding, N. and Skakkebaek, N.E. (1992) Evidence for decreasing quality of semen during the past 50 years. *British Medical Journal* **305**, 609–613.

Carrasco-Tauber, C. (1989) Pesticide productivity revisited. M.S. Thesis, University of Massachusetts, Amherst.

Carson, R. (1962) *Silent Spring.* Houghton Mifflin, Boston, MA.

Casey, P. and Vale, J.A. (1994) Deaths from pesticide poisoning in England and Wales: 1945–1989. *Human and Experimental Toxicology* **13**, 95–101.

Chen, C. (1995) Flavoring against fowl. *Cornell Countryman* April/May, p. 24.

Chevreuil, M., Carru, A., Chesterikoff, A., Boet, P., Tales, E. and Allardi, J. (1995) Contamination of fish from different areas of the river Seine (France) by organic (PCB and pesticides) and metallic (Cd, Cr, Cu, Fe, Mn, Pb, Zn) micropollutants. *Science of the Total Environment* **162**, 31–42.

Ciesielski, S., Loomis, D.P., Mims, S.R. and Auer, A. (1994) Pesticide exposures, cholinesterase depression, and symptoms among North Carolina migrant farmworkers. *American Journal of Public Health* **84**, 446–451.

Clark, R.B. (1989). *Marine Pollution.* Clarendon Press, Oxford.

Cohen, B., Wiles, R. and Bondoc, E. (1995) *Weed Killers by the Glass.* Environmental Working Group, Washington, DC.

Colborn, T., Myers, J.P. and Dumanoski, D. (1996) *Our Stolen Future: How We Are Threatening Our Fertility, Intelligence, and Survival: A Scientific Detective Story.* Dutton, New York.

Colvin, B.M. (1987) Pesticide uses and animal toxicoses. *Veterinary and Human Toxicology* **29** (Suppl. 2), 15 pp.

Committee, House of Commons Agriculture (1987) *The Effects of Pesticides on Human Health.* Report and Proceedings of the Committee, 2nd Special Report, Vol. I. Her Majesty's Stationery Office, London.

Connelly, N.A. and Brown, T.L. (1991) Net economic value of the freshwater recreational fisheries of New York. *Transactions of the American Fisheries Society* **120**, 770–775.

Cordle, M.K. (1988) USDA regulation of residues in meat and poultry products. *Journal of Animal Science* **66**, 413–433.

Crawford, L.M. and Schor, D.M. (1991) Pesticides from a regulatory perspective. In Tweedy, B.G., Dishburger, H.J., Ballantine, L.G. and McCarthy, J. (Eds) *Pesticide Residues and Food Safety: A Harvest of Viewpoints.* American Chemical Society, Washington, DC, pp. 308–312.

Cribb, D.M., Hand, D.W. and Edmondson, R.N. (1993) A comparative study of the effects of using the honeybee as a pollinating agent of glasshouse tomato. *Journal of Horticultural Science* **68**, 79–88.

Croft, B.A. (1990) *Arthropod biological control agents and pesticides.* John Wiley & Sons, New York.

Currie, R.W., Winston, M.L. and Slessor, K.N. (1992) Effects of synthetic queen mandibular pheromone sprays on honey bee hymenoptera apidae pollination of berry crops. *Journal of Economic Entomology* **85**, 1300–1306.

D'Anieri, P., Leslie, D.M. and McCormack, M.L. (1987) Small mammals in glyphosphate-treated clearcuts in Northern Maine. *Canadian Field Naturalist* **101**, 547–550.

Davis, J.R., Brownson, R.C. and Garcia, R. (1992) Family pesticide use in the home, garden, orchard, and yard. *Archives of Environmental Contamination and Toxicology* **22**, 260–266.

Dobbins, J. (1986) *Resources Damage Assessment of the T/V Puerto Rican Oil Spill Incident.* James Dobbins, Report to NOAA, Sanctuary Program Division, Washington, DC.

Dogheim, S.M., Nasr, E.N., Almaz, M.M. and El-Tohamy, M.M. (1990) Pesticide residues in milk and fish samples collected from two Egyptian Governorato. *Journal of the Association of Official Analytical Chemists* **73**, 19–21.

Eckert, J.W. (1988) Historical development of fungicide resistance in plant pathogens. In Delp, C.J. (Ed.) *Fungicide Resistance in North America*, APS Press, St. Paul, MN, pp. 1–3.

Ecobichon, D.J., Davies, J.E., Doull, J., Ehrich, M., Joy, R., McMillan, D., MacPhail, R., Reiter, L.W., Slikker, W. and Tilson, H. (1990) Neurotoxic effects of pesticides. In Wilkinson, C.F. and Baker, S.R. (Eds) *The Effect of Pesticides on Human Health*. Princeton Scientific, Princeton, NJ, pp. 131–199.

EPA (1977) *Minutes of Administrator's Pesticide Policy Advisory Committee*. US Environmental Protection Agency, Washington, DC, March.

EPA (1989) *Carbofuran: A Special Review Technical Support Document*. US Environmental Protection Agency, Office of Pesticides and Toxic Substances, Washington, DC.

EPA (1990a) *National Pesticide Survey—Summary*. US Environmental Protection Agency, Washington, DC.

EPA (1990b) *Fish Kills Caused by Pollution, 1977–1987*. Draft Report of US Environmental Protection Agency, Office of Water Regulations and Standards, Washington, DC.

FAA (1988) *Census of Civil Aircraft*. US Federal Aviation Administration, Washington, DC.

Faber, D. (1993) *Environment Under Fire*. Monthly Review Press, New York.

Fantle, W. (1994) The incredible shrinking man. *The Progressive*. October pp. 12–13.

FAO (1988) *Integrated Pest Management in Rice in Indonesia*. Food and Agriculture Organization, United Nations, Jakarta, May.

FAO (1990) *The State of Food and Agriculture 1989*. Food and Agriculture Organization of the United Nations, Rome.

FAO (1994) 1993 *Production Yearbook*. Food and Agriculture Organization of the United Nations, Rome.

FDA (1989) *Food and Drug Administration Pesticide Program Residues in Foods—1988*. Journal of the Association of Official Analytical Chemists **72**, 133A–152A.

FDA (1990) *Food and Drug Administration Pesticide Program Residues in Foods—1989*. Journal of the Association of Official Analytical Chemists **73**, 127A–146A.

FDA (1993) FDA memo from Division of Federal–State relations, 7 May 1993.

Fiore, M.C., Anderson, H.A., Hong, R., Golubjatnikov, R., Seiser, J.E., Nordstrom, D., Hanrahan, L. and Belluck, D. (1986) Chronic exposure to aldicarb-contaminated groundwater and human immune function. *Environmental Research* **41**, 633–645.

Flickinger, E.L., King, K.A., Stout, W.F. and Mohn, M.M. (1980) Wildlife hazards from furadan 3G applications to rice in Texas. *Journal of Wildlife Management* **44**, 190–197.

Flickinger, E.L., Juenger, G., Roffe, T.J., Smith, M.R. and Irwin, R.J. (1991) Poisoning of Canada geese in Texas by parathion sprayed for control of Russian wheat aphid. *Journal of Wildlife Diseases* **27**, 265–268.

Fluetsch, K.M., Sparling, D.W. (1994) Avian nesting success and diversity in conventionally and organically managed apple orchards. *Environmental Toxicology and Chemistry* **13**, 1651–1659.

Foote, R.H., Schermerhorn, E.C. and Simkin, M.E. (1986) Measurement of semen quality, fertility, and reproductive hormones to assess dibromochloropropane (DBCP) effects in live rabbits. *Fundamental and Applied Toxicology* **6**, 628–637.

Forget, F. (1991) Pesticides and the third world. *Journal of Toxicology and Environmental Health* **32**, 11–31.

Gaafar, S.M., Howard, W.E. and Marsh, R. (1985) *World Animal Science B: Parasites, Pests, and Predators*. Elsevier, Amsterdam.

GAO (1986) *Pesticides: EPA's Formidable Task to Assess and Regulate Their Risks*. US General Accounting Office, Washington, DC.

Garrido, M.D., Jodral, M. and Pozo, R. (1994) Organochlorine pesticides in Spanish sterilized milk and associated health risks. *Journal of Food Protection* **57**, 249–252.

Georghiou, G.P. (1986) *The Magnitude of the Resistance Problem. Pesticide Resistance, Strategies and Tactics for Management*. National Academy of Sciences, Washington, DC, pp. 18–41.

Georghiou, G.P. (1994) Principles of insecticide resistance management. *Phytoprotection* **75**, 51–59.

Gill, R.A. (1991) The value of honeybee pollination to society. *Acta Horticultural* **288**, 62–68.

Greig-Smith, P.W. (1990) Understanding the impact of pesticides on wild birds by monitoring incidents of poisoning. In Kendall, R.J. and Lacher, T.E. *Wildlife Toxicology and Population Modeling*. Lewis, Boca Raton, FL, pp. 301–320.

Gustafson, D.I. (1993) *Pesticides in Drinking Water.* Van Nostrand Reinhold, New York.

Hall, F.R. (1991). Pesticide application technology and integrated pest management (IPM). In Pimentel, D. (Ed.) *Handbook of Pest Management in Agriculture.* Vol. II. CRC Press, Boca Raton, FL, pp. 135–170.

Hanner, D. (1984) Herbicide drift prompts state inquiry. *Dallas (Texas) Morning News, 25 July.*

Hardy, A.R. (1990) Estimating exposure: The identification of species at risk and routes of exposure. In Somerville, L. and Walker, C.H. (Eds) *Pesticide Effects on Terrestrial Wildlife.* Taylor & Francis, London, pp. 81–97.

Harper, C.R. and Zilberman, D. (1990) Pesticide regulation: Problems in trading off economic benefits against health risks. In Zilberman, D. and Siebert, J.B. (Eds) *Economic Perspectives on Pesticide Use in California.* University of California, Berkeley.

Herrera, A., Arino, A.A., Conchello, M.P., Lazaro, R., Bayarri, S. and Perez, C. (1994) Organochlorine pesticide residues in Spanish meat products and meat of different species. *Journal of Food Protection* 57, 441–444.

Holmes, T., Nielsen, E. and Lee, L. (1888) *Managing Groundwater Contamination in Rural Areas. Rural Development Perspectives.* US Department of Agriculture, Economic Research Series, Washington, DC.

Hundley, H.K., Cairns, T., Luke, M.A. and Masumoto, H.T. (1988) Pesticide residue findings by the Luke method in domestic and imported foods and animal feeds for the fiscal years 1982–1986. *Journal of the Association of Official Analytical Chemists* 71, 875–877.

ICAITI (1977) *An Environmental and Economic Study of the Consequences of Pesticide Use in Central American Cotton Production.* Final Report, Central American Research Institute for Industry, United Nations Environment Programme, Guatemala City.

James, P.C. (1995) Internalizing externalities: Granula carbofuran use on rapeseed in Canada. *Ecology and Economics* 13, 181–184.

JAWWA (1995) DuPont, USEPA reach accord on phasing out cyanazine. *Journal of the American Water Works Association* 87, 16.

Johnson, D.W., Kish, L.P. and Allen, G.E. (1976) Field evaluation of selected pesticides on the natural development of the entomopathogen *Nomuraea rileyi* on the velvetbean caterpillar in soybean. *Environmental Entomology* 5, 964–966.

Keefe, T.J., Savage, E.P., Munn, S. and Wheeler, H.W. (1990) *Evaluation of Epidemiological Factors from Two National Studies of Hospitalized Pesticide Poisonings, USA.* Exposure Assessment Branch, Hazard Evaluation Division, Office of Pesticides and Toxic Substances, US Environmental Protection Agency, Washington, DC.

Keeling, J.W., Lloyd, R.W. and Abernathy, J.R. (1989) Rotational crop response to repeated applications of korflurazon. *Weed Technology* 3, 122–125.

Kelley, R.D. (1986) Pesticides in Iowa's drinking water. Proceedings of a Conference: *Pesticides and Groundwater: A Health Concern for the Midwest.* 16–17 October, Navarre, MN. Freshwater Foundation, pp. 121–122.

Kizer, K. (1986) *California's Fourth of July Food Poisoning Epidemic from Aldicarb-Contaminated Watermelons.* State of California Department of Health Services, Sacramento, CA.

Kovach, J. and Agnello, A.M. (1991) Apple pests—pest management system for insects. In Pimentel D. (Ed.) *Handbook of Pest Management in Agriculture.* CRC Press, Boca Raton, FL, pp. 107–116.

Kuehl, D.W. and Butterworth, B. (1994) A national study of chemical residues in fish. III. Study results. *Chemosphere* 29, 523–535.

LeBaron, H.M. and McFarland, J. (1990) Herbicide resistance in weeds and crops. In Green, M.B., LeBaron, H.M. and Moberg, W.K. (Eds) *Managing Resistance from Fundamental Research to Practical Strategies.* American Chemical Society, Washington, DC, pp. 336–352.

Legislative Counsel's Digest (1986) Legislative Assembly Bill No. 2755, State of California.

Leiss, J.K. and Savitz, D.A. (1995) Home pesticide use and childhood cancer: A case-control study. *American Journal of Public Health* 85, 249–252.

Lotti, M. (1992) The pathogenesis of organophosphate polyneuropathy. *Critical Reviews in Toxicology* 21, 465–487.

Luster, M.I., Bland, J.A. and Dean, J.H. (1987) Molecular and cellular basis of chemically induced immunotoxicity. *Annual Review of Pharmacology and Toxicology* 27, 23–49.

MacKenzie, K. and Winston, M.L. (1989) Effects of sublethal exposure to diazinon and temporal division of labor in the honeybee. *Journal of Economic Entomology* **82**, 75–82.

Mastrota, F.N. and Mench, J.A. (1995) Evaluation of taste repellents with northern bobwhites for deterring ingestion of granular pesticides. *Environmental Toxicology and Chemistry* **14**, 631–638.

Mazariegos, F. (1985) *The Use of Pesticides in the Cultivation of Cotton in Central America*. United Nations Environment Programme, Industry and Environment, Guatemala, July/August/September.

McDuffie, H.H. (1994) Women at work: Agriculture and pesticides. *Journal of Occupational Medicine* **36**, 1240–1246.

McGregor, S.E. (1976) *Insect Pollination of Cultivated Crop Plants*. US Department of Agriculture, Agricultural Research Series, Agricultural Handbook No. 496, Washington, DC.

ME and MNR (1990) *Guide to Eating Ontario Sport Fish*. Ministry of Environment and Ministry of Natural Resources, Ontario.

Mineau, P. (1988) Avian mortality in agroecosystems. I. The case against granule insecticides in Canada. In Greaves, M.P., Smith, B.D. and Greig-Smith, P.W. (Eds) *Field Methods for the Study of Environmental Effects of Pesticides*. British Crop Protection Council (BPCP), Monograph 40, Thornton Heath, London, pp. 3–12.

Mineau, P. and Collins, B.T. (1988) Avian mortality in agro-ecosystems. 2. Methods of detection. In Greaves, M.P., Smith, G. and Smith, B.D. (Eds) *Field Methods for the Study of Environmental Effects of Pesticides*. Proceedings of a Symposium Organized by the British Crop Protection Council, Churchill College, Cambridge, pp. 13–27.

Mineau, P., Boersma, D.C. and Collins, B. (1994) An analysis of avian reproduction studies submitted for pesticide registration. *Ecotoxicology and Environmental Safety* **29**, 304–329.

Minyard, J.P. and Roberts, W.E. (1991) A state data resource on toxic chemicals in foods. In Tweedy, B.G., Dishburger, H.J., Ballantine, L.G. and McCarthy, J. (Eds) *Pesticide Residues and Food Safety: A Harvest of Viewpoints*. American Chemical Society, Washington, DC, pp. 151–161.

Moses, M., Johnson, E.S., Anger, W.K., Burse, V.W., Horstman, S.W., Jackson, R.J., Lewis, R.G., Maddy, K.T., McConnell, R., Meggs, W.J. and Zahm, S.H. (1993) Environmental equity and pesticide exposure. *Toxicology and Industrial Health* **9**, 913–959.

Murphy, M.J. (1994) Toxin exposure in dogs and cats: Pesticides and biotoxins. *Journal of the American Veterinary Medical Association* **205**, 414–421.

Murray, D.L. (1994) *Cultivating Crisis: The Human Cost of Pesticides in Latin America*. University of Texas Press, Austin, TX.

NAS (1975) *Pest Control: An Assessment of Present and Alternative Technologies*. 4 volumes. National Academy of Sciences, Washington, DC.

NAS (1985) *Meat and Poultry Inspection. The Scientific Basis of the Nation's Program*. National Academy of Sciences, Washington, DC.

NAS (1987) *Regulating Pesticides in Food*. National Academy of Sciences, Washington, DC.

NAS (1993) *Malaria: Obstacles and Opportunities*. National Academy of Sciences, Washington, DC.

Nash, E.P. (1994) What's a life worth? *New York Times*, New York.

Natarajan, U. and Rajagopal, R. (1994) Economics of screening for pesticides in ground water. *Water Resources Bulletin* **30**, 579–588.

Nielsen, E.G. and Lee, L.K. (1987) *The Magnitude and Costs of Groundwater Contamination from Agricultural Chemicals. A National Perspective*. US Department of Agriculture, Economic Research Series, Natural Resources Economics Division, ERS Staff Report, AGES870318, Washington, DC.

NRDC (1993) *Pesticides in the Diets of Infants and Children*. National Academy Press, Washington, DC.

NRDC (1993) *After Silent Spring*. NRDC Publications, New York.

NYT (1988) Shell Loses Suit on Cleanup Cost. *New York Times*, New York.

OACSJC (1990) *San Joaquin County Agricultural Report 1989*. San Joaquin County Department of Agriculture, San Joaquin County, CA.

OACYC (1990) *Yolo County 1989 Agricultural Report*. Office of the Agricultural Commissioner, Yolo County, CA.

Oka, I.N. (1991) Success and challenges of the Indonesian national integrated pest management program in the rice-based cropping system. *Crop Protection* **10**, 163–165.

Oka, I.N. and Pimentel, D. (1976) Herbicide (2,4-D) increases insect and pathogen pests on corn. *Science* **193**, 239–140.

Osteen, C.D. and Szmedra, P.I. (1989) *Agricultural Pesticide Use Trends and Policy Issues*. US Department of Agriculture, Economics Research Series, Agricultural Economics Report No. 622, Washington, DC.

OTA (1979) *Pest Management Strategies*. 2 volumes. Office of Technology, US Congress, Washington, DC.

OTA (1988) *Pesticide Residues in Food: Technologies for Detection*. Office of Technology Assessment, US Congress, Washington, DC.

Pesatori, A.C., Sontag, J.M., Lubin, J.H., Consonni, D. and Blair, A. (1994) Cohort mortality and nested case-control study of lung cancer among structural pest-control workers in Florida (United States). *Cancer Causes and Control* **5**, 310–318.

Pimentel, D. (1971) *Ecological Effects of Pesticides on Non-Target Species*. US Government Printing Office, Washington, DC.

Pimentel, D. (1988) Herbivore population feeding pressure on plant host: Feedback evolution host conservation. *Oikos* **53**, 289–302.

Pimentel, D. (1995) Amounts of pesticides reaching target pests: Environmental impacts and ethics. *Journal of Agricultural and Environmental Ethics* **8**, 17–29.

Pimentel, D., Krummel, J., Gallahan, D., Hough, J., Merrill, A., Schreiner, I., Vittum, P., Koziol, F., Back, E., Yen, D. and Fiance, S. (1978) Benefits and costs of pesticide use in US food production. *BioScience* **28**, 778–784.

Pimentel, D., Andow, D., Dyson-Hudson, R., Gallahan,. D., Jacobson, S., Irish, M., Kroop, S., Moss, A., Schreiner, I., Shepard, M., Thompson, T. and Vinzant, B. (1980) Environmental and social costs of pesticides: A preliminary assessment. *Oikos* **34**, 127–140.

Pimentel, D., McLaughlin, L., Zepp, A., Lakitan, B., Kraus, T., Kleinman, P., Vancini, F., Roach, W.J., Graap, E., Keeton, W.S. and Selig, G. (1991) Environmental and economic impacts of reducing US agricultural pesticide use. In Pimentel, D. (Ed.) *Handbook on Pest Management in Agriculture*. CRC Press, Boca Raton, FL, pp. 679–718.

Pimentel, D., Acquay, H., Biltonen, M., Rice, P., Silva, M., Nelson, J., Lipner, V., Giordano, S., Horowitz, A. and D'Amore, M. (1992) Environmental and economic costs of pesticide use. *BioScience* **42**(10), 750–760.

PN (1990) Towards a reduction in pesticide use. *Pesticide News*, March.

Potashnik, G. and Yanai-Inbar, I. (1987) Dibromochloropropane (DBCP): An 8-year reevaluation of testicular function and reproductive performance. *Fertility and Sterility* **47**, 317–323.

Potts, G.R. (1986) *The Partridge: Pesticides, Predation and Conservation*. Collins, London.

PSAC (1965) *Restoring the Quality of our Environment*. President's Science Advisory Committee, The White House, Washington, DC.

Pye, V. and Kelley, J. (1984) The extent of groundwater contamination in the United States. In NAS (Ed.) *Groundwater Contamination*. National Academy of Sciences, Washington, DC.

Reuben, R. (1989) Obstacles to malaria control in India—the human factor. In W.W. Service (Ed.) *Demography and Vector-Borne Disease*. CRC Press, Boca Raton, FL, pp. 143–154.

Robertson, I.D., Naprasnik, A. and Morrow, D. (1990) The sources of pesticide contamination in Queensland livestock. *Australian Veterinary Journal* **67**, 152–153.

Rovira, A.D. and McDonald, H.J. (1986) Effects of the herbicide chlorsulfuron on Rhizoctonia Bare Patch and Take-all of barley and wheat. *Plant Disease* **70**, 879–882.

Salem, M.H., Abo-Elezz, Z., Abd-Allah, G.A., Hassan, G.A. and Shakes, N. (1988) Effects of organophosphorus (dimethoate) and pyrethroid (deltamethrin) pesticides on semen characteristics in rabbits. *Journal of Environmental Science and Health* **B23**, 279–290.

Savage, E.P., Keefe, T.J., Mounce, L.W., Heaton, R.K., Lewis, A. and Burcar, P.J. (1988) Chronic neurological sequelae of acute organophosphate pesticide poisoning. *Archives of Environmental Health* **43**, 38–45.

Sebae, A.H. (1977) *Incidents of Local Pesticide Hazards and Their Toxicological Interpretation*. Proceedings of UC/AID University of Alexandria Seminar Workshop in Pesticide Management, Alexandria.

Shaked, I., Sod-Moriah, U.A., Kaplanske, J., Potashnik, G. and Buckman, O. (1988) Reproductive performance of dibromochloropropane-treated female rats. *International Journal of Fertility* **33**,

129–133.

Shemi, D., Marx, Z., Kaplanski, J., Potashnik,. G. and Sod-Moriah, U.A. (1988) Testicular damage development in rats injected with dibromochloropropane DBCP. *Andrologia* **20**, 331–337.

Shemi, D., Sod-Moriah, U.A., Abraham, M., Friedlaender, M., Potashnik, G. and Kaplanski, J. (1989) Ultrastructure of testicular cells in rats treated with dibromochloropropane DBCP. *Andrologia* **21**, 229–236.

Spiric, A. and Raicevic, S. (1994) Residues of lindane in the adipose tissue of different animal species over a three year period (1991–1993) *Acta Veterinaria (Beograd)* **44**, 303–308.

Stickel, W.H., Stickel, L.F., Dyrland, R.A. and Hughes, D.L. (1984) DDE in birds: Lethal residues and loss rates. *Archives of Environmental Contamination and Toxicology* **13**, 1–6.

Stone, W.B. and Gradoni, P.B. (1985) Wildlife mortality related to the use of the pesticide diazinon. *Northeastern Environmental Science* **4**, 30–38.

Storey, D.A. and Allen, P.G. (1993) Economic impact of marine recreational fishing in Massachusetts. *North American Journal of Fisheries Management* **13**, 698–708.

Tanda, A.S. (1984) Bee pollination increases yield of two interplanted varieties of asiatic cotton (*Gossypium arboreum* 1.) (*Apis cerana indica, Apis mellifera*, India). *American Bee Journal 124*, 539–540.

Taylor, R.B. (1986) State sues three farmers over pesticide use on watermelons. *Los Angeles Times*, I (CC)-3-4.

Thomas, P.T. and House, R.V. (1989) Pesticide-induced modulation of immune system. In Ragsdale, N.N. and Menzer, R.E. (Eds) *Carcinogenicity and Pesticides: Principles, Issues, and Relationships*. American Chemical Society, ACS Symposium Series 414, Washington, DC, pp. 94–106.

Tinney, R.T. (1982) *The Oil Drilling Prohibitions at the Channel Islands and Pt. Reyes-Fallallon Island National Marine Sanctuaries: Some Costs and Benefits.* Report to Center for Environmental Educations, Washington, DC.

Trotter, W.J. and Dickerson, R. (1993) Pesticide residues in composited milk collected through the US pasteurized milk network. *Journal of the Association of Official Analytical Chemists International* **76**, 1220–1225.

Trumbel, J.T., Carson, W., Nakakihara, H. and Voth, V. (1988) Impact of pesticides for tomato fruitworm (Lepidoptera: Noctuidae) suppression on photosynthesis, yield, and nontarget arthropods in strawberries. *Journal of Economic Entomology* **81**, 608–614.

UNEP (1979) *The State of the Environment: Selected Topics—1979*. United Nations Environmental Programme, Governing Council, Seventh Session, Nairobi.

USBC (1990) *Statistical Abstract of the United States 1990*. US Bureau of the Census, US Department of Commerce, Washington, DC.

USBC (1994) *Statistical Abstract of the United States 1994*. US Bureau of the Census, US Government Printing Office, Washington, DC.

USDA (1989a) *Agricultural Statistics*. US Department of Agriculture, Government Printing Office, Washington, DC.

USDA (1989b) *The Second RCA Appraisal. Soil, Water, and Related Resources on Nonfederal Land in the United States. Analysis of Conditions and Trends*. US Department of Agriculture, Washington, DC.

USDA (1994) *Agricultural Statistics*. US Department of Agriculture, Government Printing Office, Washington, DC.

USFWS (1988) *1985 Survey of Fishing, Hunting, and Wildlife Associated Recreation*. US Fish and Wildlife Service, US Department of Interior, Washington, DC.

USFWS (1991) *Federal and State Endangered Species Expenditures*. US Fish and Wildlife Services. Washington, DC.

van der Leeden, F., Troise, F.L. and Todd, D.K. (1990) *The Water Encyclopedia*. 2nd edn. Lewis, Chelsea, MI.

van Ravenswaay, E. and Smith, E. (1986) Food contamination: Consumer reactions and producer losses. *National Food Review* (Spring), 14–16.

Walgenbach, F.E. (1979) Economic damage assessment of flora and fauna resulting from unlawful environmental degradation. Manuscript, California Department of Fish and Game, Sacramento, CA.

Wang, Y., Jaw, Ch. and Chen, Y. (1994) Accumulation of 2,4-D and glyphosate in fish and water hyacinth. *Water, Air and Soil Pollution* **74**, 397–403.

Ware, G.W. (1983) *Reducing Pesticide Application Drift-Losses.* University of Arizona, College of Agriculture, Cooperative Extension, Tucson, AZ.

Ware, G.W., Estesen, B.J., Cahill, W.P., Gerhardt, P.D. and Frost, K.R. (1969) Pesticide drift. I High-clearance vs. aerial application of sprays. *Journal of Economic Entomology* **62**, 840–843.

White, D.H., Mitchell, C.A., Wynn, L.D., Flickinger, E.L. and Kolbe, E.J. (1982) Organophosphate insecticide poisoning of Canada geese in the Texas Panhandle. *Journal of Field Ornithology* **53**, 22–27.

WHO (1990) *Tropical Diseases News* 31:3, World Health Organization Press Release, 28 March.

WHO (1992) *Our Planet, Our Health.* Report of the WHO Commission on Health and Environment, World Health Organization, Geneva.

WHO/UNEP (1989) *Public Health Impact of Pesticides used in Agriculture.* World Health Organization/United Nations Environment Programme, Geneva.

Wigle, D.T., Samenciw, R.M., Wilkins, K., Riedel, D., Ritter, L., Morrison, H.I. and Mao, Y. (1990) Mortality study of Canadian male farm operators: Non-Hodgkin's lymphoma mortality and agricultural practices in Saskatchewan. *Journal of the National Cancer Institute* **82**, 575–582.

Wiles, R. and Campbell, C. (1994) *Washed, Peeled—Contaminated: Pesticides Residues in Ready-To-Eat Fruits and Vegetables.* Environmental Working Group, Washington, DC.

Wilkinson, C.F. (1990) Introduction and overview. In Wilkinson, C.F. and Baker, S.R. (Eds) *The Effects of Pesticides on Human Health.* Princeton Scientific, Princeton, NJ, pp. 5–33.

Williamson, C.S. (1995) Conserving Europe's bees: why all the buzz? *TREE* **10**, 309–310.

Wills, L.E., Mullens, B.A. and Mandeville, J.D. (1990) Effects of pesticides on filth fly predators Coleoptera Histeridae Staphylinidae Acarina Macrochelidae and Uropodidae in caged layer poultry manure. *Journal of Economic Entomology* **83**, 451–457.

Winston, M.L. and Scott, S.C. (1984) The value of bee pollination to Canadian apiculture. *Canadian Beekeeping* **11**, 134.

WRI/UNEP/UNDP (1994) *World Resources: 1994–1995.* Oxford University Press, Oxford.

Zahm, S.H., Weisenburger, D.D., Babbitt, P.A., Saal, R.C., Vaught, J.B., Cantor, K.P. and Blair, A. (1990) A case-control study of non-Hodgkin's lymphoma and the herbicide 2,4-dichlorophenoxyacetic acid (2,4-D) in Eastern Nebraska, *Epidemiology* **1**, 349–356.

CHAPTER 5

Pesticide Use in Swedish Agriculture: The Case of a 75% Reduction

Olle Pettersson

The Swedish University of Agricultural Sciences, Uppsala, Sweden

SWEDEN AS A GENERAL OR A SPECIAL CASE?

Political decisions are just a few of the many forces in social and technological change. Legislative restrictions and regulations implemented on international, country, federal, state or local levels are factors which are important in the use of pesticides in agriculture. The same is true for guidelines originating from research and extension activities.

However, there are also other forces on macro- and micro-levels of society which are just as important for the practical application of pesticides. For example, the developmental work within the pesticide-producing companies, as well as economic and social factors in the market, also have great influence. Sometimes political goals are supported, sometimes they may be counteracted, by 'reality'—the sum effect of other forces.

Industrialized agriculture

As in most industrial and post-industrial countries, agriculture in Sweden has undergone major technological, environmental, economic and social changes during recent decades. Although the rate and degree of change have varied depending on the country and region, the dominating trends have generally been similar, i.e. yields have increased and production efficiency has been enhanced through the utilization of improved plant and animal varieties, together with technical and chemical means.

Pesticide use has been one of several prerequisites for this overall change in the technology, ecology and structure of agriculture. At the same time this has given rise to questions about the risks to health and the environment. Changes in technology, production intensity, crops grown and industrial structure have been accompanied by an increase in the environmental impacts of agriculture, leading to intensified environmental conflicts and debates. The conflicts and controversies, as well as the political consequences, may differ in character between countries and regions owing to social, ethical

Techniques for Reducing Pesticide Use. Edited by D. Pimentel
© 1997 John Wiley & Sons Ltd.

and political differences. They do not, and should not be expected to, simply reflect the absolute degree of pesticide use or environmental impact.

In this respect, some typically Swedish characteristics and responses may be identified which have their origin in the historical role of agriculture as well as social values and political experiences. As a result of these sociological, ethical and political factors, the use of pesticides as well as other environmental agents has been put on the agenda of politics and legislation earlier, and has been discussed more emphatically, than in many other countries. These aspects have been discussed in greater detail elsewhere (Pettersson, 1993).

Politics and regulation

In all industrialized countries, the use of individual pesticides has been regulated on the basis of the presumed hazard that each product poses to health and the environment. In Sweden, programmes aimed at reducing the overall pesticide use have also been initiated. Similar programmes have been implemented somewhat later in other countries, e.g. Norway and Denmark, Canada and The Netherlands (MANMF, 1993; Surgeoner and Roberts, 1993; Pettersson, 1994).

These aspects of pesticide policy will be discussed in this paper on the basis of the Swedish decision to reduce pesticide use. In the spring of 1986 the Swedish Government—and Parliament—decided that the use of agricultural pesticides should be reduced by 50% in 5 years from 1986 to 1990 compared with the average use in 1981–85. In 1990 the Parliament adopted a new bill which says that the use of pesticides in agriculture should be reduced by another 50% between 1990 and 1996.

This paper focuses on the strategies and outcomes of the policies implemented. The cultural and political background will also be touched upon: for instance, why did public demand for these kinds of policies first appear in Sweden, where problems concerning pesticide use and environmental impact are much less pronounced than in many other countries and agricultural areas? Questions raised about the impact on agricultural production and techniques include: do the new policies imply a new approach leading to radical changes in commercial agriculture and pesticide use? Alternatively, should they be viewed as attempts to take full advantage of what modern pesticide and agricultural technology can achieve without abandoning the present agricultural production system?

The second decision on pesticide use in Sweden raises another question how far can one go with mainly technological and administrative improvements. When will you reach the point where you cannot go further without challenging predominant specialisations, e.g. in crop rotations and other farming practices.

THE FORCE OF AGRICULTURAL PESTICIDES

Problems with weeds, arthropods and diseases have been farmer's companion through-out the history of agriculture. They can be traced back to the original conflict between mankind and nature. In contrast to natural ecosystems, cultivated areas have always been characterized by different degrees of monoculture. Although there are pronounced differences between the grain production areas of today and the more mixed crop rotations that existed before agriculture was 'industrialized', the aims and motives of the farmer have not changed qualitatively over time. The farmer has always been at war with weeds and other pests, using whatever means available to protect his crop.

A new tool

Thus, from one point of view, the pesticides used today are merely new means of achieving an old aim. They work with higher biological efficiency—in some cases they solve problems that earlier had to be accepted. However, they have become so important in the agriculture of today that they do not only represent new means; their availability is in itself an important driving force in agricultural change.

The motives for their use and hegemony is of an ecological as well as a technical and economic nature. The cost/benefit relation will differ depending on how environmental/health/aesthetic costs are assessed, but generally the benefits outweigh the costs. It should be noted, however, that there is a tendency for the benefit/cost ratio to decrease when environmental and social costs are considered (Pimentel et al., 1993).

At farm level the relative cost/benefit of a specific pesticide treatment can be compared with those of other alternative pest management techniques. However, at the societal level, the large number of interacting, dynamic factors involved make such an analysis 'soft' and difficult. For instance, the profits gained by carrying out a number of different treatments cannot simply be added up to obtain a total estimate of profits gained at a higher, e.g. national, level. The entity is something qualitatively different than the sum of all individual cases. This is true not only for external (environmental) effects. The internal (agroecological) conditions will also change as a result of pesticide treatment and regional crop specialization.

Or a new force

Pesticide use, like all technologies employed, both influences and is influenced by environmental conditions and the economic/technological infrastructure of society. Climate and other environmental factors strongly influence agricultural production as well as pest problems, thereby affecting the need for pesticides. Cultivation practises, e.g. crop rotations, and markets also affect pesticide decision-making. In addition, socio-political factors, such as past experience, societal values and cultural heritage, are important in this respect.

Consequently, a discussion concerning past, present and future trends in pest management, will need to be based on an assessment of the various *driving forces* involved in the changes being made. For example, we need to consider recent and expected technological advances, the commercial structure, i.e. farmers, companies and the market, and the political environment. From this point of view, the Swedish decisions and experiences could be regarded as a special case of outcome from general forces determining the application of a specific area of modern technology. (Figure 5.1).

NOT ONLY ENVIRONMENTAL IMPACT

In general, the types of concern raised with regard to chemical pesticides are not new. Secondary effects and conflicting objectives will become apparent regardless of the production technique being evaluated. Food quality, internal problems of the agrarian

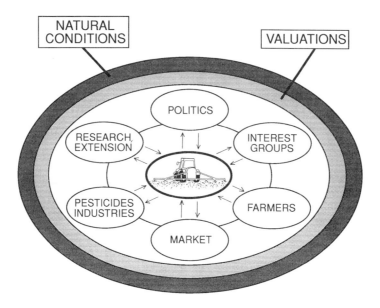

Figure 5.1   Environmental and societal factors affecting the use of pesticides in agriculture

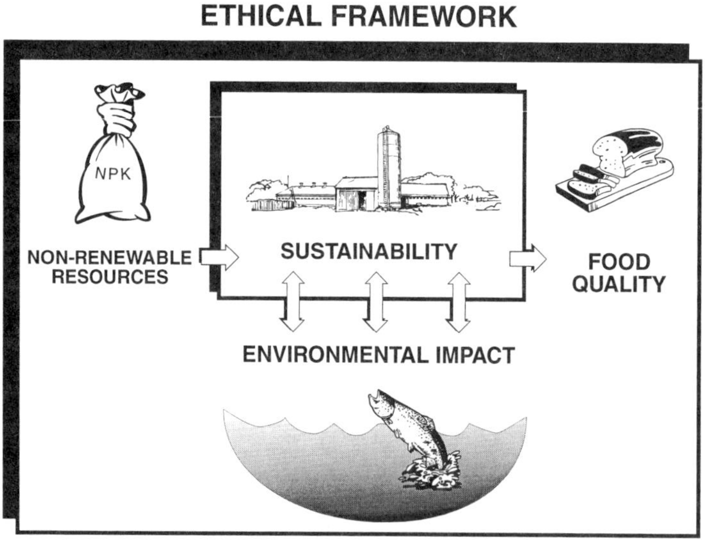

ETHICAL FRAMEWORK

Figure 5.2   Cultivation and its relation to the environment

ecosystem with regard to sustainability and production capacity, effects on the environment, and the use of non-renewable resources are all influenced by cultivation methods and technology (Figure 5.2). The aim of the following discussion is to focus on the fundamental characters and the rough dimensions of problems in connection with pesticide use in agriculture.

Quality and health

By and large, pesticides have positive effects on *product quality*. Weed reduction results in better harvesting conditions, thereby increasing the potential for improving technical and hygienic quality, speeding up the drying of grain crops, etc. The control of insects and crop diseases has the same effect. However, in many cases, the effects may be of only cosmetic character.

The general assessment with regard to *health effects* suggests that pesticide residues in food have little influence on our health compared with other hazards in connection with food safety (Gold et al., 1992). This viewpoint has its basis in toxicological assessments and epidemiological studies as well as in more qualitative discussions. For example, for the average consumer the risks posed by natural compounds in plants are several orders of magnitude larger than those posed by pesticide residues. From ecological and evolutionary points of view it is no surprise that commonly eaten plant species contain numerous substances with potentially hazardous long-term effects.

Long-term effects on the individual are of great interest for human well-being, but are not important for evolutionary selection. Our social heritage shows that human experience—based on personal empirical knowledge and not on epidemiology and natural science—has only selected out the natural toxins in foodstuffs that have more acute effects.

However, there are indications that the synthetic chemicals present in food may increase health risks above levels which may be posed by the presence of natural toxins alone. Therefore—and for many other reasons—public concern has been more focused on pesticide residues than on natural toxins. The latter are often difficult to avoid. The synthetic chemicals are perceived as added risks over which the individual has little or no control. A detailed discussion on this subject has recently been published (Culliney et al., 1993).

Sustainability of production

Among the *internal problems for the agroecosystem*, the development of pest strains resistant to pesticides is most important. Resistance has both economic and ecological consequences, and it is also a problem that may become more important with some of the modern pesticides with their specific modes of action. Adverse effects on non-target organisms, e.g. natural enemies and soil organisms, may further disturb the ecological balance. However, most often the direct impact of pesticides on soil qualities is small compared with that of other cultivation measures such as crop rotation, fertilization, application of manure or mechanical tillage.

When pesticide use results in severe side effects for soil, water or non-target organisms, it could be detrimental to natural resources. On the other hand, when managed correctly, chemical control could often be highly efficient in limiting the use of *non-renewable resources*. For example, mechanical and other alternative methods of weed control often require a considerably larger energy input in the form of tractor fuel compared with the energy required in the manufacture and application of pesticides.

Environmental impact

The *environmental effects* are of direct and indirect character. The adverse effects of fungicide (seed dressing) mercury on seed-eating birds and of DDT on predatory birds are among the classical examples of direct effects from the 1950s. The extent of this type of problem has decreased sharply, but has not entirely disappeared in most highly industrialized countries. The reduction in the severity of such problems is largely a result of changes in the assortment of pesticides used. Indirect side-effects, which exemplify a more fundamental conflict, include the decrease in floral and faunal diversity of the intensively cultivated agricultural landscape.

To summarize, the use of modern technology, including pesticides, has led to improved food product quality and enhanced agricultural production, but environmental impact from agriculture has also increased. At the same time, agricultural production depends on more input from non-renewable resources. The changes that have occurred during recent decades, including the introduction of pesticides, can also be interpreted as agriculture improving according to its original aim to produce foodstuffs. However, today the production systems have fewer positive, but more negative, side-effects. The demands for foodstuffs and the requirement placed on their quality have been dominant, whereas demands for environmental concerns have had less influence. This is true both for market influences and for government policy.

Technological development is already complex at the agroecological level of the individual crop. However, at farm level it is even more complex. The availability of pesticides is one important precondition for specialization within crop production. Thus, it will have fundamental effects on the crop rotations used, and indirectly on the landscape as well as on pest infection pressure. From being one of several means for the welfare of an individual crop, it has become a dynamic factor in agricultural change.

On the socio-economic level, for the whole agricultural sector, the overall changes will often be greater than the sum of all individual ones. The dependency of industrialized agriculture on pesticides has developed in response to a multitude of dynamic factors. Sometimes this dependence results in vicious circles where the individual inputs become increasingly profitable, but where the total result is, nevertheless, questionable. The fact that the total relative loss caused by pests in agriculture is not much lower than it used to be offers an example of this complexity (Pimentel et al., 1993). On the other hand, there are examples of pesticide use resulting in a decrease in pesticide need. The Swedish experience with herbicides represents one such exception (Figure 5.3).

POLITICS AND ECOLOGY OF PESTICIDE REDUCTION

Justifications for restricting the use of pesticides in agriculture vary from one group to another. There could be general agreement regarding certain goals, whereas others are a source of conflict. For example, most people are probably interested in avoiding the over-use of potentially hazardous substances. Thus, increasing the efficiency of a specific pesticide and thereby decreasing the dosage needed and the potential side-effects is in the interest of the general public so long as the costs are not too high.

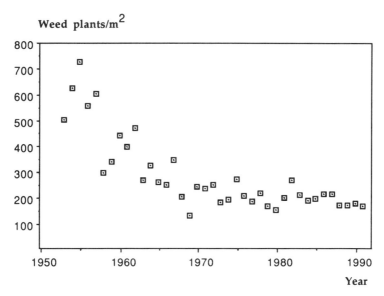

Figure 5.3   Number of weeds per square metre in Swedish field trials. *Source*: Pettersson (1994)

Of general interest

This means that most people should favour technological improvements leading to increased application precision since less pesticide would have to be applied for a given benefit. Thus, this an example of consensus. However, the pesticide manufacturer, who is interested in selling maximum amounts of a specific pesticide at a specific opportunity, may be an exception.

Improved pesticides are continually being developed and marketed. Often, they are efficient in lower doses than the older ones. Empirically, this is often accompanied with more specific knowledge on the mode of action of the compounds. This may include lower toxicity to non-target organisms and low persistence in the environment, and thus decreased health and environmental impact. Sulfonylurea herbicides are good examples of this new generation of pesticides. Thus, technological development itself results in the use of fewer kilograms of pesticides. Most people will probably agree that such improvements are desirable, including those who—for ethical reasons—are against all pesticide use.

Because many of the new compounds pose much lower risks to human health and to the environment, regulation authorities and the pesticide industry could share a common interest in promoting their use. In addition, the newly introduced products are very often more profitable for the pesticide producer than those that are withdrawn.

It is also possible, however, that the pesticides introduced are desirable from various standpoints but only work effectively at high doses. In such cases, there may be a conflict between the more specific traditional aims of decreasing risks and the more generally formulated political aims of decreasing pesticide use.

Agroecological change

In some cases agroecological changes may result in a decreased need for pesticides. Even though the mainstream of agricultural industrialization has increased dependency on pesticides, there are exceptions. For example, changes in the agricultural markets may favour crop rotations in which certain pests are much less of a problem. In addition, within a given system of crop rotation, the need for herbicides may decrease under specific circumstances.

In Swedish grain crop production, weed pressure has decreased over recent decades as a consequence of herbicide use. As discussed later, this implies that the need for herbicide efficiency has decreased. In this case, there has also been a decrease in pesticide use that has been in the common interest. Of course, the same is true for the introduction of 'alternative', 'non-chemical' methods, provided that they are safe and as efficient or profitable as the pesticides they replace, or have other desirable advantages which outweigh the higher costs.

The food market may encourage or discourage pesticide use depending on the situation. For example, certain technical and cosmetic demands on foodstuffs have encouraged the use of pesticides, whereas increasing demand for integrated pest management (IPM) and 'ecologically' or 'organically' grown products has the opposite effect.

As regards agricultural enterprises, both the farmers' attitudes and their cost/benefit calculations play an important role in the amounts and direction of pesticide use. Prices obtained for marketed products will also influence the profitability of control inputs. Market regulations in most countries have kept internal prices high, which has often had the effect of stimulating the use of pesticides, primarily insecticides and fungicides. Thus a decrease in product subsidies should tend to reduce pesticide use.

Research and development

Publicly funded research on pesticide use and effects is important in the development of pesticide-use guidelines tailored to particular crops or types of farms. Such knowledge may encourage the use of pesticides by identifying additional situations in which their use is profitable. However, research may also reveal situations in which pesticide use is not economically motivated, thereby eliminating certain cases of unprofitable use.

The national regulations stipulating which, when and how pesticides may be used have largely played a restrictive role in limiting the economically motivated use of known products. New knowledge regarding pesticide impacts, as well as changing societal values and opinions, are reflected in changes in the criteria used to assess pesticides from health and environmental standpoints. The demands placed on a given pesticide reflect a risk/benefit assessment made by society or its ruling body with regard to its use. On the other hand, general environmental levies on pesticides may be regarded as attempts at societal level to balance advantages against disadvantages.

In summary, there are many changes with regard to pesticide use that have not given rise to conflicts between different societal groups, but have nevertheless resulted in a decrease in the use of pesticides in agriculture. Other changes could be sources of conflict between farmers, the pesticide industry and the public. In addition, controversies may arise when a specific country, in its ambition to reduce pesticide use, establishes tariffs or

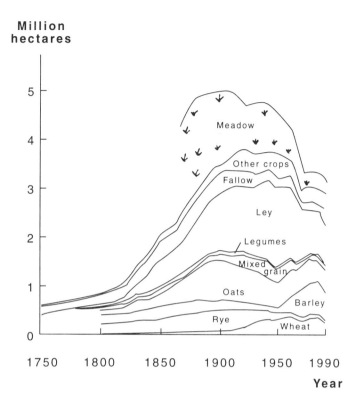

Figure 5.4 Breakdown of Swedish farmland by period

special criteria to be met for registration. As a result, the farmer may have difficulty competing in price with imported products.

When evaluating the Swedish programmes aimed at reducing pesticide use in agriculture, these aspects should be kept in mind. To what extent do the measures stimulate general, non-controversial improvements in pesticide technology and management, and to what extent do they place unique demands on the individual country's farmers?

AGRICULTURAL BASICS OF SWEDEN

Pesticide use varies between countries and regions depending on environmental conditions and the structural characteristics of the agricultural systems. The dominating trends in agricultural development in Sweden as well as in other industrialized countries have generally been similar, i.e. yields have increased and production efficiency has been enhanced through the utilization of improved plant and animal varieties, together with technical and chemical means. The hectarage required to meet the country's needs for food or the export market has been reduced over the years. Some basic statistics describing Swedish agriculture, including pesticide use, are given in Figure 5.4 and Table 5.1.

Table 5.1   Some basic facts on Swedish agriculture (1000 hectares, 1993)

Agricultural land as a percentage of total area	8 (including pasture)
Cultivated land	2872
Pasture land	576
Conversion/set-aside	368
Bare fallow, etc.	56
Wheat	304 (mainly winter wheat)
Rye	46
Barley	420 (mainly spring barley)
Oats	322
Rye wheat	35
Mixed oats/barley	25
Rape	145
Leguminous plants	55
Sugar beet	51
Potatoes	36
Cattle*	1807 (of which 525 dairy cows)
Pigs*	2277
Sheep*	470
Herbicides**	1093
Fungicides**	241
Insecticides**	15
Seed dressings**	76

* × 1000.
**Tonnes of active ingredient (sales in 1993).
Source: Statistics Sweden, 1994.

The golden mean

Sweden, as well as for example Finland and Norway, is characterized by a relatively moderate degree of agricultural specialization and non-diversification in comparison with the most intensive agricultural areas in Europe and the US, and areas in developing countries where cash crops are grown. It follows therefore that the environmental impacts and other side-effects of pesticide use will not be unendurable, or at least will be less severe than in many other regions.

The actual mix of agricultural production in terms of area, crops and different habitats is an environmental factor in itself. The influence of cultivation on the surrounding land is another. Between these two factors there is an interaction to the extent that a certain mix of production will lead to a certain type of environmental impact. Figure 5.4 illustrates the utilization of the Swedish agricultural land and its changes over time.

Arable land in Sweden comprises about 3 million hectares, which is about 7% of the country's area. Most of the agricultural and horticultural activities take place in the southern part of Sweden because of climate and soil conditions. In the southernmost part, most European crops, except grapes, corn and sunflowers, are grown. In the plain districts around the great Swedish lakes, the most frequent crops are winter wheat, oats, barley and oilseed rape. Further north there is a large proportion of forage crops, especially grass.

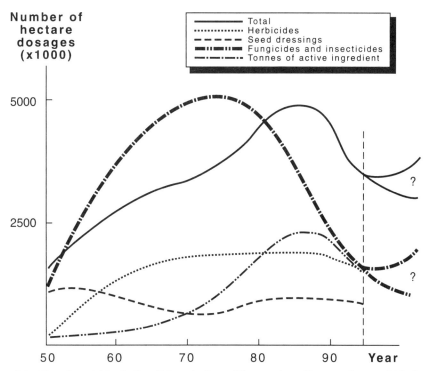

Figure 5.5   Use of pesticides in Swedish agriculture. The number of hectare dosages (thin lines) was highest in the mid-1980s, while the total national use in tonnes of active ingredient (thick broken line) peaked about 10 years earlier (simplified and generalized picture)

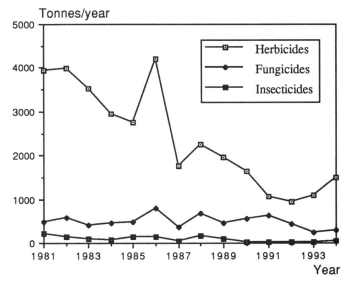

Figure 5.6   Use of pesticides in Swedish agriculture since 1980 by tonnes of active ingredient. *Source*: Swedish National Chemicals Inspectorate, 1995

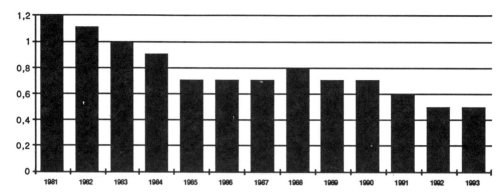

Figure 5.7 Average does of active ingredients per hectare; all use in Swedish agriculture 1981–1993. (For herbicides the quantity of active ingredients decreased from about 1.7 kg ha^{-1} to about 0.7 kg ha^{-1})

Table 5.2   The sales of pesticides in agriculture and horticulture 1981–1993 (tonnes)

Year	Seed dressings	Fungicides	Herbicides	Insecticides	Growth regulators	Total
1981–85	161	599	3536	150	82	4528
1986	199	869	4207	160	243	5678
1987	119	470	1781	63	84	2519
1988	101	662	2029	112	75	2982
1989	120	445	1871	50	35	2521
1990	97	608	1658	38	49	2450
1991	86	696	1073	27	43	1925
1992	90	479	956	36	32	1593
1993	76	283*	1105	26	41	1531
1994*	90	317	1526	46	47	2026

*The use is probably bigger than the sales.
**The sales are probably bigger than the use.
Sources: Swedish Board of Agriculture (1994); Swedish National Chemicals Inspectorate (1995).

But not uniform

All of Sweden has a relatively humid climate. In the agricultural areas the annual rainfall varies from around 600 to more than 1000 mm, half of which passes through the soil to the drainage system. The large amounts of precipitation and runoff lead to the diffuse water pollution caused by agriculture. Although nitrates account for most of the pollution, water-soluble pesticides have also shown up in some cases (Kreuger, 1992).

Thus, Sweden is not a uniform agricultural area. The southern parts contain fertile plains and resemble the granaries of Central Europe in terms of agricultural structure and yields. The sparsely populated areas in the deep forests of the northern parts of the country represent quite different conditions for agriculture—and also with respect to pesticide use. Soil types vary from heavy clays, with relatively little nutrient leaching and almost negligible soil erosion, to sandy soils exposed to wind erosion. However, the magnitude of the erosion is small in a global perspective.

Table 5.3 Area of different crops treated with pesticides (% of crop area)

Crop	Herbicides					Fungicides*					Insecticides				
	88	89	90	91	92	88	89	90	91	92	88	89	90	91	92
Cereals	83	82	77	71	75	12	15	13	15	10	47	12	12	5	23
Potatoes	41	49	44	44	50	82	92	87	88	82	19	15	19	13	17
Sugarbeet	97	98	97	97	95	9	–	–	–	–	57	48	39	25	48
Oil-seed	40	46	46	48	42	1	–	–	3	–	67	54	46	55	63
Other crops	6	6	6	4	3	–	–	–	0	0	1	2	2	1	1
All crops	50	50	48	42	42	8	9	8	9	6	30	12	11	7	16

*Seed dressings are not included.
Sources: Swedish Board of Agriculture (1994); Swedish National Chemicals Inspectorate (1995).

Table 5.4 Number of hectare dosages* used in agriculture, based on the sales of pesticides (1000 dosages)

Year	Herbicides	Fungicides	Insecticides	Growth-regulators	Total
1981	2043	1703	491	25	4262
1982	2227	1909	596	38	4770
1983	2002	1583	577	96	4258
1984	2016	1899	782	166	4863
1985	1930	1898	1026	151	5005
1986	3509	2549	1299	288	7645
1987	1359**	1365	552	94	3370
1988	1464**	1205	782	83	3534
1989	1582**	1421	504	44	3551
1990	1625**	1248	394	56	3323
1991	1381**	1237	472	49	3139
1992	1299**	1091	549	34	2973
1993	1537**	927	455	46	2965
1994†	2423	1246	803	50	4522

*One hectare dosage = one hectare sprayed once by any sort of pesticide.
**This number is probably slightly underestimated.
†Sales are probably greater than the use.
Sources: Swedish Board of Agriculture (1994); Swedish National Chemicals Inspectorate (1995).

Due to the relatively humid and cold weather, weeds tend to be more of a problem than insects and fungus diseases. This is related to the fact that compared with the rest of the European continent, herbicide use is relatively more intensive than the use of fungicides and insecticides. The use of pesticides in Sweden, expressed as amount of active ingredient and area sprayed, is shown in Figures 5.5–5.7 and Tables 5.1–5.4. Among other things, it can be seen that the boom in herbicide use in corn and soybean production in the US in the 1970a and 1980s (Osteen, 1993) has no analogy in Sweden.

SOCIAL AND POLITICAL EXPERIENCES

If the extent to which pesticides appear on a country's political agenda was determined primarily by the severity of the environmental problems that they cause, we would not have expected Sweden to be the first to pass legislation restricting pesticide use. Thus, the explanation must be sought in the socio-political sphere (Figure 5.1).

Politics and political measures have their primary origin within social values and economic and democratic structures. Consequently, there is no reason to expect relationships to exist between the absolute degree of environmental problems in a given country, on the one hand, and political discussion and political measures on the other. Instead, we must focus on the political, historical, demographic and social conditions prevailing in the country concerned.

Rural inheritance

Until fairly recently, the Scandinavian countries were predominantly peasant cultures. Most people have relatively close contacts and relationships with farming. In addition, the agricultural sector consisted mainly of free and independent farmers, and to a lesser extent of nobility and large-estate owners. The feudal system was weaker in Sweden than in continental Europe and Great Britain.

Consequently, the proportion of the population that felt affected by the changes in agriculture and the rural way of life was probably greater in Sweden than in many other parts of Europe: it is their own cultural heritage that is being changed, not just that of other people. At the same time, the Scandinavian countries are advanced welfare states and post-industrial societies. Together, this appears to have resulted in a unique combination of typically pre-industrial experience and post-industrial values.

Social engineering

In addition, Scandinavians have traditionally had a high degree of confidence in the ability of political measures to solve various societal problems. This is probably connected with Sweden's relatively stable political history, as well as the numerous successful examples of the art of 'social engineering'. Even if the trends in Sweden today are towards something more 'average European', these aspects of the past should not be neglected when analyzing agricultural, environmental and pesticide policies and opinions in Sweden. It should be noted that Denmark, Norway, Canada and The Netherlands, other countries that have implemented programmes for reduced pesticide use, have similar traditions.

THE PESTICIDE REDUCTION PROGRAMMES

Reduced pesticide use as a policy aim covers many different aspects and aims on the political and administrative levels of society. It should be stressed that the demand for reduced pesticide use in Swedish agriculture emanated from public concern about the environment and health, and thus from the socio-political sphere. This concern was formulated in general terms and did not from the very beginning deal with sophisticated distinctions between amounts of active ingredients, risk assessments for specific compounds, or the doses for different kinds of pesticides. Thus, in the initial Swedish discussions on the matter, it was not unequivocally clear which criteria or parameters should be used to evaluate the results and effects of action programmes implemented. This should come as no surprise, because by their nature, politically formulated general aims have to be less concise than specific criteria stipulated by the regulating authority.

At least three different interpretations of reduced pesticide use are relevant. First, we have the traditional formulation and ambition to reduce the *risks* posed by pesticides to human health and the environment. This has always been one of the several aims of pesticide regulation. Second, it is possible to estimate the reduction in *total* land *area* treated by pesticides. How much of the country's environment is directly or indirectly influenced by pesticides? This criterion could be complemented with data on how intensively pesticides are used on the treated areas.

Third, a vaguer aim could be to decrease the *amount of active ingredient* used. This objective has the advantage of being easily measurable, but suffers from the disadvantage of including non-relevant technological and economic factors. A decrease in pesticide use resulting from a decrease in agricultural production would be included in this category. The same applies to changes in the assortment of pesticides from high-dose to low-dose, which sometimes, but not necessarily, imply decreases in environmental and health impacts.

Goals and measures

In the spring of 1986 the Swedish Board of Agriculture, the National Chemicals Inspectorate and the Swedish Environment Protection Agency were asked by the Swedish Government to develop a programme on how to reduce the risks to health and the environment resulting from the agricultural use of pesticides. The programme has been reviewed in several papers (Bernson, 1993; Pettersson, 1993, 1994; Bernson and Ekström, 1995; Emmerman and Pettersson, 1995).

The aim of this programme was to reduce the risks to human health and the environment. The risk reduction action scheme consists of three main approaches.

(1) Change over to pesticides which are less hazardous to health and the environment.

(2) Reduce the use of pesticides. (The Government—and Parliament—decided that the use of agricultural pesticides should be reduced by 50% in the 5 years from 1986 to 1990 compared with the average use in 1981–85. In 1990 Parliament adopted a new bill which says that the use of pesticides in agriculture should be reduced by another 50% between 1990 and 1996.)

(3) Special measures to protect health and the environment.

The programme comprises several measures. Exchange of old pesticide products for new ones *with less risk* is important. The same is true for regulation measures concerning *pesticide management* on- as well as off-farm. Training and information in the *safer use of* pesticides and tighter *control of pesticide residues* in food was also emphasised, as well as reduced use of pesticides.

Different authorities are responsible for different parts of the programme. The Swedish Board of Agriculture is responsible for efforts towards reducing the use of pesticides and is also coordinator of the programme as a whole, as well as advisory work, education and research.

Different strategies

Different strategies are used in the reduction programme depending on whether it is *weeds, insect pests* or *diseases* that are being considered. For weeds, the efforts must be of a long-term character in order to reduce the use and costs as much as possible. They cannot be based on short-term changes in costs to the same extent as the strategy to control pests or diseases. Because weed pressure in grain crops in Swedish agriculture has decreased as a long-term effect of herbicide use (see Figure 5.3), it is now possible to use lower doses of herbicides and still maintain weed populations at acceptable levels without creating any increased problems in the future.

To control weeds it is therefore important to use different measures to keep down the level of weed seeds. Too low a level of weed control means that the level of weed seeds increases and so does the weed pressure, and this affects the future use of pesticides. On the other hand, too high a level of control means that the use of pesticides is unnecessarily high and therefore also expensive.

Old and new experiences from research on cereals weeds have shown that under Swedish conditions, the best yields are obtained at half of the previously recommended dose, which means a 70–75% herbicidal efficiency. Crop stress, and thus lower yields, can occur if higher doses of herbicides are used. The use of crops which are able to compete with weeds, as well as an efficient long-term weed control strategy, also make it possible to keep doses at a low level.

However, it is important to use herbicides *every year* to keep the weed pressure low. It has therefore been concluded that herbicides should be used at a low dose every year if not complemented with other methods (alternative methods or changed crop rotations). This has probably been the most important strategy for pesticide reduction.

Problems with insect pests and diseases vary between years depending on several factors. For example, the weather has a strong influence on the magnitude of the damage. However, chemical control during previous years is of less importance. Consequently, the strategy for chemical pest control is to try to adjust the control measures to what is necessary each particular year. However, for pests and diseases there can be the possibility—under certain circumstances—of reducing the pesticide dose. Measures to promote *integrated crop protection* are also used.

Training and certification

Pesticide formulations are registered in three classes: 1, 2 and 3. Class 3 pesticides are available for 'amateur use'. Since 1990 all farmers and farm workers who carry out the application of pesticides need a certificate for professional use. A 3-day course is required for those using Class 2 pesticides, and an additional day for those using Class 1 pesticides. The certificate is valid for 5 years and an additional 1-day course is required for its renewal.

During the 1988/89 and 1989/90 seasons, roughly 20 000 people participated in this training. About 30 000 people hold a certificate. The County Advisory Boards are responsible for these courses. There are now six types of courses depending on the type of production (agriculture, horticulture, green areas, seed dressing, forestry and apiculture). One-third of the course time is spent on optimizing methods to control weeds and pests and diseases. Considerable time is spent on the pesticides themselves, but there are also discussions about alternative methods.

Plant protection centres

Five regional plant protection centres have been established to promote *integrated crop protection* with chemical control being adjusted to need. The most important target groups for these centres are state, private and commercial field crop production extension officers. In total there are about 600 people involved. There is close collaboration between the centres and the staff at the Swedish University of Agricultural Sciences, and thus links to basic and applied research in the fields of plant production and pesticide use.

The work at the regional plant protection centres is concentrated on the following topics:

(1) Pest forecasting and warning services. In each district during the crop season information is collected at least once a week on 50 pests and diseases of nine crops from about

800 fields. This information is analysed at the plant protection centres.

On the same day, the main group of extension officers gets a summary of the situation in the field. This summary serves as a basis for the next morning's telephone conferences, where different problems and solutions are discussed. About 170 people are involved in these conferences each week. Afterwards, at least 4500 farmers are informed directly or indirectly by plant protection letters from the centres or from other extension officers.

(2) Strategies to combat insect pests, diseases and weeds. Each year the centres publish new information about economy of pesticide use against weeds and pests in the areas of new policies, changed prices of products, new pesticides, alternative methods, changed registrations of pesticides, etc.

(3) Developmental work. Together with the Swedish University of Agricultural Sciences, the centres give opportunities to test and further develop different forecasting methods, etc.

(4) Coordination of approaches. The centres take the initiative in different trials, follow them up, collect the results and coordinate the advisory services by making demonstration plans, participating in decision-making on different activities, participating in field courses and following up advisory activities.

(5) Dissemination of information. The centres serve teachers and authors with course materials and are a resource in the mandatory training. Results from trials, forecasting and warning activities, different investigations and new strategies are presented in courses, at conferences, and on radio and TV, etc., and by articles, information letters, books, etc.

(6) Information techniques. Special attention has been given to information techniques. For example, it is possible to put together all data collected by the forecast and warning systems very quickly, and to sort and compare the information in different ways. Information has been collected since 1988. There is also an on-going project to collect and collate old and new information in the form of articles, books, pictures, regulations, etc., about the most common pesticides, weeds, pests and diseases.

Special advisory services

An important part of the programme is to focus on the possibilities of reducing the use of pesticides, especially information on reduced herbicide dose rates. The Swedish Board of Agriculture has in various ways initiated and granted support to the advisory services at the County Boards. The aim of the activity is to implement knowledge on the risks to the environment and the possibilities of reducing the use of pesticides. Demonstration trials showing, for example, the effect of a herbicide used on a farm at full, half or a quarter dosage, are also important.

During 1993/94 the following activities were carried out in the areas of reduced dosages, techniques, flora and fauna, and the appropriate handling of pesticides (how to store pesticides, fill and clean equipment, etc.):

- Demonstration trials   About 250
- Field or farm courses   About 300, with about 7000 participants
- Other courses   About 100, with about 3500 participants
- Individual advisory service   About 550

Research and development

In connection with the pesticide reduction programme, there has been special funding of applied research which is important for the aims of the programme. Many of these research activities have been carried out at the Swedish University of Agricultural Sciences. The areas of special interest are the reduction of the risks inherent in the use of pesticides, and the more precise adjustment of chemical control each particular need (dosages, forecasting, alternative methods, etc.).

Priority is given to applied research and extension activities which can relatively quickly give solutions relevant to practical farming. To date these resources have been concentrated on the areas listed below.

Weeds Work has centred on possible ways to reduce herbicide doses and on testing new methods, as well as seeking confirmation of earlier theories. One practical task has been to develop *dosage keys* which can help the farmer to adjust the dose used. Projects on alternative methods of weed control are also being pursued.

Insect pests and diseases This work includes projects on dosages and intensity when using pesticides. The development of forecasting systems is another subject being studied. The use of weather data has become more and more important and is incorporated in various systems. There are also projects on alternative methods, including biological control.

New technique Tests have been carried out on new sprayers and assess any economic and environmental advantages. Methods which can reduce dosages as well as the risks of losses to the environment by wind have been tested and/or developed.

Flora and fauna Unsprayed and sprayed edge zones have been compared to monitor changes in weed and bird populations, etc.

A special programme for the voluntary testing of sprayers has been in operation since 1988. Grants are given to farmers or companies. Between 1988 and 1992 about 7400 sprayers were tested, which represented about 30% of the total number in agricultural use, or about 50% of the total area treated with pesticides. About 170 test examiners have been trained. Today, the test frequency is about 1400 tests per year and this number is increasing slowly.

It is difficult to calculate the costs of the programme exactly because some activities involve a distribution of resources between the authorities and the university. Special resources have also been allocated to the authorities, for example to the National Chemicals Inspectorate to re-register all pesticides and to carry out related research and development. The Swedish Board of Agriculture spent about SKr 21 million on the

programme in 1993/94. The programme is financed by the Government through environmental levies.

REDUCTION ESTIMATED AND EVALUATED

Technological, political/legislative and structural changes in the pesticide and agricultural sector have influenced pesticide use (in $kg\,ha^{-1}$ and total use in tonnes) in Swedish
agriculture. For example, the total amount of active ingredients used in agriculture in
Sweden has decreased steadily since around 1980; however, the area treated increased
during the early 1980s, since when it has been relatively stable with a possible decrease
during the 1990s (Table 5.4).

This makes it difficult to isolate and evaluate the specific impacts of the pesticide use
reduction programmes. During the 1990s, the general national agricultural policy, which
included the set-aside of agricultural land and lower grain crop prices, also influenced
pesticide use. Since 1995, Sweden has been a member of the European Union and is
influenced by its agricultural policy, GAP. This will probably increase pesticide use to
some extent, partly because of increased profitability for grain crops and partly because of
the regulations for fallow treatment. Future trends in pesticide use in Sweden will
therefore also be influenced by decisions on agricultural policy within the European
Union.

The pesticide policy of Sweden, adopted by Parliament in 1986, called for a 50%
reduction of pesticide use in agriculture within 5 years. However, the policy and actions
have not only focused on reductions in the amount of pesticides used. At the same time,
many of the old pesticides have been withdrawn in favour of new ones which are less
hazardous to human health and the environment. This re-registration procedure is
probably more important for risk reduction than the reduction in the amount of pesticides used. Empirically changing to new pesticides often has an effect on the risk as well as
on the amount of active ingredient used. A list of pesticides in Swedish agriculture that
have been suspended or restricted is given in Table 5.5.

Reduced use of pesticides

The total reduction in pesticide use in agriculture and horticulture (active ingredients) up
to 1993 has been 65% or about 3000 tonnes, compared with the average use between 1981
and 1985 (Table 5.2 and Figure 5.6). A large part of the reduction (2450 tonnes) refers to
herbicides. About 1900 tonnes of this reduction in herbicide use can be accounted for by
changes in the use of herbicides for cereals, where there has been a reduction in dose of
about 35%.

The increased use of herbicides such as sulfonylureas, which are efficient in much lower
doses than the traditional phenoxy acids, and the decrease in total cultivated area explain
about 30 and 25%, respectively. There has also been a slight decrease in the percentage of
grain crop area treated which explains about 10% of the reduction. The remaining 550
tonnes of herbicides can be accounted for by the substitution of some specific herbicides
as a result of prohibition and restrictions.

The use of fungicides and insecticides on the crops which cover the largest areas of the
country is now very close to what is really needed. In fact, on some areas the actual use is
probably lower than the economic optimum owing to the farmer's priorities. However,

Table 5.5   Pesticides in Swedish agriculture that have been suspended or restricted, 1986–1990

Removed from the market mainly because of:			Severely restricted for health or environmental reasons
Health reasons	Environmental reasons	Insufficient documentation	
Aldicarb	Aldicarb	Carbaryl	Benomyl
Bromacil	Atrazine	Chloroxuron	Captan
Carbaryl	Dicofol	Dienoclor	Carbendazim
Chlorothalonil	Lindane	Lenacil	Diquat dibromide
Cyhexatin	2-Methoxyethyl mercury acetate	Metoxuron sodium chlorate	Endosulfan
Diaminozide	Terbacil	TCA-sodium	Folpet
Dinocap	Thiram	Ziram	Simazine
1,3-Dichloro-propene	Trifluralin		Thiophanate-methyl
2-Methoxyethyl mercury acetate	Ziram		
Metoxuron			

Source: Swedish National Chemicals Inspectorate (1995).

the result as reflected in the reduction of the total amount of active ingredients used is not as successful and unequivocal as that for herbicides.

The use of pesticides could also be measured in other ways, for example, area of arable land treated with pesticides (Table 5.3), kilograms of active ingredients per hectare (Figure 5.7) or number of hectare dosages (Table 5.4).

If we examine more in detail what has happened with pesticide use in Sweden, we find that the use of pesticides measured in kilograms of active ingredient has decreased considerably since 1980 (Figures 5.5 and 5.6). From 1981 to 1990, the reduction was about 50%. The official goal was to reduce use by 50% between 1986 and 1990, where the initial value was defined as *the average for 1981–1985*. Because this procedure sets a starting point which differs from the trend value in 1986, calculations also give approximately 50% reduction for the 5-year period ending in 1990.

It should be stressed that most of the 50% reduction in pesticide use in Sweden reflects a decrease in herbicide use (Figure 5.6 and Tables 5.2–5.4). Herbicides account for most of the pesticide use in Sweden and also for the main part of the reduction. Many of the developments discussed earlier which led to decreased herbicide use are not relevant for fungicides and insecticides. Nevertheless, the overall use of insecticides and growth regulators has also decreased, primarily due to changes in the types of pesticides used.

The quality of pesticides

Another important aspect of the 1986 legislation was that regulations on the use of individual compounds were tightened, thus reducing associated risks. The new regulations include 'cut-off criteria' for determining whether pesticides are acceptable from health and environmental protection standpoints. Nowadays, pesticides can only be approved for a maximum of 5 years at a time.

During 1990, 450 of around 600 pesticide products were due for reconsideration. As a consequence, the manufacturers did not apply for continued registration for 100 products and asked that another 50 should be withdrawn. The Chemicals Inspectorate denied re-registration for further 50 products which did not meet the new criteria. Altogether, this re-registration procedure has resulted in a reduction in the number of registered pesticide products from 700 to 350. After this process, a total of about 100 active ingredients are now used in Swedish agriculture.

General aspects

Although the total amount of active ingredients applied decreased considerably, the intensity of pesticide treatment (number of dosages) increased in the beginning of the 1980s, then was relatively stable for a while and finally decreased slightly during the 1990s. The use of insecticides and fungicides in cereals intensified in Sweden during the 1970s and 1980s, as it did in many other countries. My conclusion is that thus far, the 'new' pesticide policy in Sweden should be considered to be an approximation of what is technically and economically possible. Thus, the priorities and conditions of commercial agriculture in Sweden have not changed dramatically because of the policy of pesticide use reduction.

It may be possible to get close to the goal of 75% reduction between 1986 and 1996. There are, for example, still unnecessary differences in the use of lower dose rates between different areas in Sweden. It also seems to be possible to reduce the use of fungicides on potatoes and herbicides for sugar beet.

However, we have already seen that in some areas the problems with certain weeds have increased, for example thistles in spring cereals. Sometimes there may also be conflicts between different environmental goals, for example the need for green cover during the autumn and winter in order to reduce nitrogen leakage may need an increased use of herbicides.

It has to be realized that there is also some uncertainty about the future, depending, for example, on the overall policy. It is possible that there will be some increase of the acreage for cereal production. When 'old' set-aside areas are put into production again, there will be bigger weed problems in these areas in the short term, which may mean increased use of herbicides.

During 1994 the sale of pesticides increased. To date it has not been possible to see any big changes in behaviour among the farmers. The discussion on levies during 1994 has affected sales to some extent, but not as much as it did in 1986. The main part of the increased sales is probably for storage for future use by farmers or by trading companies.

In summary, the first programme on pesticide reduction has taken full advantage of technology and administrative regulation, and has thus covered a broad range of issues. The second programme will be more restricted. It comes closer to the conflicts between the different aims involved in agricultural and pesticide policy. Therefore, it lends more urgency to the question of how much can be achieved with mainly technological and administrative improvements. When will the point be reached where it is impossible to progress without challenging the whole concept of 'industrialized' agriculture.

DISCUSSION AND CONCLUSION

During the relatively short period of agrarian history in which pesticides have played a dominating role, attitudes towards them have varied over time and between individuals.

To some, they have come to symbolize the *technological dream* that eliminates the wear-and-tear of everyday life. To most people they are probably regarded as *a necessary evil*, while still others assert that they are an *unnecessary evil*. However, *the more pragmatic attitude*, i.e. that although these aids to cultivation have their problems they can be improved and developed, seems to be quite contrary to the predominating metaphors determining opinions in post-industrial societies.

Political decisions are one of several forces driving social and technological change. Legislative restrictions and regulations implemented on federal, state or local levels are factors which have important influences on the use of pesticides in agriculture. The same is true for guidelines originating from research and extension activities. However, there are also other forces on macro- and micro-levels of society which are equally important for the practical application of pesticides. For example, the developmental work within the pesticide-producing companies as well as economic and social factors in the market also have great influence. Sometimes political goals are supported, sometimes they may be counteracted by 'reality'—the sum effect of other forces.

One question in particular arises from experiences in Sweden with regard to decision-making on the use of pesticides. Apart from regulating individual pesticides, is there also a need for public control over the total amounts of pesticides used? Control of this kind may be justified if the total negative effects of pesticide applications are greater than the sum of the adverse impacts of individual pesticides. Moreover, it could also be justified if the total impact of pesticide use differs in character from the individual impacts of single pesticides.

Personally, I believe that this problem of synergism between pesticide effects exists on the agroecological as well as on the societal level, and that there is good justification for public regulation both as regards individual compounds and total amounts used. On the other hand, I find it rather ironical that this question first surfaced on the political agenda of the countries and regions where it is least necessary, whereas things go on as usual in those parts of the world where the problem is greater.

As discussed in this paper, environmental concern and political ambitions to reduce pesticide use in Swedish agriculture must be understood primarily within the historical framework of societal values and political and social experiences. However, success in the reduction of pesticide use cannot be measured and evaluated without taking into account the specific technical and agroecological prerequisites existing in Swedish agriculture. As in many other areas of complex changes in modern society, opinion proposes, technology and markets dispose.

The success of the Swedish pesticide reduction programmes are, to some extent, dependent on the fact that technological, agroecological and market forces have been working in the same direction. Political goals have been supported by 'reality'. For these reasons, it is difficult to forecast the more exact result of the programme now working with the aim of reducing pesticide use by 50% once again, and thus by 75% over a 10-year period starting in 1986. The agricultural market and policy, as well as some agroecological factors, may be counteracting instead of supporting this goal. However if we look upon the more general aim of risk reduction by means of improved management and regulation, the changes are of a more permanent value and are not spoiled if here and there they miss the target by a few per cent.

REFERENCES

Bernson, V. (1993) The role of science in pesticide management - an international comparison. The Swedish experience. *Regional Toxicology and Pharmacology* **17**, 249–261.

Bernson, V. and Ekström, G. (1995) Swedish pesticide policies 1972–93: Risk reduction and environmental charges. *Reviews on Environmental Contamination and Toxicology* **141**, 27–70.

Culliney, T.W., Pimentel, D. and Pimentel, M.H. (1993) Pesticides and natural toxicants in foods. In Pimentel, D. and Lehman H. (Eds) *The Pesticide Question: Environment, Economics and Ethics.* Chapman & Hall, New York.

Emmerman, A. and Pettersson, O. (1995) *Reduced Pesticide Use in Swedish Agriculture.* OECD. Conference of Directors and Representatives of Agricultural Research, Agricultural Advisory Service and Higher Education in Agriculture, Paris, 4–8 September 1995, Case Study No. 4.

Gold, L.S., Slone, T.H., Stern, B.R., Manley, N.B. and Ames, B.N. (1992) Rodent carcinogens: Setting priorities. *Science* **258**, 261–265.

Kreuger, J. (1992) Occurrence of pesticides in Nordic surface waters. *Tidsskrift for Planteavls, Specialserie, Beretning,* nr S **2181**, 60–68.

MANMF (1993) *Environmental Policies in Agriculture in The Netherlands.* Ministry of Agriculture, Nature Management and Fisheries, The Netherlands, The Hague.

Osteen, C. (1993) Pesticide use trends and issues in the United States. In Pimentel, D. and Lehman, H. (Eds) *The Pesticide Question: Environment, Economics and Ethics.* Chapman & Hall, New York.

Pettersson, O. (1993) Swedish pesticide policy in a changing environment. In Pimentel, D. and Lehman, H. (Eds) *The Pesticide Question: Environment, Economics and Ethics.* Chapman & Hall, New York.

Pettersson, O. (1994) Reduced pesticide use in Scandinavian agriculture. *Critical Reviews in Plant Science* **13**(1), 43–55.

Pimentel, D., Acquay, H., Biltonen, M., Rice, P., Silva, M., Nelson, J., Lipner, V., Giordano, S., Horowitz, A. and D'Amore, M. (1993) Assessment of environmental and economic impacts of pesticide use. In Pimentel, D. and Lehman, H. (Eds) *The Pesticide Question: Environment, Economics and Ethics.* Chapman & Hall, New York.

Statistics Sweden (1994) *Statistical Yearbook of Sweden 1995.* Stockholm.

Surgeoner, G.A. and Roberts, W. (1993) Reducing pesticide use by 50% in the Province of Ontario: Challenges and progress. In Pimentel, D. and Lehman, H. (Eds) *The Pesticide Question: Environment, Economics and Ethics.* Chapman & Hall, New York.

Swedish Board of Agriculture (1994) Minskade hälso- och miljörisker vid användning av bekämpningsmedel. *Rapport 18.*

Swedish National Chemicals Inspectorate (1995) *Försålda kvantiteter av bekämpningsmedel 1994.*

CHAPTER 6

Role of Biological Control and Transgenic Crops in Reducing Use of Chemical Pesticides for Crop Protection

Heikki M.T. Hokkanen

University of Helsinki, Finland

INTRODUCTION

Biological control is one of the very few alternatives to the use of chemical pesticides in controlling insect pests, plant diseases and weeds. As a sole method of pest control in a specific target crop, however, biological control is seldom sufficient. Therefore, the requirements of biological control must be integrated with the needs and uses of other control tactics in such a way that a synergistic outcome is obtained. For example, chemical pesticides, when used, must be harmless to the key biological control agents in the target ecosystem. Biological control is a cornerstone of any sustainable pest management strategy, and it is an essential component of integrated pest management (IPM) or overall, integrated production (IP) of crops.

Biological control can be defined as the conscious use of living organisms (biocontrol agents) to restrict the population sizes of unwanted organisms (target pests). It relies on the general capacity of ecosystems to resist great changes in the population densities of any of its component species. The complex natural feedback mechanisms involved in keeping population numbers in check are utilized in various ways and to varying degrees by the different approaches to biological control. Typically, one or several factors contributing to natural control of populations are exploited. In a broad sense the natural enemies of pests, which comprise the bulk of the arsenal used in biological control, include parasitic and predatory arthropods, disease-causing organisms such as bacteria, fungi, nematodes and protozoans, and competitors such as antagonistic fungi or weakened conspecifics (e.g. hypovirulent disease strains or sterilized individuals).

The importance of biological control may best be appreciated by considering the fact that only a minute fraction of the potential pest species actually do cause economic losses at a given location and in a specific cropping season. Worldwide, for example, only a few hundred species of insects are considered as serious pests, and only about 10 000 species

Techniques for Reducing Pesticide Use. Edited by D. Pimentel
© 1997 John Wiley & Sons Ltd.

out of millions are ever recorded as pests. Natural population controls are therefore extremely important for the protection of crops, plantations, forests, livestock and health.

It may be estimated that worldwide, natural or enhanced natural controls currently satisfy at least 95% of all crop protection needs. However, the remaining pests, diseases and weeds may occasionally cause 100% crop losses if left unchecked.

IMPORTATION OF NATURAL ENEMIES

When the natural enemies present in the target ecosystem appear inefficient, new effective natural enemies may be imported and introduced, typically against a specific target pest. The aim is long-term establishment of the biocontrol agent, and permanent suppression of the pest. This is often called a one-time, or inoculative release method, or classical biological control. It has mostly been used against exotic pests, which sometimes lack effective natural enemies in their new habitat, but it can equally well be employed against native pests (Hokkanen and Pimentel, 1984, 1989). This method has proved particularly useful in relatively stable forest, orchard and pasture ecosystems, where a continuous natural enemy–pest interaction can be established (Hokkanen, 1985).

Current status and importance

To quantify the importance of, or the area under, biological control resulting from importation of natural enemies is difficult. In addition, the results obtained through classical biological control tend to be forgotten because no further action is necessary and the situation resembles effective natural control. A recent example of the magnitude of the importance of classical biological control is the protection of cassava in 34 affected African countries covering most of the continent (Herren and Neuenschwander, 1991). In another instance, some 25 million ha of rangeland in Australia were cleared of prickly pear cacti (*Opuntia* spp.) by the introduction of two insect species. This control of *Opuntia* has since been repeated in 16 other countries (Julien, 1992).

The introduction of new natural enemies has probably not solved more than a few percent of the acute pest control needs worldwide. However, the economic impact of these introductions has been far greater than this small share would indicate (Cullen and Whitten, 1995; Greathead, 1995). There have been more than 5000 introductions of beneficial species for classical biological control. About 4300 of these are insect parasitoids and predators for the control of insect pests, and some 700 are biocontrol agents (mainly insects) for the control of weeds. A small number of insect pathogens and other microbes have also been introduced with the aim of permanent establishment and continuous control of the target pest (Greathead, 1995).

Insect pests

Altogether 416 species of insect pests in a total of 1275 different control projects worldwide had been the target of classical biological control by 1992. Of the target pest species, 75 have been brought under complete control, 74 under substantial control and 15 under partial control in at least one country. Of the individual projects, 156 have been rated as a complete success, 164 as a substantial success and 64 as a partial success (Hokkanen, 1994). Considering the permanency of these successes, this is a very impressive and significant record. Examples of outstanding successes where the need for chemical control of the target pest has been virtually eliminated include:

Cottony cushion scale, *Icerya purchasi*. This insect threatened to wipe out the citrus industry in California at the end of the 19th century. In 1888 the coccinellid beetle *Rodolia cardinalis* was introduced from Australia and New Zealand, and within months the pest practically disappeared. This complete success was subsequently repeated in some 40 countries around the world. *R. cardinalis* is considered to be the single most successful biological control agent to date (Bennett et al., 1976; Greathead, 1995).

Cassava mealybug, *Phenacoccus manihoti*. Cassava, a staple food crop for over 200 million people in Africa, was devastated by the accidental introduction of the mealybug in the early 1970s. Through the largest biocontrol operation to date, the parasitoid *Epidinocarsis lopezi* has been distributed over vast areas throughout the continent, with outstanding biocontrol success (Herren and Neuenschwander, 1991).

Rhodesgrass mealybug, *Antonina graminis*. This insect is an important pest of many forage grasses. It invaded Texas in the 1940s, infesting an area of 16 million ha. A parasitoid (*Neodusmetia sangwani*) was imported from India in 1959. It was spread aerially from airplanes over all the infested area, and quickly controlled the pest completely (Caltagirone, 1981).

Walnut aphid, *Chromaphis juglandicola*. This aphid invaded California from the Old World about 100 years ago, and became a serious pest of commercial walnuts. Outbreaks over thousands of ha were routinely treated with insecticides. A parasitoid (*Trioxys pallidus*) of the aphid was imported and introduced in California in 1968. In just 3–4 years the aphid was brought under complete control throughout California (Caltagirone, 1981; Olkowski, 1989).

Further landmark examples, where the insect pest was either reduced to non-pest status or the population was lowered so that a major reduction in pesticide use and the significance of the pest resulted, include: southern green stink bug (*Nezara viridula*) in Australia, New Zealand and Hawaii; citrus blackfly (*Aleurocanthus woglumi*) in Central and South America and East and South Africa; spotted alfalfa aphid (*Terioaphis trifolii*) throughout the USA; California red scale (*Aonidiella aurantii*) and several other scale insects throughout the world; winter moth (*Operophtera brumata*) in North America; rhinoceros beetle (*Oryctes rhinoceros*) in the Pacific area (Caltagirone, 1981). Within the USA, classical biocontrol during recent decades has brought several serious agricultural pests under control, such as the cereal leaf beetle (*Oulema melanopus*), the alfalfa weevil (*Hypera postica*), the alfalfa blotch leafminer (*Agromyza frontella*) and the pea aphid (*Acyrthosiphon pisum*) in alfalfa. The savings on just those four pests owing to classical biological control were estimated (in 1989) to be about US$ 112 million every year (Morrison, 1989). It is obvious that these successes have had a tremendous impact on reducing the need to spray insecticides on the affected crops.

Weeds

For the classical biological control of weeds a total of 267 individual projects are currently listed, with 125 species of weeds as targets (Hokkanen, 1994). The success rates are clearly higher than for the control of insects, possibly owing to the more stringent procedures and generally better ecological knowledge of the system (Julien, 1989). Many of the weeds were previously sprayed with herbicides, frequently with poor control results

(Greathead and de Groot, 1993). Some of the spectacular successes are:

Prickly pear cacti, *Opuntia* spp. These cacti were brought to Australia around 1840, and by 1925 had taken over 25 million ha of rangeland. A South American moth, *Cactoblastis cactorum*, and a scale insect, *Dactylopius opuntiae*, from California cleared the whole area within 10 years of introduction, reducing the density of cacti from about 1250 to around 27 plants ha^{-1} (Julien, 1989, 1992; Hokkanen, 1994).

Water hyacinth, *Eichhornia crassipes*. This floating plant from the neotropics, claimed to be the world's worst aquatic weed, invades and blocks waterways around the tropical world. In Louisiana, some 445 000 ha of invaded waterways in 1974 were reduced to 122 000 ha by 1980 through the introduction of the weevil *Neochetina eichhorniae* from Argentina. This and other natural enemies are also used with success elsewhere to control the weed (Julien, 1989, 1992; Hokkanen, 1994).

Salvinia, *Salvinia molesta*. Salvinia, a South American floating fern, is a serious aquatic weed in Australia, Southeast Asia, the Pacific region and Africa. The beetle *Cyrtobagous salviniae* from Brazil has quickly controlled the weed in most areas; the control project in Papua New Guinea was awarded the UNESCO Science Prize in 1985 for its success (Julien, 1989, 1992; Hokkanen, 1994).

Tansy ragwort, *Senecio jacobaeae*. This European weed occupies pasture, range and forestlands in western USA, Canada, Australia and New Zealand. A flea beetle (*Longitarsus jacobaeae*) and a moth (*Tyria jacobaeae*) have given 99% control of the weed in California, excellent control in Oregon, and good or some control elsewhere (Julien, 1989, 1992; Hokkanen, 1994).

Skeleton weed, *Chondrilla juncea*. This plant was a major weed of wheat cropping in Australia, causing annual production losses of A\$25 million, and control costs of A\$5 million. In the 1970s several strains of a host-specific rust fungus (*Puccinia chondrillina*) were introduced from eastern Europe, resulting in complete control of the dominant form of the weed (Julien, 1989, 1992; Hokkanen, 1994).

Future prospects

Compared with chemical control, the success rates of classical biological control are outstanding. While only about one out of 15 000 tested chemicals ends up as a chemical pesticide meeting the requirements of efficacy and safety, approximately 12% of the individual classical biological control projects against insects have resulted in complete control of the target pest. A further 18% of projects have resulted in substantial or partial success. The figures for individual pest species are even better: 18% of the target insect species have been completely controlled in at least one country, and a total of 40% of target species have been controlled to some degree. Classical biological control of weeds has been even more successful than insect control: the success rates are approximately twice as high.

It is clear that many projects in classical biological control yield negligible economic results. On the other hand, those projects which are successful can easily cover the costs of a large research agency over a decade or more. Earlier data on the economics of classical biological control programs in California suggested a rough overall return to investment ratio of 30:1, which compares very favourably to that of about 5:1 for chemical pesticides (Pimentel, 1965). Only recently have some rigorous economic analysis been made, notably in Australia. The benefit:cost ratio for biological control in Australia was calculated at 32:1, compared with 2.5:1 for other control methods (Tisdell, 1990). Further economic data show that in classical biological control *annual* savings can exceed a *one-time* investment by almost 1000-fold; for 10 projects analysed the mean was 13-fold (see Hokkanen, 1994). On a longer-term basis, therefore, it seems likely that the early economic evaluation has been conservative rather than optimistic, confirming that investment in classical biological control is highly profitable.

Despite the proven record, classical biological control usually addresses pest problems which cannot be solved, or are very difficult to solve, with chemical pesticides. Therefore this method has played a relatively minor role in controlling insect pest, disease or weed problems in field and vegetable crops—the main targets for use of chemical pesticides (Table 6.1). On the other hand, it has an excellent record in orchard ecosystems, pastures and rangeland, and in forests, where pesticides are seldom used. In the foreseeable future this pattern probably will not change, although in some specific cases a concerted effort is being made toward classical biological control of serious agricultural pests. An example of these is the Russian wheat aphid (*Diuraphis noxia*), which currently costs some US$50 million per year in crop damage and increased insecticide use in the USA (Olkowski, 1989; Robinson, 1992, 1993). Even more attention has been paid to the possibilities of biological control of the silverleaf whitefly (*Bemisia argentifolia*), which in 1991 caused an estimated US$500 million worth of damage to vegetable crops in southern USA (Perring et al., 1993).

PERIODIC MASS RELEASES

While classical biological control with ideally just one-time introduction is sometimes the perfect solution to a pest problem, in other situations different strategies must be used. In some cases the control agent needs to be reintroduced for each season, or even more often. The seasonal inoculative method is widely used in greenhouses, but also outdoors, for example where an otherwise effective natural enemy cannot survive the winter or other seasonal/environmental conditions. Sometimes, in order to obtain effective biological control, it is necessary to manipulate the phenology of the natural enemy through mass releases at a critical time.

Several species of parasitic and predatory arthropods are utilized for insect, mite and weed control in periodic mass release in fundamentally the same way as chemical pesticides (Table 6.2). The extensive use of *Trichogramma* spp., egg parasites of many important lepidopterous pests (e.g. European corn borer *Ostrinia nubilalis* in maize, leafrollers in apple orchards and the cabbage moth *Mamestra brassicae* on vegetables) is the best example of this approach (Smith, 1996). Several other species of parasitoids are available and are used against many other pests, such as scale insects, mealybugs and house flies. A few species of ladybirds (Coccinellidae) and lacewings (*Chrysopa* spp.), which are general predators of small insects, are also commercially available. Very important to the greenhouse industry, and outdoors in apple orchards, for example, are predatory mites of the genera *Phytoseiulus*, *Amblyseius* and *Typhlodromus*. These are utilized for seasonal inoculation and natural enemy conservation programs in these crops. In the USA, at least 35 companies provide natural enemies for periodic mass release purposes (Anonymous, 1995c).

Table 6.1 Overview of current (*) or potential (o) significance of biological control methods in various target ecosystems in the practical control of pests, diseases and weeds worldwide

Target ecosystem	Target problem	Importation of natural enemies	Periodic mass releases	Enhanced natural control	Biopesticides	Transgenic crops
Field and vegetable crops	Pests	*	*/oo	*/oo	**/o	ooo
	Diseases	–	*/oo	*/o	*/oo	ooo
	Weeds	–	*	o	*/o	o
Orchards	Pests	***	**/o	**/o	**/o	oo
	Diseases	o	*/o	*/o	*/oo	oo
	Weeds	*	o	o	o	o
Protected crops	Pests	–	***	–*/oo	oo	
	Diseases	–	*/o	–	**/o	oo
Pastures rangeland	Pests	****	*	*	–	ooo
	Diseases	*/o	o	o	–	ooo
	Weeds	***	o	o	–	–
Forests	Pests	***	*	*	**/o	oo
	Diseases	o	o	o	*/o	oo
Other targets	Pests	***	**	o	**/o	–
	Diseases	–	o	o	o	–
	Weeds	**/o	o	o	*/o	–

Three symbols, highly significant in many instances; –, hardly used, or not relevant; combination (e.g. */o) indicates that the method is used but a significant potential for more impact remains

Table 6.2   Examples of the extent of biological control using periodic mass releases worldwide

Beneficial(s)	Pest(s)	Crop	Area (ha)	Region	References
Trichogramma spp.	Lepidoptera: *Helicoverpa armigera*, *Ostrinia nubilalis*, *Mamestra brassicae*, Tortricidae, many other Lepidoptera	Many crops: cotton, maize, vegetables, orchards, fruits, cereals, sugarbeet, sugarcane, pulse crops	32 000 000 17 000 000 2 000 000 335 000 20 000	World total CIS PRC USA Europe Brazil	1 2, 3 4, 5 5, 6 7 8
Habrobracon hebetor	*Agrotis segetum*, *H. armigera*	Cotton	1 800 000	CIS	2
Phytomyza orobranchia	*Orobranche cumana* (weed)	Vegetables, sunflowers, tobacco	200 000	CIS	2
Cryptolaemus, *Chrysoperla*, *Aphytis*, *Pseudaphycus*	Scales and mealybugs	Citrus, vegetables	92 000 13 000	CIS USA	2 4, 6
Predators, parasitoids	Mites, aphids	Orchards	10 000	Europe	4
Phytoseiulus persimilis	*Tetranychus urticae*	Greenhouse vegetables	5 350 5 100	CIS Western Europe	2, 3 9, 10
Encarsia formosa	*Trialeurodes vaporariorum*	Greenhouse vegetables	1 050 3 200	CIS Western Europe	2, 3 9, 10
Predators, parasitoids	Aphids, thrips, leafminers	Greenhouse vegetables	760 2 250	CIS Western Europe	2, 3 9, 10

References: 1, Smith, 1996; 2, Filippov, 1989; 3, Pospelowa and Fliess, 1987; 4, Khloptseva, 1991; 5, Hokkanen, 1994; 6, King and Powell, 1992; 7, Bigler et al., 1992; 8, Anonymous, 1996c; 9, Ravensberg, 1994; 10, van Lenteren et al., 1992.

Outdoor uses

Egg parasitoids (*Trichogramma* and others) are the most widely used arthropods for outdoor use. *Trichogramma* are introduced annually on about 32 million ha to control caterpillars and other pests on maize, fruit, cotton, sugar-beet, sugar-cane, vegetables, etc. (Smith, 1996). This figure includes repeated applications per season in some areas, and therefore the total area under biological control by *Trichogramma* is somewhat less (cf. Table 6.2). By far the largest users, with more than 20 million ha treated annually, are Russia and the CIS (Commonwealth of Independent States) republics. Over 2 million hectares per annum are treated in China, and some 350 000 ha in the USA, mainly on cotton and sugar-cane. In Europe over 10 000 ha of maize are treated annually. In orchard ecosystems, predatory mites and insect parasitoids are regularly mass-released on about 13 000 ha in the USA, and on 10 000 ha in Europe.

Outdoor uses of mass-released natural enemies have had a major impact on substitu-

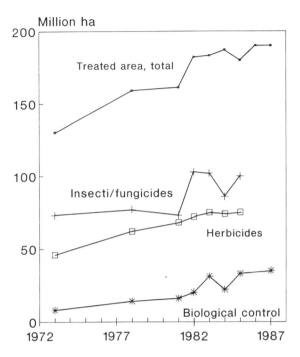

Figure 6.1   Biological control methods as a part of crop protection in the former USSR. At the end
of 1980s biological control was used on an estimated 35 million ha, i.e. about 20% of the total area
treated. *Sources*: Data from Pospelowa and Fliess (1987) and Filippov (1989)

ting for chemical pesticides only in the former Soviet Union (USSR) and in China. Even in
the USSR the total area treated with biological control agents—although immense by
any standard—represents only about 10–15% of the total area treated (Figure 6.1). If only
the use of insecticides is considered, the proportion of biological controls is about
20–30%. The driving force in obtaining such a high proportion of plant protection in the
USSR by biological means has, unfortunately, been the lack of chemical pesticides and
the funds to purchase them, rather than a preference for biological controls *per se*. The
trend of substituting the biological control systems developed during the past decades in
China and Russia by the use of chemical pesticides is already evident, as pesticides have
become more readily available in these countries (Pospelowa and Fliess, 1987). In the
Western world the trend is the opposite: biological control utilizing *Trichogramma* and
other mass-reared natural enemies is slowly increasing and has good potential, for
example in the USA to control such important pests as *Helicoverpa zea*, *Heliothis
virescens*, *Pectinophora gossypiella* and *Lygus* bugs on cotton (King and Powell, 1992).
Biological control utilizing parasitoids and commercially available predators is also
estimated to have good potential for the successful control of sweetpotato and silverleaf
whiteflies (*Bemisia* spp.) in the USA (Breene et al., 1994).
     Another important, related technique is the use of sterile insects in mass releases to
eradicate the constantly reappearing Mediterranean fruit fly (*Ceratitis capitata*) in California, as well as to control invasions by the screw-worm fly (*Cochliomyia hominivorax*) in
southwestern USA. Recently this technique was successfully applied to eradicate the

potentially devastating, first-time invasion of the screw-worm fly in Africa (Hokkanen, 1994).

Greenhouse uses

Perhaps the most spectacular successes by the periodic mass-release method have been obtained in the biological control of greenhouse pests. In Finland, for example, currently almost all pest control on greenhouse vegetables is biological. Control of the two-spotted spider mite (*Tetranychus urticae*) on cucumber is by biological means only (100% of control need), and on tomato the proportion is 94%. The greenhouse whitefly (*Trialeurodes vaporariorum*) control relies on biological means in 82% of the need on tomato, and in 62% on cucumber. Similarly, 89% of aphid control on tomato, and 50% on lettuce, is biological (all data from Vänninen, 1994). Compared with the situation about 25 years ago when biological control in greenhouses was started in Finland, the reduction in insecticide use has been about 80% (I. Vänninen, personal communication, 1996). This figure is practically the same for the most important greenhouse producer in the world, The Netherlands (80% reduction in pesticide use on greenhouse vegetables due to biological controls; J.C. van Lenteren, personal communication, 1995).

Overall, however, of the total European glasshouse area only 30% employs biological control, mainly on vegetables. Where used, this has resulted in 50–99% reduction in pesticide use (van Lenteren et al., 1992). Worldwide the figure for biocontrol in greenhouses is much smaller, about 5%. Therefore much potential remains for further reductions in pesticide use in the greenhouse sector through already well established and effective biocontrol practices.

ENHANCED NATURAL CONTROL

A potentially very important method of biological control is to enhance the activity and impact of the control agents already present in the ecosystem. A more effective use of the resident natural enemies may be obtained through habitat management, alteration of cropping practices, or by other means, but it always requires a thorough knowledge of the ecological interactions in the system. This method is compatible with, and can enhance the effect of, introduced biocontrol agents in the target agroecosystem.

When naturally occurring biological control is inadequate, the first action should be to investigate why. Often the efficiency of indigenous natural enemies is hampered by factors which are relatively easy to change. For example, the timing of pesticide applications might be modified to provide the maximum kill of the pest, while causing the minimum kill of the most important natural enemies. Similarly, the active ingredients, rates, formulations and location of pesticide application could be altered for maximal conservation of the natural enemies. Other means of conservation include the maintenance of refuges, provision of alternative prey or other food (e.g. nectariferous plants), crop rotation planning, management of crop residues, avoidance of mechanical destruction of natural enemies through minimum tillage, etc., depending on the ecology of each particular situation (Pickett and Bugg, 1996).

Management for the enhancement of natural control may be particularly important in annual crops, where the natural feedback mechanisms often have difficulties in operating effectively. The immense diversity of resident antagonists in these systems is easily overlooked, and their potential underestimated. For example, over 600 species of pred-aceous arthropods are known to be present in Arkansas cotton fields, more than 750 predator species in alfalfa in New York, and some 100 species in Florida soybean fields (Whitcomb & Godfrey, 1991). Similar, if not greater, diversity is known for insect parasitoids.

Stable equilibrium population densities for pest–enemy interactions in annual crops may in most cases be unattainable. However, in practically all annual cropping systems large numbers of generalist predators such as beetles, true bugs and spiders can be found, and in the tropics, also ants (Whitcomb and Godfrey, 1991; Lövei and Sunderland, 1996). Because generalists are not coupled with the pest species in the same way as specialists are, it might be easier to enhance their numbers in the crop in order to obtain an increased 'predator buffer' in the system. Ideally, a stable, high-density guild of predators should be formed, capable of rapid aggregation to high-density 'hot spots' of colonizing pests, and capable of switching between pest species and alternative food (Helenius, 1996).

Enhancement and skillful exploitation of indigenous biological control offers possibly the greatest potential for quickly and significantly reducing the use of chemical pesticides in agriculture. Such manipulation may be used to improve the control of all pests, and it may be utilized in synchrony with other methods of biological control, such as classical introductions or inoculative or inundative releases. Examples of the potential of such methods include the utilisation of strip intercropping (Grossman and Quarles, 1993), the conservation of epigeial predators for the control of cereal aphids through the creation of 'beetle banks' (Chiverton, 1989), conservation of the parasitoids of the pollen beetle *Meligethes aeneus* (Hokkanen et al., 1988), enhancement of *Entomophthora muscae* for the control of the onion fly *Delia antiqua* (Carruthers and Haynes, 1986), *Zoophthora* spp. mycosis of the alfalfa weevil *Hypera postica* (Brown and Nordin, 1986; Giles et al., 1994), and the enhancement of *Trichoderma* spp. and other antagonists of plant pathogens (Nelson et al., 1988). Fundamental to all these is a thorough understanding of the functioning of the particular agroecosystem to be manipulated. Learning how to conserve and maximally utilize indigenous natural enemies may be the most effective way to increase the use of biological control in agriculture.

Evidence which actually demonstrates the impact of enhanced natural control in reducing the use of pesticides is scarce. Filippov (1989) claims that habitat management for the enhancement of natural enemies has allowed the elimination of pesticide use on 10–11 million hectares of cropland in the USSR. The main crops involved in this are cabbage and cereals, where for example nectariferous plants such as anise and fennel are managed to attract parasitoids and predators, and to provide them with supplementary food (Adashkevich, 1974). Hokkanen et al. (1988, and unpublished data) have shown that under northern European conditions, simple management of the pollen beetle (*Meligethes aeneus*) parasitoid *Phradis morionellus* is likely to result in complete control of this major pest of oilseed rape. As under normal years about 50% of all insecticides in Finland and Sweden are used to control this particular pest (Mörner, 1980), the impact of enhanced natural control in this case would be extraordinary.

BIOPESTICIDES

Biocontrol agents are often released repeatedly, in great numbers, very much in the same way as chemical pesticides. The objective of this approach is a quick remedy of an acute pest problem, rather than long-term control. The agents are usually insect pathogens, and sometimes they are not even expected to multiply in the target ecosystem. The use of biopesticides is also known as the inundative release method of biological control. It is typically used in places where, and at times when, the natural spread of the biocontrol agents is not possible or sufficient, for example in annual crops and in greenhouses, but also increasingly in forests and orchards.

Biopesticides have rapidly penetrated the pesticide market, and their use is growing faster than any other type of pest control. The sales of microbial pesticides in the world currently amount to 2–3% of the sales of chemical pesticides. In terms of area treated worldwide they still lag far behind *Trichogramma*, but this is rapidly changing with the increasing use of *Bacillus thuringiensis* (*Bt*)-based products. *Bt* currently accounts for over 90% of the world use of microbial pesticides. Other bacteria, viruses and fungi are used as biopesticides against insects, weeds and plant pathogens to a small extent, but their use is expected to increase rapidly in the near future. Biopesticides are expected to increase their share of the total pesticide sales by up to 10–20% by the year 2000 (Hokkanen, 1994).

Current situation

Animal pests

For insect control, products based on bacteria, viruses, fungi, nematodes and a protozoan are at present on the market (Table 6.3). By far the most extensive is the use of several dozen different products based on the bacterium *Bacillus thuringiensis* (*Bt*). In the USA, over 450 uses and formulations are registered (Cannon, 1995). Various strains of the bacterium produce some 40–50 different endo- and exotoxins with activity against insects. The most common types of toxins kill lepidopterous larvae, and they have been used since their discovery in the 1950s for the control of at least 75 pest species (Krieg and Franz, 1989). A strain of *Bt* which kills mosquito larvae was discovered in 1977, and has been used extensively in the malaria control programs of the World Health Organization. In 1983 another pathotype was isolated, with activity against certain species of beetles. Within only a few years the strain was commercialized for the control of the Colorado potato beetle (*Leptinotarsa decemlineata*) and a few closely related species such as the eucalyptus leaf beetle *Chrysophtharta bimaculata* (Krieg and Franz, 1989; Elliott et al., 1993). Product development with *Bt* has recently produced several improved strains, such as transconjugants, which combine two or more different kinds of toxins to provide protection against a wide range of pests at the same time. Several improved strains are already sold commercially.

The world production of *Bt* has increased rapidly from about 2300 tonnes in the mid-1980s (Rowe and Margaritis, 1987) to about 5000 tonnes a few years later (Krieg and Franz, 1989) and up to 14 000 tonnes in 1992 (Marrone, 1994). This increase has far outpaced the predicted 20% annual growth for *Bt* in the early 1990s (Cannon, 1995). The trend is mainly due to a dramatic shift in North American vegetable cultivation to exploit *Bt* to avoid two problems: pesticide residues on crops and insecticide resistance in pests. At the same time, the quality of *Bt* products has improved significantly, and they now provide a predictable level of control at a competitive price (Marrone, 1994). Besides vegetables, *Bt* is now commonly employed for pest control in cotton (against *Spodoptera* and *Heliothis*), orchards (Tortricidae; *Anarsia lineatella*), maize (*Ostrinia nubilalis*) and

Table 6.3 Examples of microbial pesticides already commercialized, or close to market entry

Control organism	Target pest(s)	Remarks
Bioinsecticides		
Bacillus thuringiensis, several strains	Many Lepidoptera	Production about 14 000 t in 1992, annual growth over 20%; available since 1950s; dozens of products; main users USA, Canada, CIS, PRC
B. thuringiensis tenebrionis	*Leptinotarsa decemlineata* and some closely related leaf beetles	Available since 1990; used on 30 000 ha in CIS (1992); about 10 000 ha elsewhere
B. thuringiensis israelensis	Mosquitoes, blackflies	Available since 1980s; on > 10 000 ha annually in Europe; on large areas worldwide for control of malaria
B. thuringiensis japonensis	Scarabaeid larvae	USA, under field evaluation
B. popilliae	*Popillia japonica*	Several 1000 ha annually in north America; since 1940s
Insect viruses	About 30 spp. Lepidoptera, Hymenoptera	Nine products for six pest spp. available in Europe; in USA six products for six spp; in CIS many products for about 15 target species
NPV	*Helicoverpa* spp.	USA since 1975 on cotton, soybean, maize; several improved products approved in mid-1990s
NPV	*Autographa californica*	USA since 1981 on vegetables
NPV	*Anagrapha falcifera*, many Lepidoptera	USA, approved 1995
GV	*Cydia pomonella*	USA since 1981 on fruit trees; Europe since 1987/89 (CH/FRG)
GV	*Agrotis segetum*	Denmark since 1989
NPV	*Mamestra brassicae*	France
Beauveria bassiana	Many pests: mites, thrips, whiteflies, aphids, grasshoppers; *L. decemlineata, Ostrinia nubilalis*	USA since 1962; many new products registered in mid-1990s; most widely used in CIS
Metarhizium anisopliae	Spittle bugs	Brazil; on sugar-cane; about 50 000 ha treated annually
	Scarabaeid larvae	Australia, 1995
	Aphids, whiteflies on greenhouse crops	USA, close to market
Metarhizium flavoviride	Locusts, grasshoppers	Extensively field-tested in Africa and elsewhere; close to commercialization (IIBC/UK)
Aschersonia aleyrodis	Whiteflies	CIS
Hirsutella thompsonii	Mites	USA, on citrus (withdrawn from the market)
Verticillium lecanii	Aphids, whiteflies	Europe, in the greenhouse
Paecilomyces fumoso-roseus	Pests on greenhouse crops	USA 1995, Europe 1996
Paecilomyces lilacinus	Nematodes	Philippines; on potato etc.; Australia, under registration

Table 6.3   (*continued*)

Control organism	Target pest(s)	Remarks
Steinernema carpocapsae, S. feltiae, Heterorhabditis bacteriophora, H. megidis	Many pests in high-value crops; sciarid flies, *Otiorrhynchus* spp.; fleas of pet animals	Ornamentals, home gardens, tree nurseries; in PRC hundreds of ha annually in orchards treated against *Carposina nipponensis*, >100 000 shade trees infected annually
Biofungicides		
Trichoderma spp.	*Chondrostereum purpureum*	Reg. in France (1976), UK, Belgium, Switzerland, Sweden, and Chile (1987) for use on fruit trees
T. lignorum	*Rhizoctonia, Pythium,* etc.	CIS; on about 4000 ha annually on greenhouse crops
T. harzianum	*Fusarium, Pythium, Rhizoctonia,* etc.	USA 1995; on turf, ornamentals, etc.
	Botrytis cinerea on vines	USA/Israel, close to market
Pseudomonas putida	*Fusarium*	USA 1988; on cotton
P. syringae	Post-harvest diseases of fruit; *Boytrytis cinerea* and others	USA 1995
P. gladioli	Bacterial wilt on tomatoes, eggplant and peppers	USA 1995
Streptomyces griseoviridis	*Fusarium, Alternaria, Pythium*	Finland 1991, many other countries; USA 1995
Agrobacterium radiobacter	*Agrobacterium tumefaciens*	Australia 1989; ornamentals and fruit trees
Gliocladium virens	Damping-off diseases	USA 1992
G. catenulatum	*Pythium, Rhizoctonia*	Finland 1995, horticultural crops
Gliocladium spp.	*Phytophthora* spp.	USA, close to market
Ampelomyces quisqualis	Powdery mildew on many crops	USA 1995
Candida oleophila	Diseases of stored fruit	USA 1995
Bioherbicides		
Colletotrichum gloesporiodes	*Aeschnomene virgica*	USA, 1973; in rice and soybean used on about 4000 ha annually
Phytophthora palmivora	*Morrenia odorata*	USA, for use in citrus; currently unavailable
Other biopesticides		
Nematode Phasmarhabditis hermaphrodita	Slugs and molluscs	UK, 1994
bacterium *Salmonella enteridis*	Small rodents: *Mus, Microtus apodemus*	CIS, over 3 million ha treated annually
Haemorrhagic virus	Rabbits	Field trials in Australia
Herpes virus	Mice	Genetically modified virus to sterilize mice; product being developed in Australia

Sources: Hokkanen, 1994 (updated); recent volumes of Agrow World Crop Protection News (PJB Publications, UK); recent volumes of Biocontrol News and Information (CAB International, UK).

soybeans (*Plathypena scabra* and *Chrysodeixis includens*). The North American agricultural sector currently uses about 50% of the world production, and the forestry sector consumes a further 20–25% (*Lymantria dispar*, *Choristoneura fumiferana* and others) (Marrone, 1994). *Bt* is also gaining ground in forest protection in Europe, where instead of the usual insecticides, *Bt* was sprayed over 150 000 ha in Poland in 1994 with excellent results (Anonymous, 1995a). It is likely that neighbouring countries will follow this example in the coming years. In the USSR at the end of 1980s, *Bt* was used on over 2 million ha annually (Filippov, 1989).

Several other bacteria are utilized to a limited extent for insect pest control. The most important one dates back to before the introduction of *Bt*: *Bacillus popilliae* has been used for the control of the Japanese beetle *Popillia japonica* for over 50 years. This bacterium has a narrow host range, and unlike *Bt* it cannot be grown on an artificial medium.

Insect viruses have been tested successfully under practical conditions for the control of about 30 different pest species, but only a few have been developed to marketable products (Huber, 1995). The most widely used is the nucleopolyhedrosis virus of the pine sawfly *Neodiprion sertifer*, which is commonly used at least in the USA, UK and Finland. A granulosis virus of the codling moth *Cydia pomonella*, a serious pest on apple, is another virus product on the market in North America as well as in Europe. In East European countries and in the CIS, over a dozen different viral products for controlling insect pests, particularly in orchards, are on the market. Commercial interest in developing viruses as insecticides has lagged behind that in *Bt* because viruses kill the target insects much more slowly, typically have a narrow host spectrum, and are very costly to produce (Huber, 1995).

Of the over 750 species of fungi known to be pathogenic to insects, only a few have been commercialized, and none is widely used yet. *Beauveria bassiana* and *Metarhizium anisopliae* are the best known of these fungi. They can kill a wide range of insects, and are used, for example, to control sucking insects on vegetables and cotton (USA), grasshoppers and crickets (USA), the Colorado potato beetle (CIS), spittle bugs on sugar-cane (Brazil) and the European corn borer (France). Other commercialized fungi include *Aschersonia aleyrodis* for whitefly control (CIS), *Hirsutella thompsonii* for mite control on citrus (USA), *Verticillium lecanii* for aphid and whitefly control (Europe), and *Paecilomyces fumoso-roseus* for aphids and whitefly (USA, Europe) (Hokkanen, 1994; Anonymous, 1995m).

Insect-killing nematodes of the genera *Steinernema* and *Heterorhabditis* are produced in many countries. Nematodes have been commercially available for only about a decade, but their use is growing rapidly. At present nematodes are sold in USA, Australia, Europe and China to control many different pests, primarily on high-value crops such as greenhouse crops, strawberries, field vegetables and tree nurseries. New and more cost-effective production technologies may soon make it possible to use them on other crops as well (Ehlers and Peters, 1995).

Other biopesticides which are commercially available for controlling animal pests include an intracellular protozoan parasite of grasshoppers and locusts, *Nosema locustae*, for the control of these pests on rangeland (in the USA), and the fungus *Paecilomyces lilacinus* for the control of plant parasitic nematodes (registered in the Philippines; pending registration in Australia) (Goettel, 1993; Hokkanen, 1994). In addition, in the former USSR large areas are treated annually with a bactorodenticide containing *Salmonella enteridis*, to control small rodents such as *Microtus agrestis*, *Mus musculus* and

Apodemus silvaticus. Over three million ha are treated every year, apparently with good control results and no adverse non-target effects (Filippov, 1989).

Plant diseases

Biopesticides to control plant diseases and weeds are under intensive development. About 10 different biofungicides are currently available commercially, and many more are expected soon. Fungi of the genus *Trichoderma* inhibit the growth of several important plant pathogens such as *Pythium*, *Rhizoctonia* and *Sclerotium*. Products based on *Trichoderma* are available in several countries; in CIS they are used on about 4000 ha annually. For the biological control of *Fusarium* diseases on cotton, a product based on the bacterium *Pseudomonas putida* was commercialized in the USA in 1988. Another biofungicide, based on the fungus *Gliocladium virens*, entered the market in the USA in 1992 for the control of damping-off diseases of vegetable and ornamental seedlings in the glasshouse. In Finland, an actinomycete, *Streptomyces griseoviridis*, was discovered, developed and registered (in 1991) for the control of *Fusarium*, *Alternaria*, *Rhizoctonia* and *Pythium* on ornamentals; it has also proved to be effective on cole crops, cucumber, wheat and oil palm. Worldwide registration of the product ('Mycostop') is in progress. In Australia a biofungicide based on *Agrobacterium radiobacter* has been available since 1989 for the control of crown gall (*Agrobacterium tumefaciens*) on ornamental plants. Worldwide registration is in progress. The newest biocontrol products for the control of plant diseases contain, as their 'active ingredient': *Gliocladium catenulatum* against *Phytium*, *Rhizoctonia* and *Plicaria* ('Gliomix', Kemira Ltd., Finland); *Pseudomonas gladioli* against bacterial wilt on tomatoes, eggplants and peppers ('AM-301', Japan Tobacco); *Pseudomonas syringae* for the control of post-harvest diseases as well as *Botrytis cinerea*, *Mucor piriformis* and *Penicillium expansum* on apple, pear and citrus ('Bio-Save 10' and 'Bio-Save 11', EcoScience Corp., USA); the yeast *Candida oleophila* for the control of stored fruit ('Aspire', Ecogen Corp., USA); (Anonymous, 1995e,i,j,r).

Weeds

Despite the proven potential, only two bioherbicides have been commercialised—and neither of them is currently available. The first product, based on the fungus *Colletotrichum gloesporioides*, controlled Northern joint-vetch (*Aeschynomene virgica*) on rice and soybean in the USA. It was on the market for over 20 years, but was withdrawn in 1995 (Anonymous, 1995b). The other product was based on *Phytophthora palmivora* and it controlled milkweed vine (*Morrenia odorata*) in citrus orchards in the USA. It proved to be so effective that only one treatment was necessary over many years—therefore the demand for the product decreased and finally it disappeared from the market. A third mycoherbicide has been developed for use against *Malva pusilla* (employing *Colletotrichum gloesporioides* f.sp. *malvae*), but its future remains uncertain (Anonymous, 1995b). At least two other programs are far advanced: a *Xanthomonas campestris*-based mycoherbicide for the control of annual bluegrass (*Poa annua*) on golf courses (Japan Tobacco/Mycogen; Anonymous, 1995r), and 'rhizobacteria'-based bioherbicides for the control of such grassy weeds as wild oats (*Avena fatua*), downy brome (*Bromus tectorum*) and green foxtail (*Setaria viridis*) (Agriculture Canada; Anonymous, 1994).

The slow progress in weed control with bioherbicides is probably due to a concern about the safety of target and non-target crops and other plants. However, many opportunities exist for exploiting bioherbicides, and several new products are expected to reach the market within the next 10 years (Te Beest et al., 1992).

Future possibilities

Major constraints limiting the development and/or use of biopesticides include (i) the speed of effect (often too slow, or 'inconspicuous'), (ii) inconsistent performance, (iii) specificity (often too narrow), (iv) production costs and (v) complicated and expensive registration processes. Most of these problems can be solved with innovative research and breakthroughs in technology. Steady progress is being made in the discovery, production technology and genetic improvement of control agents used as biopesticides. This promises an accelerating rate of practical uses for this kind of biological control, and it is certain that the use of biopesticides will continue to increase very rapidly as a means of crop protection.

Discovery of new organisms

New, effective biocontrol organisms are constantly being discovered. Different strains of *Bacillus thuringiensis* appear to produce many narrow-range insecticidal toxins that only await discovery. Recent examples include a *Bt* strain which effectively kills scarabeid beetle larvae (grass grubs), and other strains which are effective against mites, corn rootworm (*Diabrotica* spp.), nematodes, adult flies and ants (Marrone, 1994).

Similar biodiversity is likely to be discovered among other microbes (e.g. Fokkema, 1995), for example some new strains of entomopathogenic fungi appear quite tolerant to environmental extremes. Thus, promising mycoinsecticides based on recently discovered strains of *Metarhizium flavoviride* and other fungi are now being intensively developed for the control of the desert locust—an unlikely target for an entomopathogenic fungus *a priori*. An experimental product ('Green Muscle') has performed well in field trials, providing a 70% reduction in grasshopper numbers in Niger 1 month after application (Anonymous, 1996b).

New species and strains of entomopathogenic nematodes are also constantly being discovered, with highly varying properties and host preferences (Ehlers and Deacon, 1996; Hominick et al., 1996). For example, the discovery and commercial development of *Steinernema scapterisci* for the control of mole crickets (*Scapteriscus* spp.) (Parkman and Smart, 1996) well illustrates this potential.

Improvements in production and delivery

Insect viruses and certain fungi are grossly underemployed as biocontrol agents, partly because many of them are currently difficult or impossible to rear on artificial media. They are also easily inactivated when exposed to sunlight, which further restricts their practical use. Fungi and nematodes also require a relatively humid microclimate to be effective. However, the production and formulation technology for biopesticides is improving, and may quickly change the ecological as well as the economic performance of many potential products. For example, the development of insect cell cultures for virus production will cut down production costs dramatically (Shieh, 1989). In the case of entomopathogenic nematodes, liquid fermentation can be developed and perfected to decrease production costs and make the price of nematodes competitive for large-scale operations (Ehlers and Peters, 1995). Currently, high costs prevent their use on other than high-value speciality crops.

Formulation technology may be crucial to the success of a biopesticide. Oil-based formulations show much better efficacy in some tests with mycoherbicides and mycoinsecticides compared with water suspensions (Womack et al., 1996), and are currently being developed, for example in a *Metarhizium flavoviride*-based locust control (Anonymous, 1996b). Innovative application methods may be equally important. One recent successful idea is to employ honey bees (*Apis mellifera*) to disperse insect viruses and entomopathogenic as well as antagonistic fungi to many different crops which are frequently visited by the bees. Examples include the control of fire blight on apple and pears (Johnson et al., 1993) and grey mold on strawberries (Peng et al., 1992); further potential targets include pests and diseases of oilseed rape, alfalfa, and clover (Gross et al., 1994; H.M.T. Hokkanen, unpublished data 1993).

Genetic engineering

Genetic improvement of biological agents is relatively new. Pesticide-resistant strains of predatory mites and a parasitoid have been obtained through traditional selection and are used in integrated control programs in orchards in northwest USA.

Only very few biocontrol agents have been genetically engineered so far, and then almost exclusively for research purposes. However, the number of experiments, and of the species involved, is increasing exponentially. Field trials have mainly involved baculoviruses, bacteria and fungi.

Currently only two commercially available, living, genetically engineered biocontrol agents exist. The first is a modified form of *Agrobacterium radiobacter* (K84) which kills the crown gall disease *A. tumefaciens* on fruit trees and roses. The product ('NoGall', BioCare, Australia) was developed in Australia and came onto the Australian market in 1989 (Anonymous, 1995f). Numerous earlier attempts to control crown gall by chemical, physical or management procedures had had limited success. Biological control with unmodified K84 is hampered because *A. tumefaciens* will readily develop resistance to the antibiotic produced by K84. This resistance is mediated by a small piece of DNA in the K84 plasmid, which transfers to the pathogen. The engineering involved removing the piece of DNA from K84 to prohibit resistance development through the same mechanism as before (Whitten, 1995).

The first viable recombinant *Bt* product ('Raven', Ecogen Inc., USA) was approved by the US EPA in 1995 (Anonymous, 1995g). This product combines two coleopteran-active proteins and gives more consistent control of Colorado potato beetle larvae than earlier *Bt* products.

Several years earlier a dead, recombinant *Bt*-based product was approved. To increase the environmental stability and effectiveness of the *Bt* toxin under field conditions, genes encoding proteins active against beetles and caterpillars are cloned into the rhizobacterium *Pseudomonas fluorescens*. After fermentation, the bacteria are killed and the cell walls are hardened chemically. The endotoxins are thereby microcapsulated, resulting in insecticides with enhanced residual activity. This product obtained full registration in the USA in 1991 (Cannon, 1995).

Genetic engineering is expected to improve the properties of insect viruses significantly. For example, the *Autographa californica* virus (AcMNPV) has been engineered to kill insects more quickly by expressing either enzymes or toxins soon after host invasion (Crook and Winstanley, 1995). Furthermore, viruses can be 'programmed' to produce insect neurohormones, which can cause rapid physiological disruptions in extremely narrowly defined target hosts. This strategy is in its early stages of development, but there is little doubt that in the near future we will have viruses with extended or specifically designed host ranges, capable of killing insects within 24–28 h. These recombinant viruses should be important for use against hosts which are not easily controlled by *Bt* (Crook and Winstanley, 1995).

As well as the already successful control of crown gall, genetic engineering may eventually help in controlling other important plant diseases (Weller et al., 1995). Antagonists such as *Trichoderma* spp. are being studied for their possible genetic manipulation, or as a source of valuable genes for disease resistance to be engineered directly into plants. To enhance the performance of plant pathogens for weed control, many aspects of the pathogen, such as increased virulence, improved toxin production, altered host range, resistance to crop protection chemicals, altered survival or persistence in soil, broader environmental tolerance, increased propagule production in fermentation systems and enhanced tolerance to formulation processes, can be targets for genetic improvements. However, engineered bioherbicides are not expected to become commercially available in the near future.

TRANSGENIC CROPS

Transgenic crop plants with resistance to pests, diseases and certain herbicides are expected to reduce the total use of chemical pesticides in world agriculture dramatically. In high-value crops such as cotton, fruits, vegetables, rapeseed, maize and potatoes, where pesticides currently play a key role in crop protection practices, the reductions may be particularly drastic (cf. Table 6.1). In other systems, such as pastures and forests, transgenic crops may provide the first possibility of controlling pests and diseases economically. They also make plant protection less dependent on weather, provide protection to plant parts difficult to reach with conventional pesticides, and have less effect on non-target organisms (Meeusen and Warren, 1989).

The potential impact of transgenic crops on reducing the global use of pesticides may best be appreciated by the example of just one crop, cotton. Despite significant gains in decreasing insecticide use due to intensive research, increased adoption of IPM and the substitution of more effective insecticides that are used at very low rates, cotton continues to account for about 29% of all insecticides used worldwide (Hearn and Fitt, 1992). As Lepidopterous pests, efficiently controlled by *Bt*-transformed cotton, account for over 50% of insecticide applications, it has been estimated that transgenic cotton alone could reduce global insecticide use by at least 10–15% (Fitt et al., 1994; Roush, 1994). Similarly, the recent introduction of herbicide- and insect-tolerant maize is expected to bring 'revolutionary' changes to the maize agrochemical market within just a few years. The maize pesticide market is estimated to exceed US$2700 million annually (Anonymous, 1995s).

During 1995 the first transgenic crops with improved crop protection properties were de-regulated in North America (Canada and/or USA), and entered commercial production. These included herbicide-tolerant cotton, soybeans and oilseed rape, insect-resistant potatoes, cotton and maize, and virus-resistant squash (Anonymous, 1995d,l,n,o,p). Australia, Europe and many other regions are expected to follow the North American lead soon.

Insect resistance

The first published reports of successful engineering of crop plants to produce insecticidal or antifeedant proteins were in 1987. The crop plants were tobacco and tomato which produced the delta endotoxin of *Bacillus thuringiensis* to make them resistant against Lepidopterous caterpillars. Since then, transgenic crop plants expressing the *Bt*-toxin have been produced in over 50 different species, including potato, cabbage, sugar-beet, rice, soybeans, maize, rapeseed sunflower, clover, apple, kiwi, walnut and poplar (see Hokkanen and Deacon, 1994). Strategies for safe and rational deployment of *Bt* genes in crop plants have been developed through international collaboration (Hokkanen and Wearing, 1994; Wearing and Hokkanen, 1995).

Another approach to provide insect resistance in transgenic crops has been to identify and transfer genes encoding various plant-derived protease inhibitors, such as the cowpea trypsin inhibitor (CpTi), tomato inhibitor I, potato inhibitor II and pea lectin (James et al., 1992; Gatehouse et al., 1994). These may provide broad-range protection across several insect orders. Recently the *Galanthus nivalis* (snowdrop) lectin (GNA) was found to provide resistance against sucking insects which do not rely on proteolysis for nutrition (Gatehouse et al., 1994). This is claimed to be the first gene to be isolated that could effectively control aphids, whiteflies, bugs and planthoppers; it has also shown potential for nematode control (Anonymous, 1996a; see also Grundler, 1996).

At least two further, novel approaches are under development for transgenic insect control. For the targeted control of the cotton bollworm (*Hellicoverpa armigera*), cotton plants are being transformed with genetic material from an insect virus (*H. armigera* stunt virus); initial success in the project has been reported (Anonymous, 1995q). 'Second generation transgenic crops' are being developed by the Monsanto corporation, based on the discovery and isolation of an enzyme, cholesterol oxidase, from a *Streptomyces* bacterium. When the gene responsible for cholesterol oxidase production was introduced into the model plants (tobacco and tomato), it provided effective protection against a wide range of insect pests. The company plans to incorporate both the *Bt*-toxin and cholesterol oxidase expression into the same plant, as these have different modes of action. Therefore greater efficacy and a much lower probability of resistance to the *Bt*-toxin is expected (Anonymous, 1995h).

Disease resistance

As for insect control, many different approaches have been proposed and are being studied to provide protection against bacteria, fungi and viruses in transgenic crops. Garcia-Olmendo et al. (1996) have outlined a general strategy for engineering plants against pathogens; Panopoulos et al. (1996) have described the options for resistance to bacteria, and Elzen et al. (1994) have done the same for virus and fungal resistance. For virus control, for example, coat protein and replicase-mediated resistances, satellite RNAs, antisense sequences, ribozymes, antibodies produced in plants, defective interfering RNAs and virus-inhibiting proteins all show promise for use in transgenic crop plants (Zaitlin, 1993). Furthermore, lignan-forming properties in transgenic plants have been proposed and are being studied as a new general approach to improving disease resistance, and to decreasing herbivore susceptibility, in non-woody plants. Various lignans and neolignans are known to have fungicidal, bactericidal and antifeeding properties (Lewis et al., 1995).

Virus resistance has already been successfully engineered into many different plants which are commercially available, and the first crop plants with a broad spectrum of fungal resistance are now being field-tested. These include transgenic carrots engineered with anti-fungal genes from tobacco resistant to powdery mildew, *Cercospora* and *Alternaria* (Anonymous, 1995t). The same technology is being extended to vegetables, oilseed rape, sugarbeet and strawberries. In strawberries, disease-resistant varieties are expected to save the growers up to US\$4000 ha^{-1} in spray and fumigation costs (Anonymous, 1995k).

Herbicide tolerance

By far the greatest number of field tests with transgenic crops have involved herbicide-tolerant plants. From the point of view of reducing the use of chemical pesticides, the rationale in developing herbicide-tolerant crop plants is to improve the efficiency of current treatments and to increase their selectivity to crop plants (Tsaftaris, 1996). If the crops tolerate a broad-spectrum herbicide, the timing of treatment can be optimised for the best control effect. Thus, for example, for tomatoes and sugarbeet several separate treatments can be replaced by just one spray, using a safer herbicide (in terms of non-target effects) than any of those previously employed (El Titi and Landes, 1990; Hauptli et al., 1990).

CONCLUSIONS

Biological control in its various forms has great potential to reduce the use of chemical pesticides, and in many cases to eliminate them completely. Classical introductions of natural enemies have done this repeatedly in the past, and will continue to be successful in specific cases, but are not likely to play a key role in intensive cropping systems. Mass releases of natural enemies and habitat management for enhancing natural control are both used extensively in some crops, where they have replaced most chemical pesticide treatments (e.g. in protected crops). It is expected that the rapid development of microbial pesticides will have the greatest impact in substituting for chemical pesticides in the near future, while the engineering and adoption of transgenic crops are likely virtually to eliminate the use of insecticides and fungicides in many crops. The long-term success of the latter strategy will depend on skillful exploitation of complementary natural regulatory factors in an integrated crop management system (c.f. Hokkanen and Deacon, 1994; Wearing and Hokkanen, 1995).

REFERENCES

Adashkevich, B.P. (1974) Number dynamics of predatory insects on nectariferous plants. In Kogan A. (Ed.) *Entomophages and Micro-Organisms in Plant Protection*. TsK KP Moldavii, Kishinev,

USSR, (in Russian).

Anonymous (1994) Rhizobacteria against weeds. *Biocontrol News and Information* **15**(3), 25N.

Anonymous (1995a) Polish forests saved. *Biocontrol News and Information* **16**(1), 1N.

Anonymous (1995b) Commercial bioherbicides. *Biocontrol News and Information* **16**(2), 21N.

Anonymous (1995c) 1996 directory of least-toxic pest control products. *IPM Practitioner* **17**(11/12), 1–37.

Anonymous (1995d) US approval for Asgrow's squash *Agrow 223*. PJB Publications, UK, p. 16.

Anonymous (1995e) Finns develop microbial products. *Agrow 224*. PJB Publications, UK, p. 16.

Anonymous (1995f) Bio-Care's biopesticide plans. *Agrow 225*. PJB Publications, Richmond, Surrey, UK, p. 18.

Anonymous (1995g) US approval for Ecogen's Raven. *Agrow 226*. PJB Publications, Richmond, Surrey, UK, p. 16.

Anonymous (1995h) Monsanto's next generation of pesticidal plants. *Agrow 227*. PJB Publications, Richmond, Surrey, UK, p. 16.

Anonymous (1995i) US approval for Ecogen's Aspire. *Agrow 227*. PJB Publications, Richmond, Surrey, UK, p. 19.

Anonymous (1995j) EcoScience biofungicides approved. *Agrow 228*. PJB Publications, Richmond, Surrey, UK, p. 21.

Anonymous (1995k) Mogens/PSI to develop disease-resistant strawberries. *Agrow 230*. PJB Publications, Richmond, Surrey, UK, p. 12.

Anonymous (1995l) Transgenic approvals for Monsanto. *Agrow 230*. PJB Publications, Richmond, Surrey, UK, p. 23.

Anonymous (1995m) Grace/Biobest extend PFR deal. *Agrow 232*. PJB Publications, Richmond, Surrey, UK, p. 19.

Anonymous (1995n) First US herbicide approvals for transgenic crops. *Agrow 234*. PJB Publications, Richmond, Surrey, UK, p. 20.

Anonymous (1995o) US deregulation of Ciba's transgenic maize. *Agrow 234*. PJB Publications, Richmond, Surrey, UK, p. 21.

Anonymous (1995p) US approvals for Ciba's *Bt* maize. *Agrow 239*. PJB Publications, Richmond, Surrey, UK, p. 22.

Anonymous (1995q) Australians pioneer new approach to bollworm control. *Agrow 239*. PJB Publications, Richmond, Surrey, UK, p. 22.

Anonymous (1995r) Japan Tobacco's biopesticide plans. *Agrow 241*. PJB Publications, Richmond, Surrey, UK, p. 24.

Anonymous (1995s) Revolution in maize market. *Agrow 245*. PJB Publications, Richmond, Surrey, UK, p. 18.

Anonymous (1995t) Mogen confirms fungal resistance. *Agrow 246*. PJB Publications, Richmond, Surrey, UK, p. 19.

Anonymous (1996a) Axis tests pest control genes. *Agrow 247*. PJB Publications, Richmond, Surrey, UK, p. 17.

Anonymous (1996b) IIBC seeks industry partner. *Agrow 248*. PJB Publications, Richmond, Surrey, UK, p. 5.

Anonymous (1996c) Biopesticide for Brazilian tomato moth problem. *Agrow 248*. PJB Publications, Richmond, Surrey, UK, p. 13.

Bennett, F.D., Rosen, D., Cochereau, P. and Wood, B.J. (1976) Biological control of pests of tropical fruits and nuts. In Huffaker, C.B. and Messenger, P.S. (Eds) *Theory and Practice of Biological Control*. Academic Press, New York, pp. 359–395.

Bigler, F., Forrer, H.R. and Fried, P.M. (1992) Integrated crop protection and biological control in cereals in Western Europe. In van Lenteren, J.C., Minks, A.K. and de Ponti, O.M.B. (Eds) (1992) *Biological Control and Integrated Crop Protection: Towards Environmentally Safer Agriculture*. Pudoc, Wageningen, pp. 95–116.

Breene, R.G., Dean, D.A. and Quarles, W. (1994) Predators of Sweetpotato whitefly. *IPM Practitioner*, **16**(8), 1–9.

Brown, G.C. and Nordin, G.L. (1986) Evaluation of an early harvest approach for induction of *Erynia* epizootics in alfalfa weevil populations. *Journal of the Kansas Entomological Society* **59**, 446–453.

Caltagirone, L.E. (1981) Landmark examples in classical biological control. *Annual Review of Entomology* **26**, 213–232.

Cannon, R.J.C. (1995) *Bacillus thuringiensis* in pest control. In Hokkanen, H.M.T. and Lynch, J.M. (Eds) *Biological Control: Benefits and Risks*. Cambridge University Press, Cambridge, pp. 190–200.

Carruthers, R.I. and Haynes, D.L. (1986) Temperature, moisture and habitat effects on *Entomophthora muscae* (Entomophthorales: Entomophthoraceae) conidial germination and survival in the onion agroecosystem. *Environmental Entomology* **15**, 1154–1160.

Chiverton, P.A. (1989) The creation of within-field overwintering sites for natural enemies of cereal aphids. Brighton Crop Protection Conference on *Weeds*. Vol. 3. British Crop Protection Council, Surrey, pp. 1093–1096.

Crook, N.E. and Winstanley, D. (1995) Benefits and risks of using genetically engineered baculoviruses as insecticides. In Hokkanen, H.M.T. and Lynch, J.M. (Eds) *Biological Control: Benefits and Risks*. Cambridge University Press, Cambridge, pp. 223–230.

Cullen, J.M. and Whitten, M.J. (1995) Economics of classical biological control: A research perspective. In Hokkanen, H.M.T. and Lynch, J.M. (Eds) *Biological Control: Benefits and Risks*. Cambridge University Press, Cambridge, pp. 270–276.

Ehlers, R.-U. and Deacon, J. (Eds) (1996) Introduction of non-endemic nematodes for biological control: Scientific and regulatory policy issues. Special Issue, *Biocontrol Science and Technology*, **6**, 291–480.

Ehlers, R.-U. and Peters, A. (1995) Entomopathogenic nematodes in biological control: Feasibility, perspectives and possible risks. In Hokkanen, H.M.T. and Lynch, J.M. (Eds) *Biological Control: Benefits and Risks*. Cambridge University Press, Cambridge, pp. 119–136.

Elliott, H.J., Bashford, R., Greener, A. and Candy, S.G. (1993) Integrated pest management of the Tasmanian eucalyptus leaf beetle, *Chrysophtharta bimaculata* (Oliver) (Coleoptera: Chrysomelidae). *Forest Ecology Management* **53**, 29–38.

El Titi, A. and Landes, H. (1990) Integrated farming system of Lautenbach: A practical contribution toward sustainable agriculture in Europe. In Edwards, C.A., Lal, R., Madden, P., Miller, R.H. and House, G. (Eds) *Sustainable Agricultural Systems*. Soil and Water Conservation Society, Ankeny, Ia, pp. 265–286.

Elzen, P.J.M., van den, Jongedijk, E., Melchers, L.S. and Cornelissen, B.J.C. (1994) Virus and fungal resistance: From laboratory to field. In Bevan, M.W., Harrison, B.D. and Leaver, C.J. (Eds) *The Production and Uses of Genetically Transformed Plants*. Chapman & Hall, London, pp. 83–90.

Filippov, N.A. (1989) The present state and future outlook of biological control in the USSR. *Acta Entomoligica Fennica* **53**, 11–18.

Fitt, G.P., Mares, C.L. and Llewellyn, D.J. (1994) Field evaluation and potential ecological impact of transgenic cottons (*Gossypium hirsutum*) in Australia. *Biocontrol Science and Technology* **4**, 535–548.

Fokkema, N.J. (1995) Biological control of foliar fungal diseases. In Hokkanen, H.M.T. and Lynch, J.M. (Eds) *Biological Control: Benefits and Risks*. Cambridge University Press, Cambridge, pp. 167–176.

Garcia-Olmendo, F., Molina, A., Segura, A., Moreno, A., Castagnaro, A., Titarenko, E., Rodriguez-Palenzuela, P., Piñeiro, M. and Diaz, I. (1996) Engineering plants against pathogens: A general strategy. *Field Crops Research*, **75**, 79–84.

Gatehouse, A.M.R., Shi, Y., Powell, K.S., Brough, C., Hilder, V.A., Hamilton, W.D.O., Newell, C.A., Merryweather, A., Boulter, D. and Gatehouse, J.A. (1994) Approaches in insect resistance using transgenic plants. In Bevan, M.W., Harrison, B.D. and Leaver, C.J. (Eds) *The Production and Uses of Genetically Transformed Plants*. Chapman & Hall, London, pp. 91–98.

Giles, K.L., Obrycki, J.J., DeGooyer, T.A. and Orr, C.J. (1994) Seasonal occurrence and impact of natural enemies of *Hypera postica* (Coleoptera: Curculionidae) larvae in Iowa. *Environmental Entomology* **23**, 167–176.

Goettel, M.S. (1993) *Directory of Industries Involved in the Development of Microbial Control Products* (Supplement No. 1). Microbial Control Division, Society for Invertebrate Pathology. Bethesda, MD.

Greathead, A. and de Groot, P. (1993) *Control of Africa's Floating Water Weeds*. CAB International, Ascot.

Greathead, D.J. (1995) Benefits and risks of classical biological control. In Hokkanen, H.M.T. and Lynch, J.M. (Eds) *Biological Control: Benefits and Risks.* Cambridge University Press, Cambridge, pp. 53–63.

Gross, H.R., Hamm, J.J. and Carpenter, J.E. (1994) Design and application of a hive-mounted device that uses honey bees (Hymenoptera: Apidae) to disseminate *Heliothis* nuclear polyhedrosis virus. *Environmental Entomology* 23, 492–501.

Grossman, J. and Quarles, W. (1993) Strip intercropping for biological control. *IPM Practitioner* 15(4), 1–11.

Grundler, F.M.W. (1996) Engineering resistance against plant-parasitic nematodes. *Field Crops Research,* 45, 99–109.

Hauptli, H., Hatz, D., Thomas, B.R. and Goodman, R.M. (1990) Biotechnology and crop breeding for sustainable agriculture. In Edwards, C.A., Lal, R., Madden, P., Miller, R.H. and House, G. (Eds) *Sustainable Agricultural Systems.* Soil and Water Conservation Society, Ankeny, Ia, pp. 141–156.

Hearn, A.B. and Fitt, G.P. (1992) Cotton cropping systems. In Pearson, C.J. (Ed.) *Ecosystems of the World: Field Crop Ecosystems.* Elsevier, Amsterdam, pp. 85–142.

Helenius, J. (1996) Enhancement of predation through within-field diversification. In Pickett, C.H. and Bugg, R.L. (Eds) *Enhancing Natural Control of Arthropod Pests through Habitat Management.* Ag Access, Davis, CA, pp. 111–150.

Herren, H.R. and Neuenschwander, P. (1991) Biological control of cassava pests in Africa. *Annual Review of Entomology* 36, 257–283.

Hokkanen, H.M.T. (1985) Success in classical biological control. *CRC Critical Reviews in Plant Science* 3, 35–72.

Hokkanen, H.M.T. (1994) Pest management; biological control. *Encyclopedia of Agricultural Science.* Vol. 3. Academic Press, San Diego, CA, pp. 155–167.

Hokkanen, H.M.T. and Deacon, J. (Eds) (1994) OECD Workshop on Ecological Implications of Transgenic Crop Plants Containing *Bacillus thuringiensis* Toxin Genes. *Biocontrol Science and Technology* 4(4), Special Issue.

Hokkanen, H.M.T. and Pimentel, D. (1984) New approach for selecting biological control agents. *Canadian Entomologist* 116, 1109–1121.

Hokkanen, H.M.T. and Pimentel, D. (1989) New associations in biological control: theory and practice. *Canadian Entomologist* 121, 829–840.

Hokkanen, H.M.T. and Wearing, C.H. (1994) The safe and rational deployment of *Bacillus thuringiensis* genes in crop plants: Conclusions and recommendations of OECD workshop on ecological implications of transgenic crops containing *Bt* toxin genes. *Biocontrol Science and Technology* 4, 399–403.

Hokkanen, H.M.T., Husberg, G.-B. and Söderblom, M. (1988) Natural enemy conservation for the integrated control of the rape blossom beetle *Meligethes aeneus* F. *Annales Agriculturae Fenniae* 27, 281–294.

Hominick, W.M., Reid, A.P., Bohan, D.A. and Briscoe, B.R. (1996) Entomopathogenic nematodes: Biodiversity, geographical distribution and the Convention on Biological Diversity. *Biocontrol Science and Technology,* 6, 317–331.

Huber, J. (1995) Opportunities with baculoviruses. In Hokkanen, H.M.T. and Lynch, J.M. (Eds) *Biological Control: Benefits and Risks.* Cambridge University Press, Cambridge, pp. 201–206.

James, D.J., Passey, A.J., Easterbrook, M.A., Solomon, M.G. and Barbara, D.J. (1992) Progress in the introduction of transgenes for pest resistance in apples and strawberries. *Phytoparasitica* 20 (suppl.), 83–87.

Johnson, K.B., Stockwell, V.O., McLaughlin, R.J., Sugar, D., Loper, J.E. and Roberts, R.G. (1993) Effect of antagonistic bacteria on establishment of honey bee-dispersed *Erwinia amylovora* in pear blossoms and on fire blight control. *Phytopathology* 83, 995–1002.

Julien, M.H. (1989) Biological control of weeds worldwide: Trends, rates of success and the future. *Biocontrol News and Information* 10(4), 299–307.

Julien, M.H. (1992) *Biological Control of Weeds. A World Catalogue of Agents and their Target Weeds.* CAB International, Wallingford.

Khloptseva, R.I. (1991) The use of entomophages in biological pest control in the USSR. *Biocontrol News and Information* 12(3), 243–246.

King, E.G. and Powell, J.E. (1992) Propagation and release of natural enemies for control of cotton insect and mite pests in the United States. *Crop Protection* **11**, 497–506.

Krieg, A. and Franz, J.M. (1989) *Lehrbuch der biologischen Schädlingsbekämpfung*. Paul Parey, Berlin/Hamburg.

Lewis, N.G., Kato, M.J., Lopez, N.P. and Davin, L.B. (1995) Lignans: Diversity, biosynthesis and function. In Siedl, P.R., Gottlieb, O.R. and Kaplan, M.A.C. (Eds) *Chemistry of the Amazon*. ACS Symposium, Series 588, pp. 135–167.

Lövei, G.L. and Sunderland, K.D. (1996) Ecology and behavior of ground beetles (Coleoptera: Carabidae). *Annual Review of Entomology (1996)*, 231–256.

Marrone, P.G. (1994) Present and future use of *Bacillus thuringiensis* in integrated pest management systems: An industrial perspective. *Biocontrol Science and Technology* **4**, 517–526.

Meeusen, R.L. and Warren, G. (1989) Insect control with genetically engineered crops. *Annual Review of Entomology* **34**, 373–382.

Mörner, J. (1980) Samordnad oljeväxtodling—problem och möjligheter. *Växtskyddsrapporter, Jordbruk* **12**, 21–29. (Swedish University of Agricultural Sciences).

Morrison, J. (1989) Biological control turns 100 this year. *Agricultural Research* **37**(3), 4–8.

Nelson, E.B., Harman, G.E. and Nash, G.T. (1988) Enhancement of *Trichoderma*-induced biological control of *Pythium* seed rot and pre-emergence damping-off of peas. *Soil Biology and Biochemistry* **20**, 145–150.

Olkowski, W. (1989) Update: Biological control of aphids. What's really involved? *IPM Practitioner* **11**(4), 1–9.

Panopoulos, N.J., Hatziloukas, E. and Afendra, A.S. (1996) Transgenic crop resistance to bacteria. *Field Crops Research*, **45**, 85–97.

Parkman, J.P. and Smart, G.C., Jr. (1996) Introduction of *Steinernema scapterisci* in Florida, USA. *Biocontrol Science and Technology*, **6**, 413–419.

Peng, G., Sutton, J.C. and Kevan, P.G. (1992) Effectiveness of honey bees for applying the biocontrol agent *Gliocladium roseum* to strawberry flowers to suppress *Botrytis cinerea*. *Canadian Journal of Plant Pathology* **14**, 117–129.

Perring, T.M., Cooper, A.D., Rodriguez, R.J., Farrar, C.A. and Bellows, T.S., Jr. (1993) Identification of a whitefly species by genomic and behavioural studies. *Science* **259**, 74–77.

Pickett, C.H. and Bugg, R.L. (Eds) (1996) *Enhancing Natural Control of Arthropod Pests through Habitat Management*. Ag Access, Davis, CA, in press.

Pimentel, D. (1965) Restoring the quality of our environment. In *President's Science Advisory Committee 1965*. The White House, Washington, DC, pp. 227–291.

Pospelowa, G. and Fliess, H. (1987) *Biologischer Pflanzenschutz in der Sovietunion. Giessener Abhandlungen zur Agrar- und Wirtschaftsforschung des Europäischen Ostens, Band 147*. Duncker & Humboldt, Berlin.

Ravensberg, W.J. (1994) Biological control of pests: Current trends and future prospects. Brighton Crop Protection Conference, *Pests and Diseases* 1994, pp. 591–600.

Robinson, J. (1992) Predators and parasitoids of Russian wheat aphid in central Mexico. *Southwestern Entomologist* **17**, 185–186.

Robinson, J. (1993) Studies in host-plant resistance to Russian wheat aphid (*Diuraphis noxia Kurdjumov*). Ph.D. Thesis, University of Helsinki.

Roush. R.T. (1994) Managing pests and their resistance to *Bacillus thuringiensis*: Can transgenic crops be better than sprays? *Biocontrol Science and Technology* **4**, 501–516.

Rowe, G.E. and Margaritis, A. (1987) Bioprocess developments in the production of bioinsecticides by *Bacillus thuringiensis*. *CRC Critical Reviews in Biotechnology* **6**, 87–127.

Shieh, T.R. (1989) Industrial production of viral pesticides. *Advances in Virus Research* **36**, 315–343.

Smith, S.M. (1996) Biological control with *Trichogramma*: Advances, successes, and potential of their use. *Annual Review of Entomology* **41**, 375–406.

Te Beest, D.O., Yang, X.B. and Cisar, C.R. (1992) The status of biological control of weeds with fungal pathogens. *Annual Review of Phytopathology* **30**, 637–657.

Tisdell, C.A. (1990) Economic impact of biological control of weeds and insects. In Mackauer, M., Ehler, L.E. and Roland, J. (Eds) *Critical Issues in Biological Control*. Intercept, Andover, pp. 301–316.

Tsaftaris, A. (1996) The development of herbicide-tolerant transgenic crops. *Field Crops Research*,

45, 115–123.

van Lenteren, J.C., Minks, A.K. and de Ponti, O.M.B. (Eds) (1992) *Biological Control and Integrated Crop Protection: Towards Environmentally Safer Agriculture.* Pudoc, Wageningen.

Vänninen, I. (1994) *Pests and Pesticide Usage on Greenhouse Cultivations. Results of a Questionnaire Survey from 1992.* Agricultural Research Centre of Finland, Jokioinen, Report 7/94, pp. 1–30. (in Finnish, with a summary in English).

Wearing, C.H. and Hokkanen, H.M.T. (1995) Pest resistance to *Bacillus thuringiensis*: Ecological crop assessment for *Bt* gene incorporation and strategies of management. In Hokkanen, H.M.T. and Lynch, J.M. (Eds) *Biological Control: Benefits and Risks.* Cambridge University Press, Cambridge, pp. 236–252.

Weller, D.M., Thomashow, L.S. and Cook, R.J. (1995) Biological control of soil-borne pathogens of wheat: Benefits, risks and current challenges. In Hokkanen, H.M.T. and Lynch, J.M. (Eds) *Biological Control: Benefits and Risks.* Cambridge University Press, Cambridge, pp. 149–160.

Whitcomb, W.H. and Godfrey, K.E. (1991) The use of predators in insect control. In Pimentel, D. (Ed.) *CRC Handbook of Pest Management in Agriculture.* Vol. II. CRC Press, Boca Raton, FL, pp. 215–241.

Whitten, M. (1995) An international perspective for the release of genetically engineered organisms for biological control. In Hokkanen, H.M.T. and Lynch, J.M. (Eds) *Biological Control: Benefits and Risks.* Cambridge University Press, Cambridge, pp. 253–260.

Womack, J.G., Eccleston, G.M. and Burge, M.N. (1996) A vegetable oil-based invert emulsion for mycoherbicide delivery. *Biological Control* **6**, 23–28.

Zaitlin, M. (1993) Molecular strategies for the control of plant virus diseases. *Acta Horticulturae* **336**, 63–68.

CHAPTER 7

Host-Plant Resistance to Insect Pests

Helmut F. van Emden

The University of Reading, Reading, UK

INTRODUCTION

Host-plant resistance (HPR) to insect pests would appear to offer many advantages over other methods of pest control. It should usually be fully environmentally friendly, in contrast particularly to toxic pesticides. It is compatible with pesticide use, in contrast to many forms of biological control. In contrast to cultural control, it can be expected to fit in with the farmer's crop management scheme; it is often merely the replacement of one seed stock by another. As it requires no special knowledge, equipment or skill in use, it is very attractive in many developing countries. Indeed, HPR was the early focus of pest control in the chain of International Agricultural Research Institutes set up by the Consultative Group for International Agricultural Research (CGIAR). More recently, the development of direct gene transfer techniques, which avoid some of the problems and long time-scales of traditional plant breeding, has made HPR attractive in developed countries as well.

Emphasis on genetic HPR has rather overshadowed non-genetic expressions of HPR. Examples of changes in the susceptibility of plants to pests with age and developmental stage, high and low temperature, light intensity, fertilizer, herbicides, insecticides and plant growth regulators are given by van Emden (1987). Environmentally induced resistance is an aspect of HPR in its own right, since resistance can be obtained by treatments to susceptible genotypes as well as, and certainly far more quickly than, by plant breeding. The subject is also important to genetically based HPR. The changes involved can enhance HPR mechanisms already possessed by plants. Recently, it has also been found (R. Cole, unpublished data, 1995) that stress (e.g. wind, lack of water) greatly increases secondary compound concentrations in field-grown compared with glasshouse-grown plants. This probably explains the decreased suitability to *Myzus persicae* of Chinese cabbage plants brushed regularly during early growth (van Emden et al., 1990). HPR studies in atypical environments such as glasshouses can therefore be very misleading. In addition, the stability of any resistance needs to be confirmed in different environments (especially of soil and temperature) at an early stage of any research work.

Techniques for Reducing Pesticide Use. Edited by D. Pimentel
© 1997 John Wiley & Sons Ltd.

GENETIC PLANT RESISTANCE

Although biotechnological techniques now allow the transfer of foreign genes, much development of HPR involves enhancement of a property already possessed by the plant. This is shown by the changing suitability of plants with age and environment, as mentioned earlier. Even so-called 'susceptible' varieties present the insect with problems (Southwood, 1973) of tough tissues, sites protected with toxic and unpalatable secondary compounds and limited nitrogen availability. Indeed, most plants are resistant to most insects. 'Host selection' in many insects restricts them to a very narrow range, and rejection rates as high as 50–95% are shown by small insects arriving on 'susceptible' host plants by passive dispersal (e.g. Müller, 1958, with *Aphis fabae*).

Farmers have known for centuries that different varieties of a crop can vary considerably in their susceptibility to pests. The many local varieties grown only in small areas in the tropics reflect past selection by farmers for those able to produce high yields in spite of intense pest and disease pressure.

HPR has long provided control of plant pathogens and eelworms. Early cultural and biological control methods and the few agrochemicals which were available had little effect on such problems, although they did reduce insect pests. Some commercial use of insect-resistant crop varieties dates back to the middle (e.g. apples and woolly aphid) or end (e.g. grape rootstocks and phylloxera) of the 19th century. A definitive treatise on HPR to insects was published over 40 years ago (Painter, 1951). Nevertheless, the real increase in HPR studies with respect to insect pests is a phenomenon of only the last 20 years, coinciding with pressures on agriculture to effect a considerable reduction in the use of agrochemicals and the emphasis on HPR at CGIAR's International Agricultural Research Institutes.

The classification of plant resistance

Painter's (1951) classification distinguished three categories of HPR, namely 'non-preference', 'antibiosis' and 'tolerance', and these are still useful today. The only change has been the introduction by Kogan and Ortman (1978) of the term 'antixenosis' (resistance to colonisation) to replace non-preference, so that all these three categories are properties of the plant. Non-preference is a characteristic of the insect.

Antixenosis. This refers to plant properties which reduce colonisation by pests seeking food or oviposition sites. It is 'host-plant resistance to pest arrival'. However, antixenosis and non-preference are not strictly synonyms. Antixenosis, unlike non-preference, includes the concept of non-acceptance in a no-choice situation (van Marrewijk and de Ponti, 1975).

Antibiosis. This is resistance to biological processes such as growth and reproduction. It is therefore resistance which reduces some aspects of insect performance on the plant. Such aspects are survival, growth, generation time, fecundity and longevity.

Tolerance. This is the ability of a plant variety to show a reduced damage response (often measured as yield) compared with another variety when both varieties suffer the same insect pest burden. Tolerance is therefore resistance to damage rather than to the insect, yet host-plant resistance remains the umbrella term, if only by accepted usage.

A later classification of plant resistance devised by van der Plank (1963) had its origin in Flor's (1942) 'gene-for-gene' hypothesis for resistance to plant diseases, which was proposed a decade before Painter's book. Major genes for plant resistance to a pathogen are matched by corresponding genes in the pathogen for overcoming that resistance. Where HPR is based on a single gene (monogenic) or just a few genes (oligogenic), resistance and susceptibility to different races of the pathogen or pest are clear-cut and race-specific. van der Plank (1963) termed this 'vertical resistance', by analogy with the ups and downs which would be shown by a histogram for the susceptibility of different gene combinations to different races of a pest or pathogen. By contrast, polygenic resistance depends on the percentage of the pathogen and host alleles at the appropriate gene loci which are 'virulent' or 'resistant'. Such outcomes are quantitative and non-specific for race, resulting in what van der Plank termed 'horizontal' HPR.

What is the ideal type of plant resistance? Using van der Plank's classification, polygenic horizontal resistance will last the longest and be effective against many races of the pest. Using Painter's (1951) classification, it would be tempting to propose tolerance, since there is no pressure for selection of an adapted race. However, farmers growing tolerant varieties would allow the pest to multiply and create a problem for other farmers using different varieties. Paradoxically, antibiosis is the preferred category of resistance. It is more effective than antixenosis in a no-choice situation for the pest, and Painter's three categories of resistance rarely occur in isolation. Antibiosis is often accompanied by some antixenosis and may be combined with tolerance in the course of a crop improvement programme.

The true mechanisms of plant resistance are morphological, physiological, anatomical or chemical, but Painter (1951) was rather dismissive of the practical importance of understanding them. However, there are three ways in which knowledge of the mechanism may be useful:

(1) an assay for the resistance mechanism may be quicker than waiting for, and then measuring, pest populations;

(2) it may point to the category of resistance and even indicate the genetic complexity;

(3) predictions about the stability of the resistance may be possible.

The major mechanisms are shown in Figure 7.1 linked to the three categories of resistance and presented in the approximate order in which an insect encounters them. Most are self-explanatory, and a fuller discussion with examples can be found in van Emden (1987). However, the classification of mechanisms presented in Figure 7.1 is not clear-cut. Many examples of 'hairiness' are actually based on chemicals produced by glandular hairs, 'waxiness' often relates to HPR based on high levels of diketones in the wax of less waxy varieties, and 'hardness' may involve the presence of silica grains in the tissues rather than hardness of the plant cells. Four other groups of mechanisms which perhaps require further explanation are 'morphology', 'necrosis', 'phenology' and 'extrinsic resistance'.

Morphology. An example of morphology is the incorporation into commercial cotton varieties of narrow twisted bracts ('Frego bract') under the boll and the nectary-free characteristic (no leaf nectaries). Both confer resistance to the bollworm *Heliothis*.

Necrosis. Necrosis means the death of the plant or part thereof, apparently the antithesis of resistance (hypersensitivity). The local collapse and death of cells around the stylet tips of a sucking insect brings together enzymes and substrates for a polyphenol reaction which inhibits or deters further penetration. Even whole-plant necrosis can be

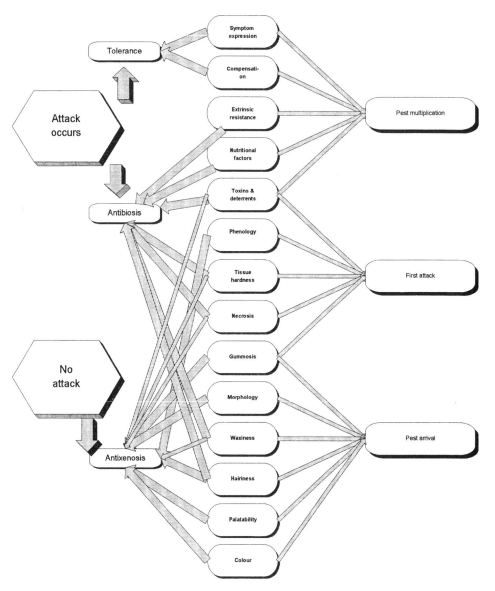

Figure 7.1   A classification of plant resistance mechanisms based on host-plant influences on insect populations. The mechanisms link right to stage of pest infestation and left to categories of host-plant resistance (thick arrows, most success; thin arrows, fair success)

used in densely sown crops such as cereals, where inter-plant competition allows yield compensation over a wide range of plant densities.

Phenology. Phenological resistance, often termed 'pseudoresistance', is the reduction of attack by a variety, or its escape because the time of sowing or speed of maturation means that it has reached a resistant stage at the normal time of insect attack. If grown at a different time, a pseudoresistant variety will be just as susceptible to the pest as other varieties.

Extrinsic resistance. Extrinsic resistance does not stem from the plant–pest interaction. It occurs only because an unrelated source of mortality (usually biological control) is enhanced on a variety. This will be discussed more fully later as part of the integration of HPR into integrated pest management (IPM).

The practical development of genetic plant resistance

Developing pest-resistant crop varieties requires three elements: sources of genes for resistance, screening techniques and methods for transferring the selected genes into agronomically adapted varieties (i.e. with desirable agronomic characters, including good yield potential).

Sources of genes for plant resistance to insect pests

The genetic range of most of the world's major crops has been assembled in 'germplasm banks' where plant material of a particular crop, collected regionally or world-wide, is held in long-term storage and regenerated as necessary to maintain viability. The International Board for Plant Genetic Resources (IBPGR) was established in 1973 to assemble material and to allocate it to germplasm banks at the different CGIAR International Research Institutes. The IBPGR list (first published in 1984) is of accessions for more than 60 crops held at over 100 centres (IBPGR, 1984). Another list, now somewhat dated, is available for 22 major world crops only (Harlan and Starks, 1980). A database of world germplasm collections is held at the Information Science/Genetic Resources Program of the University of Colorado in the USA.

Germplasm banks assemble examples of genetic variation from a wide variety of sources. Many accessions result from major expeditions to collect examples of local crops used for centuries by peasant farmers. One of the reasons for establishing IBPGR was the fear that the genetic diversity of these local varieties would be lost. The germplasm banks also assemble commercial varieties before they become obsolete and, being at research centres with active international plant breeding programmes, they also collect the many breeders' lines tested at different stages in these programmes.

Very often, genes for resistance are found in the 'wider gene pool' of wild ancestors of present day crops or even of plants totally unrelated to the crop. Use of this wider gene pool is difficult, if not impossible, in traditional plant breeding, but is easier now that crossing can be replaced by direct transfer of genes.

In the case of the wild relatives of modern crops, there are good reasons why they are targeted in the search for resistance genes. Genetics is a science stemming from the start of the 20th century. Before then, farmers were unaware that low-yielding types discarded during selection may have had valuable pest-resistance properties. Secondly, the early stages of human efforts at, for example, turning a minute grass seed into a 'grain' involved selection for spot characteristics rather than for resistance. Thirdly, much of the resistance of wild plants is based on toxins or strong flavours which were dangerous or unpalatable to humans. For example, wild lettuce is extremely bitter and unpleasant to human taste. Table 7.1 gives some quantified examples. That it was necessary for humans to breed out these compounds does limit how far they can be restored in crops, although the concentrations required to deter insects may not be high. Alternatively, even with toxins, it may be possible to protect some plant organs (e.g. leaves) without the same levels of chemicals

Table 7.1 Contrast between some wild and cultivated plants in levels of secondary compounds

Chemical (units)	Plant	Quantity	Reference
2-tridecanone (μg cm^{-2} leaf surface)	*Lycopersicon esculentum* (tomato)	0.1	Kennedy, 1986
	L. hirsutum	44.6	
Allylisothiocyanate (μg g^{-1} dry weight)	*Sisymbrium officinale*	3660.0	van Emden, 1990
	Brussels sprout	167.0	
	January King cabbage	847.0	

being present in the organs destined for market (e.g. tubers or seeds). Finally, the pests attacking our crops may not be identical to those of the past. New races of pest species or new species may have been imported from abroad or arisen as a result of changes in agronomic practices.

There are several ways of creating new genetic diversity. Paramount among these is 'random out-crossing'. This is such a routine method that male-sterile lines of agronomically adapted varieties have been bred for many crops. If such lines are not available, flowers of an adapted variety have to be emasculated by hand. Either way, plants of an adapted variety which are unable to produce pollen are planted in rows at intervals between several rows of single seeds of as many varieties as possible, e.g. from a germplasm bank. Each seed produced on the male-sterile line must represent a cross between that line and an unknown male parent from elsewhere in the field, i.e. it is potentially a new genetic combination! New variation can also be created by mutation. Techniques include the gamma-irradiation of seed and the soaking of seeds in certain chemicals, including colchicine and dimethyl sulphate. However, only a few mutated seeds from such treatments are likely to be viable.

Another approach to creating new variation is to exploit the considerable somaclonal variation which can be shown in plants derived from callus in cell and tissue culture. For example, sugar cane resistance to the borer *Diatraea saccharalis* was found in one of 2000 somaclonal variants derived from a susceptible cultivar.

Screening methods

Many hundred to several thousand accessions for one crop may be available in a germplasm bank, but each can usually only be supplied as a few seeds. Replication of each accession is therefore usually impracticable; nor is it necessary if the first screens are 'negative', i.e. they aim to eliminate obvious susceptibility to the pest rather than select for resistance. Large numbers of accessions can then be screened, each as a single unreplicated row. A known susceptible variety (the 'susceptible check') is planted at intervals and each accession can be compared visually with the nearest check row and rated on a simple visual damage scale (often 1–5).

The first negative screen will usually eliminate the majority of accessions as susceptible. Most of the others will have escaped attack by chance, but they will also include any truly resistant types. A second negative screen of the retained accessions will thus eliminate many more, as will a third, and so on. The process of negative screening continues until the number of accessions left is small enough to make replicated blocks, several metres square, possible. These accessions will already have shown low damage scores on several occasions.

Insects or damage (e.g. per cent pods bored, per cent defoliation) can be counted in the replicated trials with some precision, using established methods. Chiarappa (1971, with later supplements) gives a particularly extensive range of protocols, as do Panda and Khush (1995) for both field and glasshouse work. An alternative approach is to compare the yields of accessions in adjacent half-plots, with and without insecticide protection. The lower the difference between these two treatments, the greater is the likely resistance of the accession.

Two further steps are necessary before any accessions are selected for a breeding programme. One is to include promising accessions in 'multi-location trials', spanning as far as possible the range of soil types and climates in which the crop is grown. The other is to identify how far the three categories of antixenosis, antibiosis and tolerance account for the resistance observed in the field. This usually has to be done in the glasshouse with different known races of the pest, or with pest populations collected from different areas and introduced artificially. Another important check is that the resistance is not a 'novel association' phenomenon, i.e. antixenosis and antibiosis do not disappear once the pest has been reared for a few generations on the resistant variety.

The transfer of genetic resistance

Although the degree of partial HPR that can be found in commercial cultivars may be immediately useful in an IPM programme, it is usually necessary to transfer the resistance into agronomically desirable cultivars. The strategies for such transfer have traditionally involved grafting, or various well-established plant breeding protocols involving steps such as self-fertilisation, hybridisation and back-crossing. These protocols are outside the scope of this chapter, but a useful summary can be found in Everson and Gallun (1980).

Modern techniques of genetic transformation make possible the transfer into a different genetic background (an agronomically adapted cultivar of the relevant crop) of a foreign gene derived from an unrelated plant or even from a far more genetically distant organism such as a micro-organism or an animal. Genes can often be transferred to dicotyledons indirectly using a pathogenically disarmed root-parasitic bacterium (especially *Agrobacterium tumefaciens*) or directly to more difficult subjects (e.g. cereals) by culturing plasmid DNA with protoplasts of the crop variety in special media by using high-voltage external electrical fields (electroporation), by firing tungsten or gold particles coated with DNA into a plant's genome using a special high-velocity gun (biolistics), or by micro-injection of DNA.

Genetic transformation offers not only the possibility of circumventing natural barriers to genetic exchange, but also the lure of transferring the desirable gene (e.g. for resistance to a pest) alone, without other characteristics which might reduce the agronomic desirability of the new cultivar. Indeed, several major multi-national companies have begun genetic engineering of pest- and pathogen-resistant plants as an important new expansion of their crop-protection provision.

Most novel techniques have problems, and transgenic plant resistance to pests is no exception. Stringent risk assessment is imposed by governments fearful that the transgenes might escape by invasion or pollination into unintended environments, such as natural and semi-natural vegetation. There are also potential problems in the use of transgenic HPR in IPM, and these will be discussed later.

Even without such problems, scientists involved in genetic transformation do not find the techniques as promising and speedy as those outside the subject often assume.

Considering that transgenic plants were first reported as late as 1984 (Herrera-Estrella et al., 1984), the 19 crop–pest associations for which Panda and Khush (1995) list examples of transgenic HPR represent rapid progress. However, most examples have involved the same gene source, that coding for the production of toxins in the bacterium *Bacillus thuringiensis*.

LIMITATIONS AND PROBLEMS OF PLANT RESISTANCE AS A PEST CONTROL MEASURE

Problem-trading

There is always the possibility that whatever anatomical, physiological or chemical change confers greater resistance to organism A may increase the susceptibility of the plant to organism B.

Examples in practice are numerous. An early discovery was that the hairy cottons resistant to leafhoppers were, unfortunately, especially susceptible to aphids. Like the non-hairy cottons, the smooth-leaved soybeans resistant to *Heliothis* provided varieties very vulnerable to leafhoppers.

These examples all relate to problem-trading between different insect pests. It is equally important to anticipate problem-trading between pests and fungal diseases. A practical commercial case occurred in California when new wilt-resistant alfalfa varieties were released and immediately proved very susceptible to the spotted alfalfa aphid (*Therioaphis maculata*) (van den Bosch and Messenger, 1973).

The reaction of plant breeders to problem trading is to suggest pyramiding genes for resistance to both (or several) organisms. This slows down breeding programmes and requires finding new sources of resistance which can be expressed together, unlike the incompatibility of smooth and hairy foliage in cotton mentioned earlier.

Yield drag (yield penalty)

Some widely accepted theories of ecology, including the co-evolution theory of Ehrlich and Raven (1964) and the apparency theory of Feeny (1976), require a trade-off between the effectiveness of plant defences and the energetic cost to the plant of that defence. This thesis was nicely summarised by Hodkinson and Hughes (1982) 'Energy or other resources which the plant diverts for defence cannot be used for growth and reproduction'. An example of evidence for yield drag is the ICRISAT (1980) screening of 31 pigeon pea varieties for resistance to pod borers. The negative correlation between the yield achieved by these varieties under insecticide protection and the level of resistance to borer was very high ($r = 0.96$), and leads to the prediction that 90% resistance incurs a yield drag of 31%.

An important practical consequence, where yield drag of HPR can be established, is that the minimum level of HPR required in an IPM context will be most compatible with yield aspirations.

However, in spite of the reasonable basis for the thesis that maximum yield and HPR are incompatible, it is by no means certain that yield drag is inevitable. A moment's

Table 7.2 Costs of raw materials fore the formation of various antiherbivore defence compounds (minimum, medium and maximum costs for each chemical class, selected from Gershenzon, 1994)

Chemical class	Compound and source	Conc. (% dry wt.)	Cost (mg glucose per g tissue)
Terpenoids			
Diterpene	Abietic acid in needles of *Larix laricina*	0.1	3
Sesquiterpene	Caryophyllene in leaves of *Hymenaea courbaril*	0.5	19
Triterpene	Papyriferic acid in juvenile twigs of *Betula resinifera*	1.3	307
Phenolics			
Isoflavenoid	Daidzein in seeds of *Phaseolus mungo*	0.1	2
Condensed tannins	Linear procyanidin polymer in leaves of *Betula allegheniensis*	1.2	25
Flavonoid	Apigenin in leaves of *Isocoma acradenia*	5.0	103
Alkaloids			
Quinolizidine	Lupanine in leaves of *Lupinus polyphyllus*	0.2	7
Pyridine/pyrrolidine	Nicotine in leaves of *Nicotiana sylvestris*	0.5	18
Purine	Caffeine in leaves of *Camellia sinensis*	2.5	72
Other nitrogenous defences			
Proteinase inhibitor	Inhibitor II from leaves (24 h after wounding) in *Lycopersicon esculentum*	0.004	0.1
Glucosinolates	3,4-Dihydroxybenzylglucosinolate in leaves of *Bretschnedera sinensis*	0.7	13
Non-protein amino acids	Mimosine in mature leaves of *Leucaena leucocephala*	1.5	42

reflection will show that the phenomenon of problem trading (see above) is at variance with the yield drag hypothesis. Simply put, if HPR to organism A is inversely correlated with HPR to organism B, yield cannot be negatively correlated with resistance to both organisms.

Gershenzon (1994) has compiled a table from the literature of costs for various antiherbivore defence compounds (Table 7.2). In terms of mg glucose per g plant tissue, the costs range from 0.1 for a proteinase inhibitor to 307 for a triterpene. Many workers have argued that these costs are small, and would only incur yield drag where carbon dioxide, nitrogen or radiation are limiting factors.

However, modelling the theoretical yield costs of the combinations of different defences to different organisms (Table 7.3 and Figure 7.2) shows that negative relationships between yield and HPR may remain hidden. Taking six hypothetical varieties, random numbers were used to assign yield drags of 0, 10, 20, 30, 40 and 50% for HPR to each of three pest organisms. Thus for each organism, large costs of resistance were built into the model, and the actual overall yield of the six varieties was then calculated. For two of the three organisms, no significant negative relationship could be established between resistance and yield for the varieties.

In summary, it would appear that yield drags may well occur for certain types of HPR,

Table 7.3 Hypothetical relationship for six varieties between randomly assigned yield drags caused by host-plant resistance to three organisms and the overall potential yield of each variety

Variety	I	II	III	IV	V	VI
% yield drag due to resistance to organism:						
A	50	0	20	40	30	10
B	50	20	0	10	40	30
C	10	0	40	50	20	30
Yield as per cent of an equal potential for each variety after the combined effects of the three component yield drags						
	17.5	80.0	43.2	30.0	33.6	44.1

Regressions of yield (y) on per cent resistance (x) to each organism.
A: $y = 67.4 - 1.04x$ ($r^2 = 0.838$, $P < 0.01$).
B: $y = 50.5 - 0.36x$ ($r^2 = 0.102$, $P > 0.05$).
C: $y = 53.0 - 0.48x$ ($r^2 = 0.167$, $P > 0.05$).

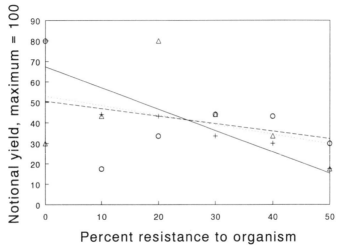

Figure 7.2   Yield of six varieties against per cent resistance to three organisms (A, +; B, △; C, ○), each of which imposes a randomly assigned yield drag per variety (data of Table 7.5)

even though they cannot always be demonstrated. The matter remains sufficiently open and important to warrant further research.

Health hazard

A potential human health problem exists, especially if the resistance is based on a new or on higher levels of an allelochemical with unknown toxicity to humans or animals. New allelochemicals for a crop are particularly likely to be a problem with transgenic plants. Natural products may well be carcinogenic and mutagenic (Ames et al., 1990). Nor is palatability always checked. The famous example is the US potato variety 'Lineup' which was taken off the market because consumers found the flavour of raised levels of glycoaldehyde, bred in for resistance to Colorado beetle, distasteful and even mildly toxic.

Biotypes

The high selection pressure on a pest population of strongly resistant varieties can lead a biotype adapted to the plant resistance mechanism to increase in abundance from a previously low gene frequency. Such breakdown of plant resistance by an adapted biotype has long been a serious problem with plant pathogens (e.g. rust in cereals).

The famous example with insects is the release of 'IR26', a rice variety resistant to brown planthopper, developed at the Rice Research Institute in the Philippines and planted over large areas of southeast Asia. The resistance was based on a single gene, and adapted biotypes appeared within 2–3 years. Yet against such examples can be set others where HPR to insects has lasted for a long time. The resistance to phylloxera of American rootstocks used for grafting with European grape varieties lasted in Europe from the late 19th century until a new adapted phylloxera biotype was found in a small area of Germany in 1994 (Anonymous, 1994), and wheat varieties resistant to Hessian fly in the USA have been used successfully for over 50 years, in spite of the existence (known since 1930) of biotypes with some adaptation to the resistant varieties.

Biotype problems have arisen remarkably rarely with HPR to insects. Table 7.4 lists reported cases. It is almost certainly not complete, but many of the examples relate to laboratory testing of different sources of the pest rather than to the breakdown of HPR in the field.

The relative rarity of the biotype problem with HPR to insects probably has several explanations. Antibiosis is often associated with antixenosis, so that some insects rejecting the plants may find a host plant elsewhere, perhaps outside the crop. Secondly, traditional plant breeding rarely produced varieties with single mechanisms of resistance. Finally, insect generation times are long when compared with those of plant pathogens, giving fewer opportunities per season for selection of adapted biotypes.

Just as there are strategies for managing the resistance of pests to insecticides, so there are others for managing biotypic breakdown of HPR, especially if the latter is based on single genes. These include pyramiding and various forms of deployment (in time, space or as mixtures) of varieties with similar agronomic characters but differing in the resistance genes they carry.

Deleterious effects on natural enemies

Some secondary plant compounds in resistant varieties show toxicity to parasitoids while within their hosts or to predators (Herzog and Funderburk, 1985). Table 7.5 gives some examples. The transfer of allelochemicals has so far been the principal approach of transgenic HPR, so these examples sound a warning. However, Herzog and Funderburk (1985) also quote several examples where no deleterious effects on natural enemies occurred. There is also one example (of DIMBOA in wheat and the coccinellid *Eriopis connexa* preying on *Rhopalosiphum padi*) which shows that strong allelochemically based resistance may paradoxically protect the natural enemy. The aphid reduced its feeding rate on high DIMBOA varieties and became less toxic to the predator (Martos et al., 1992).

Even without allelochemicals as the basis for HPR, deleterious effects on parasitoids may still be found, since their size and fecundity is usually correlated with the size of the prey, which tends to be smaller on resistant varieties. Jan Salim and H.F. van Emden

Table 7.4   Insect biotypes and plant resistance

Insect	Crop	No. of biotypes	Reference
Acryrthosiphon pisum (pea aphid)	Alfalfa	4	Cartier et al., 1965; Frazer, 1972, Auclair, 1978
	Pea	5	Frazer, 1972
	Pea	3	Markkula and Roukka, 1971
Amphorophora rubi (*Rubus* aphid)	Raspberry	4	Keep and Knight, 1967
Aphis craccivora (groundnut aphid)	Cowpea	2	Ansari, 1984
Brevicoryne brassicae (cabbage aphid)	Sprouts	2	Lammerink, 1968
	Sprouts	7	Dunn and Kempton, 1972
Dysaphis devecta (rosy leaf curling aphid)	Apple	3	Briggs and Alston, 1969
Eriosoma lanigerum (woolly aphid)	Apple	3	Briggs and Alston, 1969
	Apple	3	Sen Gupta, 1969
	Apple	?	Knight et al., 1962
Mayetiola destructor (hessian fly)	Wheat	9	Everson and Gallun, 1980
Nephotettix cinctipes (green rice leafhopper)	Rice	2	Sato and Sogawa, 1981
Nephotettix virescens (green leafhopper)	Rice	3	Heinrichs and Rapusas, 1985; Takita and Hashim, 1985
Nilapavarta lugens (brown planthopper)	Rice	4	Pathak and Saxena, 1980
Pachydiplosis oryzae (rice gall midge)	Rice	4	Pathak and Saxena, 1980
Rhopalosiphum maidis (corn leaf aphid)	Sorghum and maize	5	Painter and Pathak, 1962; Wilde and Feese, 1973
Schizaphis graminum (greenbug)	Wheat	8	Wood et al., 1969; Puterka et al., 1988
Sitobion avenae	Wheat	3	Lowe, 1981
Therioaphis trifolii (spotted alfalfa aphid)	Alfalfa	9	Nielson and Don, 1974; Manglitz et al., 1966
Viteus vitifolii (phylloxera)	Grapes (USA)	2	Stevenson, 1970
	Grapes (Europe)	2	Anonymous, 1994

(unpublished data, 1994) have reared the parasitoid *Aphidius rhopalosiphi* on the aphid *Metopolophium dirhodum* on susceptible and partially resistant wheat varieties for 10 generations. The size of the parasitoid was reduced by 11% in the first generation and the reduction had reached 15% by F_{10}. Equivalent figures for fecundity were 21 and 56%, respectively. However, even after seven generations, a return of the culture to the susceptible variety immediately restored parasitoid performance. Nevertheless, such results confirm that there is little merit in seeking higher levels of HPR than necessary, especially in an IPM context.

Spread of plant diseases

Pests on resistant plants can be expected to be more restless, and arriving pests to test more plants. This would suggest that the spread of plant diseases vectored by insects

Table 7.5 Examples of plant-derived compounds toxic to natural enemies *via* their prey

Prey	Natural enemy	Plant (allelochemical)	Reference
Homoptera (Aphididae)			
Aphis magnoliae	*Coccinella septempunctata*	Elder (sambunigrin)	Hodek, 1967
Aphis nerii	*Propylaea 14-punctata*	*Asclepias curassavica* and *Nerium oleander* (cardenolides)	Rothschild et al., 1970
Aphis sambuci	*Chrysoperla carnea* *C. septempunctata* *Metasyrphus corollae*	Elder (sambunigrin)	Philippe, 1972 Hodek, 1956 Ruzicka, 1975
Macrosiphum albifrons	*Carabus problematicus*	Lupin (quinolizidine alkaloids)	Wink and Römer, 1986
Rhopalosiphum padi	*Eriopis connexa*	Maize (DIMBOA)	Martos et al., 1992
Lepidoptera			
Anticarsia gemmatalis	*Nomuraea rileyi*	Soybean (?)	Oliveira, 1981
Heliocoverpa zea	*Bacillus thuringiensis* *Hyposoter exiguae*	Soybean Tomato (α-tomatine)	Bell, 1978 Campbell and Duffey, 1979
	Microplitis croceipes	Soybean (?)	Powell and Lambert, 1984
Manduca sexta	*Cotesia congregata* and *Hyposoter annulipes*	Tobacco (nicotine)	Barbosa et al., 1986; Thorpe and Barbosa, 1986
Pseudoplusia includens	*Copidosoma truncatellum*	Soybean	Herzog and Funderburk, 1985

(primarily in a non-persistent manner) might be more rapid on resistant than on suscep-tible crop varieties (as proposed by van Marrewijk and de Ponti, 1975).

The only experimental evidence for a positive effect of HPR on disease spread comes from a cage experiment by Atiri et al. (1984) on cowpea mosaic virus vectored by *Aphis craccivora*. In contrast, all the field evidence shows that HPR can effect some control of insect-vectored plant diseases. This has been shown for sugarbeet and yellows virus (Lowe, 1975), for broad beans and bean yellow mosaic virus (Hagel et al., 1972) and for raspberries and mosaic virus (Knight et al., 1959).

HOST-PLANT RESISTANCE IN THE CONTEXT OF INTEGRATED PEST MANAGEMENT (IPM)

HPR is unlikely to be effective against all the organisms attacking a crop. Its use in IPM will therefore be either as an effective target-specific measure against one or more pests or as part of a diversified attack strategy against a single pest.

Use as a target-specific measure

High levels of plant resistance have been used in IPM programmes to replace broad-spectrum insecticides for the control of single major pests of a crop (e.g. Hessian fly in wheat in the USA). Decreased insecticide use allows more biological control to operate, not only of the pest which is the target for the plant resistance, but also of other pests. Thus the crop system begins on an 'up escalator', with further reductions in pesticide inputs progressively becoming possible.

In the integrated approach, therefore, the target for HPR will often by the key pest which accounts for the bulk of insecticide use. However, there is an alternative strategy. This is to develop plant resistance against those target pests which make it possible to delay pesticide application for as long as possible at the beginning of the season, as in developing HPR for insect pests of cowpeas (Singh, 1978).

Use as part of a diversified attack strategy against a single pest

'Overdosing' is a charge usually levelled at those relying on agrochemicals for their pest control. Yet the charge can equally be levelled at any over-use of one pest control method to the point where there are undesirable side-effects.

Attention has already been drawn to the possibilities for overdosing plant resistance. Strong plant resistance, especially if allelochemically based, is the form of HPR most likely to incur yield drag, damage biological control and lead to the appearance of adapted biotypes of the pest. To these side-effects must be added the possible increase in tolerance to insecticides of pests on plants highly resistant to the pest because of allelochemicals (see below). However, positive synergism between plant resistance and other control measures employed in IPM will be emphasised in this section. A key principle of IPM is that adaptation to any single pest control measure can be delayed or avoided by diversifying the selection pressures against which the pest must counter-adapt. Clearly there is little point in combining a pest control measure with other measures if that one measure is already effective on its own, and this argues for partial rather than high levels of plant resistance. Further, if such partial plant resistance is to be combined with other measures for IPM, then it is most sensible to look to those measures where the combined effect with plant resistance is likely to be greater than the sum of the combined effects. Two measures which have often shown positive synergism with partial HPR are insecticides and biological control.

Partial plant resistance and insecticides

Toxicity to pesticides of an organism is a function of body weight. Since one effect of antibiotic plant resistance is to reduce the body size of pests, we might expect them to become more susceptible to pesticide. Some comparisons of the susceptibility to pesticide of sucking and chewing pests on antibiotic and susceptible plants are given in Figure 7.3. Most of the data come from laboratory studies, but there is also one result from a pesticide application in the field. The reductions in LC50 vary from about 25 to 75%. However, the effect remains almost as strong if the data are corrected for differences in body weight (LD50). Some stress of HPR on the pest, such as poorer nutrition, appears to be far more important than body weight changes.

There are also examples where pests on resistant plants are more resistant to pesticides (e.g. Kea et al., 1978, on soybeans; Heinrichs et al., 1984, on rice). The soybean resistance to lepidopterous larvae is known to be allelochemically based. Brattsen et al. (1977)

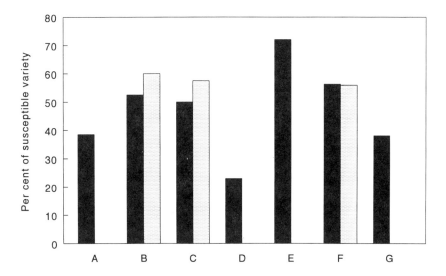

Figure 7.3   (A–F) LC50 (black) and LD50 (stippled) to insecticides of pests on resistant varieties as per cent of that on susceptible ones: (A) *Myzus persicae* and malathion on chrysanthemums (Selander et al., 1972); (B) *Myzus persicae* and malathion on Brussels sprouts (Mohamad and van Emden, 1989); (C) *Metopolophium dirhodum* and malathion on wheat (Attah and van Emden, 1993); (D) *Plutella xylostella* and abamectin and (E) cypermethrin on cabbage (Abro and Wright, 1989); (F) *Phaedon cochliariae* on brassicas (Smith, 1990); (G) Survival of *Empoasca* spp. on resistant cowpeas in the field after spraying 50 g a.i. (grams of active ingredient) ha^{-1} dimethoate as per cent of the survival on a susceptible variety (Raman, 1977)

explained such examples in terms of an induction by secondary plant metabolites of insecticide-detoxifying mixed function oxidases in the pest.

Partial plant resistance and biological control

The possible deleterious effects on natural enemies of toxic allelochemicals present in resistant plants have already been mentioned. Positive synergism between HPR and biological control shows as 'extrinsic resistance', referred to earlier as one of the mechanisms of HPR.

Some examples of extrinsic resistance are highly crop-specific. One concerns the nectary-free cotton varieties in Australia (Adjei-Maafo, 1980), on which a phytophagous ladybird (*Cryptolaemus* sp.) which uses the nectar from the nectaries on normal varieties as its main food source changes its food habits on nectary-free varieties and becomes a predator on the immature stages of bollworm.

Another example comes from the USA and concerns the 'leafless' varieties of pea bred for resistance to mildew, and showing only about 8% intrinsic resistance to pea aphids. However, the difference in aphid populations on normal and leafless varieties in the field is far greater when ladybird predators are present. These find it hard to keep their footing on the smooth leaves of normal plants, and a high percentage fall. On the leafless plants, in contrast, the enlarged stipules and tendrils provide a sure footing, and predation of the aphids is more intense (Kareiva and Sahakian, 1990).

Additionally, there are generalised effects on biological control that are the conse-

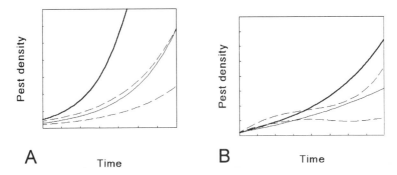

Figure 7.4   Pest population curves on susceptible (solid lines) and partially resistant (broken lines) varieties with (lower line) and without (upper line) biological control. *Sources:* (A) Theoretical relationships as proposed by van Emden and Wearing (1965); (B) data for *Schizaphis graminum* and *Lysiphlebus testaceipes* from Starks et al. (1972)

quences of a slower rate of increase on partially resistant crop varieties, and the fewer and smaller pest individuals that result.

Synergistic effects were first proposed by van Emden and Wearing (1965). Their simple model suggested that the combination of partial plant resistance and biological control, both quite insufficient on their own to prevent the pest population rising, could together exert economic control (Figure 7.4A). The first experimental evidence that this could occur came completely by accident. Wyatt (1970) was screening chrysanthemum varieties for resistance to the aphid *Myzus persicae* and finding no variety sufficiently resistant to reduce the aphid to economically acceptable levels. However, the accidental appearance in the glasshouse of the parasitoid *Aphidius matricariae* did result in acceptable levels of biological control, but only on partially resistant varieties. The next confirmation of van Emden and Wearing's (1965) model came when Starks et al. (1972) published graphs from an aphid:parasitoid system on sorghum varieties which were almost identical to the model (Figure 7.4B). When 12 *Schizaphis graminum*, with or without one parasitoid (*Lysiphlebus testaceipes*), were introduced per plant, the aphids were only controlled in the combination of resistant variety with parasitoid.

van Emden (1987) elaborated on the van Emden and Wearing model by identifying three possible outcomes of biological control:partial HPR interaction, using Hassell's (1975) density-dependent population model as the starting point (Figure 7.5). Hassell's model was used to produce population growth curves with two differing potential growth rates (S, hypothetical susceptible variety; R, hypothetical variety with a low level of resistance), as well as a curve (S+) for the susceptible variety with an arbitrary and low level of density-dependent mortality (biological control). One possible outcome of biological control on the resistant variety is shown by the line R + p, which reflects the same proportional mortality from biological control as on the susceptible variety. The other two possibilities are described below.

(1) Less than proportional mortality on the resistant variety. The line R + models this as the expected mortality from the density-dependence model.

(2) More than proportional mortality on the resistant variety. The line R + a models this as the occurrence of the same absolute numerical mortality on both varieties.

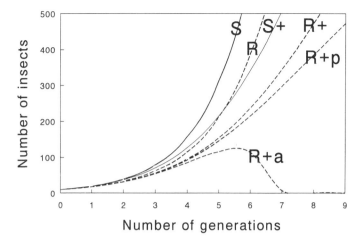

Figure 7.5   Theoretical insect population growth curves arising from the possible interactions between partial plant resistance and biological control on susceptible and resistant varieties (S and R, respectively) without biological control; S+ and R+, the same growth curves, but with density-dependent mortality; R+p and R+a, curves on the resistant variety with, respectively, identical proportional and numerical mortality from biological control as on the susceptible variety

The latter represents the greatest synergism between partial HPR and biological control, and leads to a rapid decline of the population. Table 7.6 shows data from a number of field or glasshouse studies where the degree of partial plant resistance and the amount of biological control on the susceptible variety can be calculated. The predictions of the expected population on the resistant variety on the extreme assumption of the same absolute mortality on susceptible and resistant varieties are not very different from the observed values, indicating strong synergism between HPR and biological control.

Some explanations for such dramatic synergism have been proposed, as described below, although sampling twice with an interval between will itself lead to some magnification of the difference between pest populations on the two varieties, even in the absence of biological control.

Numerical responses of natural enemies   Many natural enemies, particularly parasitoids, respond positively to host-plant odours from a distance. Any lower pest densities on resistant varieties will therefore not decrease the numbers of natural enemies arriving. It is important that resistant varieties should liberate the same odours at about the same concentrations as other varieties. There is the danger that varieties with a resistance based on allelochemicals may not be attractive to natural enemies. van Emden (1978) reported that a Brussels sprout variety partially resistant to *Brevicoryne brassicae* carried more aphids than the susceptible variety in glasshouse experiments when the parasitoid *Diaeretiella rapae* was present. The parasitoid normally develops on plants containing mustard oils, and hardly visited the aphid-resistant variety with lower concentrations of such volatiles; parasitisation of the aphids was therefore very low.

Functional responses of natural enemies   The fact that pests on resistant varieties are likely to be smaller than those on susceptible ones means that a predator will eat more

Table 7.6 Expected per cent reductions in pest populations on resistant varieties (on the assumption of equal numerical mortality on susceptible and resistant varieties) and experimentally observed reductions

Crop plant	Per cent plant resistance	Percent biological control on susceptible plant	Reduction		Reference
			Expected	Observed	
Barley	44	40	92	98	1
Cereals	26	70	80	64	2
Wheat	38	23	80	86	3
Wheat	12	44	70	93	3
Brassicas	14	20	40	92	3
Brassicas	18	14	35	64	4
Brassicas	30	16	22	43	5

Calculated from data in the following references.
1. Starks et al. (1972), *Schizaphis graminum* parasitised by *Lysiphlebus testaceipes*.
2. Lykouressis (1982), *Sitobion avenae* parasitised by *Aphelinus abdominalis*.
3. Gowling and van Emden (1994), *Metopolophium dirhodum* parasitised by *Aphidius rhopalosiphi* and *Brevicoryne brassicae* and natural predation.
4. Dodd (1973), *Brevicoryne brassicae* and natural predation.
5. Bogahawatte (1993), *Plutella xylostella* and natural parasitisation, especially by *Cotesia plutellae*.

individuals on the resistant variety before satiation. Another effect of the smaller size of pests on partially resistant varieties is that they may be less able to escape from parasitoids or predators by kicking or running away.

Another aspect of synergism in the functional response relates to the dislodgement of prey from plants by natural enemies. That the proportion of prey dislodged is higher on resistant than on susceptible varieties, probably because the pests are more restless there, has been shown for both caterpillars and aphids (Bogahawatte, 1993; Gowling and van Emden, 1994).

Interactions in relation to biotypes adapted to the resistance

If pest biotypes with different levels of adaptation to the plant resistance co-exist on the resistant variety, then only those poorly adapted to the resistance will suffer from any synergism with a second mortality factor. Where there is positive synergism between partial plant resistance and biological control, HPR-adapted biotypes will be selected faster in the presence of biological control than in its absence (Gould et al., 1991). This applies equally to other synergistic combinations between control measures, and some workers have used Gould et al.'s paper to throw doubt on IPM in general. However, the practical contrast is not between partial plant resistance alone and in combination with another factor, but between the same level of control given either by high plant resistance alone or by the combination of less plant resistance with the other factor. Adaptation will be considerably faster to the high level of plant resistance alone (Gould et al., 1991).

The 3-way interaction between plant resistance, biological control and insecticides

Positive synergism between partial plant resistance and insecticides takes the form (see earlier) that it may be possible to achieve the same kill on a resistant variety with a lower

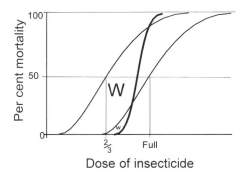

Figure 7.6   Theoretical selectivity of pesticide application on a resistant as compared with a susceptible plant variety. Thick curve, dose–mortality response of natural enemy; thin lines, dose–mortality response of pest at (right) full dose on the susceptible variety and (left) reduced dose on the resistant variety. Small and large 'w' show the respective selectivity windows

dose of insecticide than would be required on a susceptible variety.

The potential then exists for greater selectivity of kill in favour of natural enemies, because the slope of the dose–mortality response of the natural enemy will often be steeper than that for the pest. This implies less variation in insecticide tolerance in carnivores than in herbivores, which Plapp (1981) attributed to the need for herbivores to possess an armoury of oxidative enzymes evolved for metabolising plant toxins and also effective for detoxifying insecticides.

The differing steepness of the dose–mortality response (Figure 7.6) means that the ratio of per cent kill of pest to natural enemies (and therefore selectivity) can increase with reduction in pesticide dose, even on susceptible varieties. Partial plant resistance will often allow the insecticide dose to be reduced by one-third, moving the pest's dose–mortality curve to the left, as shown in Figure 7.6. The curve for a natural enemy should move to the left by only a small amount if pests are smaller, and then only for parasitoids. The displacement of the two curves with partial plant resistance should therefore widen the window of selectivity in favour of the natural enemy.

CONCLUSIONS

Plant resistance is a valuable part of the armoury of control measures other than insecticides. However, like insecticides, it is not free from side-effects, and it may be unwise to look on HPR as a sole control measure to replace insecticides. It is multi-component IPM, which may include some insecticide use, which is the true alternative. This chapter has sought to emphasise two aspects of the role of HPR in IPM.

(1) In a multi-component pest management system, HPR has unique abilities to synergise with biological control and insecticides.

(2) There are two types of plant resistance in terms of their compatibility with IPM. HPR can contribute to pest mismanagement (van Emden, 1991) as well as to pest management. On the one hand, high levels of plant resistance, especially if based on a single allelochemical, are inimical to IPM in that they are most likely to lead to the appearance of adapted biotypes, will damage biological control and may induce

insecticide-detoxifying enzymes in the pest. Unfortunately, this is the kind of HPR which is most likely to result from genetic transformation approaches. However, this is not inevitable if the potential problems are consciously avoided. Transgenic plants can have several mechanisms, and allelochemicals can be concentrated in particular organs or at particular times in the phenology of the plant. Also, the resistance system can be made pest-inducible rather than always being present. More partial, polygenic 'horizontal' resistance has quite other properties. Pests on such varieties will be easier to kill with insecticide and the impact of biological control will probably be increased.

Unfortunately, the impressive resources available to develop HPR in the private sector, and which lead to profits for plant-breeding companies including multi-national companies, are likely to produce the first type of plant resistance, not the second. Individuals or organisations which supply multi-component IPM packages are not often in a position to develop suitable varieties; hence the emphasis in this chapter on the identification of suitable levels of resistance in varieties which are already commercially available.

REFERENCES

Abro, G.H. and Wright, D.J. (1989) Host plant preferences and the influence of different cabbage cultivars on the toxicity of abamectin and cypermethrin against *Plutella xylostella* Lepidoptera: Plutellidae. *Annals of Applied Biology* **115**, 481–487.

Adjei-Maafo, I.K. (1980) Effects of nectariless cotton trait on insect pests, parasites and predators with special reference to the effects on the reproductive characters of *Heliothis* spp. Ph.D. Thesis, University of Queensland.

Ames, B.N., Profet, M. and Gold, L.S. (1990) Dietary pesticides (99.99% all natural). *Proceedings of the National Academy of Sciences, USA* **87**, 7777–7781.

Anonymous (1994) Die Rückkehr der Reblaus, *Profil, October 1994*, p. 11.

Ansari, A.F. (1984) Biology of *Aphis craccivora* (Koch) and varietal resistance of cowpeas. Ph.D. Thesis, University of Reading.

Atiri, G.I., Ekpo, E.J.A. and Thottappilly, G. (1984) The effect of aphid-resistance in cowpea on infestation and development of *Aphis craccivora* and the transmission of cowpea aphid-borne mosaic virus. *Annals of Applied Biology* **104**, 339–346.

Attah, K.P. and van Emden, H.F. (1993) The susceptibility to malathion of *Metopolophium dirhodum* on two wheat species at two growth stages, and the effect of plant growth regulators on this susceptibility. *Insect Science and its Application* **14**, 101–106.

Auclair, J.L. (1978) Biotypes of the pea aphid *Acyrthosiphon pisum* in relation to host plants and chemically defined diets. *Entomologia Experimentalis et Applicata* **24**, 212–216.

Barbosa, P., Saunders, J.A., Kemper, J., Trumbule, R., Olechno, J. and Martinat, P. (1986) Plant allelochemicals and insect parasitoids. Effect of nicotine on *Cotesia congregata* (Say.) (Hymenoptera: Braconidae) and *Hypsoter annulipes* (Cresson) (Hymenoptera: Ichneumonidae). *Journal of Chemical Ecology* **12**, 1319–1328.

Bell, J.V. (1978) Development and mortality in bollworms fed resistant and susceptible soybean cultivars treated with *Nomuraea rileyi* or *Bacillus thuringiensis*. *Journal of the Georgia Entomological Society* **13**, 50–55.

Bogahawatte, C.N.L. (1993) Glasshouse and field studies on diamondback moth *Plutella xylostella* (Lepidoptera, Yponomeutidae) on host-plant resistance in *Brassica* and biological control. Ph.D. Thesis, University of Reading.

Brattsen, L.B., Wilkinson, C.F. and Eisner, T. (1977) Herbivore–plant interactions: Mixed function oxidases and secondary plant substances. *Science, New York* **196**, 1349–1352.

Briggs, J.B. and Alston, F.H. (1969) Sources of pest resistance in apple cultivars. *Report of East Malling Research Station, 1996*, pp. 170–171.

Campbell, B.C. and Duffey, S.S. (1979) Tomatine and parasitic wasps: Potential incompatibility of plant antibiosis with biological control. *Science, New York* **205**, 700–702.

Cartier, J.J., Isaak, A., Painter, R.H. and Sorensen, E.L. (1965) Biotypes of pea aphid *Acyrthosiphon pisum* (Harris) in relation to alfalfa clones. *Canadian Entomologist* **97**, 754–760.

Chiarappa, L. (1971) (+ later supplements) *Crop Loss Assessment Methods*. Commonwealth Agricultural Bureaux, Farnham Royal (loose-leaved).

Dodd, G.D. (1973) Integrated control of the cabbage aphid (*Brevicoryne brassicae* (L.)). Ph.D. Thesis, University of Reading.

Dunn, J.A. and Kempton, D.P.H. (1972) Resistance to attack by *Brevicoryne brassicae* (L.) on Brussels sprouts. *Annals of Applied Biology* **72**, 1–11.

Ehrlich, P.R. and Raven, P.H. (1964) Butterflies and plants: A study in coevolution. *Evolution* **18**, 586–608.

Everson, E.H. and Gallun, R.L. (1980) Breeding approaches in wheat. In F.G. Maxwell and P.R. Jennings (Eds) *Breeding Plants Resistant to Insects*. Wiley, New York, pp. 513–533.

Feeney, P. (1976) Plant apparency and chemical defense. *Recent Advances in Phytochemistry* **10**, 1–40.

Flor, H.H. (1942) Inheritance of pathogenicity in *Melampsora lini*. *Phytopathology* **32**, 653–659.

Frazer, B.D. (1972) Population dynamics and recognition of biotypes in the pea aphid (Homoptera: Aphididae). *Canadian Entomologist* **104**, 1729–1733.

Gallun, R.L. (1977) The genetic basis of hessian fly epidemics. In P.R. Day (Ed.) *The Genetic Basis of Epidemics in Agriculture. Annals of the New York Academy of Sciences*, **287**, 1–400.

Gershenzon, J. (1994) The cost of plant chemical defense against herbivory: A biochemical perspective. In E.A. Bernays (Ed.) *Insect–Plant Interactions*. Vol. 5. CRC Press, Boca Raton, FL, pp. 173–205.

Gould, F., Kennedy, G.G. and Johnson, M.T. (1991) Effects of natural enemies on the rate of herbivore adaptation to resistant host plants. *Entomologia Experimentalis et Applicata* **58**, 1–14.

Gowling, G.R. and van Emden, H.F. (1994) Falling aphids enhance impact of biological control by parasitoids on partially aphid-resistant varieties. *Annals of Applied Biology* **125**, 233–242.

Hagel, G.T., Silbernagel, M.J. and Burke, D.W. (1972) Resistance to aphids, mites, and thrips in field beans to infection by aphid-borne viruses. *Bulletin of the USDA Agricultural Research Service*, No. 33-139, 4 pp.

Harlan, J.R. and Starks, K.J. (1980) Germplasm resources and needs. In F.G. Maxwell and P.R. Jennings (Eds) *Breeding Plants Resistant to Insects*. Wiley, New York, pp. 253–273.

Hassell, M.P. (1975) Density-dependence in single-species populations. *Journal of Animal Ecology* **44**, 283–295.

Heinrichs, A.E., Fabellar, L.T., Basilio, R.P., Tu Cheng Wen and Medrano, F. (1984) Susceptibility of rice planthoppers *Nilapavarta lugens* and *Sogatella furcifera* (Homoptera: Delphacidae) to insecticides as influenced by level of resistance in the host plant. *Environmental Entomology* **13**, 455–458.

Heinrichs, A.E. and Rapusas, H.R. (1985) Cross virulence of *Nephotettix virescens* (Homoptera; Cicadellidae) biotypes among some rice cultivars with the same major resistance gene. *Environmental Entomology* **14**, 696–700.

Herzog, D.C. and Funderburk, J.E. (1985) Plant resistance and cultural practice interactions with biological control. In M.A. Hoy and D.C. Herzog (Eds) *Biological Control in Agricultural IPM Systems*. Academic Press, Orlando, FL, pp. 67–88.

Hodek, I. (1956) The influence of *Aphis sambuci* L. as prey of the ladybird beetle *Coccinella septempunctata* L. *Acta Societatis Zoologicae Bohemoslovaca* **20**, 62–74 (in Czech.).

Hodek, I. (1967) Bionomics and ecology of predaceous Coccincllidae. *Annual Review of Entomology* **12**, 79–104.

Hodkinson, I.D. and Hughes, M.K. (1982) *Insect Herbivory*. Chapman & Hall, London, 77 pp.

Herrera-Estrella, L., Van Den Broeck, G., Maenhaut, R., Van Montagu, M., Schell, J., Timko, J.M. and Cashmore, A. (1984) Light-inducible and chloroplast associated expression of a chimaeric gene introduced into *Nicotiana tabacum* using a Ti plasmid vector. *Nature, London* **310**, 115–120.

IBPGR (1984) *Institutes Conserving Crop Germplasm: The IBPGR Global Network of Genebanks*. IBPGR, Rome, 25 pp.

ICRISAT (1980) *Annual Report 1978–79*. ICRISAT, Patancheru, 288 pp.

Kareiva, P. and Sahakian, R. (1990) Tritrophic effects of a simple architectural mutation in pea plants, *Nature, London* **345**, 433–434.

Kea, W.C., Turnipseed, S.G. and Carner, G.R. (1978) Influence of resistant soybeans on the susceptibility of lepidopterous pests to insecticides. *Journal of Economic Entomology* **71**, 58–60.

Keep, E. and Knight, R.L. (1967) A new gene from *Rubus occidentalis* L. for resistance to strains 1, 2 and 3 of the *Rubus* aphid, *Amphorophora rubi* Kalt. *Euphytica* **16**, 209–214.

Kennedy, G.G. (1986) Consequences of modifying biochemically mediated insect resistance in *Lycopersicon* species. In M.B. Green and P.A. Hedin (Eds) *Natural Resistance of Plants to Pests*. American Chemical Society, Washington, DC, pp. 130–141.

Knight, R.L., Briggs, J.B., Massee, A.M. and Tydeman, H.M. (1962) The inheritance of resistance to woolly aphid, *Eriosoma lanigerum* (Hsmnn.), in the apple. *Journal of Horticultural Science* **37**, 207–218.

Knight, R.L., Keep, E. and Briggs, J.B. (1959) Genetics of resistance to *Amphorophora rubi* (Kalt.) in the raspberry. *Journal of Genetics* **56**, 261–280.

Kogan, M. and Ortman, E.F. (1978) Antixenosis—a new term proposed to define Painter's 'non-preference' modality of resistance. *Bulletin of the Entomological Society of America* **24**, 175–176.

Lammerink, J. (1968) A new biotype of cabbage aphid (*Brevicoryne brassicae* (L.)) on aphid-resistant rape (*Brassica napus* L.). *New Zealand Journal of Agricultural Research* **11**, 341–344.

Lowe, H.J.B. (1975) Infestation of aphid-resistant and susceptible sugar beet by *Myzus persicae* in the field. *Zeitschrift für Angewandte Entomologie* **79**, 376–383.

Lowe, H.J.B. (1981) Resistance and susceptibility to colour forms of the aphis *Sitobion avenae* in spring and winter wheats (*Triticum aestivum*). *Annals of Applied Biology* **99**, 87–98.

Lykouressis, D. (1982) Studies under controlled conditions on the effects of parasites on the population dynamics of *Sitobion avenae* (F.). Ph.D. Thesis, University of Reading.

Manglitz, G.R., Calkins, C.O., Walstrom, R.J., Hintz, S.D., Kindler, S.D. and Peters, L.L. (1966) Holocyclic strains of the spotted alfalfa aphid in Nebraska and adjacent states. *Journal of Economic Entomology* **59**, 636–639.

Markkula, M. and Roukka, K. (1971) Resistance of plants to the pea aphid *Acyrthosiphon pisum* Harris (Hom. Aphididae). III. Fecundity on different pea varieties. *Annales Agriculturae Fenniae* **10**, 33–37.

Martos, A., Givovich, A. and Niemeyer, H. (1992) Effect of DIMBOA, an aphid resistance factor in wheat, on the aphid predator *Eriopis connexa* Germar (Col.: Coccinellidae). *Journal of Chemical Ecology* **18**, 469–479.

Mohamad, B.M. and van Emden, H.F. (1989) Host-plant modification to insecticide susceptibility in *Myzus persicae* (Sulz.). *Insect Science and its Application* **10**, 699–703.

Müller, H.J. (1958) The behaviour of *Aphis fabae* in selecting its host plants, especially different varieties of *Vicia faba*. *Entomologia Experimentalis et Applicata* **1**, 66–72.

Nielson, M.W. and Don, H. (1974) A new virulent biotype of the spotted alfalfa aphid in Arizona. *Journal of Economic Entomology* **67**, 64–66.

Oliviera, E.B. (1981) Effects of resistant and susceptible soybean genotypes at different phenological stages on development, leaf consumption and oviposition of *Anticarsia gemmatalis* Hubner. M.S. Thesis, University of Florida.

Painter, R.H. (1951) *Insect Resistance in Crop Plants*. Macmillan, New York, 520 pp.

Painter, R.H. and Pathak, M.D. (1962) The distinguishing features and significance of the four biotypes of the corn leaf aphid, *Rhopalosiphum maidis* (Fitch). *Proceedings of the 11th International Congress of Entomology, Vienna, 1960*, Vol. 2, (Austrian Entomological Society, Vienna, pp. 110–115.

Panda, N. and Khush, G.S. (1995) *Host-Plant Resistance to Insects*. CAB International, Wallingford, 431 pp.

Pathak, M.D. and Saxena, R.C. (1980) Breeding approaches in rice. In F.G. Maxwell and P.R. Jennings (Eds) *Breeding Plants Resistant to Insects*. Wiley, New York, pp. 421–455.

Philippe, R. (1972) Biologie de la reproduction de *Chrysopa perla* (L.) (Neuroptera, Chrysopidae) en fonction d'alimentation imaginale. *Annales de Zoologie, Ecologie Animale* **4**, 213–227.

Plapp, F.W., Jr. (1981) Ways and means of avoiding or ameliorating resistance to insecticides. *Proceedings of Symposia, 9th International Congress of Plant Protection, Washington, 1979*, Vol. 1. Burgess, Minneapolis, MN. pp. 244–249.

Powell, J.E. and Lambert, L. (1984) Effects of three resistant soybean genotypes on development of

Microplitis croceipes and leaf consumption by its *Heliothis* spp. hosts. *Journal of Agricultural Entomology* 1, 169–176.

Puterka, G.J., Peters, D.C., Kerns, D.L., Slosser, J.E., Bush, L., Worrall, D.W. and McNew, R.W. (1988) Designation of two new greenbug (Homoptera: Aphididae) biotypes G and H. *Journal of Economic Entomology* 81, 1754–1759.

Raman, K.V. (1977) Studies on host-plant resistance of cowpeas to leafhoppers. Ph.D. Thesis, University of Reading.

Rothschild, M., von Euw, J. and Reichstein, T. (1970) Cardiac glycosides in the oleander aphid *Aphis nerii*. *Journal of Insect Physiology* 16, 1141–1145.

Ruzicka, Z. (1975) The effects of various aphids as larval prey on the development of *Metasyrphus corollae* (Dipt., Syrphidae). *Entomophaga* 20, 393–402.

Sato, A. and Sogawa, K. (1981) Biotypic variations in the green rice leafhopper, *Nephotettix virescens* (Uhler) (Homoptera: Deltocephalidae). *Japanese Journal of Applied Entomology and Zoology* 16, 55–57.

Selander, J.M., Markkula, M. and Tiittanen, K. (1972) Resistance of the aphids *Myzus persicae* (Sulz.), *Aulacorthum solani* (Kalt.) and *Aphis gossypii* Glov. to insecticides and the influence of the host plant on this resistance. *Annales Agriculturae Fenniae* 11, 141–145.

Sen Gupta, G.C. (1969) The recognition of biotypes of the woolly aphid, *Eriosoma lanigerum* (Hausmann), in South Australia by their differential ability to colonise varieties of apple rootstock, and an investigation of some possible factors in the susceptibility of varieties to these insects. Ph.D. Thesis, University of Adelaide.

Singh, S.R. (1978) Resistance to pests of cowpea in Nigeria. In S.R. Singh, H.F. van Emden and T.A. Taylor (Eds) *Pests of Grain Legumes: Ecology and Control*. Academic Press, London, pp. 267–279.

Smith, L.J. (1990) The host plant modification of malathion tolerance of *Phaedon cochleariae* and *Myzus persicae* including reference to the third trophic level. Ph.D. thesis, University of Reading.

Southwood, T.R.E. (1973) The insect/plant relationship—an evolutionary perspective. In H.F. van Emden (Ed.) *Insect/Plant Relationships*. Blackwell Scientific, Oxford, pp. 3–30.

Starks, K.J., Muniappan, R. and Eikenbary, R.D. (1972) Interaction between plant resistance and parasitism against the greenbug on barley and sorghum. *Annals of the Entomological Society of America* 65, 650–655.

Stevenson, A.B. (1970) Strains of the grape phylloxera in Ontario with different effects on the foliage of certain grape cultivars. *Journal of Economic Entomology* 63, 135–138.

Takita, T. and Hashim, H. (1985) Relationship between laboratory-developed biotypes of green leafhopper and resistant varieties of rice in Malaysia. *Japanese Agriculture Research Quarterly* 19, 219–223.

Thorpe, K.W. and Barbosa, P. (1986) Effects on consumption of high and low nicotine tobacco by *Manduca sexta* (Lepidoptera: Sphingidae) on survival of the gregarious endoparasitoid *Cotesia congregata* (Hymenoptera: Braconidae). *Journal of Chemical Ecology* 6, 1329–1337.

van den Bosch, R. and Messenger, P.S. (1973) *Biological Control*. Intext, New York, 137 pp.

van der Plank, J.E. (1963) *Plant Diseases: Epidemics and Control*. Academic Press, New York, 349 pp.

van Emden, H.F. (1978) Insects and secondary plant substances—an alternative viewpoint with special reference to aphids. In J.B. Harborne (Ed.) *Biochemical Aspects of Plant and Animal Coevolution*. Academic Press, London, pp. 309–323.

van Emden, H.F. (1987) Cultural methods: The plant. In A.J. Burn, T.H. Coaker and P.C. Jepson (Eds) *Integrated Pest Management*. Academic Press, London, pp. 27–68.

van Emden, H.F. (1990) Limitations on insect herbivore abundance in natural vegetation. In J.J. Burdon and S.R. Leather (Eds) *Pests, Pathogens and Plant Communities*. Blackwell Scientific, London, pp. 15–30.

van Emden, H.F. (1991) The role of host-plant resistance in insect pest mis-management. *Bulletin of Entomological Research* 81, 123–126.

van Emden, H.F., Macklin, R.J. and Staunton-Lambert, S. (1990) Stroking plants to reduce aphid populations. *Entomologist* 109, 184–188.

van Emden, H.F. and Wearing, C.H. (1965) The role of the host plant in delaying economic damage levels in crops. *Annals of Applied Biology* 56, 323–324.

van Marrewijk, G.A.M. and de Ponti, O.M.B. (1975) Possibilities and limitations of breeding for pest resistance. *Mededelingen van de Fakulteit Landbounwwetenschappen Rijkuniversiteit Gent* **40**, 229–247.

Wilde, G. and Feese, H. (1973) A new corn leaf aphid biotype and its effect on some cereal and small grains. *Journal of Economic Entomology* **66**, 570–571.

Wink, M. and Römer, P. (1986) Acquired toxicity—the advantages of specializing on alkaloid-rich lupins to *Macrosiphon albifrons* (Aphidae). *Naturwissenschaften* **73**, 210–212.

Wood, Jr., E.A., Chada, H.L. and Saxena, P.N. (1969) Reactions of small grains and grain sorghum to three greenbug biotypes. *Report of the Oklahoma Agricultural Experiment Station*, No. 618, 7 pp.

Wyatt, I.J. (1970) The distribution of *Myzus persicae* (Sulz.) on year-round chrysanthemums. II. Winter season: The effect of parasitism by *Aphidius matricariae* Hal. *Annals of Applied Biology* **65**, 31–42.

CHAPTER 8

Disease Resistance in Plants: Examples of Historical and Current Breeding and Management Strategies*

Cesare Gessler

Swiss Federal Institute of Technology, Zurich, Switzerland

INTRODUCTION

Agriculture is, and will continue to be, the second most important human activity keeping mankind alive on this planet. From past, present and future perspectives, the goal of agriculture is to increase and optimize the production of food, fibre, stimulants and other plant and animal products. Major constraints, as well as soil and water, were and are plant pests, weeds and plant diseases. World food yield reduction through pathogens is estimated to be of the order of 20% (Russell, 1978). Until barely two centuries ago, ignorance of plant diseases was absolute and led to mysticism. Disease avoidance strategies knew only repent and quit sinning (Bible, Amos 4:9 'I have smitten you with blasting and mildew ... yet you have not returned to me said the Lord'). The Romans put the responsibility of the dreaded rust of cereals into the hands of two gods, Robigo and Robigus. Nevertheless, rational studies of plant disease did occur sporadically. Cleidemus (400 BC) described diseases on grapes, figs and olives. Theophrastus (300 BC) observed that the amount of disease was higher in low (probably more humid) spots than on high ground (Horsfall and Cowling, 1977). Albertus Magnus (1200–1280) conceived the idea that mistletoe was a parasitic plant and that wood decay fungi were exhalations of humid ingredients which condense and solidify in the cold air. However, the association of a fungus with a plant disease occurred only after the invention of the microscope (in 1667 R. Hooke observed teliospores of cereal rust), but erroneous interpretation still led to the mystical idea that diseases are spontaneously generated and not the expression of a parasitic interaction. Micheli (about 1726, Italy) showed that a saprophytic fungus was able to reproduce from spores, and Tillet, working on brining treatment (previously improved by Pluchet) of wheat, clearly showed that artificial inoculation of wheat seeds with the black dust of smut led to smutted crops. Control was possible by seed treatment with copper sulphate, as Pluchet had already shown with natural infections. Prévost, in

---

* This chapter is a revision of a paper entitled "Results of plant breeding during the last decade in relation to resistance against pathogens" by C. Gessler, published in *Acta Horticulturae*, Feb. (355), pp. 35–62, 1994.

Techniques for Reducing Pesticide Use. Edited by D. Pimentel
© 1997 John Wiley & Sons Ltd.

Table 8.1   Evolution of knowledge of plant pathogens

Roman culture,	ca. 700BC	Cereals gods of rust Robigo and Robigus
Cleidemus,	400 BC	Diseases on grape, figs and olives
Theophrastus,	300 BC	The amount of disease was higher in low areas than on high ground
Bible, Amos 4:9		'I have smitten you with blasting and mildew ... yet you have not returned to me said the Lord'
Albertus Magnus,	1200–1280	Mistletoe was a parasitic plant and wood decay fungi were exhalations of humid ingredients which condense and solidify in the cold air
R. Hooke,	1667	A fungus was associated with a plant disease: teliospores of cereal rust
Micheli,	1726, Italy	Saprophytic fungus can reproduce from spores
Pluchet,	1746	Seed treatment with copper sulphate
Tillet,	1755	Artificial inoculation of wheat seeds with the black dust of smut led to crops with smut
Prévost,	ca. 1800	Plant diseases were caused by organisms reproducing themselves and parasite plants. Germination of spores and its inhibition by copper sulphate
DeBary,	ca. 1850	Established the living nature of plant parasites
Bailey,	1892	'Enemies often progress or develop as rapidly as do the host plant. I imagine that by the time we are able to breed scab-proof varieties our scab-fungus will have developed a capability to attack more uncongenial hosts'.
Biffin,	1907	Resistance to yellow rust of wheat controlled by a single recessive gene
Orton,	1909	Rational use of resistance and therefore the resistance breeding
Vavilov,	1920	Suggests the use of wild resistance sources

the early 1800s, again advanced the idea that plant diseases were caused by organisms reproducing themselves and parasitising plants. He even devised a method for observing germination of spores and their inhibition by copper sulphate. DeBary's work and the tragic consequences of the *Phytophthora* epidemic in the middle of the last century clearly established the living nature of plant parasites, finally opening the way for rational elaboration of strategies to avoid and control plant diseases (Table 8.1).

Prior to DeBary, recurrent losses probably went unremarked, and only exceptional epidemics leading to tragic famines were noted. However, empirically effective control strategies were developed from the Middle Ages, mostly based on crop rotation. The first truly rational control attempts involved chemicals. It is not surprising that the first great success in controlling epidemics is attributed to the use of sulfur dust to control powdery mildew of grapes in Europe in the autumn of 1854. Chemicals still are, and will be to a large extent in the future, the most convincing way to control pathogens. They fit in our science-orientated society: a cause is identified, an action can be taken and the success is visible. Quarantine, hygiene or sanitation (use of clean seed or propagation material) could only be taken into account once the infectious living nature of pathogens was established, and even then their real usefulness was recognized only slowly and applied generally only in this century. They have the disadvantage that their effects are much less evident and that they are mostly long-term strategies. Less susceptible plants, on the other hand, have been used unconsciously since the dawn of agriculture, as nature allows better survival of fitter plants which were selected automatically, while pathogens were a regular constraint. Later, deliberate selection of plants (cultivars) giving the best yield and/or best quality may often have also included deliberate selection for less susceptible plants. The rational use of resistance, and therefore resistance breeding, came into use only at the beginning of this century (Orton, 1909), and the search for resistance sources in natural ecosystems (wild resistance sources) began even later (Vavilov, 1920). A great stimulus to breeders derived from the recognition that a particular resistance to yellow rust of wheat was controlled by a single recessive gene (Biffin, 1907). To a large extent, breeding programs do (and today have to) consider the resistance aspect. To the scientists involved it has always been clear that the benefit to producers and the environment from the use of pathogen-resistant plants is extremely high, and is superior *per se* to other strategies. On the other hand, we cannot forget that the usefulness of a new cultivar will depend primarily on its yielding capacity and on the quality (appearance, content, etc.) of the product. A resistant cultivar is unlikely to be widely grown unless it has these qualities. Minor faults may often be compensated by the presence of resistance; this leads to economical considerations, and we can postulate that a resistant cultivar will be successful only if the benefits outweigh the disadvantages, a point which is often overlooked and leads back to the chemical control of pathogens. The advantage gained through resistance is often expressed only as a difference from a standard cultivar resulting from the lower cost of chemical control (product, application and equipment costs). As this is often low compared with the value of the crop, it is evident that resistant cultivars need to have almost all of the quality aspects of the susceptible control. Moreover, while other methods of control of a particular disease (e.g. crop rotation, quarantine or hygiene) are effective and are traditionally practised with no evident drawbacks, it is not useful to devote efforts to breeding for resistance against such diseases.

Definitions

Before venturing into breeding for resistance against pathogens, a few terms have to be defined and types of resistance clearly stated in order to have a common language. For this paper, an infectious disease is defined as a harmful alteration of the normal physiological state of a particular individual of a species by a pathogen of a particular species or *forma specialis*. A pathogen of a particular species can be (and is, in most cases) a non-pathogen on other plant species. All members of a non-host species are characterized by immunity or resistance against the non-pathogen. On the other hand, not all genotypes constituting the host species may be susceptible; some may be completely resistant, and their resistance is called host resistance and the pathogen is said to be avirulent on them (Andrivon, 1993). The interaction of the two organisms is then incompatible. Virulence therefore implies that the pathogen is able to infect and to reproduce on a particular host. The host is then said to be susceptible and the interaction compatible.

Aggressiveness defines the amount of disease produced by a particular individual of the pathogen on the total range of hosts that it can infect; this non-specific attribute is quantitative. In addition, individuals may differ in their ability to infect a host for which they have the same specific virulence. In other words, they may differ quantitatively in their specific virulence. Quantitative differences in aggressiveness and virulence are difficult to distinguish experimentally. Fitness relates to the maintenance or increase in relative frequency of a particular pathogen genotype in the gene pool during evolution.

Parasitic fitness is therefore a basic attribute of a pathogenic strain within a population of pathogens (Andrivon, 1993).

Resistance of a host cultivar can be of two types: general, which allows less disease (development of the pathogen) than a susceptible cultivar by all genotypes of the pathogen, and is often referred to as horizontal resistance (elsewhere in the literature, general resistance is sometimes referred to as resistance of a particular host against several diseases); and specific, which allows no or less disease compared with a susceptible control cultivar by those pathogen genotypes that lack virulence specific to the resistance in question, and is often referred to as vertical resistance. The pathogen genotype that does have the appropriate virulence allowing it to induce normal disease is known as the pathotype. Different pathogen isolates with the same pathotype belong to the same physiologic race or, more commonly, race.

Inheritance of resistance mostly follows a Mendelian pattern, as does inheritance of virulence in the pathogen. Inheritance analysis can show two patterns: discontinuous variation or continuous variation. In the first case, the segregating progeny will segregate typically as susceptible and resistant individuals; this pattern is often found for differential resistance. General resistance, on the other hand, although often not exclusively, shows continuous variation in the segregating progeny, the individuals ranging from susceptible to resistant with a majority being intermediate.

From these data the underlying genetic base of the resistance can be of two types. The first type is monogenic (dominant or recessive, also referred to as the major gene), whose expression may be affected by other genes (modifiers). The majority of examples of differential resistance can be attributed to the effect of a single gene, as can the capacity of a particular pathotype to be virulent or avirulent. This led Flor (1942, 1956, 1971) to postulate the gene-for-gene theory: each pair of resistance/susceptibility alleles in the host has a matching pair of virulence/avirulence alleles in the pathogen. In other words, for each allele of resistance in the host, a particular allele of avirulence is present in the pathogen. Another theoretical model, the matching allele model, was recently proposed (Frank, 1994), but no experimental data are available to prove its fit to reality.

Resistance genetics focused its attention on sharply defined phenotypes, so differential resistance is also called qualitative resistance and results are expressed as compatible and incompatible interactions.

Considerably less information is available on quantitative interactions, even though they are arguably more common. The genetic basis underlying resistance may also be additive, where each gene contributes only partially to the level of resistance. Here, the expression of resistance is increased by the presence of more genes, also referred to as minor genes. As it is assumed that each minor gene has little effect and that many genes contribute (perhaps unequally) to this resistance (second type: polygenic resistance), it is also called quantitative resistance. However, resistance recognized as quantitative is often assumed to be polygenic, which may be incorrect in the absence of a genetic test since, as indicated above, a single gene may give quantitative resistance that segregates with a continuous distribution.

Resistance genes at different loci can control the same resistance mechanisms and can be overcome by the same race; these are duplicate genes and can be considered as one in breeding and effect. Conversely, resistance genes that are overcome by different races can be described as functionally different. Ephemeral resistance can be defined as resistance (vertical or horizontal) that selects rapidly for the pathogen genotype(s) that can overcome it. General or horizontal resistances are usually considered to be durable, since no races are known that can overcome them. On the other hand, Johnson (1981) defines durable disease resistance as resistance that lasts at least for the period during which the resistant cultivar is widely cultivated. Differential resistance can thus be durable because either virulence is extremely rare and/or it is linked to lower fitness (seldom the case over longer periods). Alternatively, planting strategies involving mixtures with functionally different resistances may inhibit virulent races from reaching epidemic levels (Wolfe, 1983, 1985, 1992, 1993; Gessler and Blaise, 1994).

To understand the complexity of a host–pathogen interaction and its analysis, the state of the genotypes should be known. In a diploid or dicaryontic organism it can be homozygous, meaning that the two alleles of a gene are identical, or heterozygous, where the two alleles are dissimilar. Often allelic relationships are not so simple, since the expression of a given allele may be affected by other non-allelic genes (e.g. epistatic effects). Many plant pathogens, such as the ascomycetes, are haploid with a single set of chromosomes carrying a single allele for each gene.

With new biomolecular methods allowing isolation and cloning of plant disease-resistance genes, it is necessary to state clearly what type of genes are envisaged. Plant resistance genes can not only be classified as described above, but also in relation to their function. Most genes of the potential ephemeral type are genes which allow the plant to recognize the pathogen (R-genes) and then activate a cascade of responses encoded by defence–response genes, such as the production of phytoalexins. These responses are often similar between plant species. Recognition genes often fit the gene-for-gene concept. Being potentially ephemeral, they are not desirable in breeding. Defence–response genes, on the other hand, cannot be used (or only in exceptional cases) as they code *per se* for non-specific phytotoxic products or cell letal functions (see Dixon et al., 1995; Innes, 1995; Young, 1995). Which type of genes (R-genes and/or defence–response genes) can and should be used in the construction of transgenic disease-resistant plants has to be considered very carefully.

THE CASE OF APPLE (*MALUS* × *DOMESTICA*)

Apple is not the most important fruit crop, but it is grown almost world-wide in temperate zones and requires massive fungicide inputs under intensive production systems. Increasing public pressure is demanding production systems with reduced inputs of pesticides. Integrated production (IP) systems are more and more popular in Europe. Although the fungicides used are decreasingly harmful to the environment and consumers, the input of pesticides is still questioned. In low-input systems high-quality apples are difficult or impossible to produce. Less susceptible cultivars or even resistant cultivars can be a step towards both production systems.

Genetics of the interaction scab–apple

Commercial apple cultivars have never been analyzed for segregation of genes for differential resistance against scab caused by *Venturia inaequalis*. Because of the variability of the scab population and the high frequency of virulence genes overcoming specific resistances of the commercial cultivars, breeders opted to use material resistant against all strains of *V. inaequalis* present at their location. This included other *Malus* species, often from other continents (Asia). For example, *M. baccata* Borkh. originates from the Himalayas, Siberia and eastern Asia. *M. floribunda* is known only as a cultivated species from Japan, probably originating from interspecies crosses from *M. kaida* × *M. baccata*, *M. ringo* × *M. spectatabilis* × *M. baccata* or others (Hegi, 1963). *M. micromalus* comes from China.

Resistances in other *Malus* species were studied intensively and this indicated the existence of a large pool of resistances (Williams and Kuc, 1969). The available data allow various interpretations: (a) Vf-resistance from *M. floribunda* 821 is due to a single gene (Williams and Kuc, 1969, p. 226); (b) the original level of resistance in *M. floribunda* 821 was due to a group of closely linked quantitative genes; (c) resistance could be due to a qualitative gene giving a class 3 reaction type, closely linked with one or more quantitative genes always inherited to the modified back-cross progeny and due to a gene giving a class 1 reaction, not detected in the back-cross progeny (Williams and Kuc, 1969, p. 227). Crosby et al. (1992), reviewing the success of breeding apples for resistance against scab, list 48 cultivars resistant to scab, of which 37 carry the Vf-resistance. As Rousselle et al. (1974) and Gessler (1989) did earlier, they hypothesized that the phenotype of the Vf-resistance is enhanced or augmented by additional genes. The fraction of a segregating population carrying the Vf gene from a cross between a Vf carrier and a susceptible parent shows continuous gradation of the resistance reaction from no symptoms to restricted sporulation and even to full susceptibility.

To my knowledge, 19 independent resistance loci have been described to date (Bagga and Boone, 1968a). So far, only six resistances at different loci from six *Malus* species are named (Vf, Va, Vb, Vbj, Vr and Vm). As Dayton and Williams (1967) noted, it is still not clear whether all genes are truly different or whether some are identical genes that were transferred to non-homologous chromosomes (or loci) by aberration during the evolutionary development of *Malus*. On the other hand, there may be as many mechanisms of resistances as there are resistance genes present (Bagga and Boone, 1968a) and, correspondingly, the same number of different virulence genes. This can be established only after the corresponding virulences have been found and the resistance carriers have been tested accordingly.

The cultivar 'Nova Easygro' should carry the Vr-resistance since it was bred from a Russian seedling. This resistance is described as different and independent from all other resistances mentioned above (Dayton et al., 1953; Dayton and Williams, 1967). However, it was possible to show by the use of molecular DNA markers for the Vf locus that 'Nova Easygro' also carries the Vf gene (N.F. Weeden, personal communication, 1995; Gianfranceschi et al., 1996). The Vm resistance is different from the other single resistances as it is overcome by Race 5 (Williams and Brown, 1968).

Bagga and Boone (1968b) studied 41 scab-apples for crab resistance and found 25 independent loci rendering resistance. Some of them must be functionally different from each other because Williams and Kuc (1969, p. 228) found that they remained resistant to an inoculum consisting of a combination of isolates. Again we cannot deduce that the seven crab-apple selections which were resistant to the isolate used by Bagga but susceptible to the undefined inoculum used by Williams all had identical, or up to seven different, resistance genes. Similarly, we cannot deduce that the remainder had the same functional resistance but at different loci. Other examples where interpretation is impossible can be found in the literature. Recently (Gessler et al., 1993; Sierotzki et al., 1994a,b) it was shown that three popular cultivars considered to be susceptible have differential and

Table 8.2   Major steps leading to scab resistance in apple

Aderhold,	1899	Particular apple cultivars seem to be more resistant than others. The observations were not consistent. Resistance/susceptibility of a single cultivar changes with the years. Accommodation of the scab fungus to the host
Aderhold,	1902	Recommendation to avoid spread of scab isolates and to test the behaviour of various cultivars at various sites over several years
Crandall,	ca. 1910	Collection of *Malus* species (see Crandall, 1926)
Wallace,	1913	Use of resistant apple cultivars offers little promise as a means of scab control, since varieties known to be fairly resistant changed to susceptible over the years
Crandall,	1914	Cross *Malus floribunda* 821 (Vf-resistance) × Rome Beauty (see Crandall, 1926 and Kellerhals, 1989)
Crandall,	1926	Full-sib cross between two F1 selections and selection of No. 26829-2-2
Hough and PRI-program,	1945	Developing scab resistant cultivars using Crandall's No. 26829-2-2 (see Hough et al., 1953 and Kellerhals, 1989)
Dayton et al.,	1970	Scab-resistant cultivar Prima (Vf)
Crowe,	1971	Scab-resistant cultivar Easygro (Vr from Russian seedling No. 12740-7A) (Crowe, 1975)
Williams et al.	1972	Scab-resistant cultivar Priscilla (Vf)
Williams et al.	1975	Scab-resistant cultivar Sir Prize (Vf)
Lespinasse et al.	1977	Scab-resistant cultivar Florina (Vf) (Lespinasse et al., 1985)
Norelli and Aldwinckle,	1993	Transgenic apple rootstock M26 with increased resistance to fire blight (see Norelli et al., 1994)

ephemeral resistances, each being active against populations of the pathogen collected from the other two cultivars. We may now assume that speculation that a gene-for-gene relationship exists (Boone, 1971) is also correct and valid for these hidden resistance genes in commercial cultivars. The total number of functionally different resistances present in wild *Malus* species and in cultivated commercial cultivars, almost all recognized as ephemeral, may be many times greater than expected.

Based on these assumptions, a set of old cultivars has been inoculated under controlled conditions with conidia from monoconidial isolates made from single lesions collected randomly and sparsely in various orchards. The results (Koch et al., 1996; Table 8.3) show that all cultivars carry resistances against particular isolates. Strongly sporulating lesions were observed when the inoculum originated from the same host cultivar. The isolates of an orchard with many cultivars carried more virulences than the isolates of an orchard with only a few cultivars.

The data described above indicate that many functionally different ephemeral resistance genes are present in old cultivars and other *Malus* species. A corresponding number of pathotypes is probably present in the world population of the scab pathogen, with the limitation, often found in other host–pathogen systems, that a number of the resistance genes have the same function, so that the number of different virulences may be lower.

Strategies to improve the durability of ephemeral resistance

By incorporating any resistance (or part of the immunity) into *Malus* × *domestica* we exert selection on the scab pathogen which leads to the emergence of a specific race (or

Table 8.3 Frequency (%) of *Venturia inaequalis* isolates capable of causing abundantly sporulating lesions or flecks with slight sporulation (in parentheses) on a set of eight cultivars. Monosporic isolates were obtained from single lesions randomly collected in an orchard planted with 26, 3 or 1 cultivars (adapted from Koch et al., 1996)

Origin of isolates	Ananas Reinette	Boskoop	Champagner Reinette	Glockenapfel	Golden Delicious	Gravensteiner	Jonathan	Klarapfel
Orchard with 26 cultivars								
Ananas Reinette (5 isolates)	100	20	80	(40)	(60)	20+(40)	(20)	20
Boskoop (4 isolates)	(75)	75+(25)	50+(50)	(75)	0	0	25	(50)
Champagner Reinette (10 isolates)	10+(30)	20+(30)	90+(10)	(10)	10+(10)	(10)	(20)	0
Glockenapfel (8 isolates)	25+(37)	12+(37)	25+(12)	87+(12)	12+(25)	20+(40)	(25)	(37)
Orchard with 3 cultivars								
Golden Delicious (15 isolates)	0	0	6.7	6.7	93+(6.7)	6.7	0	0
Jonathan (7 isolates)	28	14	0	0	57	0	29+(57)	0
Gravensteiner (4 isolates)	25	0	0	0	25	75	0	0
Orchard with 1 cultivar								
Golden Delicious	0	0	0	0	82+(9)	0	0	0

races), as Parisi et al. (1993) and Roberts and Crute (1994) showed. Substituting susceptible cultivars by cultivars with Vf or any other single resistance may not be a long-term
strategy as long as we maintain the concept of monoculture; we can predict that the
resistance will be completely overcome and the cultivars will then be dependent only on
their background resistance.

Populations of pathogens are genetically variable and therefore respond to selection
pressure. In natural systems, one apple tree represents a single genetic entity. Mankind
changed this situation drastically by creating large and dense, genetically uniform host
areas (monoculture). Seemingly, the Greeks propagated apple clones by grafting, and the
Romans cultivated particular cultivars in orchards and had disease problems with them.
Until the beginning of this century, the planting system consisted of single trees in
meadows and non-arable land, most of the trees originating from seedlings (Morgan and
Richards, 1993). Today's intensive orchard forms planted with only a limited number of
cultivars were adopted only during this century, mostly thanks to the systematic selection
of rootstocks with particular growth characteristics (Hatton, 1939). Parallel to the
disappearance of the sparse planting system of trees of unequal genome, the disease
situation worsened. With the introduction of dense orchards, chemical disease control
was introduced. Even today we can observe that scab does not cause a total loss on single
standing trees of 'older' cultivars (cultivars not popular anymore). Under similar conditions, however, scab can cause total or near total loss in an unsprayed, intensive single-
cultivar orchard. Even if the destabilizing effect of increasing aggregation of plants with
the same genome is poorly substantiated (Zadoks, 1993), we can postulate that this
change drastically shifted the relationship between the host (now a single entity) and the
parasite. Previously, variability was required to enable the pathogen to infect more than a
single entity (by sexual ascospores), followed by asexual reproduction from a successful
infection. Now, however, uniformity gives an advantage to the pathogen. Scab can
damage trees to such an extent that trees are defoliated early and some individuals are no
longer able to survive competition from others, leading to a thinning of the close stand.
However, humans have intervened by reducing pathogen damage through protection by
pesticides. As the efficacy of the pesticides and of pesticide scheduling improved, host
density was maintained or increased. At this stage, the pathogen population also re-
sponds to the fungicides, and resistant strains or sub-populations are selected. We are
now dependent to such an extent on the use of pesticides to maintain an equilibrium
between pathogen and host that we have forgotten how this situation arose; the current
unnatural situation is now considered natural.

Can we return to the earlier system of dispersed trees of various cultivars (as described
above for apple)? I am not advocating this because it is unrealistic. Can we therefore
continue on our current road? Again this is unrealistic, or at least short-sighted. Few
classes of fungicides have been developed, and the probability of finding new products is
diminishing. Moreover, even with the best strategy of use, fungicide resistance is likely to
emerge. Indeed, by definition, no chemical strategy is sustainable because of either
resistance or toxicity to humans or other organisms. In the co-evolution of a host and a
pathogen in a particular environment, the main stabilizing factors are, to different extents,
distances between hosts, inter-cropping and the genetic variability of the host population.
As the first two factors cannot easily be altered, the latter remains.

A possible strategy is cultivar mixing (Wolfe, 1983; Blaise and Gessler, 1994), where
each cultivar needs to have a different resistance. A simple mixture of only three cultivars

greatly limits the production of primary lesions, but we have now to prove or demonstrate the validity of the theory for this host–pathogen system. However, we do not know the resistance-gene composition of all cultivars, and old cultivars do not correspond to our needs. Therefore modern cultivars need to be analyzed accordingly or new cultivars need to be bred.

Breeders have two options. The first is to breed a set of cultivars each having a functionally different resistance. These unrelated cultivars, each with one or more resistances, may differ so much that in an appropriate planting system that adaptation of the fungus is difficult if not impossible. In other words, occurrence of a super-race (a single isolate able to overcome all mixtures of resistance components) may be much less likely in appropriate cultivar mixtures than in one cultivar with the same total number of resistances, even without considering the possibility of stabilizing selection *sensu* Vanderplank (1982; see Schaffner et al., 1992, for similar data on the response of populations of the barley mildew pathogen to large-scale use of cultivar mixtures).

The recognition of functionally different resistances is possible if the resistances are ephemeral, but only through the use of differential races. Such a selection scheme would be very cumbersome. An alternative would be to transfer the genome segment carrying the resistance from one cultivar into the target cultivar. The target cultivar could be a popular commercial cultivar. By incorporating functionally different resistances, the concept of near-isogenic lines (NIL) could be adopted. Although this cannot yet be done, it may well be possible in the foreseeable future, but it is not clear whether this will be acceptable to consumers. Moreover, by incorporating the resistances into a single genetic background, we may simplify the problem for the pathogen of overcoming the resistances (Wolfe, 1993) and thus select a fit super-race even in a resistance-mixture concept.

The second option is to breed cultivars that contain the greatest possible number of resistance genes (pyramiding). The presence or absence of each gene could be recognized by serial testing of the progenies of correctly selected parents with inoculum of races carrying the appropriate virulences. Alternatively, the races could be mixed in the inoculum to reach the same conclusion in fewer tests, except that genes giving incomplete resistance may be missed. This system could again be used for resistances known to be ephemeral, i.e. if we have the corresponding virulent races which would allow identification of a second (or third?) resistance. However, selection for super-races may be stronger than in the NIL concept, particularly for a parasite such as *V. inaequalis* which reproduces sexually each year. There has been much discussion about the advantages and disadvantages of both strategies (Wolfe, 1993), mostly by experts on annual plants. The choice of strategy (multi-cultivar mixture, NIL or pyramiding resistance genes) may not be relevant to the apple system because the constraints on success may be of a completely different nature.

A more elegant approach than selecting for resistance in the glasshouse or field is to identify or mark the genome segment carrying the resistance information and to select progeny by the presence of such markers. The success of marker-assisted breeding is based on the assumption that resistances at different loci are also functionally different. Since the resistance genes are often not easily identified, it may be better to concentrate on marker-assisted breeding, particularly because of its high feasibility and cost–benefit ratio. An alternative to classical breeding is the transformation of a desired cultivar through incorporation of genes leading to resistance. Two types of genes can be considered. (a) The traditional resistance genes which allow the host to recognize the pathogen

(R-genes). Such genes, as already mentioned, are (potentially) ephemeral, and the strategies stated above would be needed to render these resistances durable. (b) Defense response genes (reviewed by Dixon et al., 1995), which may well lead to stable resistance, as, for example, would the constitutive expression of one or more phytoalexins. An example is the introduction of the gene encoding the lytic protein attacin E which is active against a broad spectrum of gram-negative bacteria. Aldwinckle and Norelli recently obtained a transgenic apple rootstock expressing this gene and being significantly more resistant to fire blight (*Erwinia amylovora*) than the parent root stock M26 (Norelli et al., 1994). For apples, a strategy based on defense–response genes must be considered very carefully as regards the side effects on the plant itself and toxicity to consumers.

COFFEE

Coffee is, in global trading, the highest revenue cash crop, with US\$11–12 billion year^{-1}. Several countries are highly dependent on this crop. Coffee originates from tropical Africa, an important center of origin being Ethiopia. The genus *Coffea* includes about 90 species. Commercially important are *C. arabica* and *C. canephora*. *C. arabica* has excellent quality. *C. canephora*, on the other hand, is more resistant to diseases (*C. canephora subspecies robusta*). *C. arabica* is tetraploid and self-fertile while *C. canephora* and most other species are diploid and cross-pollinating. The basic number of chromosomes is 11 (Van der Graaff, 1986; Lutzeyer et al., 1993). The most important diseases are coffee-rust (caused by *Hemileia vastatrix*) and coffee-berry disease (caused by *Colletotrichum coffeanum*).

Breeding

As with apple, breeding a high-quality perennial crop is time-consuming and success is slow (see Muller, 1986; Van der Graaff, 1986). The market for resistant plants or seeds is small and resistance has to be durable. Coffee breeding was therefore of no interest to private breeders. Coffee is bred in public institutions in producer countries. Great regional differences can be observed among these institutions. The CENICAFE (Centro National de Investigationes de Café de la Federacion Nacional de Cafeteros de Colombia) can be regarded as the leading institution for research and breeding, together with the CIFC (Centro de Investigação das Ferrugens do Cafeeiro) in Portugal (Oeiras). These institutions generally carry a big deficit; international collaboration is often hindered by economic and political problems. Sinking world prices for coffee do not favour investment in research.

Coffee rust

Coffee rust struck in 1868 in Ceylon (Sri Lanka), completely devastating the large British-owned plantations. Since then, coffee rust has spread to all coffee-planting areas, fortunately (for producers and coffee drinkers) with less impact, thanks to less favourable climates and/or the use of fungicides.

Host–pathogen interaction

About 30 races of the pathogen have been described so far at CIFC based on the interactions of dominant host resistance genes and pathogen virulence genes. The resistance genes can give complete protection, but most have been overcome; a few single genes and some combinations are still effective against the complete population of *H. vastatrix*. An important source of resistance are the Timor hybrids (probably spontaneous hybrids between *Coffea arabica* and *C. canephora*), which possess the genes Sh 6–9. These hybrids are often used in breeding.

Horizontal resistance can be found in *C. arabica* and in *C. canephora*. In *C. canephora* the pathogen response ranges from almost susceptible to immunity. Breeding strategies include the use of transgressive segregation in interspecies crosses of *C. arabica* with the introduction of resistance from *C. canephora* and the use of these resistances in *C. canephora* itself by back-cross breeding and the pyramiding of the vertical resistance gene in new selections. Male-sterile lines can be used to produce hybrids. As may be expected, introduction of general resistance is problematic because the resistance may be completely or partially lost during back-crossing.

Impact

Resistant cultivars of *C. arabica* from breeding programs are only slowly finding their way into cultivation. In 1986, the rust control strategy was generally based on the replacement of *C. arabica* by resistant *C. robusta* (Van der Graaf, 1986), a strategy which leads to lower quality since *C. robusta* is inferior to *C. arabica*.

However, in Colombia more than 200 000 ha are currently planted with the cultivar mixture 'Colombia' which is uniform for high quality, dwarf stature and high yield, but heterogeneous for a range of rust-resistance genes defined through collaboration with CIFC in Portugal (Lutzeyer et al., 1993). The lines derive from crosses between the tall, rust-resistant population of 'Hibrido de Timor' and high-quality dwarf varieties of *C. arabica* (e.g. 'Caturra'). Appropriate lines are selected in F5 and composed in mixtures on the basis of agronomic trials in Colombia and resistance screening in Portugal. Mixtures of different compositions are released in different years depending on the current structure of the pathogen population (Moreno-Ruiz and Castillo-Zapata, 1990). This highly dynamic use of the available resistance genes in a mixture strategy gave durable resistance for some 20 years. The use of pathogen-resistant cultivars allows important savings in pesticide costs, amounting in Colombia alone to US$28 million (calculated after the official plant protection recommendations).

Coffee berry disease

The coffee berry disease (CBD) originated in Africa; it can be devastating in higher zones of Kenya. The situation regarding resistance against CBD is very unclear. Until now no race has been detected which is able to overcome any resistance, and the resistances are assumed to be general. The mode of inheritance is unclear, but selection of resistant cultivars seems to be promising. Recently in Kenya a new cultivar ('Ruiru 11') has been introduced with good resistance against CBD and partial resistance against rust, and this cultivar is being propagated. It is also a hybrid originating from the Timor hybrids and *C.*

arabica. However, multiplication is cumbersome and several decades may be needed before such varieties are widely distributed.

The use of 'Ruiru 11' would avoid all the fungicide treatments which are now necessary to control CBD and rust, and which are a large proportion of the variable costs (28%) in Kenya, amounting to US$560 ha^{-1} (Omondi, 1994; Opile and Agwanda, 1993). Theoretically, Kenya could save about US$10–19 million in direct fungicide costs on its 35 000 ha of coffee. In Ethiopia, CBD-resistant cultivars cover the highest surface percentage of any coffee-producing country, although this only amounted to 5% in 1986.

Even though encouraging rust- and CBD-resistant cultivars are available, the area covered by disease-resistant cultivars world-wide is still extremely small.

FRENCH BEANS (*PHASEOLUS VULGARIS*)

The importance of beans is less evident from commercial statistics, since beans are produced as high-quality food for own or local consumption. They are an important source of protein and also vitamins. As a basic food the crop plays an essential role in many Third World countries. Indeed, in some regions of Africa it is the most important food source. It is consumed as green pods or as seed, and different cultivars have been selected for each purpose. The genetic variability is immense (Cepts and Debouck, 1991), with more than 40 000 registered entries at the gene resource bank of CIAT (Centro International de Agricoltura Tropical; Hidalgo, 1991). The most important diseases include anthracnose caused by *Colletotrichum lindemuthianum*, which is important in colder humid climates and can be transmitted by seed, and rust caused by *Uromyces appendiculatus* (Beebe and Corrales, 1991; Singh, 1991).

Breeding

A leader in breeding and research on beans is CIAT, which has the global mandate for the crop. Breeding at the local level is also important in the USA, Italy, France and The Netherlands, etc.

The breeding strategy is negative selection: in the absence of major genes, minor genes should be accumulated in breeding populations. Beans are intrinsically easy to breed since generation times are short and the plant is a typical outbreeder (cross-pollinating; $n = 11$). On the other hand, the crop has to be highly adapted to regional requirements, therefore breeding for resistance is often of secondary importance.

Anthracnose

C. lindemuthianum was used as an example to demonstrate the presence of different races in a fungal pathogen population. Initially there was thought to be little variability in the fungus, but many different races have now been identified. However, the possibility of obtaining durable resistance has been described as good because of the slow rate of spread of the fungus and the rarity of sexual recombination (see Beebe and Corrales, 1991; Singh, 1991). Thirteen vertical resistances have been described, which cover a large range of the world population of *C. lindemuthianum* (ARE, MEXIQUE1, MEXIQUE2, TO, TU, etc.). The corresponding genes are described as complementary dominant, simple dominant or double dominant, and even some recessive. More recently, resistance sources have been found which give resistance against all known races of *C. lindemuthianum*. Even if these resistances are controlled by single genes they cannot be regarded as vertical. Attempts to incorporate partial general resistance are problematic because such resistance is often not expressed in seedlings in the glasshouse. Yield has been static over the last 40 years, but the cultivars used have changed, especially in intensive agriculture, due to mechanization and market requirements. Disease-resistant cultivars are now available for almost any requirements (Silbernagel and Hannan, 1988; Allavena and Ranalli, 1989; Beebe and Corrales, 1991; Singh, 1991).

Bean rust

Various dominant monogenic vertical resistances against rust have been described and a gene-for-gene relation has been postulated (Christ and Groth, 1982a,b). The bean rust population is genetically highly variable because sexual recombination frequently occurs. It is therefore likely that vertical resistance has little future in resistance breeding. Through the use of an international rust nursery, the variability of the pathogen in relation to virulence is being analyzed to find resistance sources which are effective against the whole pathogen population. Habtu and Zadoks (1995) analyzed components of partial resistance (latent period, infection efficiency, sporulating capacity, infectious period and pustule size). The contribution of the various components to the resistance varies from cultivar to cultivar. Some of these components could be used in selection for partial resistance. Accumulation of genes contributing to such a resistance could be feasible with a method of screening for each component. This type of resistance could be truly horizontal.

Impact

The impact of resistance breeding can be regarded as high in intensive agriculture. Without resistance, production would be seriously hampered. However, there is still much room for improvement in sustainable agriculture. Trials in Rwanda showed that through the use of fungicides the average local yield of 873 kg ha^{-1} could be increased by 343 kg ha^{-1}. Disease represented the second most important constraint after the lack of fertilizer (increment through fertilization 571 kg ha^{-1}; Voss and Graf, 1991). Yield increment is vital for countries with dense populations such as Rwanda. Resistant cultivars are particularly important since money for fungicides is not available. Several cultivars were developed in the 1950s and 1960s with ARE resistance, but they were overcome in the early 1970s by a specific race (Messiaen, 1981). This virulence is now widely distributed in South America. Locally important cultivars are available with other resistance genes.

Progress in breeding still continues. In 1975/76 almost no generally resistant material was available, but by 1983/84, 33 out of 100 breeding populations were resistant in all localities where tests were made (Van Schoonhoven and Voyset, 1991). The old strategy of small farmers to mix cultivars (Panse, 1988) has probably helped to extend the durability of vertical resistances. Breeding needs to adopt a holistic point-of-view and cannot concentrate on a single cultivar and pathogen.

WHEAT (*TRITICUM AESTIVUM*)

Globally, wheat is one of the most important food crops, and is planted on 220 million ha; almost 600 million tonnes were produced in 1990 (ca. 120 kg per capita; FAO, 1990). Productivity varies from 0.4 t ha^{-1} (Somalia) and 1.2 t ha^{-1} for Africa to 4.8 t ha^{-1} in Europe, with The Netherlands reaching the highest yield with 8 t ha^{-1}. The climatic and pedologic conditions contribute most to this variability, but other major determinants are inputs such as fertilizer and pesticides. Western Europe planted about 7.6% of the world surface area of wheat in 1987 and produced 15.7% of the world production; this required 78.2% of all fungicides used world-wide on wheat (corresponding to US$430 million in 1987; Verreet, 1991).

Resistance breeding therefore has different goals and different impacts in various countries. In Western Europe it can lead to a more ecological production system, and in Third World countries to more stable and higher yields.

Breeding

Wheat-breeding programs are present in almost all developed countries as well as the major Third World countries. In addition, CIMMYT (Centro Internacional de Mejoramiento de Maiz y Trigo, Mexico) is financed internationally, and has the largest collection of genetic material and an exceptionally productive breeding program. Cultivars selected at CIMMYT now cover almost a quarter of the world wheat-growing area (> 50.7 million ha; Singh and Rajaram, 1992). Breeding concept (Simmonds and Rajaram, 1988) as well as general knowledge (Saunders, 1991) are among the most advanced for all crops.

Wheat includes several species, of which bread wheat (*Triticum aestivum*, hexaploid; $n = 7$, self-pollinating) is dominant. Local varieties, durum wheat (*Triticum durum*), *Triticum spelta* and wild relatives (grasses) can be used as sources of disease resistance (Dosba et al., 1982). Many resistance genes are catalogued and already introgressed in breeding lines. The most important diseases are rusts (especially leaf-rust caused by *Puccinia recondita*), mildew (*Erysiphe graminis*), and leaf and glume blotch (*Septoria* spp.) (Verreet, 1991), but other diseases of more local importance are also considered in resistance breeding (yellow rust, eyespot, stem rust, etc.).

Powdery mildew

Fifteen major genes (Pm1 ... Pm15) giving resistance against one or more races of the powdery mildew fungus have been identified so far (McIntosh, 1988; Fried et al., 1993), all of which have proved to be ephemeral. The cultivar 'Walter', used in Switzerland, is an excellent example: with its increasing popularity the frequency of the corresponding virulence in the mildew population also increased to levels rendering the resistance ineffective (Winzeler et al., 1990). Similar boom and bust cycles are well documented for other nations and other cereals and diseases. Minor genes conferring quantitative resistance, described as slow-mildewing, are widely used in breeding and have given sufficient protection at lower levels of production. However, they may not compete with the use of fungicides at high production levels. Selection in the breeding material has to avoid the trap of vertical resistance (Bartos et al., 1990), and breeders try to expose their material to inoculum of all known virulences (Winzeler et al., 1990).

Leaf rust

Resistance breeding against leaf-rust could follow the same strategy as that for powdery mildew, but the situation is more complex. Genetic variability in the leaf-rust population is high, partly in response to the 34 known major genes for resistance (Lr1, Lr2, ...). Several single genes and combinations are known to have been overcome by the pathogen. In North America more than 100 distinct races have been identified using differential cultivars. The cultivar 'Mironovskaja 808' carrying Lr3 became susceptible just 2 years after its introduction in 1968 (Bartos et al., 1990). In former Czechoslovakia the gene Lr26 became ineffective in the same year that varieties with the gene ('Kavkaz', 'Aurora') were introduced. A long list of such events is described (Russell, 1978). Zadoks (1972) pointed to the probable presence of minor genes inducing a slow-rusting effect. Breeding for slow rusting has to overcome obstacles such as epistatic effects of major genes in the material. Moreover, some are expressed only in adult plants and others are temperature-dependent for their expression (Fried et al., 1993).

For particular resistance-gene combinations no corresponding pathotype has yet been found. For example, the cultivar 'Frontana' is still resistant (it has Lr13, Lr34 and two further unidentified resistance genes; Singh and Rajaram, 1992). Similarly, no pathotype overcoming the genes Lr9 and Lr24 has yet been identified. Breeding for resistance against leaf-rust is still based mainly on major genes and adopts the strategy of pyramiding the best combination of genes. However, even with conventional pedigree selection procedures a good resistance level can be selected from parents with an intermediate resistance level (Pieters et al., 1991).

Glume blotch

Glume blotch, caused by *Septoria nodorum*, is often mentioned in conjunction with *S. tritici* (leaf blotch) since they both infect the leaves, but damage due to glume blotch may be more serious since it can move from the leaves to the glumes. Resistance against this disease may be indirect, based on the distance between leaves, and between the flag-leaf and the spike. Tall cultivars with long internodes are less affected; this resistance is a form of disease escape based on morphology. Under artificial infection conditions, the cultivar may be susceptible. Direct resistance against glume blotch in the leaves and glumes is controlled by different genes (Fried and Meister, 1987). Both are polygenic and additive (Ecker et al., 1989; Wilkinson et al., 1990). Selection is cumbersome, since the development of the disease has to be observed over long periods because it is highly weather-dependent (Brönnimann, 1968). It is almost impossible to determine the resistance level in single plants early in the selection program. The current Swiss breeding program is based on careful determination of the resistance level of possible parents and selection of the progeny late in the program (F6 generation). Development of genetic markers for quantitative trait loci (QTL) could advance *Septoria* breeding greatly.

Stem rust

Stem rust caused by *Puccinia graminis* sp. *tritici* presents analogous problems to leaf rust. It is absent or only a minor problem in some countries due to eradication of the haploid host barberry. Resistance is also based on major vertical genes. Noteworthy is the explication of virulence frequency observations made in North America by Vanderplank (1982). In particular cases the frequency of individuals carrying a particular virulence gene combination is significantly lower than expected. Therefore, Vanderplank postulated a stabilizing effect on the pathogen population due to an epistatic effect which caused two virulences to dissociate. If this could be proved it may lead to a new strategy in breeding.

Impact

Wheat resistance breeding has often produced cultivars resistant against some diseases with clear success in practice, but in most cases the pathogens were able to overcome those resistances sooner or later; in other cases the popularity of resistant cultivars decreased, not because of susceptibility but because new and more productive cultivars were released. Zwatz (1992) reported the progress of disease resistance in cereals in Austria. Noteworthy is the change of the average response to diseases of all cultivars used in Austria between 1975 and 1990: for leaf rust it changed from an average of 6.9 to 5.6 (on a scale 1–9, with 1 indicating completely immune and 9 indicating highly susceptible), which is a clear success. For other diseases the change was less obvious, and in the case of glume blotch, there was a slight increase in susceptibility (from 6.84 to 6.91), possibly due to a gradual height reduction of the popular varieties. Similar results have been reported from other countries (Zimmermann and Strass, 1991). Using even partially resistant cultivars the use of fungicides is economically no longer justified, neither to increase nor to stabilize yield (risk insurance; Zwatz, 1990).

In Switzerland, the disease resistance range of the recommended cultivars also shows clear progress, although in the late 1980s the situation has been clouded by the dominance of a single cultivar, rather susceptible to leaf rust, and to the intensification of production. Therefore no practical effect can be demonstrated other than a clear increase in the use of fungicides. Fortunately, new and highly productive cultivars with good resistance qualities are slowly replacing the susceptible cultivars. The potential for savings is high, since US$12–15 million are spent on fungicides in Switzerland per year on the 100 000 ha planted with wheat (Winzeler et al., 1990). Multi-line cultivars (NIL) carrying various resistances were developed (Fried et al., 1992), but have not been used in practice for regulatory reasons.

TOMATO (*LYCOPERSICON ESCULENTUM*)

The modern Western European kitchen is almost unimaginable without the tomato. Tomato production is associated with the technical revolution in agronomy. The demands of consumers (from industrial use to local consumption) has led to intensive breeding of new cultivars. The tomato has an important role not only in the developed world but also in Third World countries; the total production in 1992 was 70.4 million t on 2.9 million ha. Yield ranges from a high of 433 t ha^{-1} in The Netherlands (Western

Europe 37 t ha^{-1}, USA 58 t ha^{-1}) to a low of 18 t ha^{-1} for Asia (FAO, 1992). Tomato is a self-fertilizing crop with a high degree of cross-pollination under natural conditions; it is easy to breed and to maintain as a pure stand. Tomato is highly susceptible to a great number of pathogens, which can seriously limit production.

Breeding

Disease resistance breeding has been a clear goal for a long time. Resistances to the major diseases are controlled by single, mostly dominant genes, e.g. resistance to *Fusarium* wilt (genes I-1 to I-3), *Verticillium* wilt (Ve), leaf mould (*Fulvia fulva* (= *Cladosporium fulvum*) genes Cf1–Cf5), *Septoria* etc. Resistance to bacterial wilt (*Pseudomonas solanacearum*) and bacterial spot (*Xanthomonas campestris* pv. *vesicatoria*) is polygenically inherited (Callow, 1993). Cultivars which are resistant in a particular area may be susceptible in another. For example, the cultivars 'Venus' and 'Saturn' are resistant in the USA, but susceptible in Mexico. Most single-gene resistances are known to be race-specific. The original source of resistance is usually an individual of a related wild species found in the Andes. As most cultivars used in commercial planting today have resistances against several pathogens, the main diseases can be controlled by the appropriate choice of cultivar. Even if most of the resistances are vertical and the pathogen population adapts accordingly, it is easy to change to cultivars carrying another resistance. Virulence changes in pathogen population are relatively slow (*Fusarium* and *Verticillium* wilt are soil-borne) and often clearly hampered by the closed environment (glasshouses) in high-production systems. Breeding in this case is predominantly one step ahead of the pathogen. New techniques (transgenic plants) may even increase the advantage of the breeders over the various pathogens. Nevertheless, not all problems due to diseases have been resolved as new problems can arise quickly.

Impact

In the late 1970s and early 1980s plastic tunnels increasingly replaced field planting of tomato in southern Switzerland. Tunnels are now being replaced by glasshouses. All popular cultivars in 1994 carried resistance against TMV virus, *Verticillium*, *Fusarium* (mostly F2), the nematode *Meloidogyne incognita* and corky root rot caused by *Pyrenochaeta lycopersici*. Some cultivars also had resistance against *Fusarium* crown and root rot (*F. oxysporum* sp. *radicis-lycopersici*) and *Cladosporium* leaf mould. *Cladosporium* is often a problem, and appropriate studies showed that locally some of the Cf-genes are ineffective. However, the presence of cultivars with many efficient resistances is not due to farmers deliberately selecting resistant cultivars to avoid disease problems. It is dictated by the quality requirements of the market. Farmers do not acknowledge resistance; for them it is an irrelevant surplus. The disease problems that still occur, such as grey mould (*Botrytis cinerea*), black mould (*Alternaria alternata* f. sp. *lycopersici*) and in particular cases leaf mould, are due to lack of resistance in the cultivars and are resolved by the use of fungicides. Problems due to wilts which occurred with particular cultivars in the early 1980s were resolved by using grafts on resistant roots.

CONCLUSIONS

The arguments for increased efforts to breed resistant cultivars and their increased use can be separated into two groups: the first is more valid for the developed world, where production currently covers the needs of its population, and the second for a large group of countries—with the majority of the human population—which cannot produce enough to satisfy their needs.

The requirements of consumers and environmental concerns impose strong pressure to change intensive production systems towards systems with lower pesticide inputs. Resistance breeding is regarded as highly relevant, since widespread adoption of disease-resistant cultivars would solve many of the concerns expressed by environmental movements and consumers. However, a crucial question for the success of resistant cultivars is the faith of the producer in the marketing quality of the cultivars and in the superiority of the cultivar because of its resistance. If the producer is not convinced of the durability of the resistance, it will be difficult to substitute well-known susceptible cultivars by unfamiliar resistant cultivars.

Moreover, farmers will switch to resistant cultivars only if there is a pathogen problem without an easy solution. If resistant cultivars are already in use, or the problem is not pressing, the choice of cultivar is dictated by other reasons. Only clear awareness of the risk of a particular pathogen becoming a problem could influence a possible 'wrong choice'.

The responsibility of scientists, administrators and politicians in countries with insufficient food production is to favour strategies to increase and maintain this production with the constraint that the increase should not be detrimental to the environment (sustainable agriculture). As, for economic reasons, inputs can be increased only to a limited extent, all possible measures should be taken to help stabilize and increase yields. Inputs to control diseases are very cost-effective under conditions of high financial yield. For example, a 10% loss in wheat under the Swiss production system is equivalent to a loss of 300–400 ECU ha^{-1}, but only to 200 kg ha^{-1} or about 22 ECU in low production systems due to price and yield differences. Under such conditions only resistant cultivars are cost-effective; they may also help to avoid sudden, unexpected epidemics due to particular weather patterns. In developing countries, it is even more important to strive for durability of disease resistance, or at least to minimize the impact of a loss of resistance. The choice of breeding strategies depends more on the crop, the availability of a resistance type and a source, and the pathogen than on use in intensive or sustainable agriculture (see also Singh, 1986; Callow, 1993). In all cases, the goal of breeding strategies should be, as far as possible, to produce cultivars with durable resistance.

The usefulness of resistance in controlling pathogen epidemics and possible approaches to increase durability have been discussed elsewhere (Wolfe and Gessler, 1992). One approach is to use quantitative resistance, inherited oligo- or polygenically. It has been observed that the durable resistance of some cultivars are often of this type, but the assumption that any cultivar with resistance inherited this way is durable is dangerously optimistic (Wolfe, 1993). Moreover, as the above examples show, this type of resistance is rarely available and often involves time-consuming breeding strategies and selection procedures.

The use of major resistance genes has often led to rapid breakdown of the resistance. Here the assumption that major genes are in all cases ephemeral is too pessimistic. Which type of resistance should be used has to be considered case by case.

Apple culture appears to be a classic case, since it carries the complexity which we should also expect in other systems. The Vf-resistance introgressed into apple was tested carefully over 40 years in scab-favourable situations all over the world; breakdown of the resistance was not expected (Crosby et al., 1992). However, it occurred (Parisi et al., 1993), and made breeders and producers doubt the relevance of resistant cultivars. As documented by more than 100 years of observation and research, the breakdown of resistance in apple is a natural and common phenomenon, being so frequent (specific resistances in commercial cultivars) that we do not notice it except in special cases (e.g. the Vf-resistance).

Commercial cultivars differ from each other in their level of resistance; some of those differences were never affected regardless of the area covered by a single cultivar. In these cases the resistances can truly be called general, but some cultivars that are moderately susceptible in one area are highly susceptible in others and are ranked accordingly (Gessler, 1994; compare Govi, 1996 to Götz and Silbereisen, 1989).

Parents for specific crosses should be cultivars with truly general resistance even in breeding steps where strong resistance is introgressed from an other source. Many problems currently seem insoluble (e.g. recognition of several functionally different resistances in a single individual), but new biotechnological aids are now being developed. Some of these, such as gene transfer, are being questioned and may even delay widespread application of biotechnology because of patent protection. Also they may have negative effects due to their own success, reducing variability in the crops instead of increasing it, and thus increasing dependence on inputs (e.g. herbicide-resistant cultivars) and the risk of breakdown (large areas with the same genotype). On the other hand, some biotechnological methods can greatly increase the efficiency of breeding. In tomatoes produced for processing, traditional breeding has rather reduced the genetic variability and increased reliance on inputs. Some argue that genetic engineering would reverse this trend (Hauptli et al., 1990).

However biotechnology can be of more immediate advantage by integration with classical breeding. Plant genome maps can associate the information on traits of interest with DNA markers. Once these traits can be identified, they can be selected, manipulated and complemented (see Allen, 1994).

The problem of recognizing different resistances in a single individual due to major genes can be overcome with modern genomic markers closely linked to the R-genes. Quantitative trait loci (QTL) markers are much more difficult to find, but not impossible. They can be useful for selecting individuals with as many QTLs as possible and with major gene resistance. The use of QTL markers could also avoid the *Vertifolia* effect, defined by Vanderplank (1963, 1968) as loss of general resistance in the process of breeding for vertical resistance. The erosion of general resistance is due to the unrecognized loss of additive minor genes during back-crossing.

Instead of pyramiding R-genes in a single cultivar, several cultivars adapted for planting in mixtures, each with a functionally different resistance, can be bred and selected. As Wolfe (1992) points out, different resistances should be kept in separate cultivars in order to maximize varietal diversity and as a consequence to maximize diversity within the pathogen population. Planting strategies oriented towards the use of cultivar mixtures can stabilize yield and increase durability compared with mono-culture (Wolfe, 1993).

The use of such complementary strategies is imperative in a sustainable agriculture (low- or no-input production systems). Their use may also be advantageous in intensive agriculture (see Wolfe, 1993). With the appropriate cultivar combination, mixtures can be devised that impede the pathogen from reaching intolerable epidemic levels. Genetic uniformity in a crop (field) is much more risky than diversity. Natural, successful 'planting systems' are usually based on genetic variability; our concept for the next century should follow this lesson. Enhancing sustainability, increasing production and decreasing pesticide input are possible by coordinated action on the entire cropping system. Breeders who have worked traditionally on a particular set of problems in a crop must increasingly consider modification of cultivars to fit within the overall cropping pattern of a particular climate and society. Of prime importance is the maintenance of genetic variability, in order to have a reservoir of adaptability that acts as a buffer against harmful environments (FAO, 1993b), agricultural practices and other changes.

OUTLOOK

Modern agriculture will have to increase production of essential food crops substantially over the next 50 years. Production has more than doubled in the last 35 years, and this pace should be maintained. The main factors for these increases were newly cultivated land, and pesticide and chemical fertilizer use (FAO, 1993a). Pesticides will also be one of the main factors of increment in agricultural production in the near future. Strong arguments can be developed in favour of pesticides and chemical fertilization even with respect to sustainable agriculture (Deichner, 1995). However, considering consumer perceptions in the industrialized nations, the current information status of farmers in the Third World, and economic factors, plant resistance against pathogens should be favoured. Breeding can and will produce such plants. Under particular conditions transgenic resistant plants may also be developed. The problem arising will not be the availability of resistant cultivars, but their introduction in practice, since a new cultivar is often associated with a change in characteristics. Moreover, even if this can be accomplished, plant resistance can only be of importance if strategies granting durability of resistance can be implemented. The primary goal of plant breeding and agronomy should be the development of strategies—appropriate to crop and local situation—rendering resistance durable. Otherwise, plant pathogens will cause sporadic and erratic high losses, or regular losses when no fungicides are used. We recognize the high benefit/cost ratio of plant resistance against pathogens. This should be an incentive to develop durable resistance. However, as the beneficiaries (farmers) are not directly and visibly the same organizations and persons as those carrying the costs (mostly state-financed agencies), this argument generally fails. Scientists cannot count on support in proportion to benefit to the same extent as do their colleagues involved in pesticide research and application.

ACKNOWLEDGEMENTS

I thank Dr. B. Koller for editing this paper. Research mentioned in this paper was supported by the Swiss National Foundation for Scientific Research grants 31-29928.90 and 31-36271.92, grant NF-Biotechnology Module 6 5002-034614, the Swiss Federal Office for Education and Science grant No. A6010.

REFERENCES

Aderhold, R. (1899) Auf welche Weise können wir dem immer weiteren Umsichgreifen des Fusic-ladium in unseren Apfelkulturen begegnen und welche Sorten haben sich bisher dem Pilz gegenüber am widerstandsfähigsten gezeigt? *Pomologische Monatshefte*, XLV, Heft 12 266–272.

Aderhold, R. (1902) Aufforderung zum allgemeinen Kampf gegen die Fusicladium- oder sog. Schorfkrankheit des Kernobstes. Kaiserliches Gesundheitsamt. *Biologische Abteilung für Land- und Forstwirtschaft* **1**, 4 pp.

Allavena, A. and Ranalli, P. (1989) Miglioramento genetico e nuove varietà di fagiolo. *Informatore Agrario* **45**, 49–55.

Allen, F.L. (1994) Usefulness of plant genome mapping to plant breeding. In Gresshoff, P.M. (Ed.) *Plant Genome Analysis*. CRC Press, Boca Raton, FL, 247 pp.

Andrivon, D. (1993) Nomenclature for pathogenicity and virulence: The need for precision. *Phytopathology* **83**, 889–890.

Bagga, H.S. and Boone, D.M. (1968a) Genes in *Venturia inaequalis* controlling pathogenicity to crabapples. *Phytopathology* **58**, 1176–1182.

Bagga, H.S. and Boone, D.M. (1968b) Inheritance of resistance to *Venturia inaequalis* in crabapples. *Phytopathology* **58**, 1183–1187.

Bartos, P., Maly, J. and Parizek, P. (1990) Dauerhaftigkeit gegen Blattkrankheiten-Zweckmässig-keit und Wege zu ihrer Erreichung. *Arbeitstagung der Saatzuchtleiter innerhalb der Vereinigung österreichischer Pflanzenzüchter*, 163–172.

Beebe, S.E. and Corrales, M.P. (1991) Breeding for disease resistance. In van Schoonhoven, A. and Voysest, O. (Eds) *Common Beans: Research for Crop Improvement* CAB International, CIAT London, pp. 561–618.

Biffin, R.H. (1907) Studies on the inheritance of disease resistance. *Journal of Agricultural Science, Cambridge* **2**, 109.

Blaise, Ph. and Gessler, C. (1994) Cultivar mixtures in apple orchards as a means to control apple scab? *Norwegian Journal of Agricultural Sciences* **17** (Suppl.), 105–112.

Boone, D.M. (1971) Genetics of *Venturia inaequalis*. *Annual Review of Phytopathology* **9**, 297–318.

Brönnimann, A. (1968) Zur Kenntniss von *Septoria nodorum* Berk. dem Erreger der Spelzenbräune und einer Blattdürre des Weizen. *Phytopathologische Zeitschrift* **61**, 101–146.

Callow, G. (1993) Tomato, *Lycopersicon esculantum* Miller. In: Callow, G. and Bergh, B.O. (Eds) *Genetic Improvement of Vegetable Crops*. Pergamon Press, Oxford, 833 pp.

Cepts, P. and Debouck, D. (1991) Origin, Domestication and Evolution of the Common Bean (*Phaseolus vulgaris* L.). *Common Beans; Research for Crop Improvement*. CIAT London, pp. 7–54.

Christ, B.J. and Groth, J.V. (1982a) Inheritance of virulence to three bean cultivars in three isolates of the bean rust pathogen *Vromyces phaseoli*. Phytopathology **72**, 767–770.

Christ, B.J. and Groth, J.V. (1982b) Inheritance of resistance in three cultivars of beans to the bean rust pathogen and the interaction of virulence and resistance genes *Vromyces phaseoli*. Phytopathology **72**, 771–73.

Crandall, C.S. (1926) Apple breeding at the University of Illinois. Illinois Agric. Exp. Stn. bull. **275**, 341–600.

Crosby, J.A., Janick, J., Pecknold, P.C., Korban, S.S., O'Connor, P.A., Ries, S.M., Goffreda, J. and Voordeckers, A. (1992) Breeding apples for scab resistance, 1945–1990. *Fruit Varieties Journal* **46**(3), 145–166.

Crowe, A.D. (1975) "Nova Easygro" apple. Fruit Varieties Journal. **29**, 76.

Dayton, D.F., Shay, J.R. and Hough, L.F. (1953) Apple scab resistance from R12740-7A, a Russian apple. *Proceedings of the American Society for Horticultural Science* **62**, 334–340.

Dayton, D.F. and Williams, E.B. (1967) Independent genes in *Malus* for resistance to *Venturia inaequalis*. *American Society for Horticultural Science* **92**, 89–94.

Dayton, D.F. and Williams, E.B. (1970) Additional allelic genes in *Malus* for scab resistance of two reaction types. J. Amer. Soc. Hort. Sci. **95**, 735–736.

Dayton, D.F., Mowry, J.B., Hough, L.F., Bailey C.H., Williams, E.B., Janik, J., and Emerson F.H. (1970) Prima – an early fall red apple with resistance to apple scab. Fruit Vars Hortic. Dig. **24**, 20–22.

Deichner, K. (1995) Agrarchemie ist Umweltschutz. Von der Moral der Agrarchemie und ihren Kritiker. BASF AG. 67117, Limburgherhof, 83 pp.

Dixon, R.A., Paiva, N.L. and Bahttacharyya, M.K. (1995) Engineering disease resistance in plants: An overview. In Singh, R.P. and Singh, U.S. (Eds) *Molecular Methods in Plant Pathology*. CRC Press, Boca Raton, FL, 523 pp.

Dosba, F., Doussinault, G., Jahier, J. and Trottet, M. (1982) Utilisation d'espèces sauvages dans l'amelioration de l'état sanitaire du blé tendre: *Triticum aestivum*, In: La Sélection des Plantes pour la résistance aux maladies. Colloques franco-israélien, Bordeaux mars 1982. Serie Colloques de l'INRA 11, INRA Publ. Versailles, Paris France 231 pp. 75–85.

Ecker, R., Dinoor, A. and Cahaner, A. (1989) The inheritance of resistance to *Septoria* glume blotch in common bread wheat, *Triticum aestivum*. *Plant Breeding* 102, 113–121.

FAO (1990) *Production Yearbook*, Rome.

FAO (1992) *Production Yearbook*, Rome.

FAO (1993a) *Production Yearbook*, Rome.

FAO (1993b) Safeguarding the diversity of plant genetic resources. In *World Agriculture*. Vol. 1. Sterling Publications Limited, Hong Kong, pp. 11–14.

Flor, H. (1942) Inheritance of pathogenicity in *Melampsora lini*. *Phytopathology* 32, 653–669.

Flor, H. (1956) The complementary genic system in flax and flax rust. *Advances in Genetics* 8, 29–54.

Flor, H. (1971) Current status of the gene-for-gene concept. *Annual Review of Phytopathology* 9, 275–296.

Frank, S.A. (1994) Recognition and polymorphism in host parasite genetics. *Philosophical Transactions of the Royal Society of London* 346, 283–293.

Fried, P.M. and Meister, E. (1987) Inheritance of leaf and head resistance of winter wheat to *Septoria nodorum* in a diallel cross. *Phytopathology* 77, 1371–1375.

Fried, P.M., Streckeisen, Ph., Winzeler, H., Winzeler, M., Forrer, H.R. and Saurer, W. (1992) Development of a multiline for cereal rust and powdery mildew resistance. *Vortäge Pflanzezüchtung* 24, 295–298.

Fried, P.M., Barben, H., Keller, S., Müller, M.D., Winzeler, H., Winzeler, M. and Weisskopf, P. (1993) Expertise betreffend Möglichkeiten des Einsatzes biotechnologischer Methoden zur Erhöhung der Resistenz gegen Krankheiten und Schädlingen wichtiger Kulturpflanzen in der Schweiz. Report to the Swiss National Research Funds, 90 pp.

Gessler, C. (1989) Genetics of the interaction *Venturia inaequalis*–*Malus*: The conflict between theory and reality. In Gessler, C., Butt, D. and Koller, B.(Eds) *Integrated Control of Pome Fruit Diseases*. II. OILB-WPRS Bulletin XII/6, pp. 168–190.

Gessler, C. (1994) Biology and biotechnology in strategies to control apple scab. *Norwegian Journal of Agricultural Sciences* 17 (Suppl.), 337–353.

Gessler, C. and Blaise, Ph. (1994) Differential resistance in apple against scab and its use in breeding and in orchard planting strategies to control the disease. In Development of plant breeding Vol 1. Schmidt, H. and M. Kellerhals (eds) Progress in temperate fruit breeding. Kluwer Academic Press. 470 pp, 99–104.

Gessler, C., Eggenschwiler, M. and Sierotzki, H. (1993) Vertikale gegen Schorf in anfälligen Apfelsorten. *Schweizerische Landwirtschaftloche Forschung* 32(3), 401–410.

Gianfranceschi, L., Koller, B., Seglias, N., Kellerhals, M. and Gessler, C. (1996) Molecular selection in apple for resistance to scab caused by *Venturia inaequalis*. Theor. Appl. Genet., 199–204.

Götz, G. and Silbereisen, R. (1989) *Obstsorten-Atlas*. Eugen Ulmer, Stuttgart, Germany, 365 pp.

Govi, G. (1966) Sensibilità varietale del melo alle infezioni di Venturia inaequalis. *Phytopathologia Mediterranea* 5, 145–146.

Habtu, A. and Zadoks, J.C. (1995) Components of partial resistance in *Phaseolus* beans against an Ethiopian isolate of bean rust. *Euphytica* 83(2), 95–102.

Hatton, R.G. (1939) Rootstocks work of East Malling. *Scientia Horticultural* 7, 7–16.

Hauptli, H., Katz, D., Thomas, B.R. and Goodman, R.M. (1990) Biotechnology and crop breeding for sustainable agriculture. In Edwards, A., Lal, R., Madden, P., Miller, R.H. and House, G. (Eds) *Sustainable Agricultural Systems*. Soil and Water Conservation Society, Ankeny, IA, pp. 141–156.

Hegi, G. (1963) *Illustrierte Flora Mitteleuropas (1905–1931)*. Vol. IV/2. Carl Hansen, München, pp. 745–754.

Hidalgo, R. (1991) *CIAT'S World Phaseolus Collection. Common Beans: Research for Crop Improvement.* CIAT London, pp. 163–197.

Horsfall, J.G. and Cowling, E.B. (1977) The sociology of plant pathology. In Horsfall, J.G. and Cowling, E.B. (Eds) *Plant Disease.* Vol. I. Academic Press London.

Hough, L.F., Shay, J.R. and Dayton, D.F. (1953) Apple scab resistance from *Malus floribunda*. Proc. Americ. Soc. Hort. Sci. **62**, 341–347.

Innes, R. (1995) *Arabidopsis* as a model host in molecular plant pathology. In Singh, R.P. and Singh, U.S. (Eds) *Molecular Methods in Plant Pathology.* CRC Press, Boca Raton, FL, 523 pp.

Johnson, R. (1981) Durable resistance: Definition of, genetic control, and attainment in plant breeding. *Phytopathology* **71**, 567–568.

Kellerhals, M. (1989). Breeding of pome fruits with stable resistance to diseases. In Gessler, C., Butt, D. and Koller, B.(Eds) *Integrated control of Pome Fruit Diseases.* II OILB-WPRS Bulletin XII/6, pp. 116–129.

Koch, Th., Kellerhals, M. and Gessler, C. (1996) Virulence analysis of *Venturia inequalis* populations. In: Polesny, F., Müller, W. and Olszak, R.W. (eds). International Conference on Integrated Fruit Production, Cedzyna Poland 28 August–2 September 1995. OILB-WPRS Bulletin 19/4, 400–401.

Lespinasse, Y., Olivier, J.M., Lespinasse, J.M. and LeLezec, M. (1985) Florina Quérina la résistance du pommier à la tavelure. Arboriculture Fruitiére **378**, 43–47.

Lutzeyer, H.-J., Pülschen, L., Compart, W. and Scholaen, S. (1993) Neue Erkenntnisse über Pflanzenschutz in Plantagenkulturen dargestellt am Beispiel Kaffee. *Forschungsbericht des Bundesministeriums für wirtschaftliche Zusammenarbeit und Entwicklung*, Band 107, pp. 148.

McIntosh, R.A. (1988) Catalogue of gene symbols for what. 7th International Wheat Genetics Symposium, 1988. Cambridge, Vol. 2, Cambridge, UK, pp. 1225–1322.

Messiaen, C.M. (1981) *Les Variétés Resistantes.* INRA, Paris, pp. 215–232.

Moreno-Ruiz, G. and Castillo-Zapata, J. (1990) The variety Columbia: A variety of coffee with resistance to rust (*Hemileia vastatrix* Berk. & Br.). CENICAFE Publication, Chinchin-Cladas, Columbia, 27 pp.

Morgan, J. and Richards, A. (1993) *The Book of Apples.* Ebury Press, London, 304 pp.

Muller, R.A. (1986) General scheme of studies conducted by IRCC concerning varieties of coffee resistant to orange leaf rust (*Hemileia vastatrix*). FAO Plant Production and Protection Paper, Vol. 70, pp. 74–79.

Norelli, J.L., Aldwinckle, H.S., Destéfano-Beltràn, L. and Jaynes, J.M. (1994) Transgenic 'Malling 26' apple expressing the attacin E gene has increased resistance to *Erwinia amylovora*. In Schmidt, H. and Kellerhals, M. (Eds) *Progress in Temperate Fruit Breeding.* Kluwer, Dordrecht, 470 pp.

Omondi, C.O. (1994) Resistance to coffee berry disease in Arabica coffee variety 'Ruiru 11'. Plant-breed. P. Parey Berlin **112**(3) p. 256–259.

Opile, W.R. and Agwanda, C.O. (1993) Propagation and distribution of cultivar Ruiru 11: a review. Kenya-Coffee **58**, 1496–1508; 14 ref.

Orton, W.A. (1909) Yearbook of Agriculture, US Dept. Agr. Gov't Printing Office pp. 453–464

Panse, A. (1988) Unterschungen an Linienmischungen von *Phaseolus vulgaris* in den Tropen. Dissertation, Technical University of München.

Parisi, L., Lespinasse, Y., Guillaumes, J. and Krüger, J. (1993) A new race of *Venturia inaequalis* virulent to apples with resistance due to the Vf gene. *Phytopathology* **83**, 533–537.

Pieters, R., Aalders, A. and Van-der-Beek, J. (1991) Practical breeding for horizontal resistance in wheat to brown rust. *FAO Plant Protection Bulletin* **39**, 35–42.

Roberts, A.L. and Crute, I.R. (1994) Apple crab resistance from *Malus floribunda* 821 (Vf) is rendered ineffective by isolates of *Venturia inaequalis* from *Malus floribunda*. In Schmidt, H. and Kellerhals, M. (Eds) *Progress in Temperate Fruit Breeding.* Kluwer, Dordrecht, 470 pp.

Rousselle, G.L., Williams, E.B. and Hough, L.F. (1974) Modification of the level of resistance to apple scab from the Vf gene. In *Proceedings of the XIXth International Horticultural Congress*, Vol. III. Warsaw, International Society for Horticultural Science, pp. 19–26.

Russell, G.E. (1978) *Plant Breeding for Pest and Disease Resistance. Studies in the Agricultural and Food Sciences.* Butterworth, London, Boston, 485 pp.

Saunders, D.A. (Ed.) (1991) *Wheat for the Non-traditional Warm Areas.* CIMMYT, Mexico DF, 545 pp.

Schaffner, D., Koller, B., Müller, K. and Wolfe, M.S. (1992) Response of populations of *Ersiphe graminis* f. sp. *hordei* to large-scale use of variety mixtures. In Zeller, F.J. and Fischbeck, G. (Eds) *Cereal Rusts and Mildews*. Vorträge für Pflanzenzüchtung Heft, Vol. 24, pp. 317–319.

Sierotzki, H., Eggenschwiler, M., McDermott, J. and Gessler, C. (1994a) Specific virulence of isolates of *Venturia inaequalis* on 'susceptible' apple cultivars. *Norwegian Journal of Agricultural Sciences*, Supplement No. 17: 83–93.

Sierotzki, H., Eggenschwiler, M., Boillat, O., McDermott, J. and Gessler, C. (1994b) Detection of variation in virulence toward susceptible apple cultivars in natural populations of *Venturia inaequalis*. Phytopathology **84**, 1005–1009.

Silbernagel, M.J. and Hannan, R.M. (1988) Utilisation of genetic resources in the development of commercial bean cultivars in the USA. In Gepts, P. (Ed.) Genetic Resources of *Phaseolus* beans. Kluwer, Dordrecht, pp. 561–600.

Simmonds, N.W. and Rajaram, S. (Eds) (1988) *Breeding Strategies for Resistance to the Rust of Wheat*. CIMMYT, Mexico DF, 151 pp.

Singh, D.P. (1986) *Breeding for Resistance to Diseases and Insect Pests. Crop Protection Monographs*. Springer, Berlin, Heidelberg, 222 pp.

Singh, S.P. (1991) *Bean Genetics. Common Beans: Research for Crop Improvement*. CIAT London, pp. 199–286.

Singh, R.P. and Rajaram, S. (1992) Durable resistance to *Puccinia recondita tritici* in CIMMYT bread wheats: Genetic bases and breeding approaches. *Vorträge Pflanzezüchtung* **24**, 239–241.

Van der Graaff, N.A. (1986) *Coffee. Breeding for Durable Resistance in Perennial Crops*. FAO Plant Production and Protection Paper, Vol. 70, Rome pp. 49–74.

Vanderplank, J.E. (1963) *Plant Diseases: Epidemics and Control*. Academic Press, New York.

Vanderplank, J.E. (1968) *Disease Resistance in Plants*. Academic Press, New York, 206 pp.

Vanderplank, J.E. (1982) *Host–Pathogen Interaction in Plant Disease*. Academic Press, New York, 207 pp.

Van Schoonhoven, A. and Voysest, O. (1991) *Common Beans. Research for Crop Improvement*. CIAT London.

Vavilov, N.I. (1920) "The Origin, Variation, Immunity and Breeding of cultivated plants" Waltham, Mass., Chronica Botanica (1949–1959). (Translated from Russian by K.S. Chester).

Verreet, J.A. (1991) Grundlagen des integrierten Pflanzenschutes gegen Pilzkrankheiten in Weizenanbausystemen. *Nachrichtenblatt Deutscher Pflanzenschutzdienst* **43**(6), 119–132.

Voss, J. and Graf, O. (1991) On-Farm Research in the Great Lakes Region of Africa. *Common Beans. Research for Crop Improvement*. CIAT London, pp. 891–929.

Wallace, E. (1913) Scab disease of apples. *New York Cornell Agricultural Experimental Station Bulletin* **335**, 543–642.

Wilkinson, C.A., Murphy, J.P. and Rufty, R.C. (1990) Diallel analysis of components of partial resistance to *Septoria nodorum* in wheat. *Plant Disease* **74**, 47–50.

Williams, E.B. and Brown, A.G. (1968) A new physiologic race of *Venturia inaequalis*, incitant of apple scab. *Plant Disease Reporter* **52**(10), 799–801.

Williams, E.B. and Kuc, J. (1969) Resistance in *Malus* to *Venturia inaequalis*. *Annual Review of Phytopathology* **7**, 223–246.

Williams, E.B., Janik, J., Emerson, F.H., Hough, L.F., Bailey, C.H., Dayton, D.F., and Mowry, J.B. (1972) Priscilla — a fall red apple with resistance to apple scab. Fruit Vars Hortic. Dig. **26**, 34–35.

Williams, E.B., Janik, J., Emerson, F.H., Dayton, D.F., Mowry, J.B., Hough, L.F., and Bailey, C.H., (1975) Sir Prize apple Hort Science **10**, 281–282

Winzeler, M., Streckeisen, Ph., Winzeler, H. and Fried, P.M. (1990) Züchtung auf dauerhafte Mehltauresistenz bei Weizen und Dinkel. In *Arbeitstagung der Saatzuchtleiter innerhalb der Vereinigung österreichischer Pflanzezüchter, 1990.* pp. 173–179.

Wolfe, M.S. (1983) Genetic strategies and their value in disease control. In Kommwdahl, T. and Williams, Ph. (Eds) *Challenging Problems in Plant Health*. Phytopathology Society, St. Paul, MN, pp. 461–473.

Wolfe, M.S. (1985) The current status and prospects of multiline cultivars and cultivar mixtures for disease resistance. *Annual Review of Phytopathology* **23**, 251–273.

Wolfe, M.S. (1992) Barley diseases: Maintaining the value of our varieties. In Munck, L. (Ed.) *Barley Genetics VI*. Vol. II. Munksgaard International Publishers, Copenhagen pp. 1055–1067.

Wolfe, M.S. (1993) Can the strategic use of disease-resistant hosts protect their inherent durability? In Jacobs, Th. and Parleviet, J.E. (Eds) *Durability of Disease Resistance*. Kluwer, Dordrecht, pp. 83–96.

Wolfe, M.S. and Gessler, C. (1992) Resistance genes in breeding: Epidemiology. In Boller, T. and Meins, F. (Eds) *Plant Gene Research. Genes Involved in Plant Defense*. Springer, New York pp. 3–23.

Young, N.D. (1995) Isolation and cloning of plant disease resistance genes. In Singh, R.P. and Singh, U.S. (Eds) *Molecular Methods in Plant Pathology*. CRC Press, Boca Raton, FL, 523 pp.

Zadoks, J.C. (1972) Modern concepts of disease resistance in cereals. In Lupton, F.G., Jenkins, G. and Johnson, R. (Eds) *Proceedings of the 6th Congress of Eucarpia*. Eucarpia, Cambridge, UK, 1971, p. 98.

Zadoks, J.C. (1993) The partial past. In Jacobs, Th. and Parleviet, J.E. (Eds) *Durability of Disease Resistance*. Kluwer, Dordrecht, pp. 11–22.

Zimmermann, G. and Strass, F. (1991) Ertrag und Qualitità von Winterweizensorten bei unterschiedlicher Fungizidbehandlung. *Bayrisches Landwirtschaftliches Jahrbuch* **68**(5), 635–648.

Zwatz, B. (1990) Weitere Ergebnisse aus der langjährigen Versuchsreihe zur Extensivierung durch sortenspezifische Minimierung der Chemotherapie. *Arbeitstagung der Saatzuchtleiter innerhalb der Vereinigung österreichischer Pflanzenzüchter, 1990.* pp. 155–162.

Zwatz, B. (1992) Progress in disease resistance in cereals and strategy in plant protection in Austria. *Vorträge Pflanzenzüchtung* **20**, 241–245.

CHAPTER 9

What is Durable Resistance? A General Outline

Jan E. Parlevliet

Wageningen Agricultural University, Wageningen, The Netherlands

INTRODUCTION

Plants employ a great variety of defence mechanisms to cope with the multitude of organisms that try to exploit them. These defence mechanisms can be classified as avoidance, resistance or tolerance mechanisms. Avoidance is mainly used against animal parasites as these have sensory abilities. Tolerance, i.e. enduring the parasite while sustaining relatively little damage, does not seem to play a significant role with pathogens. In breeding against pathogens it is resistance that is used almost exclusively. Resistance mechanisms, i.e. reducing the growth and/or development rate of the pathogen, are nearly always of a biochemical nature.

Soon after resistance to pathogens was introduced it became apparent that pathogen populations could adapt to such resistances; the resistance 'broke down'. However, not all resistance broke down, and differences in the durability of resistance became apparent.

MEASURING RESISTANCE

Ideally one should measure the amount of pathogen present at a given moment compared with the amount present on or in an extremely susceptible control cultivar. The larger the difference in amount the larger the difference in susceptibility/resistance. To measure the amount of pathogen is not normally possible because the pathogen is either invisible or only partially visible. However, one can evaluate the direct or indirect effects of the pathogen on the host if the pathogen itself is not visible. In this respect it is possible to group the pathogens roughly into three types.

(1) Pathogens that are partially visible, such as the ectoparasitic powdery mildews (most of the pathogen visible) and the rusts.

(2) Pathogens whose direct effects can be assessed; the pathogen itself is not visible but its presence is recognizable by discoloured tissue, e.g. *Ascochyta* spp.

(3) Cases where the amount of pathogen has to be assessed through the indirect effects on

Techniques for Reducing Pesticide Use. Edited by D. Pimentel
© 1997 John Wiley & Sons Ltd.

the host. These are the true disease symptoms, e.g. wilting with vascular pathogens (*Fusarium oxysporum*), leaf rolling, stunted growth, leaf mosaic, etc., caused by viruses. When the disease symptoms are restricted to the indirect effects of the pathogen the plant is diseased. Signs of direct effects show in the tissue invaded by the pathogen, mostly through discolouration. Tissues or plant parts not invaded have a normal appearance. These plants are affected, not diseased.

Experience has shown that assessing the amount of tissue affected by pathogens of types (1) and (2) gives a good estimate of the amount of pathogen present. Assessing resistance through assessing the amount of tissue affected relative to that of a highly susceptible control is therefore a good method.

On the other hand, the true disease symptoms, the indirect effects caused by disease inciting pathogens such as viruses, form a much less reliable way of assessing the amount of pathogen present. The relationship between the severity of such disease symptoms and the amount of pathogen present may vary from reasonable to poor.

The amount of tissue affected is in general a good estimator of the amount of pathogen present. The amount of pathogen present, however, is not only dependent on the level of resistance of the host cultivar. Other factors, such as those described below, may and do interfere with it.

(i)   Interplot interference. Screening for resistance is generally carried out in small adjacent plots. A fairly resistant entry may receive an abundance of inoculum if it has a highly susceptible neighbour. The amount of pathogen on the fairly resistant entry can then be enlarged very considerably, especially with airborne pathogens, resulting in an underestimation of the level of resistance of that entry. With barley leaf rust, where the affected leaf area consists of a myriad tiny individual sporulating infections, this interplot interference is very pronounced, as shown in Table 9.1. In adjacent plots, with interplot interference operating, the range of resistance was only 16-fold. In isolated plots, when interplot interference was absent, the range was about 1000-fold. Interplot interference may not only underestimate the partial resistance, but may also cause the ranking order to change as it does for powdery mildew in barley. Single-row plots gave a ranking order of the cultivars different from the one obtained in large isolated plots (Nørgaard Knudsen et al., 1986). The authors concluded that for a reliable selection for partial resistance, plots of about $1.4\,m^2$ were advisable. However, not all airborne pathogens cause interplot interference. For stripe rust in wheat no interplot interference of any significance could be observed, and even single hill plots next to spreader rows gave a representative assessment of resistance (Parlevliet and Danial, 1992). Interplot interference in airborne pathogens appears to vary greatly. It seems that this interference is greater the more the total disease is based on more and smaller individual infections.

(ii) Earliness. If the cultivars differ considerably in earliness, the period of exposure to the pathogen varies greatly because assessment is usually done at the same time for all varieties. Resistance to head blight caused by *Fusarium* in wheat is considerably overestimated in late cultivars for this reason. The same is true for *Septoria* leaf and glume blotch in wheat; the later the cultivar, the lower the blotch scores.

(iii) Inoculum density. This may obscure small but real differences in resistance if high

Table 9.1   Percentage leaf area of seven spring barley cultivars affected by *Puccinia hordei* in plots well isolated from one another (no interplot interference) and in adjacent single-row plots (strong interplot interference). Means over 2 years relative to the extremely susceptible L94, set at 100%

Cultivars	Isolated plots	Adjacent plots
L94	100	100
Mamie	75	92
Sultan	23	56
Zephyr	16	39
Volla	3	27
Julia	0.5	17
Vada	0.1	6

Source: Parlevliet and Van Ommeren, 1975

inoculum densities are used. At too low a density escapes can be confounded with resistance.

(iv) Plant habit. This may also affect the assessment of resistance. Dense crops and short plants tend to increase the amount of tissue affected, and loose crops and tall plants tend to decrease it. This is probably due to micro-climatic effects. Short wheat cultivars are more affected by *Septoria* leaf and glume blotch than tall cultivars.

EFFECT OF FARMING SYSTEM ON THE DURABILITY OF RESISTANCE

The more inoculum of a pathogen is present, the greater the chances that new variants (new races) can arise. In Western Europe, potato viruses are kept at a reasonably low level by a tightly controlled system of seed potato production. The race-specific resistance genes to potato viruses A, X and Y do not show any sign of losing their effectiveness. In a few cultivars this resistance has been effective for over 60 years. The good control of virus levels has most probably contributed to the durability of these resistance genes.

In flax in North America, the race-specific major genes against flax rust (*Melampsora lini*) lasted some 5 years in the period 1930–1950 when flax was still an important crop. From 1950 onward the resistance genes lasted much longer: from 10 to over 13 years. In this period the flax acreage decreased and farmers went over to resistant cultivars much faster. In The Netherlands, the race-specific major genes were used from 1962 onward. In 1962 the crop was still of some importance, but it rapidly became a minor crop. In the period 1962–1989, 13 cultivars were introduced carrying major gene resistance. Four of them were grown for a period of 16–20 years. Resistance in these cultivars did not break down, with the exception of one cultivar, 'Berber', whose resistance became ineffective after 2 years of use despite the fact that it was rarely grown (less than 2% of the flax acreage in the 4 years of its existence). The smaller the acreage of flax became and the higher the percentage of resistant cultivars grown, the more durable the resistance became.

Furasium oxysporum f.sp. *pisi* is a serious pathogen of peas. At least four races are recognized and resistance to these races is governed by single dominant genes (Haglund and Kraft, 1979; Meindert, 1981). In Western Europe most *Pisum sativum* cultivars are resistant to the prevailing Race 1 of the pathogen, and this resistance has been effective for more than 30 years. As with flax, the area covered by the crop has been limited in all those

Table 9.2   Host–pathogen systems with at least 10 race-specific resistance genes and many races known (Group A pathogens)

Pathogen	Host	
Fungi		
Puccinia graminis f.sp. *tritici*	Wheat	B, S, air-borne
Puccinia coronata	Oats	B, S, air-borne
Puccinia recondita f.sp. *tritici*	Wheat	B, S, air-borne
Puccinia sorghi	Maize	B, S, air-borne
Melampsora lini	Flax	B, S, air-borne
Ustilago nuda	Barley	B, S, air-borne
Erysiphe graminis f.sp. *hordei*	Barley	B, S, airborne
Phytophthora infestans	Potato	HB, S, air-borne/splash-borne
Bremia lactucae	Lettuce	B, S, air-borne/splash-borne
Fulvia fulva	Tomato	B, S, air-borne
Colletotrichum lindemuthianum	Bean	HB, S, seed-borne/splash-borne
Pyricularia oryzae	Rice	HB, S, air-borne
Rhynchosporium secalis	Barley	HB, S, air-borne/splash-borne
Bacteria		
Xanthomonas campestris pv. *oryzae*	Rice	HB, S, splash-borne/water-borne

B, biotrophic, HB, hemibiotrophic, S, specialist.

years and more and more cultivars have become resistant, probably reducing the inoculum present.

The durability of resistance in such cases therefore depends on the farming system, and may change when the system changes.

EFFECT OF THE PATHOGEN ON THE DURABILITY OF RESISTANCE

Pathogens can be grouped in various ways based on aspects such as:

- taxonomic position (viruses, bacteria and fungi representing major taxons);
- parasitic way of life (biotrophs, i.e. pathogens that exploit the living cells (rusts, powdery mildews, viruses); hemibiotrophs, i.e. pathogens that are biotrophic at the lesion perimeter, but necrotrophic near the lesion centre (*Rhynchosporium secalis* in barley, *Phytophthora infestans* in potato); necrotrophs, i.e. pathogens that exploit the host tissue after killing it with toxins (*Septoria* leaf and glume blotch in wheat);
- degree of host specialization (specialists such as the formae speciales of *Puccinia graminis* and *Erysiphe graminis* in cereals versus generalists; viruses, for instance, are often less specialized in their host range than many fungi);
- way of dispersal (often expressed as soil-borne, air-borne, splash-borne or seed-borne pathogens).

There is a relationship between these aspects and the frequency with which non-durable resistance, based on major, race-specific genes, occurs. The pathogens in which new races develop easily and against which several to many race-specific resistance genes occur are very often specialized, biotrophic or hemibiotrophic, air-borne or splash-borne fungi. Table 9.2 gives a representative sample of these (Group A pathogens).

There is another large group of pathogens (Group B) where very few races of each are

Table 9.3   Host–pathogen systems with some races and a few race-specific resistance genes known
(Group B pathogens)

Pathogen	Host	
Fungi		
Fusarium oxysporum f.sp. *lycopersici*	Tomato	V, S, soil-borne
Fusarium oxysporum f.sp. *pisi*	Pea	V, S, soil-borne
Fusarium oxysporum f.sp. *tracheiphilum*	Cowpea	V, S, soil-borne
Fusarium oxysporum f.sp. *niveum*	Watermelon	V, S, soil-borne
Fusarium oxysporum f.sp. *conglutans*	Cabbage	V, S, soil-borne
Fusarium oxysporum f.sp. *vasinfectum*	Cotton	V, S, soil-borne
Trichometasphaeria turcica	Maize	N, S, air-borne
Cochliobolus carbonum	Maize	N, S, air-borne
Ascochyta pisi	Pea	HB, S, seed-borne/splash-borne
Plasmopara halstedii	Sunflower	HB, S, soil-borne
Peronospora tabacina	Tobacco	B, S, air-borne
Bacteria		
Xanthomonas campestris pv *vesicatoria*	Bell peper	HB, S
Viruses		
Tobacco mosaic virus	Tomato	B, MS
Tobacco mosaic virus	Bell peper	B, MS
Virus X and Virus Y	Potato	B, MS
Peanut mottle virus	Peanut	B, MS
Bean common mosaic virus	Bean	B, MS
Bean yellow mosaic virus	Bean	B, G
Soybean mosaic virus	Soybean	B, MS
Pea seed-borne mosaic virus	Pea	B, MS
Pea seed-borne mosaic virus	Lentil	B, MS

V, vascular wilt fungi; S, specialist; MS, moderately specialized; G, generalist; B, biotrophic; HB, hemibiotrophic;
N, necrotrophic.

known, and where the number of resistance genes described in each host is restricted
(Table 9.3). In these host–pathogen systems the adaptation of the pathogen population to
introduced resistances is usually slower or much slower, than in Group A. In some cases
no changes appear to occur in the pathogen population, the resistance is durable (potato
virus X in Western Europe). Meindert (1981), for example, concluded that all known
resistances to pea viruses are race-specific in nature, but these resistances can be used
because they appear to last a long time.

No races of the third group of pathogens (Group C) are known, and the resistance used
has been effective from the start. Examples of pathogens that, despite widespread exposure, did not respond to monogenic resistance by the formation of a new race are the fungi
Cladosporium cucumerinum and *Corynespora melonis* in cucumber, *Periconia circinata* in
sorghum, *Pseudocercosporella herpotrichoides* in wheat, *Helminthosporium victoriae* in
oats, the bacterium *Xanthomonas campestris* pv *glycines* in soybean, and bean common
mosaic virus in the common bean (the I-gene).

Apparently there are large differences in the ease with which pathogens adapt to
introduced resistances, and it is difficult to deduce what to expect from the taxonomic
position alone. *Xanthomonas campestris* pathovars are found among all three groups of
pathogens mentioned above. However, it seems clear that the *X. campestris* taxon is a
rather artificial one, probably representing a complex of species. The fact is that by far the

most pathogens belonging to Group A are fungi (Table 9.2). Viruses do not occur in this group, but are very common in the second group (Table 9.3).

DURABLE HOST RESISTANCE. DOES IT EXIST?

It is durable host resistance to the pathogens represented in Table 9.2 which is of most importance. The discussion here is therefore concentrated on questions such as:

• Does durable host resistance to these pathogens exist?
• If so, how is it characterized?
• How should one select for it?

All host–pathogen systems in Group A are characterized by a large number of major resistance genes in the host population and many races in the pathogen population. The major genes are nearly always race-specific and cause a hypersensitive-type response or a low-infection-type response to biotrophic fungi such as rusts and powdery mildews. When durable resistance exists in such host–pathogen systems it is another type of resistance.

When cultivars classified as susceptible are compared with each other, considerable differences in the level of susceptibility often become apparent. The pathosystem barley–barley leaf rust (*Puccinia hordei*) demonstrates this. There are at least nine major genes (Pa-genes) known which are race-specific, of the hypersensitive type, and which are not durable when used on a large area (Parlevliet, 1983b). Cultivars not carrying such major genes are designated as susceptible. They do have a susceptible infection-type response, but the amount of leaf rust in the field varies greatly among these cultivars. Table 9.1 shows that it is quite inappropriate to classify cultivars like 'Julia' and 'Vada' as susceptible. In the absence of interplot interference, which is the farmer's situation, they never become seriously affected. Even in the breeders' type of plot with interplot interference (very strong in this pathogen) the cultivar differences remain quite clear. This partial resistance is widespread at various levels in barley, and despite its wide exposure over a long period no signs of erosion have yet been observed. This partial resistance is due to the collective effect of longer latency periods, reduced infection frequencies and lower rates of sporulation (Parlevliet, 1979). The cultivar 'Vada', for instance, compared with L94, has a latent period which is almost twice as long, an infection frequency about half as great and a sporulation rate which is about 40% of that of L94 in adult plants. These three components of partial resistance to *P. hordei* in barley appear to be largely pleiotrophic (Parlevliet, 1986) and governed by polygenes (Parlevliet, 1978).

Partial or quantitative resistance not, governed by race-specific major genes, has been reported from various host–pathogen systems such as wheat–*P. graminis*, f.sp. *tritici*, wheat–*P. recondita* f.sp. *tritici*, maize–*P. sorghi*, maize–*P. polysora*, barley–*Erysiphe graminis* f.sp. *hordei*, potato–*Phytophthora infestans* (named field resistance), lettuce–*Bremia lactucae*, barley–*Rhynchosporium secalis*, maize–*Trichometasphaeria turcica*, rice–*Xanthomonas campestris* pv *oryzae*, potato–virus X, potato–virus Y, groundnut–*Puccinia arachides*, common bean–*Fusarium solani* f. sp. *phaseoli*, common bean–*Rhizoctonia solani*, common bean–*Xanthomonas phaseoli* and faba beans–*Botrytis fabae*. This suggests a general situation with on the one hand race-specific, non-durable major genes, and on the other a kind of partial resistance, sometimes called field resistance or residual resistance, and assumed to be race-non-specific and durable. The

Table 9.4   Mean resistance values of cultivars in which the resistance 'broke down' at entering (first year) the Dutch recommended cultivar lists and just before they were removed from that list (last year). Period 1960–1991. Resistance values on a scale of 1 (extremely susceptible) to 10 (completely resistant)

Crop	Pathogen	n	First year	Last year	Range*	Most susceptible** exotic
Winter wheat	Stripe rust	9	8.22	4.28	3–6	1
Winter wheat	Leaf rust	6	7.50	4.33	4–5	1
Winter wheat	Powdery mildew	6	7.17	4.33	3–6	1
Spring barley	Powdery mildew	9	8.06	4.56	4–6	1

*Range of resistance values after resistance broke down.
**Resistance value of most susceptible genotypes, always unadapted exotics.

existence of partial resistance can also be shown in another way. When newly released cultivars carry an effective race-specific major gene against one of its important pathogens, the cultivar will get a high score for resistance on the recommended cultivar list. Over the years many cultivars 'lost' their resistance. They became susceptible, but the level of susceptibility was always higher than that of extremely susceptible unadapted (to the region) exotic genotypes (Table 9.4).

All these cultivars apparently carry some residual resistance, which is of a quantitative nature. It is introduced accidentally by the breeder through his choice of parents for crossing. The parents are either fully resistant or nearly so due to race-specific major genes (which will break down later, after release), or susceptible, but never very susceptible. Through those moderately susceptible parents the breeder introduces partial or quantitative resistance, which after the break-down of the race-specific major gene, becomes visible as residual resistance (Table 9.4).

The inheritance of partial or quantitative resistance has been investigated in some detail in only a limited number of host–pathogen systems. Partial resistance is often recessive and the result of the cumulative effects of several genes with small to intermediate effects. The number of genes tends to be fairly small in most cases: two in maize to *P. sorghi* (Kim and Brewbaker, 1977); two–three in wheat to *P. recondita* f.sp. *tritici* (Broers and Jacobs, 1989; Jacobs and Broers, 1989); three–five in wheat to *P. graminis* f.sp. *tritici* (Knott, 1988). Black (1970) reported polygenic inheritance of field resistance to *Phytophthora infestans* in potato, as did Habgood (1974) to *Rhycosporium secalis* in barley. Western European barley cultivars carry from one to five or six minor genes for partial resistance to barley leaf rust (Parlevliet, 1978), but more minor genes appear to exist (Parlevliet et al., 1985). In none of these cases were there indications that there was a large number of minor genes.

As already mentioned, this partial or quantitative resistance is often assumed to be race-non-specific, but this is not true. In several host–pathogen systems small race-specific effects appear to be present (Parlevliet, 1979). Table 9.5 gives an example of such small race-specific effects. The overall impression is that of a race-non-specific resistance. 'Julia', however, had too high a level of barley leaf rust with isolate 18, and 'Berac' with isolate 22. The former effect was significant, the latter was not. The too high level on 'Berac' with isolate 22 was observed in two consecutive years (Parlevliet and Van Ommeren, 1985). Despite the occurrence of such small race-specific effects and an

Table 9.5   Percentage leaf area affected in three barley cultivars, partially resistant to *P. hordei*, exposed to five isolates of *P. hordei*. All plots were well isolated from each other to avoid interplot interference

Cultivar	Isolate				
	11-1	18	1-2	22	24
Berac	8.1	6.7	3.1	5.0**	0.9
Julia	4.5	12.1*	1.8	1.1	0.6
Vada	0.8	0.5	0.6	0.2	0.1

*If race-specific effects had been absent, this value would have been ca. 3.2.
**If race-specific effects had been absent, this value would have been ca. 2.0.
Source: Parlevliet, 1977

Table 9.6   Quantitative resistance (resistance to infection) of some potato cultivars to four potato viruses expressed on a scale of 1 (extremely susceptible) to 10 (zero percentage of infected plants) (Joosten, 1988)

Cultivar	Potato virus			
	Leaf roll	A	X	Y
Arkula	8	8	8	9
Agria	5.5	9	9	9.5
Kennebec	5	9	5	8
Doré	7.5	2	6	2
Primura	6	3	5	3.5
Saskia	7	–	4	8

extensive exposure over millions of hectares during several decades, no erosion of partial resistance to barley leaf rust has been observed in Western Europe (Habgood and Clifford, 1981).

This partial or quantitative resistance is not the general resistance it is often assumed to be. It is always at least pathogen-specific. Partial resistance in wheat or barley to a specific species of rust is not effective for any other rust species. Quantitative resistance in potato to one virus (leafroll, virus A, virus X, virus Y) does not operate with any other viruses (Table 9.6), and as well as this pathogen specificity, small race-specific effects may occur, as shown above. No case of erosion or loss of partial or quantitative resistance has been reported as yet. So the conclusion must be that durable forms of resistance exist to the pathogens that so easily adapt to major resistance genes (Group A). This resistance tends to present itself as a quantitative type of resistance. The reverse of this conclusion, however, is not correct. Several cases of monogenic, incomplete resistance are known, which are indistinguishable from the other forms of quantitative resistance and which rapidly become neutralized by the pathogen population.

BREEDING FOR DURABLE RESISTANCE

The approach may vary with the type of pathogen. Pathogens classified as Group C (no races known, no resistance breaking reported) or Group B (few races identified, resistance breaking uncommon or not yet known to occur) can be dealt with in a straightforward way. Any resistance observed can be selected for and used. Pathogens of Group A require a different approach.

Resistance to pathogens of Groups B and C

Although the selection is said to be straightforward, this does not mean that it is a simple job. In cases where major genes are involved it is usually not difficult, but when the resistance is of a quantitative nature, various problems may arise in trying to distinguish the more resistant entries from the less resistant ones.

Soil-borne pathogens tend to be heterogeneously distributed. Homogeneous exposure to the pathogen is often not easy, and the accuracy of disease assessment is often far from satisfactory. In wheat, no satisfactory screening method is yet available against take-all (*Geaumannomyces graminis*). For resistance to eyespot, *Pseudocercosporella herpotrichoides*, the situation is better, but screening is still not easy. In peas it is not yet clear how many races exist to *Fusarium oxysporum* f.sp. *pisi*.

Soil-borne pathogens that spread upward to the leaves and ears offer different problems in screening for resistance. Resistance to such pathogens (*Septoria nodorum, Septoria tritici, Fusarium culmorum*) is often confounded with earliness and tallness.

Quantitative resistance to viruses carries a similar problem: homogeneous exposure to the virus is difficult, especially if an animal vector (aphids, nematodes) is involved, and screening methods that identify small differences in resistance are not readily available. Resistance to barley yellow dwarf virus in wheat and oats are quantitative in nature and screening is not easy, in barley it is easier as two major genes have been identified. In potatoes the quantitative resistance to the leaf roll, the X, the A and the Y viruses is expressed as a reduced frequency of plants becoming infected (showing symptoms) when exposed. To obtain a fairly accurate assessment of the differences in frequency of diseased plants, it is essential that a sufficient number of plants of each variety is tested: some 20–30 per season per variety. This precludes screening for this resistance in the early stages of the selection program when the number of plants per entry (clone) is still small.

Good screening methods that can identify small differences in resistance are absolutely essential in order to obtain satisfactory levels of resistance, and such screening methods are not available for many pathogens in these two groups.

Resistance to pathogens of Group A

Selection for partial or quantitative resistance in the absence of major resistance genes is also straightforward, and in fact easier in many cases than in the situations described above. This is because many of the pathogens in this group are air-borne (Table 9.2), and it is much easier to obtain a uniform exposure to air-borne inoculum, while assessment of the amount of tissue affected is also more reliable with biotrophic fungi. However, there are only a very few host–pathogen systems where the host is free, or mostly free, of effective major genes. Groundnut–*P. arachidis*, barley–*P. hordei* and maize–*P. sorghi* are examples of this type. Selection for increased levels of partial resistance is relatively easy in such cases (Parlevliet et al., 1985; Parlevliet and Van Ommeren, 1988). Three cycles of recurrent selection in two heterogeneous barley populations increased the level of partial resistance to *P. hordei* from an average level of susceptibility similar to that of cultivar 'Zephyr', see Table 9.1, to a level considerably better than that of 'Vada'. The large

Table 9.7 Percentage of host tissue affected if cereal cultivars carrying different race-specific resistance (R) genes are exposed to a mixture of rust races carrying different virulence (a) genes

| R-genes | a-genes of rust races | | | | Percentage of tissue affected |
	a2 (30)*	a1a3 (40)	a1a4 (25)	a2a3a4 (5)	
–	+**	+	+	+	70
R1	–	+	+	–	50
R3	–	+	–	+	40
R2R4	–	–	–	+	20

*Percentage of each race in the initial inoculum in parentheses.
**a, +, host cultivar susceptible for that race; –, host cultivar not susceptible for that race.

increase in resistance was obtained by removing only the 30% most susceptible entries at all selection stages (Parlevliet and Van Ommeren, 1988).

However, there are very few host–pathogen systems in this group where major, non-durable resistance genes are not present. In several crops they are abundantly present, making selection for partial resistance far from easy. How should one select for durable resistance in these host–pathogen systems? Before a selection strategy is laid out, one has to take a few facts into consideration.

(i)   The major genes present in the population from which to select may vary from totally ineffective to completely effective to the races present in the area. Some of the major genes are effective to some races and ineffective to other races. The latter genes may easily give the impression that partial resistance is present when the pathogen population consists of a mixture of races. This situation is demonstrated in Table 9.7.

The cultivars show different levels of affected tissues where the lesions have a susceptible infection type, but the differences in the level of host tissue affected are solely due to the major genes. So the conscious use of race mixtures should be avoided if one wishes to select for partial resistance. Instead one should use one race with as wide a virulence range as possible (Parlevliet, 1983a).

(ii) There are major, race-specific resistance genes with an incomplete effect. In the field such genes are difficult to distinguish from partial resistance based on minor genes. There is no clear-cut difference between the low-infection (major gene) type of resistance and the high-infection (minor gene) type of resistance.

(iii) There is no guarantee that the resistance selected is durable. Only time and exposure can give us firm evidence.

(iv) Resistance is not the only trait to be improved. One has to select for a variety of other agronomic traits as well.

Before suggesting a strategy, it is worthwhile discussing a selection experiment and a practical example.

The experiment, carried out in Wageningen, The Netherlands, consisted of recurrent selection for partial resistance to two pathogens in barley, barley leaf rust, *Puccinia hordei* and powdery mildew, *Erysiphe graminis* f.sp. *tritici* (Parlevliet and Van Ommeren, 1988).

Table 9.8   Mean leaf area affected in two barley populations (A and B) to barley leaf rust and powdery mildew after one (S1) and three (S3) cycles of recurrent selection, relative to the leaf area affected in the starting populations, S0 (set at 100%). Populations A and B were merged after the second cycle by intercrossing the selected lines. Each cycle of recurrent selection consisted of a single plant (F2) followed by F3 line selection. The selected F3 lines were intercrossed

Cycle	Leaf rust		Powdery mildew	
	A	B	A	B
S0	100	100	100	100
S1	35	58	105	141
S2	7	10	38	71
S3	5		26	

Source: Parlevliet and Van Ommeren, 1988

Two unrelated heterogeneous populations were produced, A and B. Against leaf rust only partial resistance was present, and during the whole selection procedure the same single race was used. Against powdery mildew Population A carried partial resistance while Population B carried many major genes as well as partial resistance. The powdery mildew to which the barley populations to be selected were exposed was the natural mixture of races that varies from year to year. Each cycle of recurrent selection consisted of single plant selection in F2 followed by line selection in F3. The selection consisted of a mild selection against susceptibility; the 30% most affected plants or lines were removed for each of the pathogens. This left the possibility of selecting for other agronomic traits as well among the remaining plants or lines at each stage. After two cycles the selected lines were intercrossed between the populations. Table 9.8 summarizes the results.

The gain in resistance to leaf rust was more than five times the gain in resistance to powdery mildew. The gain in resistance to the leaf rust was solely due to increased partial resistance, the gain in resistance to powdery mildew probably also contained some major, race-specific genes owing to the effect shown in Table 9.7. This experiment showed two aspects very clearly. (i) Selection for partial resistance in the absence of major genes is far easier than selection for partial resistance in the presence of major genes when exposed to a mixture of races. (ii) Even a very mild selection pressure (only the 30% most affected units were removed) appeared to be very effective. Owing to this mild selection it was possible to select for other traits as well. The yield after S3 was 26% higher than the mean yield of the starting populations, a gain obtained by selecting for increased ear weight and higher tillering separately in these early generations.

What strategy should one follow with the above-mentioned facts in mind? Assuming the selection is carried out in the field only, there are two possibilities. (i) The pathogen is introduced in the screening trials. In this case a single, widely virulent race should be used, or one should rely on the natural inoculum, which might be a mixture of races.

That selection against susceptibility can be very effective is supported by experience accumulated in the soybean–*Pseudomonas glycinea* pathosystem in the USA (Wilcox, 1983). This bacterial blight occurs throughout the soybean-growing areas. Major gene resistance and the existence of eight races have been reported. Soybean breeders in the USA have never put much emphasis on breeding for a high level of resistance. They have eliminated the highly susceptible lines continuously during the breeding procedures. The current cultivars are relatively free from bacterial blight during most growing seasons, whereas unselected accessions from abroad quite often become heavily infected in the same breeders' nurseries.

Based on the information discussed above, the following strategy is suggested to avoid major race-specific resistance genes as far as possible and to accumulate partial resistance genes which, it is hoped, are considerably more durable: *in segregating genetically heterogeneous populations, selection should aim at removal of the most susceptible entries (selection against susceptibility rather than for resistance) and removal of those entries that do not show any disease. The latter are assumed to carry major genes, which should not be selected. Among the remainder selection for other traits is carried out and the selected entries are then recombined. This is recurrent selection against susceptibility and against presumed major genes.*

DURABLE USE OF NON-DURABLE RESISTANCE GENES

If non-durable major genes are used in combination it may be more difficult for the pathogen to build up races with a wider virulence spectrum (complex races). Such genes can be exploited in combination in two ways:

- through guided distribution in space and time (gene deployment);
- through physical combination in the same genotype (multiple gene barriers).

Gene deployment in a wide sense

The genes to be used occur in different cultivars. The cultivars carrying the different resistance genes can be distributed in different ways:

- they occur within the same field, sown as a unit (the multiline or the cultivar mixture);
- they occur as different cultivars in different fields within the same farm (gene deployment at the farm level);
- they occur as cultivars recommended for different regions within the same epidemiological area (gene deployment at regional level.

The multiline and cultivar mixture approach

The basic idea is of course to confront the pathogen with a greater diversity of resistance genes. The greater diversity could make it more difficult for the pathogen to adapt to such cultivars. However, this assumption may not be true, as several pathogens of Group A have been shown to be able to produce highly complex races without loss of fitness (Parlevliet, 1981). At the same time it is far from easy to exploit multilines and cultivar mixtures. The production of a multiline requires quite a number of resistance genes that are still effective, while maintenance and multiplication pose other, practical problems, which are increased if breeders' rights and/or official registration are required. So, on the whole the disadvantages seem to outweigh the advantages to a considerable extent.

Some of the disadvantages of the multiline approach do not apply to the cultivar mixture approach. Maintenance, multiplication and registration do not create additional problems because only existing cultivars are used. The limitations of this approach are determined by the number of adapted cultivars available and the diversity in resistance genes present in these cultivars. If uniformity in the quality of the harvested product is not essential, cultivar mixtures could provide a solution in certain cases.

Gene deployment at the farm level

In farming systems with large farms it is possible to advise farmers to use more than one cultivar of the same crop. A farmer could plant one cultivar on one part of the acreage intended for that crop and a second cultivar on the other part. These cultivars should carry different resistance genes that are still effective. Both in England and The Netherlands the recommended cultivar lists give information on which wheat cultivars should be combined in this way in the case of stripe rust, and which barley cultivars in the case of powdery mildew. If this is done on a wide scale it could reduce the rate of development of more complex races. However, in areas where most farmers are smallholders, this approach is unrealistic.

Gene deployment at the regional level

If the cultivars recommended in different regions of an epidemiological area carry consistently different resistance genes it would certainly reduce the rate at which complex races could develop. However, it is next to impossible to realize this as the breeders in that area must make and keep agreements over the use of the resistance genes, and cultivars intended for one region should not be grown in an adjacent region. Most epidemiological areas cover several countries and often several breeders are involved. Good agreements that are kept over decades are most unlikely in such situations, and often are not even possible, as in most pathosystems only a few of the resistance genes are defined and described satisfactorily.

Multiple gene barrier

Two or three fully effective race-specific genes together appear to be a barrier not easily broken. Several winter wheat cultivars remained resistant to stripe rust for periods of 15 or more years in Western Europe because of such a multiple gene barrier ('Felix', 'Manella', 'Arminda'). The resistance ultimately broke done ('Felix', 'Manella') as the resistance genes involved were also used singly in other cultivars. Through these other cultivars the pathogen could develop races with virulence to all these genes in a stepwise fashion.

If such combinations of resistance genes could be used exclusively in combination, the resistance would be highly durable, but in order to arrange this, one has to know all the resistance genes involved, and all the breeders in the whole epidemiological area would have to stick rigorously to the agreement made over the use of these genes. To bring together two fully effective major genes in an agronomically successful cultivar is far from easy. In the cases mentioned it was an accidental occurrence, not a planned event.

What are the difficulties in bringing together two major resistance genes? They should be fully effective to all races in the intended region. This means that when tested in that region the genes cannot be distinguished. Genotypes carrying $R1$ cannot be differentiated from genotypes carrying only $R2$, or $R1$ and $R2$ together. This makes the selection of genotypes carrying both resistance genes together with the required agronomic traits a highly complicated affair.

Since the resistance genes are often not satisfactorily identified, the breeders cannot make agreements about the sets of genes used in combination even if they were willing to do so. This, together with the complexities of bringing two such genes together, makes it unlikely that this approach will be applied, except accidentally. However, when sufficient RFLP markers have become available, the application of a multiple-gene barrier will become much easier as tracing an individual resistance gene becomes much easier.

INTEGRATED DISEASE MANAGEMENT

The concept of IPM was initiated by entomologists, and although they suggested that resistance should be one of the components of integrated pest management, they rarely integrated it in their research projects. This is not unexpected, as resistance to animal parasites is not always easy to integrate into a coherent set of control measures.

In the control of diseases the situation is different. Pathologists have been applying integrated disease control for a long time without using the term (Zadoks, 1989). A few examples will illustrate this.

(i)   Potato–*Synchytrium endobioticum* (wart disease). This soil-borne pathogen is considered to be very serious in many potato-growing areas. In The Netherlands it was so threatening that in the 1930s a law was introduced requiring new cultivars to be completely resistant to this pathogen. This measure, together with measures such as a prohibition on growing susceptible potatoes on infested soils, kept the pathogen under control for a long time. It was so effective that the law requiring new cultivars to be resistant was abandoned in 1974. In the 1970s a second race appeared in a small area in the north east of the country, a race which is still slowly spreading, partly due to the fact that cultivars resistant to this new race are still rare. The committee that decides which cultivars can enter the list of national cultivars considers any level of resistance to either Race 1 or Race 2 as a highly desirable characteristic. The measures taken in The Netherlands did keep this disease restricted to an area in the northeast of the country and to two races. The durability of race-specific resistance has been very considerable owing to this integrated set of measures. This is in sharp contrast to areas (e.g. Eastern Europe) where such measures were not taken, and where many races have been identified (Langerfeld, 1984; Saltykova and Yakovleva, 1976; Stachewicz et al., 1977).

(ii) Potato–potato cyst nematodes (*Globodera rostochiensis*, *Globodera pallida*). These are also very serious soil-borne parasites. In The Netherlands a series of control measures, including resistance, were introduced several decades ago. On nematode-free soils the growing of susceptible cultivars is allowed only once in 4 years. If resistant cultivars are alternated with susceptible ones, once in 3 years is acceptable. On infested soils resistant cultivars only are allowed once in 3 years, or once every 2 years if combined with soil fumigation to kill the nematodes. This integrated set of measures was meant to keep the population levels of the nematodes below the damage threshhold. Success was only partial. The population growth of the nematodes was severely reduced, but not enough. Over the years a slow but steady increase was observed in the number and size of nematode-infested areas and in the number of races. Without integrated management the present situation would have been reached decades earlier, and resistance-breeding would have been completely unable to keep up with the parasite.

(iii) Potato–potato viruses. Viruses such as potato leaf roll virus, potato virus X, potato Virus A, potato virus S, potato virus M and potato virus Y_N form a very serious threat of potatoes in almost all growing areas. To keep these viruses at low, non-damaging levels it is essential that the seed potatoes from which commercial cultivation starts are free, or almost free, from these viruses. Most of the viruses are transmitted by aphids. In The Netherlands, a set of measures ensures satisfactory control. To prevent transmission by aphids, the foliage of all seed potato crops has to be killed or removed before a specified date, which is announced yearly and based on aphid catch data. Furthermore, maintenance of the cultivars, based on a compulsory and specified program of progeny testing in which all progeny showing virus infection are completely removed, ensures a complete virus-free start of the multiplication phase. During multiplication checks for the presence of viruses are continued, and seed potato lots that show a higher incidence than the one allowed (very low) cannot be used as seed potatoes. To reduce the transmission of especially persistent viruses, such as the leaf roll virus, within the seed potato field, systemic insecticides are effective (Hille Ris Lambers, 1980) and are still often used. Resistance plays an important role in the certification program described above. From experience it has been found that it is almost impossible to keep the virus levels of a cultivar sufficiently low when that cultivar is very susceptible, e.g. cultivar 'Dore' for viruses A and Y_N (Table 9.6). The committee in charge of the admission of new cultivars to the list of national cultivars has therefore set minimum levels of resistance for potato viruses X, Y_N, A and leaf roll virus.

In the management of diseases the pathogen-reducing effects of resistance are generally additive, or even multiplicative, to the pathogen-reducing effects of other measures such as crop rotation, seed or plant certification, and pesticides. This statement is based on experience, as very few (if any) experiments have been carried out to study the effect of quantitative disease resistance on other control measures. Because of this additive or multiplicative effect, quantitative resistance is a powerful tool in integrated disease management. Any increase of the level of quantitative resistance adds to the reduction of the pathogen level and so to a reduction in the disease level.

In the first two examples the resistance used was, and is, predominantly of the race-specific, non-durable type. The integrated system works as long as the resistance is effective, but breaks down when a new race of the parasite appears. With the potato cyst nematodes in particular this has been a problem. For a more durable integrated disease management, non-durable, race-specific resistance genes should not be used. This is because they obscure the quantitative, more durable resistance, which in the long run is more effective, and which is nearly always available, albeit often at too low a level.

Because of this additive/multiplicative effect of resistance, experiments on the effects of integrated control measures should specify the level of quantitative resistance of the cultivars used. This is rarely done because of the often qualitative way of thinking of the investigator; a cultivar is classified as either susceptible or resistant. All cultivar–virus combinations shown in Table 9.6 below an 8 or 9 would, in such thinking, be classified as susceptible. Integrated control experiments with the leaf roll virus, for example, would give quite different results if 'Kennebec' was used instead of 'Dore'.

REFERENCES

Black, W. (1970) The nature and inheritance of field resistance to late blight (*Phytopththora infestans*) in potatoes. *American Potato Journal* **47**, 279–288.

Broers, L.H.M. and Jacobs, Th. (1989) The inheritance of host-plant effect on latency period of wheat leaf rust in spring wheat. II. Number of segregating factors and evidence for transgressive segregation in F3 and F5 generations. *Euphytica* **44**, 207–214.

Habgood, R.M. (1974) The inheritance to *Rhynchosporium, secalis* in some European spring barley cultivars. *Annals of Applied Biology* **77**, 191–200.

Habgood, R.M. and Clifford, B.C. (1981) Breeding barley for disease resistance: The essence of compromise. In Jenkyn, J.F. and Plumb, R.T. (Eds) *Strategies for the Control of Cereal Diseases*. Blackwell Scientific, Oxford, pp. 15–25.

Haglund, W.A. and Kraft, J.H. (1979) *Fusarium oxysporum* f.sp. *pisi*, race 6: Occurrence and distribution. *Phytopathology* **69**, 818–820.

Hille Ris Lambers, D. (1980) Integrated control of aphid-borne viruses of potatoes. In Minks, A.K. and Gruys, P. (Eds) *Integrated Control of Pests in The Netherlands*. PUDOC, Wageningen, pp. 59–66.

Jacobs, Th. and Broers, L.H.M. (1989) The inheritance of host-plant effect on latency period of wheat leaf rust in spring wheat. I. Estimation of gene action and number of effective factors in F1, F2 and back-cross generations. *Euphytica* **44**, 197–206.

Joosten, A. (1988) *Geniteurslijst voor Aardappelrassen, 1988*. COA, RIVRO, Wageningen.

Kim, S.K. and Brewbaker, J.L. (1977) Inheritance of general resistance in maize to *Puccinia sorghi*. *Schw. Crop Science* **17**, 456–461.

Knott, D.R. (1988) Using polygenic resistance to breed for stem rust resistance in wheat. In Simmonds, N.W. and Rajaram, S. (Eds) *Breeding strategies for resistance to the rusts of wheat*. CIMMYT, Mexico, pp. 39–47.

Langerfeld, E. (1984) *Synchytrium endobioticum* (Schilb.) *perc. Mitt. Biol. Bundesanstalt für Land- und Forstwirtschaft, heft 219*. Dahlem, Berlin, 142 pp.

Meindert, J.P. (1981) Genetics of disease resistance in edible legumes. *Annual Review of Phytopathology* **19**, 189–209.

Nørgaard Knudsen, J.C., Dalsgaard, H.H. and Jøgensen, H.J. (1986) Field assessment of partial resistance to powdery mildew in spring barley. *Euphytica* **35**, 233–243.

Parlevliet, J.E. (1977) Evidence of differential interaction in the polygenic *Hordeum vulgare–Puccini hordei* relation during epidemic development. *Phytopathology* **67**, 776–778.

Parlevliet, J.E. (1978) Further evidence of polygenic inheritance of partial resistance in barley to leaf rust, *Puccinia hordei*. *Euphytica* **27**, 369–379.

Parlevliet, J.E. (1979) Components of resistance that reduce the rate of epidemic development. *Ann. Rev. Phytopathol.* **17**, 203–222.

Parlevliet, J.E. (1981) Stabilizing selection in crop pathosystems; An empty concept or a reality? *Euphytica* **30**, 256–269.

Parlevliet, J.E. (1983a) Can horizontal resistance be recognized in the presence of vertical resistance in plants exposed to a mixture of pathogen races? *Phytopathology* **73**, 379.

Parlevliet, J.E. (1983b) Race-specific resistance and cultivar-specific virulence in the barley leaf rust pathosystems and their consequences for the breeding of leaf rust resistant barley. *Euphytica* **32**, 367–375.

Parlevliet, J.E. (1986) Pleiotrophic association of infection frequency and latent period of two barley cultivars partially resistant to barley leaf rust. *Euphytica* **35**, 267–272.

Parlevliet, J.E. and Danial, D.L. (1992) How does interplot interference affect the field assessment for resistance in cereals to rusts and powdery mildews? In *Proceedings of the 8th European Mediterranean Cereal Rust and Powdery Mildews Conference*, Weihenstephan, September, Th. Mann Verlag, Gelsenkorchen-Buer, Germany pp. 289–291.

Parlevliet, J.E. and Van Ommeren, A. (1975) Partial resistance of barley leaf rust, *Puccinia hordei*. II.

Relationship between field trials, microplot tests and latent period. *Euphytica* **24**, 293–303.

Parlevliet, J.E. and Van Ommeren, A. (1985) Race-specific effects in major genic and polygenic resistance of barley to barley leaf rust in the field and how to distinguish them. *Euphytica* **34**, 689–695.

Parlevliet, J.E. and Van Ommeren, A. (1988) Accumulation of partial resistance in barley to barley leaf rust and powdery mildew through recurrent selection against susceptibility. *Euphytica* **37**, 261–274.

Parlevliet, J.E., Leijn, M. and Van Ommeren, A. (1985) Accumulating polygenes for partial resistance in barley to barley leaf rust, *Pyccinia hordei*. II. Field evaluation. *Euphytica* **34**, 15–20.

Saltykova, L.P. and Yakovleva, V.I. (1976) Races of potato wart, *Synchytrium endobioticum* (Schilb.) Perc. established in the USSR and Czechoslovakia. *Mikologiya i Fitopatologiya* **10**, 503–507 (*Plant Breeding Abstracts*, 1979, **49**, 38).

Stachewicz, H., Burth, U., Eiternick, G., Effmert, M., Demny, L. and Ballhausen, S. (1977) Die Prüfung der Kartoffelkrebses (*Synchytrium endobioticum* (Schilb.)) Perc. in der Deutschen Demokratischen Republik. *Nachrichten für den Planzenschuts in der DDR* **33**, 170–173.

Wilcox, J.R. (1983) Breeding soybeans resistant to diseases. *Plant Breeding Reviews* **1**, 183–235.

Zadoks, J.C. (1989) Does IPM give adequate attention to disease control, in particular at the farmer level? *FAO Plant Protection Bulletin* **37**, 144–150.

CHAPTER 10

Benefits of Minimum Pesticide Use in Insect and Mite Control in Orchards

Torgeir Edland

Norwegian Crop Research Institute, Ås, Norway

INTRODUCTION

Norway is a relatively small fruit-producing nation. Apples, pears, plums and cherries are grown commercially as far north as Valldal, Møre og Romsdal, at 62°18′ N. However, the main areas of fruit growing are located along the Oslo fjord, around some lakes in the eastern part of the country and in the fjord districts of the west (Figure 10.1).

According to Kvåle (1995) the total area of commercial orchards is approximately 3300 ha, which is only about 0.3% of the total agricultural area. The annual fruit production, including crops produced in private gardens, is estimated to be about 51 100 tons, with apples as the most important fruit crop (65%).

The average crop yield per hectare is normally quite low compared with more southern countries. In modern apple orchards, the crop yield varies between 12 and 15 tons ha^{-1}. During recent years the annual import of apples has been somewhere between 40 000 and 46 000 tons, while our commercial production has been only 9000–13 000 tons per year.

Since 1960 the number of apple, pear and plum trees has been decreasing, while cherry has increased. The total number of fruit farms has also decreased considerably during the last 30–40 years, while those with at least 1000 trees each have increased from 170 farmers in 1959 to 480 farmers in 1989. Many of the fruit farms are situated in hilly areas, too steep for other agricultural crops. Therefore, although small, fruit production is still important, since there are very few, if any, alternative possibilities for employment.

The average temperature during the summer season (May–September) ranges from 12.5 to 13.5°C, with the highest temperatures in the fruit areas of eastern Norway. Eastern Norway has a continental type of climate with moderate rainfall, relatively high summer temperatures and cold winters. The fruit areas of western Norway may be characterised by a more oceanic type of climate with moderate summer temperatures, fairly high precipitation and mild winters. Because of the different climatic situation, the fruit orchards in western Norway normally suffer from considerable disease problems, especially scab, whereas the orchards in the eastern part of the country have more problems with insect and mite pests.

Techniques for Reducing Pesticide Use. Edited by D. Pimentel
© 1997 John Wiley & Sons Ltd.

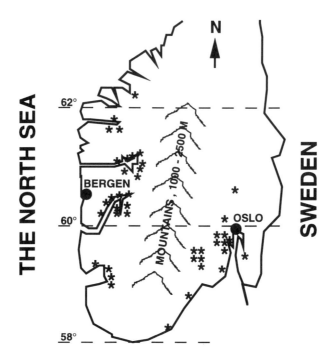

Figure 10.1   Commercial fruit growing in Norway. The asterisks indicate the most important
fruit-growing regions

THE PEST SITUATION

The agro-ecosystems in Norway are characterised by relatively few and moderate plant
protection problems. In general, the number of pest species is smaller than in the more
southern parts of Europe. Owing to a shorter growing season and lower temperatures,
many multivoltine insects have only one annual generation.

However, a large number of different insects and mites have been recorded as occasional feeders on fruit trees. Approximately 80 species may have pest status. Several
species are distributed in a few limited areas or regions only. On the other hand, some
species occur with periodically heavy outbreaks, causing massive pesticide use to protect
the crop from devastation.

According to Alford (1984), 190 different arthropod species may attack fruit trees in
Great Britain. In The Netherlands, Frankenhuyzen (1988) reported about 130 species
which feed on fruits. Evidently he has deleted many unimportant pests that are commonly
distributed in Europe. In Norwegian orchards, more than 150 different species feed on
various parts of the trees (Edland, 1991). Only half of these are considered as permanent
or occasional pests. More than 100 of the listed species in the British and Dutch reviews
are not recorded from fruit trees in Norway.

On the other hand, the summer fruit tortrix moth, *Adoxophyes orana* (Fischer von
Röslerstamm), regarded as one of the most injurious leaf rollers on top fruit in many
European countries, has been caught in pheromone traps in several localities in eastern
Norway, but its larvae have never been recorded as feeders on fruit trees. Other species,

especially among the lepidopterans, which are common pests on Norwegian fruit trees are not pests in more southern latitudes. For instance, the noctuid, *Eupsilia transversa* (Hufnagel), one of our most important pests on both pome and stone fruits, has seldom been mentioned as a pest in any other country. Likewise, the dun-bar moth, *Cosmia trapezina* (L.), sometimes described as beneficial because it may prey on various pest caterpillars (Alford, 1984), may reach high densities causing significant foliage damage.

As will be discussed below, in Norway the sizes of fruit pest populations fluctuate considerably from year to year, their importance also changes. Thus, in the period 1915–30, the pear thrips, *Taeniothrips inconsequens* (Uzel), was a most destructive pest on pear and stone fruits in western Norway (Schøyen and Jørstad, 1944). Since 1950, this insect has caused no trouble. In contrast, the rose thrips, *Thrips fuscipennis* Haliday, suddenly attacked apple in 1993–94 and destroyed the entire crop in unsprayed plots of some commercial orchards (Edland et al., 1995). This insect had been a common pest on roses for more than 70 years in Norway, but not on fruit trees until recently.

At least 12 species of injurious mites occur in our orchards, a few of them as troublesome pests (Edland, 1983). The fruit tree red spider mite, *Panonychus ulmi* (Koch), did not appear as a serious problem until farmers started their massive use of pesticides (Fjelddalen, 1952). During the last 15 years, the apple rust mite, *Aculus schlechtendali* (Nalepa), has frequently been the greatest economic hazard to the apple crop, causing drastic deterioration of the fruits through russeting. This increased damage was probably due to the reduced use of sulphur in most spray programmes.

In Norway the fauna of beneficial arthropods, still insufficiently researched, seems to be efficient in keeping many orchard pests at a low level. The number of species is probably lower in the northern regions than in warmer southern countries, but the abundance of each species is normally fairly high. The most important natural enemies among predators seem to be certain species of heteropterous bugs, especially among anthocorids and mirids, and also various predatory mites.

More than 50 species of predatory mites, belonging to the family Phytoseiidae, have been recorded on different plants in Norway. This mite family is widely distributed throughout the country and as far north as Alta, located at the same latitude as Murmansk, Russia, and the northern shorelines of Alaska. Investigations carried out during the last 10 years show that at least 18 phytoseiid species live on fruit trees in Norway (Edland, 1993). In similar studies, Tuovinen (1993) found 12 species occurring on apple trees in Finland, while Miedema (1987) recorded 20 species on fruit trees in The Netherlands. Most of the phytoseid species recorded in our orchards exploit both spider mites and rust mites as food, and some species have proved to be efficient predators on important mite pests. Thus, they have great potential as a valuable biocontrol agent for utilization in integrated pest management (IPM) programmes in orchards.

THE 'OLD SPRAY SCHEDULE'

During the first half of this century, comprehensive changes took place in the fruit industry world wide. Together with progress made in growing technology, new and often serious problems with pests and diseases occurred. For example, in Great Britain the apple capsid, *Plesiocoris rugicollis* (Fallén), the common green capsid, *Lygocoris pabulinus* (L.), and the fruit tree red spider mite, *Panonychus ulmi*, which until the 1920s were unknown to the growers, became some of the most troublesome pests attacking fruit

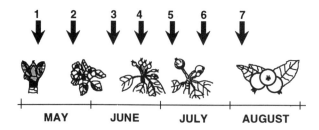

Figure 10.2   Example of a traditional spray schedule, commonly used on apple in the 1950–60s, in Norwegian orchards. The arrows indicate insecticide and acaricide applications for controlling insects and mites

by 1936 (Massee, 1954). Because of greater demands for both higher yields and more perfect-looking fruits of better quality, more effort was put into plant protection measures. This resulted in increased pesticide use in many countries.

In Norway, pesticide use in commercial orchards became common and widespread in the pre-war years. However, the drastic increase did not occur until after the war, when synthetic pesticides were introduced. In the 1950s and early 1960s, when it was generally accepted that chemical treatments were the correct and only way of controlling pests, a heavy spray programme, consisting of 6–8 annual applications, was commonly used against insects and mites in orchards (Figure 10.2).

The stages of bud, flower and fruit development in apple for each of the seven sprays indicated in Figure 10.2, and the assumed target pests at each application, were as follows. (Stages in parentheses follow the European system, according to IOBC/WPRS, 1980.)

(1) Green cluster (D), chewing pests, especially the winter moth, *Operophtera brumata* (L), and allied species, leaf rollers that hibernate as small larvae, and the apple blossom weevil *Anthonomus pomorum* (L).

(2) Pink pub (E2), chewing pests, mainly the leaf rollers that hibernate as eggs, some noctuids, e.g. *Eupsilia transversa* (Hufnagel), and other lepidopterous pests. Some sucking insects such as the apple sucker, *Psylla mali* (Schmidberger), were also controlled by this spray.

(3) Petal-fall (H), sucking pests, especially the fruit tree red spider mite, *Panonychus ulmi* (Koch), various aphids and the apple capsid, *Plesiocoris rugicollis* (Fallén).

(4) Small fruitlet (I), apple fruit moth, *Argyresthia conjugella* Zeller. It was long assumed that this key pest of our apple orchards had its main flight around mid-summer. A spray with a fairly persistent insecticide was applied against the adult moths to prevent oviposition.

(5) Fruitlet, early July (J1), apple fruit moth. Since the mid-summer spray did not give satisfactory results against this pest every year, a second spray was usually applied about 3 weeks later against eggs and young larvae to achieve complete control.

(6) Fruitlet, late July (J2), mites and aphids. In orchards receiving five insecticidal sprays in spring and early summer, most beneficial arthropods were greatly diminished. Consequently, in the absence of natural enemies, populations of phytophagous mites

and certain aphids often increased rapidly, and more pesticide use was necessary to prevent serious damage.

(7) Full grown fruit, early August (K), leaf rollers. Some species that hibernate as small larvae may feed on the surface of maturing fruits and cause severe damage. In certain years, such late damage has occurred in some areas and up to 80% of the apple fruits have been destroyed. Therefore it became common to apply a spray with a short-lasting insecticide in early August to prevent such damage.

In general, this spray schedule was used by most farmers for many years. However, in certain years one of the pre-blossom or one of the summer treatments was omitted, resulting in six annual applications, while in other years an extra spray was applied against mites and aphids in late June, amounting to eight sprays per year on some farms.

REDUCTION IN THE USE OF PESTICIDES

The routine, heavy spray programme, commonly used during the first two decades after World War II, was not only expensive, but also had undesirable environmental impacts. Therefore, an extensive research programme on IPM was started in 1962, when we asked the following question: Is it really necessary to spray so frequently every year and in all fruit areas? The investigations provided important information about major pests and their natural enemies, on their biology and ecology, and about the effects and side-effects of different chemicals. Based on this new knowledge, the use of pesticides was reduced. By 1974, farmers who had adopted IPM in their orchards, used only 0–3 pesticide application a year without significant damage to the crop (Edland, 1989a). The important information which made this reduction possible is discussed below.

Geographical distribution of pest species

An analysis of the arthropod fauna in sprayed and unsprayed orchards showed that orchard fauna was more diverse and complex than expected, and that some important pest species, generally assumed to be widely distributed in southern Norway, did not exist at all in some areas. Thus, among the 15–20 leaf roller species which have been recorded on fruit trees in Norway, the fruit tree tortrix moth, *Archips podana* (Scopoli), and the bud moth, *Spilonota ocellana* (Denis & Schiffermüller), which are normally responsible for late damage on the fruit surface, were found not to occur in the western fruit areas. Therefore, the August spray could safely be omitted in this part of the country. During the last 20 years further field investigations have been conducted using pheromone traps, and these showed that the same two species were absent or very rare in many fruit areas of eastern Norway. Consequently, the August spray has been omitted in most orchards for many years.

Seasonal development

The life cycles of major pests were investigated in detail. This work greatly improved our knowledge so that applications of control measures could be better timed. It also showed that various pest species that develop two or three annual generations in more southern countries have only one generation in Norway.

For several years the oviposition period of the apple fruit moth, *Argyresthia conjugella*, was studied at different altitudes in various parts of Norway. Contrary to the assumption that this insect had its peak of oviposition around mid-summer, the investigation showed that the moth deposits approximately 80% of its eggs in July, with a clear peak in the first half of the month. Because only a few eggs are laid in June, spray No. 4 (Figure 10.2) was omitted. In most years complete control of the apple fruit moth is achieved by a single spray (azinphos-methyl) applied in late June to early July. That means that spray No. 5 should normally be applied about 1 week earlier than shown in Figure 10.2.

Population fluctuation

In western Norway four species of winter moths (*Operophtera brumata* (L), *O. fagata* (Scharfenberg), *Erannis defoliaria* (Clerck) and *Agriopis aurantiaria* (Hübner) occur together in regular outbreaks, with peaks at 12–15-year intervals. During each outbreak, which normally lasts for 2–4 years, the larvae of these geometrids often cause complete defoliation of their host trees. In extreme cases, the larval density has been assessed at more than 4000 per beating sample of 33 apple branches (Edland, 1981). Moreover, during the outbreaks numerous newly hatched larvae are dispersed by wind from the surrounding forest slopes to the orchards, making pre-blossom sprays necessary (Edland, 1971). However, during the periods between the outbreaks the population densities of the winter moths are normally very low, often less than 10 larvae per beating sample. This is far below the economic threshold. Since other pests, including leaf rollers and noctuids, rarely exceed the thresholds so early in the year, spray No. 1 (Figure 10.2) was omitted in most of the 10–12 years between each moth outbreak.

Assessment of the degree of pest attack/economic thresholds

After three sprays had been omitted from the schedule without significant damage to trees and fruits, many farmers asked if further reduction in pesticide use was possible. Therefore, some of them started a programme of systematic examination of the trees. When they observed none or only a few pests and very little damage, they decided to omit spray No. 2 (Figure 10.2), hoping that the first post-blossom treatment would give sufficient protection. Then when a subsequent examination during late blossom showed that the pest situation remained unchanged, spray No. 3 was frequently omitted.

In 1970, to facilitate this monitoring programme, practical methods were developed for assessing fruit pests and economic thresholds. Based on field tests, the beating method, developed by Steiner (1967), proved to be a suitable method for assessing attacks of beetles and caterpillars, and for estimating the density of many beneficial insects. 'Visual control', according to IOBC/WPRS (1980), was a practical method for assessing attacks of small pests species such as mites and aphids, and of mining insects. Today both methods are widely used in our orchards.

These investigations proved that the economic thresholds for many fruit pests are considerably higher in Norway than further south in Europe owing to the differences in the developmental rate of pests and their host plants. Using winter moth as an example, the blossom period for apple in central Europe takes place in late April to early May, while in Norway it occurs about 1 month later (Figure 10.3). However, the difference in larval development and time for pupation is considerably less. Thus, in central Europe the

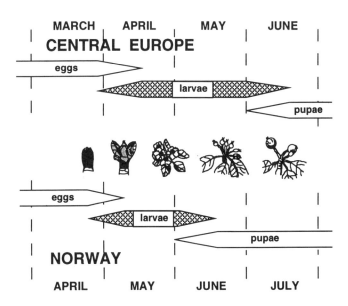

Figure 10.3   Differences in the rate of development of winter moth, *Operophtera brumata*, and its host, apple, in central Europe and Norway

caterpillars of *Operophtera brumata* are not full-grown until the fruitlets have been growing for some time. To avoid too much damage to the fruits, the economic threshold for this pest is frequently set at 8–12 caterpillars per beating sample (100 branches) before blossom, and at 12–15 after (IOBC/WPRS, 1980). In Norway, the caterpillars of this pest are normally full-grown by petal fall and they usually disappear from the trees before the fruitlets start to grow (Figure 10.3). Thus, under normal climatic conditions there is no damage to the fruitlets and 30–40 caterpillars per beating sample (33 branches) can be tolerated (Hesjedal et al., 1993). Results of tests indicate that the threshold can be higher than 50 caterpillars per beating sample, indicating that the need for pesticide treatments is even lower than is usually believed.

A similar difference in thresholds exists for most of the leaf rollers that hibernate as young larvae, e.g. *Hedya nubiferana* (Haworth), *Spilonota ocellana* and *Archips podana*. Also for various noctuids, such as *Orthosia* spp., a threshold can be tolerated in Norway which is 6–12 times higher than in central Europe. These findings have made it possible to reduce chemical treatments around blossom time still further (Figure 10.2).

The apple-grass aphid, *Rhopalosiphum insertum* (Walker), is a common species in Europe, often abundant on apple but rarely a serious pest. In central Europe its economic threshold is set at 60 colonies per 100 leaf/flower rosettes before and shortly after blossom (IOBC/WPRS, 1980). In Norway the threshold is set higher, i.e. 80% of the spurs with attack (Hesjedal et al., 1993). However, such heavy infestation very seldom occurs. Unnecessary spray against this aphid may disturb or destroy the natural enemies and result in significant damage by other aphid species, e.g. the rosy apple aphid *Dysaphis plantaginea* (Passerini). This is illustrated in Figure 10.4, which shows the relationship between two aphid species and their natural enemies in three different ecological situations. Figure 10.4a shows apple trees infested with both *R. insertum* and *D. plantaginea*. The number of *R. insertum* increases rapidly in May and approaches the threshold (*i*) at

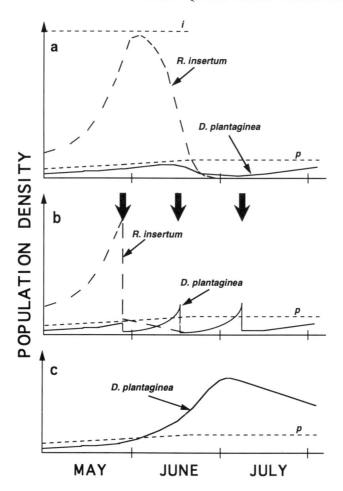

Figure 10.4   Relationship between two aphid species, *R. insertum and D. plantaginea*, and their natural enemies in apple orchards in three different ecological situations. The dotted lines, *i*, and *p*, indicate the economic thresholds for *R. insertum* and *D. plantaginea*, respectively. The arrows indicate aphicide applications. See text for further explanation

early blossom. The high aphid density attracts many natural enemies, which deposit their eggs in the aphid colonies. By early June, at the height of the blossom period, *R. insertum* starts emigrating to grasses, and disappears from the apple trees by the end of the month. The nymphs and larvae of various aphid predators remain on the apple trees, keeping the number of *D. plantaginea* below the economic threshold (*p*), so no pesticide treatment against aphid is needed. Figure 10.4b shows same situation as above, but here a spray is applied against *R. insertum* just before blossom. This treatment greatly reduces the number of aphids, but also reduces the number of natural enemies. In the absence of aphid predators, *D. plantaginea* has favourable conditions and exceeds the threshold in mid-June and early July, so two additional aphicide treatments are required. In Figure 10.4c *R. insertum* is absent on apple, and does not attract predators. Thus, *D. plantaginea* will increase in number and become abundant in late June. Natural enemies will not reduce the aphid population until July. To avoid economic damage, two sprays should be applied.

Thus, since the presence of R. *insertum* often results in a more efficient control of other pests by natural enemies, a moderate infestation of this species is considered as a benefit in IPM programmes.

Forecasting systems

Apple fruit moth, Argyresthia conjugella

In many areas of Fennoscandia the apple fruit moth is regarded as a key pest of apple. Its natural and preferred host plant, however, is the commonly distributed wild mountain ash *Sorbus aucuparia* L. In those years when berry production is too small to support all the egg-laying moths, a number of them emigrate to apple for oviposition. Then injury to the apple crop becomes severe, sometimes resulting in total destruction of the fruit. On the other hand, when sufficient numbers of mountain ash berries are available for the egg-laying females within an area, oviposition is completely limited to these berries, while apples avoid infestation.

During the 1970s a system for warning about apple moth infestation was developed (Edland, 1974). This system is based on various field observations, analyses and calculations made at different localities within a specific area. At each locality, about 20 'reference trees' of mountain ash are selected, and all the berry/flower clusters are counted every autumn and spring. In the autumn, samples of berries are collected in the neighbourhood of the reference trees to estimate the percentage of attack and the amount of larvae killed by the parasitoid *Microgaster politus* Marshall. This braconid wasp, widely distributed throughout the country, is considered the most important natural enemy regulating the population density of A. *conjugella*.

Based on the data obtained in the autumn, the number of moths expected to emerge the following spring is estimated. As each moth requires about five berries to limit its oviposition to the mountain ash, it is possible to calculate a prognosis to show whether attack on apples is expected. The calculation is made by the formula

$$P = \frac{a \cdot b \cdot c}{d \cdot 20 \cdot 100}$$

where P is the prognosis value, a is the percentage attack on the berries, b is the percentage unparasitized larvae, c is the number of clusters in the autumn, d is the number of clusters in the spring and 20 is a constant.

When $P < 1.0$ there is a surplus of berries and no attack on apple will occur, so all sprays against this pest should be omitted. When $P > 1.0$ insufficient berries are available for the moths, and attack on apple must be expected. Then an insecticidal treatment is recommended. In most years of moderate infestation, one spray of azinphos-methyl, applied in late June to early July, gives full protection against attack, but in years when the mountain ash has no clusters, or very few (d is very low), the prognosis value (P) will be very high. Then the oviposition period on apple may last for several weeks, and two sprays will be required to achieve full control of the pest.

The situations in 1994 and 1995 are good examples. In 1994, the *Sorbus* trees failed to crop in all regions, so *c* approached 0 in most areas. Consequently, *P* became extremely high (somewhat between 10 and 1000), indicating severe attacks on apple in all localities. In unsprayed orchards, 90–100% of the apple crop was destroyed by the apple fruit moth that year. Due to the shortage of *Sorbus* berries, very few *Argyresthia* larvae developed in 1994, resulting in a minimum of emerging moths the following year. Thus, for 1995 *abc* was very low. This year, the *Sorbus* produced a very high number of flowers/berries clusters, resulting in an unusually high *d* and a very low *P* in all regions. In most apple orchards, no sprays were applied against this pest in 1995, and no attack was recorded.

This warning system, which has been applied in our fruit areas for more than 20 years, has been tested under different ecological and geographical conditions at approximately 150 localities. It has proved to give reliable forecasts every year, and has been a very useful tool in making decisions about spray programmes. During recent years prognoses for 100–125 localities have been worked out annually, which covers most areas of our commercial apple growing. The system has shown that sprays are needed only every 3–5 years. In all other years, spray No. 5 (Figure 10.2) is omitted by practically all growers. Thus, this warning system has resulted in a significant reduction in pesticide applications (80–90%) for controlling *A. conjugella*.

Winter moths, Operopthera brumata *and allied species*

During the periods of outbreak, winter moths cause severe damage on fruit trees. Consequently, pesticide use is normally high during such periods, causing undesirable environmental impact. A reliable warning system, projecting the start and end of an outbreak, helps avoid unnecessary sprays.

A project was carried out in an outbreak area of western Norway in 1973–1980 to provide reliable data on the timing of moth outbreaks. A strong correlation between the number of male moths captured in light traps during the late autumn and the larval density on the fruit trees during the following spring showed that use of such traps provided information upon which to base forecasts (Edland, 1981). However, because light traps are costly and time-consuming to operate, and because many farmers have difficulty in distinguishing between winter moths and other geometrids captured in the traps, the system has not come into practical use.

Subsequent investigations using pheromone traps for *O. brumata* indicated that these traps are well suited for use in a warning system. However, the farmers found the price of the traps too high, and most were busy monitoring for other pests during the spring. Therefore, the beating method and/or visual control were the most acceptable methods, and the sprays around blossom time (Nos. 2 and 3, Figure 10.2) are being kept at a low level.

Codling moth, Cydia pomonella (L)

The codling moth, a key pest in most European countries, is not considered a serious apple pest in Norway, although it is distributed in the eastern and southern areas and in a small part of western Norway. In very warm seasons, however up to 80% of the apple crop may be destroyed by this pest in unsprayed orchards. This occurred in 1976, when monthly mean temperatures were 0.6–2.2°C higher than normal, and a second generation developed (Edland, 1977). That year larvae of the second generation destroyed about 50% of the apple crop in an orchard just before harvest time.

The codling moth requires high evening temperatures, at least 15.6°C at sunset, for successful oviposition. In many years low evening temperatures prevent egg-laying, and no chemical treatments are needed. However, to ensure adequate control a warning system was developed. In cooperation with the Institute for Plant Protection in Fruit Crops, Dossenheim, Germany, a Californian model 'Bugoff 2' was tested in Norway as well as in Germany and Italy (Blago and Edland, 1991). After 3 years of investigation, a model was developed to improve forecasts for egg-laying and egg-hatching activity under Nordic climatic conditions.

This model includes data on the number of adult moths captured in pheromone traps when the apple trees have reached petal fall stage, and a record of the sunset temperature which notes whether it has been 15.6°C or higher in 3 of the 7 preceding days. When all the criteria are positive, the beginning of egg-laying can be predicted. The developmental time for the eggs was found to be 90 degree-days above 10°C. Information about daily mean temperatures allows a forecast of when the eggs will start to hatch in different areas. The most efficient time for applying sprays can thus be determined. Use of this system has resulted in significant reduction in chemical use, especially in areas where one or more annual safety sprays were a customary practice.

Utilization of natural enemies

Studies on beneficial insects and mites have shown that many natural enemies are widely distributed in our fruit-growing areas. When several pesticide applications were eliminated from the spray schedule (Figure 10.2), the environmental conditions were greatly improved for the survival of many beneficial species. Thus, some important predators (feeding on sucking pests), including ladybirds, e.g. *Coccinella septempunctata* L. and *Adalia bipunctata* (L.), some lacewings, e.g. *Chrysoperla carnea* (Stephens), and the common flower bug, *Anthocoris nemorum* (L.), showed an appreciable and rapid increase in number. They provided an efficient biological control of aphids and spider mites, and made it possible to omit spray No. 6 (Figure 10.2).

It seems quite clear that a more effective utilization of natural enemies is of fundamental importance in developing and maintaining sound and efficient IPM approaches. Achieving a higher density and improved activity among natural predators and parasites is considered to be the main reason for the decrease in pest attacks and the reduced need for pesticides in Norwegian fruit orchards.

Among the natural enemies, several predatory bugs have proved to be efficient agents in keeping both sucking and chewing pests at a low level. In orchards where IPM has been practised for some years, they tend to be numerous, e.g. more than 100 adults and nymphs per beating sample (beating of 33 branches). The common flower bug, *Anthocoris nemorum*, is normally the most common and abundant species. As well as being an effective predator on sucking pests, e.g. spider mites, aphids and psyllids, it often causes great reductions in the number of leaf rollers and other lepidopteran pests by feeding on eggs and small larvae. After heavy use of insecticides in pear to control *Lygocoris pabulinus* and other capsids that often cause severe fruit damage, the common pear psylla, *Psylla pyri* (L), became resistant to most commonly used organophosphorus compounds.

Today this pest is being controlled in many pear orchards by *A. nemorum*, which has proved more efficient than any insecticide.

Several mirid bugs have also become useful predators in many moderately sprayed apple orchards. The following species have been recorded as commonly distributed and abundant on fruit trees in Norway: *Blepharidopterus angulatus* (Fallén), *Malacocoris chlorizans* (Panzer), *Campylomma verbasci* (Meyer-Duer), *Atractotomus mali* (Meyer-Duer), *Psallus ambiguus* (Fallén), *Orthotylus marginalis* Reuter, *Pilophorus perplexus* (Douglas & Scott), and a few *Phytocoris* spp. Although it is known that in other countries some of these species cause damage on apple (Alford, 1984), there is no record of this in Norway. However, a few of the species may cause some damage on pear fruits (Hesjedal, 1989).

Another group of natural enemies which are becoming important for controlling the fruit tree red spider mite, *P. ulmi*, and the rust mites, *Aculus* spp., are the phytoseiid mites. Fauna studies have showed that about 20 different species may live on fruit trees (Edland, 1993). As will be discussed later, at least six of these species have sometimes occurred in high densities on fruit trees heavily treated with various fungicides at normally recommended dosages. Although they all proved to be effective predators on phytophagus mites, they are very sensitive to many insecticides. Thus they are normally scarce in commercial orchards, even in those relying on an IPM programme.

Even though it has been possible to delete many sprays from the schedule shown in Figure 10.2, no commercial fruit growers have omitted all sprays against insects and mite pests. Increasing knowledge about the pest species, their biology and ecology, and of chemical and biological control agents, however, has resulted in a significant reduction in pesticide use. As mentioned, as early as 1974 some farmers had reduced the number of pesticide applications against insects and mites from 6–8 to 0–3 per year. During subsequent years, with a better understanding of pest problems and the control achieved by the IPM approach, an increasing number of Norwegian farmers have found IPM appropriate to their farming systems.

A survey performed in different fruit-growing areas in 1988 confirmed these changes (Edland, 1989b). For this study, 16 commercial apple growers were selected, all of whom had maintained IPM in their orchards for several years. Their pesticide use, given in normally recommended dosages, varied from 0.4 to 2.0 applications (average 1.1 sprays per farm) against insect and mite pests, and from 1.9 to 8.5 applications (average 3.4 sprays) against diseases, for the whole season. Inspections of the orchards at harvest showed no damage, or only insignificant damage, on the crop and trees from insects and mites, and low damage from diseases. On one farm, which was sprayed only twice against diseases, 2% of the total crop was destroyed by scab. This was caused by infection in a plot of the scab-susceptible cultivar 'Vista Bella', where approximately 15% of the apples were infested. Such cultivars, because they are so sensitive to certain pests, should not be grown in an integrated production system.

According to a recent survey, a majority of Norwegian farmers are using IPM in their apple production (Kvåle and Hovland, 1996). Specifically, 47.3% of the farmers surveyed applied 1–2 annual sprays and 37.7% applied 3–4 annual sprays against insects and mites. Only in orchards with outbreaks, and others with special problems, were higher numbers of sprays used.

Use of insecticides at low dosages

In addition to fewer applications per season, significant reductions in the use of pesticides have been achieved by using lower concentrations or dosages. Research work on this topic started after a visit to some research centres in Switzerland and Germany in 1970. There it had been demonstrated that satisfactory control on aphids and some leaf rollers was possible after treatment with certain insecticides at from 1/5 to 1/2 of the recommended concentration of pesticide (M. Baggiolini, personal communication, 1970).

During the subsequent years, 13 insecticides applied at different concentrations, were tested for their effectiveness in controlling seven species of aphids in 15 orchards in eastern Norway (Edland, 1976). Systemic organophosphorus and carbamate insecticides proved to be superior to the non-systemic organophosphates, especially at the lower concentrations. Among the systemic compounds, only small differences appeared. Demeton-S-methyl at a dosage of 0.78 g a.i. (grams of active ingredient) per 100 l water (= 1/32 normal concentration) gave 90–100% control against green apple aphid, *Aphis pomi* Degeer, and rosy apple aphid, *Dysaphis plantaginea*, and gave 75% control against mealy plum aphid, *Hyalopterus pruni* (Geoffroy), when used at 0.2 g a.i. per 100 l (= 1/128 normal concentration). Against the same species, dimethoate gave a somewhat poorer effectiveness, but was slightly more effective than demeton-S-methyl in controlling cherry blackfly, *Myzus cerasi* (Fabricius). Ethiofencarb had similar effects to demeton-S-methyl against most aphid species. An exception was peach–potato aphid, *Myzus persicae* (Sulzer), probably an O–P resistant strain, against which demeton-S-methyl gave only 0–10% control at 1/32 and 1/16 normal concentration, respectively, while ethiofencarb showed 83% kill at 3.1 g a.i. per 100 l and 25% at 1.6 g a.i. per 100 l. Pirimicarb, which was not included in the tests, has in later experiments proved to give control similar to that of ethiofencarb.

The non-systemic compounds, e.g. azinphos-methyl, bromophos, diazinon, fenthion, malathion and parathion, were generally less effective than the systemic compounds against aphid pests, especially at low dosages (1/8 normal concentration and lower). In addition, they all showed a trend towards being more harmful to natural enemies. At all treatment levels that controlled aphids, most adult beneficial species disappeared as well. Relatively low dosages of the systemic compounds, however, allowed many larvae and nymphs of various predatory species to survive. These continued feeding on surviving aphids, and eventually provided complete and long-lasting control. Moreover, late larval and pupal stages of certain aphid parasites, e.g. *Ephedrus plagiator* (Nees) and *Praon volucre* (Haliday), developed normally after treatment with demeton-S-methyl and dimethoate at 1/16 and lower concentrations, whereas the number of aphids parasitized was smaller in plots treated with 1/8 normal concentration.

Both experimental tests and practical experience have demonstrated that systemic compounds, used at reduced dosages, are the most suitable for controlling aphids in an IPM programme. The concentration should be chosen according to the severity of infestation, the abundance of natural enemies, and the probability of migration and reinfestation after treatment. If an air-mist sprayer is used, delivering 300 l ha^{-1} or less, the application should be done in cloudy weather or during the night. Otherwise, on warm days with bright sunshine and windy weather, much of the pesticide evaporates before reaching the target. In Norwegian fruit areas, with daylight for most of the nights during the summer season, it has become a customary practice to apply most of the sprays, even fungicides, at night. Thus, failure to obtain satisfactory effects with low pesticide dosages seldom occurs. Generally, low dosages are less persistent than higher ones. During the growing season immigration of aphids into our fruit orchards does not occur, or only to a small extent. Therefore, persistent chemical treatments are not necessary unless very few natural enemies are present on the trees.

For the 15 years after 1970, when the use of DDT in Norwegian agriculture was prohibited by law, only non-systemic organophosphorus compounds were available for controlling chewing pests in orchards. These chemicals are all broad-spectrum insecticides, and applied in normally recommended dosages they are usually very harmful to most natural enemies. Research work to test the effects and side-effects of different dosages was therefore initiated. Since the widely used azinphos-methyl, even in normal concentrations, i.e. 37.5–50 g a.i. per 100 l mixture, often gives poor control of capsids, we tested it on different heteropterous species to see if lower dosages could give selective effects. At a rate of 7.5–10 g a.i. per 100 l this insecticide caused little damage to adults and nymphs of *Anthocoris nemorum*, but gave fully satisfactory control of many important caterpillars such as winter moth, leaf rollers and noctuids. However, these moderate dosages still seemed to have some detrimental impact on natural occurring phytoseiid mites.

In subsequent experiments, dosages of 5.0 g and even 2.6 g a.i. azinphos-methyl per 100 l have given satisfactory control of leaf rollers and small winter moth larvae, but varying and often unsatisfactory effects on large noctuid larvae, e.g. *Eupsilia transversa*. The rate of 3.8 g a.i. per 100 l, commonly used by IPM farmers today, had very little effect on six different non-resistant phytoseiid species. By comparison, fenthion at a rate of 5.4 g a.i. per 100 l (= 1/10 normal concentration) caused some reduction in the phytoseiid density (Edland, 1994). Experiments carried out in western Norway have shown that fenthion, even at 1.8 g a.i. per 100 l (= 1/30 normal concentration) gives sufficient control of mirid bug damage on pear, but is very detrimental to anthocorid and mirid predatory bugs (Hesjedal, 1986).

The organochlorine compound endosulfan has been registered in Norway for more than 30 years. It was primarily recommended as an acaricide for use against pear leaf blister mite, *Phytoptus pyri* Pagenstecher, and other gall mites, as well as against strawberry mite, *Tarsonemus pallidus fragariae* Zimmermann. It is an efficient acaricide against apple rust mite, *Aculus schlechtendali*, and one single spray in early blossom is usually enough to protect the apple crop from fruit russeting.

In experimental plots to test different pesticides against rust mites, endosulfan also gave excellent control against certain lepidopterous pests, even when applied at lower temperatures, whereas repeated applications seemed to have few detrimental effects on anthocorids and other predatory bugs (T. Edland, unpublished data, 1983–1988). Therefore, experiments were performed in the fruit areas of eastern and western Norway to obtain more knowledge about the selectivity of endosulfan used in different dosages. This pesticide, applied at a rate of 3.6 g a.i. per 100 l (= 1/10 normal concentration) gave satisfactory control of the larvae of winter moth *O. brumata*, and the noctuid *Orthosia gothica* (L.). In addition, applied at a rate of 7.2 g a.i. per 100 l it had no visible effects on anthocorid and mirid bugs (Meadow and Edland, 1990).

Endosulfan at moderate dosages (18.0 g a.i. per 100 l) has also proved to be an effective compound for controlling severe attacks of rose thrips, *Thrips fuscipennis*, on apple. With severe infestations of rose thrips, two pre-blossom applications are frequently needed

(Edland et al., 1995), but such treatments do not seem to reduce the number of anthocorid bugs. Although, endosulfan is not an efficient pesticide for controlling some important fruit pests such as apple fruit moth, codling moth and various leaf rollers, it is now regarded as a suitable pesticide for IPM in Norwegian fruit orchards.

In orchards which have effective natural enemy complexes, chemical applications should be low enough to ensure that part of the pest population survives to provide sufficient food resources for its natural enemies. This may be achieved by using certain pesticides at low dosages, and, if necessary, limiting the treatment to the most severely infested trees or parts of the orchard.

Norwegian IPM programmes in which pesticides are applied at low dosages, especially against aphids and many lepidopteran pests, have proved beneficial in many ways. They are providing efficient and long-lasting control measures as long as there is a sufficient density of natural enemies. Also, they are the safest and cheapest control measures for the farmers. Because they cause little contamination and harm to the environment, populations of beneficial species are increased. Further, these programmes minimize the risk of developing pesticide resistance and decrease the chance of undesirable residues accumulating on the fruits.

PESTICIDES PERMITTED AND RECOMMENDED IN NORWEGIAN FRUIT GROWING

Based on experimental research and various field surveys performed in different fruit areas over many years, specific dosages for controlling the various pest species/groups are now officially established for Norway (Svendsen, 1995). For each pesticide, the lowest dose rate which is sufficient to reduce the infestation or plant damage below the economic threshold is listed. Fifteen pesticides are currently registered for controlling insects and mites on Norwegian fruit crops. However, bromophos, fenitrotion and endosulfan have recently been withdrawn and will not be allowed after 1996–97. Owing to their severe toxicity to natural enemies, no synthetic pyretroids have been registered for use in our orchards, although they are widely used on berry, vegetable and other agricultural crops.

The guidelines for integrated pome fruit production, prepared by Hesjedal and Edland (1991), are based on the general principles, guidelines and standards for such production in Europe. Pesticides permitted in Norway, showing the permitted dosages for various insecticides and acaricides, are listed in Table 10.1. The dosages of the products permitted without restrictions are considered harmless to all major natural enemies. Clofentezine and diflubenzuron at normally recommended dosages give effective and selective control of spider mites and certain lepidopterous pests, respectively. All the other products may be used without restrictions, but only at reduced dosages. As discussed earlier, they are effective in controlling aphids and various caterpillars, while sulphur at this low rate gives excellent control of rust mites.

The products permitted with restrictions might be used at considerably higher dosages, but only for control of certain pests resistant to other measures, or in special situations. Thus, azinphos-methyl at 574–765 g a.i. ha^{-1} is applied when edge-spraying is used to control the apple fruit moth, whereas 383 g is the maximum permitted rate for an all-over coverage spray against the same pest species (cf. page 215). Moreover the high dose rates of azinphos-methyl, diazinon and fenthion are permitted for use against pests which occur only occasionally, such as mussel scale, *Lepidosaphes ulmi* (L.), and brown scale,

Table 10.1   Insecticides and acaricides permitted for integrated production of pome fruits in Norway

Pesticide	Dosages (g a.i. ha^{-1})	
	Permitted without restrictions	Permitted with restrictions
Azinphos-methyl	77–153	383–765
Chinomethionat		188
Clofentezine	150–200	
Demeton-S-methyl	13–50	
Diazinon		345
Diflubenzuron	144–288	
Endosulfan	54–107	536–714
Ethiofencarb	26–103	
Fenthion		803
Pirimicarb	15–50	
Sulphur	2400	

Source: after Hesjedal and Edland, 1991

Parthenolecanium corni (Bouché), some beetles, e.g. garden chafer, *Phyllopertha horticola* (L.), and clay-coloured weevil, *Otiorhynchus singularis* (L.).

For control of certain fruit-damaging tortricids, e.g. codling moth, *Cydia pomonella*, plum fruit moth, *Cydia funebrana* (Treitschke), and fruitlet-mining tortrix moth, *Pammene rhediella* (Clerck), azinphos-methyl at 383 g a.i. ha^{-1} is usually persistent enough to protect the fruit with one annual application. Fenthion at 803 g a.i. ha^{-1} has proved highly effective against thrips and cherry fruit fly, *Rhagoletis cerasi* L. However, such treatment is very detrimental to many beneficials and we are therefore searching for alternatives.

At present the highest dose rate of endosulfan, 714 g a.i. ha^{-1}, is recommended only for control of severe attack by pear leaf blister mite, *Phytoptus pyri*, an occasional pest on pears. This product is commonly recommended in a lower dosage (268 g a.i. ha^{-1}) for control of thrips and apple blossom weevil, *Anthonomus pomorum*, because other insecticides against these pests would have a more detrimental impact on natural enemies.

Chinomethionat is still widely used to control various species of spider mites, mainly *Panonychus ulmi*, and rust mites. Neither 2–3 applications of 188 g a.i. ha^{-1} nor 5 applications of 141 g a.i. ha^{-1} have shown any negative effects on phytoseiid mites (Edland, 1994). However, because some fruit growers have observed that chinomethionat suppresses the density of anthocorid bugs, they are now being advised to limit its use in orchards.

Dimethoate, occasionally applied at 15 and 45 g a.i. ha^{-1} against aphids and capsids, respectively, in conventional fruit production, is not permitted in integrated fruit growing in Norway. This insecticide is more harmful to many beneficial species, especially predatory bugs, than most other registered chemicals.

USE OF PHYTOSEIID MITES AS BIOCONTROL AGENTS IN ORCHARDS

As mentioned earlier, a large number of phytoseiid mites have been recorded in various parts of Norway. Even in regions with very cool climates, such as in the mountains and in the far north of the country, some species are found at high densities. In well-kept

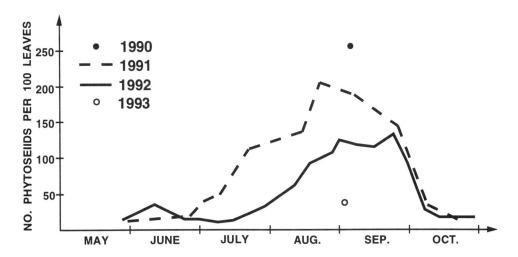

Figure 10.5 Average number of phytoseiid mites recorded in an experimental plot in eastern Norway in early September in 1990 and 1993, and at different intervals in 1991 and 1992

commercial orchards, however, these predators are sparse. In most cases, this is related to the use of pesticides, either by toxicity to the predators or because the pesticides eliminate the prey (Cranham, 1979).

In orchards with plenty of food for the phytoseiids, these predators may occur at high densities even if many annual sprays are applied. In an experimental apple orchard treated with a full dosage of azinphos-methyl in 1989, the spider mite, *Panonychus ulmi*, increased to outbreak levels the following summer. However, after some weeks the spider mite population diminished and reached an acceptable level in August. Leaf samples collected in early September showed a dense and even distribution of phytoseiid mites throughout the orchard, with an average of 2.5 mites per leaf (Figure 10.5). During subsequent years, this orchard was used for testing different fungicides at recommended dosages and a few insecticides at low dosages (Edland, 1994). In some plots, up to 10 annual sprays with fungicides were applied. Surprisingly, the phytoseiids developed well in all treatment areas, and kept the phytophagous mites far below the economic threshold. Thus, the decreased density of the predators from 1990 to 1993 shown in Figure 10.5 occurred because there was a lack of sufficient prey.

In this orchard study, 10 different phytoseiid species were recorded from about 220 apple trees. About 10 000 specimens were identified, showing that more than 95% of the total number belonged to six different species. Since all of them were dominant on certain trees during the experimental period, and since both the spider mites and the rust mites diminished to a low level, it was concluded that all six species are effective predators against harmful mite species. On the other hand, since all the naturally occurring phytoseiids are highly sensitive to most insecticides, and to organophosphorus (OP) insecticides in particular, their practical value as a biocontrol agent is limited, even in successful IPM programmes.

Following its success in England, an OP-resistant strain of *Typhlodromus pyri* Scheuten was brought from Horticultural Research International, East Malling, UK, and released in 1987 in a few apple orchards in Norway. Ever since 1979, when this strain was introduced into England from New Zealand, it has provided excellent control of both fruit tree red spider mites and apple rust mite in British IPM orchards, and has virtually abolished the need for chemical treatments (Solomon, 1988).

In Norway, this strain was first successfully established in a commercial orchard west of the Oslo fjord. There it has resisted the cold winters in eastern Norway and is now being spread to other fruit areas, where it will be tested for its predatory ability and its survival under different ecological conditions. In 1990, it survived successfully after a full-dosage treatment of azinphos-methyl applied against the apple fruit moth, and in the subsequent years it gave significantly better control of spider mites than any of the best new acaricides being tested in heavily infested experimental plots (Edland and Berle, 1994). However, it is still unclear if this OP-resistant strain is efficient enough to keep rust mites below the injury level on apple varieties that are most susceptible to fruit russeting.

Phytoseiid mites that can tolerate important pesticides which are used to control many pests are of great benefit in the efforts to produce large yields of high quality. In addition, their presence will reduce the need for acaricides, thereby stimulating populations of other natural enemies and decreasing the use of certain insecticides.

OTHER STRATEGIES FOR SUPPRESSING PEST ATTACK AND DECREASING PESTICIDE USE

In agriculture, many key pests can be regulated and kept at acceptable economic levels by using specific control techniques. Most of these biological and biotechnological techniques, apparently without any serious negative side-effects on natural enemies, have made it possible to replace broad-spectrum insecticides from the spray schedule. Thus, they are well suited for an IPM programme. However, most of these techniques are expensive, complicated to organize and maintain, and not well suited to small countries like Norway with modest and scattered production.

Blommers (1992) presents a most interesting review of the progress and prospects for selective control measures in Dutch fruit orchards. By use of one or two applications of the insect growth regulator fenoxycarb around blossom time against various leaf rollers, supplemented, if necessary, with a few applications of diflubenzuron and selective aphicides and acaricides during the season, good control of major pests has been achieved without any significant harm to the beneficials. In fact, these treatments resulted in a surplus of parasitoids such as the eulophid *Colpoclypeus florus* (Walker), which is a parasite of the summer fruit tortrix moth, *Adoxophyes orana*. According to Blommers, this might explain why this formerly most important leaf roller pest tends to become rare after a few years of IPM in the orchards.

In Norway, several insect growth regulators or inhibitors have been tested in experiments against a few lepidopterous pests and mites. Diflubenzuron at 25 g a.i. per 100 l water showed good effects against eggs of the noctuid *Xylena vetusta* (Hübner), and at 15–20 g a.i. per 100 l it was also effective against small larvae of noctuids (*Orthosia*) and geometrids (*Operophtera*) (Meadow, 1990). Flucycloxuron, when applied at full dosage (17 g a.i. per 100 l) just after egg-hatching, gave good and long-lasting effects against spider mite, *Panonychus ulmi* (Edland and Berle, 1994). Diflubenzuron, now available on the Norwegian market, is recommended for controlling noctuids and geometrids, codling moth and other fruit-feeding tortricids, e.g. *Pammene rhediella* and some leaf miners, e.g. *Lyonetia clerkella* (L.) and *Phyllonorycta blancardella* (Fabricius). Other growth regulators and some special combinations of selective pesticides, which will probably be badly needed when endosulfan is eliminated, are not likely to be registered in Norway, because the market is too small. Future lack of such efficient but environmentally harmless, agro-chemicals, which are commonly recommended and used abroad, might greatly disadvantage small countries.

In years with severe attacks of apple fruit moth, *Argryresthia conjugella*, diflubenzuron has shown some, but insufficient, control even when the apple orchards received 2 or 3 applications. Therefore, other solutions for sound management of this key pest have to be explored. During the development of the aforementioned forecasting system, we observed that the untreated control plots normally avoided attack by this moth, even when unsprayed apple trees were heavily infested in private gardens near the experimental fields. This indicated that azinphos-methyl has a repellent effect on the female moths, preventing them from immigrating into the sprayed orchards. In later experiments this was clearly demonstrated. In orchards where only the edges, i.e. the outer 1–3 rows, depending on the tree size, and the ends of each row were sprayed, the inner untreated part of the orchard avoided infestation.

This form of edge-spraying has been applied with success for many years by some growers. A prerequisite, of course, is that the edge-spray is applied before the moths start their migration, and that an insecticide persistent enough to last for the critical period of weeks is applied. In Norway, azinphos-methyl at a dosage of 38.3 g a.i. per 100 l mixture has proved to give full control. Although this strategy for pest management is detrimental to natural enemies inhabiting the edge rows, the beneficials in the major part of the orchard avoid negative side-effects. They will prevent increasing attack by aphids and mites later in the season and during the subsequent year. However, in years when the mountain ash crop fails completely, as was the case in 1994, both the earliest and latest emerging moths migrate to apple for oviposition. In such years the period of migration usually lasts for several weeks, and one single application of edge-spraying is usually not sufficient to protect the apple orchards.

Because there are no selective insecticides for controlling *Argyresthia conjugella*, and the selective technique of edge-spraying does not work efficiently enough in severe outbreak situations, we have been searching for another strategy, which hopefully may provide a permanent solution. On the basis of more knowledge of the ecology of *Argyresthia conjugella*, and a better understanding of the relationship between this moth species, its natural and preferred host-plant *Sorbus aucuparia* L, and its key natural enemy, the braconid *Microgaster politus*, it is possible that a kind of habitat manipulation could be a safe strategy in managing the pest. This hypothesis has been discussed in detail elsewhere (Edland, 1995), and only the main points will be dealt with here.

As described before, the *Argyresthia* moth attacks apple only in years when the mountain ash produce insufficient berry crops. The number of *Sorbus* berries on wild-growing trees fluctuates greatly from year to year, resulting in a similar fluctuation in the number of *Argyresthia* moths as well as in the number of parasites. Some ornamental types of *Sorbus aucuparia*, however, produce large numbers of berries every year. If these types were spread in sufficient numbers in the surroundings of the orchards, the number of *Arygresthia* and its key parasite would be greatly stabilized.

In 1991, field tests were begun to test the hypothesis that with this simple manipulation of the natural habitat, a sufficient equilibrium between *Argyresthia* and its parasites would take place at a level that prevents emigration and subsequent damage to apple. In cooperation with other research institutions, 11 different exotic *Sorbus* clones and one native one were selected to test whether all the 12 clones would produce satisfactory berry crops every year in all areas. Furthermore, the attractiveness of all the clones for oviposition by *Argyresthia* would be investigated, as well as any possible repellent effect on its key parasite.

The first results in 1994 were encouraging. All the clones, represented by three trees in each of 3–6 replicates, produced a large and even berry crop on the majority of trees. In contrast, all the trees of the wild–native type, represented by three trees in each of 30 replicates, produced no crop at all. The percentage attack of *Argyresthia* in the mountain ash varied greatly, but on some trees of three or four clones up to 17–49% of the berries were infested. The parasitization of the *Argyresthia* larvae was highest on trees with the highest berry infestation. Thus, from one replicate with 36.4% attacked berries, 12% of the *Argyresthia* larvae were parasitized by the *Microgaster* parasitoid. If this approach to habitat manipulation continues to be successful, there will be one less important pest to control, and all sprays against this target insect can cease. As with other biocontrols considerable savings in application costs will occur, and detrimental side-effects on natural enemies and other undesirable environmental impacts can be avoided.

KNOWLEDGE AND GOOD COOPERATION WITH THE FARMERS ARE AN INDISPENSABLE CONDITION FOR FURTHER PROGRESS IN IPM

Substantial knowledge about the fauna, the interactions between pests and natural enemies, and the effects and side-effects of different pesticides and other control agents are of fundamental importance for progress in plant protection. Likewise, development of practical and reliable forecasting systems and economic thresholds, and improvements in spraying machinery and other technical equipment needed in modern fruit growing, are absolutely essential to sound pest management strategies to minimize the use of undesirable chemicals.

However, even the most extensive scientific knowledge, the best warning systems and thresholds, and the possibility of using selective pesticides and other sound techniques are worthless if the farmers are not motivated to use and maintain the best measures that are available. To achieve this aim, a close and trusting cooperation between the advisors/scientists and the practical growers is indispensable.

Undoubtedly, many farmers possess thorough knowledge and valuable experience of plant protection problems. Their observations of pest infestations and damage made in frequent inspection in their own orchards may provide the scientists and advisors with significant information for use in forecasting and advisory work. Today, several fruit growers are able to recognise their important pests, and deal with them in the most effective way by an intelligent and thoughtful use of the various techniques available.

Most growers have a good general impression of the importance of beneficials, predators, parasites and pollinators, and many of them are aware of the need to measure and assess their influence on various pests under different growing and ecological conditions. In an attempt to help the growers in this respect, the first course in IPM was arranged in 1976. In most years since 1982 we have arranged annual courses, and both research workers, advisors and commercial fruit growers have participated.

Each course lasts for 1–5 days, and both summer and winter courses have been given. The programme varies. For the short summer courses, lasting for 1 or 2 days, practical training in recognition of pests and natural enemies, and in selection and use of monitoring methods are emphasised, while the winter courses are mainly lectures and workshops.

The summer courses, lasting for 3–5 days, are usually arranged during or shortly after the blossom period, when many different species of pests and beneficials are available in the orchards. These courses, which have been considered by most participants as the most valuable approach, emphasize theory, practical training, workshops and examinations.

In 1988–89, a video entitled 'Integrated Pest Management in Apple Cultivation' was prepared by A. Rein (1990) in cooperation with entomologists and plant pathologists. This video gives a most useful and practical introduction to IPM in orchards in the areas of assessment of the pest situation, use of different methods, evaluation of needs for control measures, use of low dosages and other selective means, etc. In addition, it gives an instructive survey of the commonly occurring pests and beneficials, showing how they act in nature. Most of them are pictured in different developmental stages, showing their distinguishing marks, and enabling the growers to quickly recognise the pests and beneficials in the field. Therefore, this video has become a most valuable contribution to the course work.

Up to now more than 50% of our commercial fruit growers have participated in the IPM courses. Undoubtedly, this activity has contributed to increased knowledge and greater self-confidence in making decisions on *if*, *when* and *how* sprays should be applied. Some of the growers have also started up special working groups in their home areas, where current plant protection problems are discussed at frequent meetings during the season. This has resulted in a more correct use of pesticides and a sustainable way of fruit production.

SUMMARY

The agro-ecosystems in Norway are characterised by relatively few and moderate plant protection problems. However, a great number of different insects and mites have been recorded as occasional feeders on fruit trees, and about 80 species have pest status. On the other hand, many natural enemies, especially predatory bugs and phytoseiid mites, have proved to be efficient in keeping many orchard pest populations at low levels.

In commercial orchards, pesticide use increased greatly with the introduction of synthetic pesticides. Thus, in the 1950s and early 1960s heavy spray programmes of 6–8 annual applications were commonly used against insects and mites. This routine spray programme was not only expensive, but also caused many environmental problems.

Extensive investigations of orchards and pesticide use started in 1962. This research provided important information about the major pests and their natural enemies, ecology and natural control, and about the direct and indirect effects of different chemicals. In addition, practical methods were developed for assessing fruit pests and determining the economic thresholds at which to treat.

Based on this new knowledge, technologies were developed to reduce the use of pesticides. By 1974, farmers who had adopted IPM in their orchards used only 0–3 sprays a year, which was a reduction of 60–80% in insecticide and acaricide application. The reduced sprays were used only when needed, and since they were not part of a routine schedule, they were applied at any time when necessary.

A special forecasting system for assessing the key pest, *Argyresthia conjugella*, was developed in the 1970s and subsequently was widely used in Norwegian fruit areas. It has resulted in significant pesticide reduction (80–90%) from two annual applications to just one every 3–5 years for this pest. A similar reduction has been achieved by utilization of a forecasting system for the codling moth.

In addition to fewer applications per season, a considerable reduction in use of pesticides has been achieved by using lower dosages (1/5–1/60 of normally recommended dosages). In Norwegian IPM programmes, low pesticide dosages, especially against aphids and various caterpillars, have proved beneficial in many ways. They provide efficient and long-lasting control because the natural enemies are conserved in the orchards. Also, low dosages are the safest and cheapest control measures for the farmers, because they cause little environmental contamination and populations of natural enemies are increased. Further, these programmes minimize the risk of developing pesticide resistance and reduce the chance of undesirable residues on the fruits.

Based on research and field surveys over many years, eight different pesticides are currently permitted without restriction, while five are restricted, in the integrated production of pome fruits in Norway. The permitted dosages of these pesticides are shown in Table 10.1.

During the last decade, phytoseiid mites have been used successfully as biocontrol agents in orchards. In particular, an OP-resistant strain of *Typhlodromus pyri* has proved to be efficient against both fruit tree red spider mite and rust mites. This strain is now being spread to various fruit areas in Norway. Predatory mites that tolerate important pesticides, which are sometimes needed in IPM programmes, are of great benefit, as they greatly reduce the need for acaricides, thereby encouraging populations of other natural enemies and further decreasing the use of acaricides and certain insecticides. Clearly the effective utilization of natural enemies is of fundamental importance in developing and maintaining sound and efficient IPM programmes. The event of a higher density and improved activity among natural predators and parasites is considered to be the main reason for the decreased pest attack and reduced need for pesticides in Norwegian fruit orchards.

Although a few selective pesticides, such as diflubenzuron, are now available for use in orchards, only broad-spectrum chemicals are efficient against severe attack by our key pest *Argyresthia conjugella*. Therefore, the new approach of habitat manipulation is under investigation, which hopefully may provide permanent pest control without the use of pesticides.

A recent survey has shown that the majority of Norwegian fruit farmers are using IPM with success, and it is estimated that their average use of insecticides and acaricides is equal to two normal dosage applications per year, which is a reduction of 70–75% over the past 25 years. In order to maintain this low level of pesticide use, the introduction and establishment of new or resistant pests must be avoided, and selective, efficient and environmentally friendly pesticides must be available for Norwegian fruit growers.

ACKNOWLEDGEMENTS

This paper is largely based on results obtained in several research projects financed partly by the Norwegian Research Council (1962–1991). The author is indebted to many commercial fruit growers and to the Agricultural University of Norway for providing orchards for conducting experiments. During recent years, most of the research work has been performed in orchards belonging to civ.ing. Odd Berle, Svelvik, who has also prepared the figures. A special thanks to Dr

Kåre Hesjedal, Ullensvang Research Centre, for fruitful and stimulating cooperation over many years, and to all my technical assistants for valuable help in my 30 years of work.

The author wishes to express his gratitude to Professor David Pimentel, Cornell University, USA, for reviewing and correcting the manuscript, and for all valuable help and advice during its preparation.

REFERENCES

Alford, D.V. (1984) *A Colour Atlas of Fruit Pests, Their Recognition, Biology and Control*. Wolfe, London, 320 pp.

Blago, N. and Edland, T. (1991) Vorhersage der Eiablage von *Cydia pomonella* in Europa: Erste Ergebnisse eines weiterentwickelten Prognosemodells. *Zeitschrift für Pflanzenkrankheiten und Pflanzenschutz* **98**, 378–384.

Blommers, L. (1992) Selective package and natural control in orchard IPM. *Acta Phytopathologica et Entomologica Hungarica* **27**(I), 127–134.

Cranham, J.E. (1979) Managing spider mites on fruit trees. *Span* **22**, 28–30.

Edland, T. (1971) Wind dispersal of the winter moth larvae *Operophtera brumata* L. (Lep. Geometridae) and its relevance to control measures. *Norsk entomologisk Tidsskrift* **18**, 103–107.

Edland, T. (1974) Prognosis for attack of apple fruit moth (*Argyresthia conjugella* Zell.). A preliminary report on methods and results. *Gartneryrket* **64**, 524–532 (with English summary).

Edland, T. (1976) Effectiveness of insecticides at different concentrations on aphids and natural enemies. *Forskning og Forsøk i Landbruket* **27**, 683–699 (with English summary).

Edland, T. (1977) Eple- og plommeviklar. *Gartneryrket* **67**, 444–449.

Edland, T. (1981) *Forecasting attacks by the winter moth and allied species in orchards, and evaluation of control requirements. Växtskyddsrapporter, Jordbruk* No. 16, pp. 43–53 (with English summary).

Edland, T. (1983) Midd i frukthagar, *Frukt og Bær* 1983 60–71.

Edland, T. (1989a) Integrated pest management in Norwegian orchards. Noragric Occasional Papers, Series C, Development and Environment No. 3, pp. 65–74.

Edland, T. (1989b) Kvantifisering av kjemikalbruken nytta mot soppsjukdomar og skadedyr i frukthagar. Aktuelt fra SFFL No. 3, pp. 297–305.

Edland, T. (1991) Drivhuseffekten kan føre til drastisk forverring av skadedyrproblema i norsk fruktdyrking. *Gartneryket* **81**(11), 19–22.

Edland, T. (1993) *Taksonomi: Rovmidd (Acari: Phytoseiidae), Ein Oversikt over ulike Middgrupper i Noreg, og Omtale/Bestemmingsnøklar for Slekter og Arter av Rovmidd som Lever på Frukt og Bær.* Norwegian Plant Protection Institute, Course Book, 130 pp.

Edland, T. (1994) Side-effects of fungicide and insecticide sprays on phytoseiid mites in apple orchards. *Norwegian Journal of Agricultural Sciences* Suppl. 17, 195–204.

Edland, T. (1995) Integrated pest management (IPM) in fruit orchards. In Hokkanen, H.M.T. and Lynch, J.M. (Eds) *Biological Control: Benefits and Risks*. Cambridge University Press, Cambridge, pp. 44–50.

Edland, T. and Berle, O. (1994) Observasjonar og forsøk med frukttremidd. *Gartneryrket* **84**(7), 20–22.

Edland, T., Helleland, I. and Berle, O. (1995) Trips i norske frukthagar. *Gartneryrket* **85**,(7), 36–41.

Fjelddalen, J. (1952) Midder på frukttrær og bærvekster. Biologi og bekjemping. *Frukt og Bær* **5**, 56–72 (with English summary).

Frankenhuyzen, A.van (1988) *Schadelijke en Nuttige Insekten en Mijten in Fruitgewassen*. Nederlandse Fruittelers Organisatie, 's-Gravenhage/Plantenziektenkundige Dienst, Wageningen, 285 pp.

Hesjedal, K. (1986) Effects of pesticides in different concentrations in mirids and anthocorids in orchards. *Forskning og Forsøk i Landbruket* **37**, 213–217 (with English summary).

Hesjedal, K. (1989) Økonomisk viktige tegearter i frukthagen. Informasjon fra SFFL No. 5, pp. 164–174.

Hesjedal, K. and Edland, T. (1991) *Retningsliner for integrert produksjon av kjernefrukt*. Planteforsk, 10 pp.

Hesjedal, K., Edland, T., Stensvand, A., Hovland, B., Nybøle, K. and Amundsen, T. (1993) *Notathefte for dokumentert og integrert produksjon av kjernefukt.* Planteforsk, 32 pp.

IOBC/WPRS (1980) *Visuelle Kontrollen im Apfelanbau.* No. 2, 3rd edition. Generalsekretariat WPRS, 96 pp.

Kvåle, A. (1995) *Fruktdyrking.* Landbruksforlaget, Oslo, 208 pp.

Kvåle, A. and Hovland, B. (1996) Usage of pesticides in Norwegian apple production. *IOBC/WPRS Bulletin* **19**(4), 413–414.

Massee, A.M. (1954) *The Pests of Fruits and Hops.* Crosby Lockwood & Son, London, xvi + 325 pp.

Meadow, R. (1990) An evaluation of two selective insecticides for control of noctuids (Lepidoptera: Noctuidae) and geometrids (Lepidoptera: Geometridae) in Norwegian apple orchards. Noctvids (Lepidoptera: Noctvidae) and geometrids (Lepidoptera: Geometridae) in Norwegian apple orchards-damage, economic injury levels and some alternatives for control. Doctor Scientiarum Theses, Agricultural University of Norway, Number 13, 36–51.

Meadow, R. and Edland, T. (1990) The effect of reduced dosages of azinphos-methyl, endosulfan and fenthion on noctuid larvae (Lepidoptera: Noctuidae) and geometrid larvae (Lepidoptera: Geometridae) and on adults and nymphs of predatory bugs (Hemiptera: Heteroptera) in Norwegian apple orchards. Noctvids (Lepidoptera: Noctvidae) and geometrids (Lepidoptera: Geometridae) in Norwegian apple orchards-damage, economic injury levels and some alternatives for control. Doctor Scientiarum Theses, Agricultural University of Norway, Number 13, 22–35.

Miedema, E. (1987) Survey of phytoseiid mites (Acari: Phytoseiidae) in orchards and surrounding vegetation of northwestern Europe, especially in the Netherlands. Keys, descriptions and figures. *Netherlands Journal of Plant Pathology* **93**, Suppl. 2, 1–64.

Rein, A. (1990) *Integrated Pest Management in Apple Cultivation* (Eds Edland, T. and Hesjedal, K.). An instruction film produced by Agroinform A/S, Gvarv, Norway, in 10 languages, approx. 2 hours.

Schøyen, T.H. and Jørstad, I. (1944) *Skadedyr og Sykdommer i Frukt-og Bærhagen.* Aschehoug, Oslo, 140 pp.

Solomon, M.G. (1988) Managing predators in apple orchards in the UK. *Proceedings of the XVIIIth International Congress of Entomology,* Vancouver, BC, 3–9 July 1988. Abstracts and Author Index, p. 322.

Steiner, H. (1967) Die Anwendungsmöglichkeiten der Klopfmethode bei Arbeiten über die Obstbaumfauna. *Entomophaga, Mémoire HS* No. 3, pp. 17–20.

Svendsen, S. (1995) (Ed.) *Plantevern: Kjemiske og Biologiske Midler 1995–1996.* Planteforsk–Plantevernet/Landbruksforlaget, 238 pp.

Tuovinen, T. (1993) Phytoseiid mites (Acari: Gamasina) in Finnish apple plantations with reference to integrated control of phytophagous mites. *Agricultural Science in Finland* **2**, Suppl. 1, 1–215.

CHAPTER 11

Cultural Controls for Crop Diseases

David Marshall

Texas Agricultural Experiment Station, Dallas, TX, USA

INTRODUCTION

This chapter uses definition, classification, example and analysis for an objective examination of cultural controls for plant diseases. These cultural practices have been, are now and will continue to be major factors in plant health management. By way of definition, cultural practices are the manipulations and decisions needed in crop production to eliminate or reduce initial disease inoculum or to help suppress the rate of epidemic disease development. All of the methods and practices discussed in this chapter can be accomplished at the farm level. This is important, because the best possible stewards of our invaluable land resources are farmers. The crop decisions that farmers make and the crop management practices they use have a direct impact on the sustainability of agriculture.

Why should we want to be concerned about the cultural methods farmers can use to reduce disease on their crops? The answer lies in keeping agriculture viable and sustaining the most basic of human needs, food and fibre. It is easy for agricultural scientists, particularly in developed countries, to get caught-up in the language and daily activities of their own speciality area. Too much time, energy and resources is spent convincing each other that certain agricultural practices are better than others. It is easy to lose sight of the fact that a great many people, particularly in Third World countries, suffer from undernutrition. Even though the immediate causes of undernutrition may be poverty, wealth inequalities, overpopulation and illness, rather than the lack of food (Foster, 1992), there is no guarantee that food supplies will continue to be adequate in the future. In fact, global grain production has declined in recent years (Brown, 1994). This need to produce food should be paramount in our efforts to develop effective, efficient and sustainable disease control practices. The total supply of food and fibre in the world is mainly a function of crop land area and yield per unit of land. Of these two factors, the economics of food production generally favour an increase in output per unit area rather than an increase in the amount of land used to produce agricultural products. Scientific and technological advances have served to increase the efficiency of food production. However, adverse climatic conditions, plant diseases and plant pests can restrict plants from attaining their biologic yield potential. The control or management of plant diseases can be separated into three broad categories: biological (including genetic), cultural and

Techniques for Reducing Pesticide Use. Edited by D. Pimentel

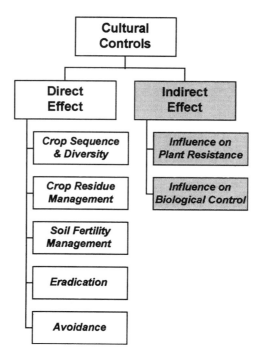

Figure 11.1   Cultural control organizational chart outlining the practices that have direct or indirect effects on the initial amount of disease or the rate of disease increase

chemical. Direct cultural controls include crop sequence (rotation) and diversification, crop residue management, soil fertility management, eradication and disease avoidance (Figure 11.1). Cultural controls also have an important indirect effect on host-plant resistance and biological control methods. These indirect effects (shaded in Figure 11.1) will be mentioned only briefly at the end of the chapter. Even though the topics in this chapter are discussed independently, most, if not all, of them are most effective when used in an integrated programme of crop production and pest control.

Matters of scale are important considerations for cultural controls. Some of the controls are aimed at an organismal level, while others are important at a population, a community or even a biosphere level. A control method that may be useful on a small scale may not be useful or practical on a large scale. Similarly, a short-term solution to a problem may not be what is needed to solve the problem in the long run (i.e. short-term solutions may not be sustainable). All good agricultural practices should have a solid theoretical framework. In theory, we must be able to relate our understanding of a biological phenomenon to the biological levels that adjoin the phenomenon (Marshall, 1991). Our work may be deemed too trivial if we cannot relate it to a higher organizational level. Similarly, if we cannot relate a phenomenon to a lower organisational level, the work may be merely descriptive.

CROP SEQUENCE AND DIVERSITY

Crop sequence or rotation

Crop sequence (rotation) is the planned succession of crops in a field over time. Crop sequence may be a more descriptive term than crop rotation because the important aspect of time is implicit in crop sequence, whereas crop rotation must use the time parameter as part of its definition. Nevertheless, the two terms will be used interchangeably here. The practice of a planned cropping sequence probably first began early in agricultural history when farmers 'rested' some fields after crop production on those fields declined. Planned crop sequences grew more out of a need to conserve water and soil rather than to control diseases. Often these fields were fallowed, and domesticated livestock may have been allowed to graze on the volunteer plants and weeds that emerged (King, 1911; Fussell, 1965). Much of the arable land in the world's temperate zones were farmed for thousands of years to cool- or warm-season cereals alternating with a leguminous or other 'break' crops every 1–3 years (Tivy, 1990). Sequences of crops tended to keep diseases, weeds and insect pests somewhat in check, while at the same time adding organic matter to the soil. During the late 1800s in the USA, farmers' alliances were first formed to spread information and exchange experiences on crop rotation and diversification. Later, these alliances became political, and eventually formed the first national independent political party, known as the People's or Populist Party (Hesseltine, 1962). It has only been since about the 1950s that crop production has intensified to the point where continuous monoculture is not uncommon. With intensification came simplification, mainly expressed as a reduced number of crop species in cropping sequences (Fiddian, 1973). Continuous cultivation of potato, wheat and rye tended to increase minor root pathogens in The Netherlands, thus emphasizing the importance of break crops (Schippers et al., 1985, 1987). In the absence of alternate crops, letting fields lie fallow can be an important component of a cropping sequence, particularly for water conservation in low rainfall, non-irrigated areas.

In the Pacific northwest of the USA, the manipulation of soil moisture is an overriding factor in controlling wheat diseases (Cook, 1986). This is because water stress predisposes the crop to infection by soil-borne fungi. Therefore, water management can clearly be an important aspect of crop rotations that include fallow periods. Water management in a crop rotation scheme also interacts with seedbed preparation, in that planting in ridges can help some crops to channel water into the area where it is most needed.

As a disease control measure, a planned sequence of crops is much more effective against pathogens that are primarily soil-borne rather than air-borne. Within the soil milieu, root-inhabiting pathogens are inclined to be affected more by crop sequences than soil inhabitants. This is because root-inhabiting pathogens tend to be somewhat crop-specific and poor competitors relative to soil-inhabiting pathogens (Kommedahl and Todd, 1991). In addition, pathogens that produce primary inoculum on the debris or stubble of the previous crop may also be affected by crop sequencing (Nusbaum and Ferris, 1973). For example, the impact of sheath blight of rice (Belmar et al., 1987) and *Septoria* diseases of wheat can be minimized with proper crop rotations (Holmes and Colhoun, 1975; Garcia and Marshall, 1992; Shaner and Buechley, 1995). A primary effect of crop sequencing is to interrupt the pathogen's life cycle (Power and Follett, 1987) with the aim of reducing the initial population density (Marshall, 1991). Kommedahl and

Todd (1991) list approximately 64 fungal, 19 nematode, 1 viral and 16 bacterial diseases in which a planned cropping sequence had a major effect on the control of a disease.

Because the soil consists of a multitude of microorganisms, a change from the primary crop of interest to a very different break crop may cause some microorganisms to proliferate, others to decline, and still others will be unaffected. Fungal populations have been found to be richer and more variable under crop sequences as opposed to continuous monoculture (Williams and Schmitthenner, 1962). Thus, a planned crop sequence will have a better chance of being successful if it is aimed at a single disease problem (Piening and Orr, 1988). Care must be taken, however, to avoid break crops that may promote the population increase of a previously undetected or minor pathogen on the primary crop (Broadfoot, 1934, Chinn, 1976). Oats serve as a host for *Fusarium* root rot in the Pacific northwest of the USA, and should not be planted in wheat–fallow or barley–fallow rotations (Cook, 1980). Another example of this occurred in some parts of Texas in 1991, when Austrian winter pea was used as a break crop to add organic matter and decrease population levels of the root rotting *Fusarium* and *Bipolaris* spp. in a continuous wheat-cropping system. After the pea crop had been incorporated into the soil and wheat planted the next year, the young wheat plants began to yellow and die. The wheat was being attacked by a *Pythium* species which had always been in the soil, primarily as a root nibbler at low population levels. The pea crop served its purpose in increasing the organic matter of the soil and decreasing the population levels of *Fusarium* and *Bipolaris*, but it was a receptive host for an unsuspected soil-borne pathogen. Luckily, the peas were tried in only a few fields (Marshall, 1993, unpublished data). One of the most underrated phenomena in agriculture is the introduction or intensification of new and unsuspected problems associated with changes in agricultural practices (Hunter and Leake, 1933).

The primary effect of a non-host crop in a cropping sequence is to reduce the level of primary inoculum of the target pathogen in the soil. Thus, cropping sequence as a practice is a preventative measure rather than a curative measure of controlling disease. Fry (1982) described the two pathogen-related characteristics of broad host range and long-lived survival structures as factors that could tender planned crop sequences ineffective. It may be difficult to find a break crop that is adapted to an area and which is a non-host for fungal pathogens such as *Sclerotium rolfsii* and *Rhizoctonia solani*, or the nematode *Pratylenchus penetrans*. Other pathogens may have long-lived survival structures, such as fungal sclerotia or nematode cysts, which could remain semi-dormant in the soil for many years, waiting for a cue from its host to become active once more and initiate infection. The cotton root-rot pathogen *Phymatotrichum omnivorum* has a wide host range (dicots) and very long-lived sclerotia. Cotton planted into fields that had not had cotton or any other dicotyledenous crop for 10 years may still suffer from cotton root rot.

A pathogen's competitive saprophytic ability, the population level when its primary host was last used, its survival structures and its host range all contribute to the amount of time a primary crop must be rested from a field. Economics (particularly in the short term) also enters into the picture, because it may be more profitable to grow a primary crop that may lose some yield to a disease, rather than an unprofitable secondary crop, or allowing a field to go fallow. Because the disease suppression effects of crop sequences may not be realized until years after their initial implementation, the profitability and effectiveness of some crop sequences must be measured over long time-intervals. (Cook, 1981) found that soil cropped continuously to wheat, or wheat in rotation with a grass mixture, became suppressive to take-all (caused by *Gaeumannomyces graminis tritici*) after

7 years. Break crops of potato, oats, alfalfa or beans resulted in the soil becoming conducive to take-all. In fact, the bean crop in the sequence served to maintain inoculum levels of *G. g. tritici*. However, in Victoria, Australia, higher wheat yields were obtained when either non-cereals (rapeseed, lupine, pea or medic) or fallow preceded a wheat crop. The non-cereal break crops were more effective than fallow in reducing the effects of take-all on wheat (Kollmorgen et al., 1985). Medic and pea rotations were also effective in South Australia for reducing take-all on wheat (Rovira and Venn, 1985).

Soybean production in the USA often benefits from planned crop sequences. However, the yield benefits are not necessarily the result of disease control. Brown stem rot (caused by *Phialophora gregata*) can be controlled by various soybean–corn rotations under some circumstances (Dunleavy and Weber, 1967; Kennedy and Lambert, 1981), but not in others (Whiting and Crookston, 1993). It has not always been possible to pinpoint a cause for the beneficial effects of rotations on soybean yield (Crookston et al., 1991). Reasons for this could include complex parameter interactions or imprecise detection methods for biotic factors. However, it must be realized that there is not necessarily a direct relationship between a specific disease and the yield of the attacked crop, even if the disease in question was apparently the only one present. Crop loss assessments must be made under conditions where the only variable is different levels of the disease in question.

A type of cropping sequence that should be given more attention and formalization is *cultivar rotation* or *sequence*. This is the use of cultivars of the same crop that vary in their reaction to diseases, and which are diversified over time and space. For example, population levels of *Cylindrocladium crotalariae* were reduced with 3 years continuous monoculture with a resistant variety as compared with a susceptible variety (Black et al., 1984). In the integrated control of *Phytophthora* root rot of soybean, rotation to tolerant cultivars play a major role (Schmitthenner and Van Doren, 1985). Examples can also be found for tobacco (Melton and Powell, 1991).

Crop diversity

A practice of modern agriculture that has enhanced the destructive potential of disease is the use of genetically similar crops in continuous monoculture. This type of agriculture serves to select pathogens for increased adaptability on the crop. It also provides the pathogen with a substrate that is continuous over time and space. Marshall (1991) separated all the methods of using crop diversification into two groups: multiple cropping, which is the diversification of more than one type of crop over space and/or time, and mixture cropping, which is the pre-plant blending of different genera, species or cultivars grown as a single unit and aimed primarily at spatial diversification.

For the majority of the time that agriculture has been in existence, crops were spaciogenically diverse (Marshall, 1989) within fields, between fields and over regions (Jasny, 1944; Harlan, 1992). The intensification of agriculture has given rise to uniformity in agricultural practices and in the crops that are produced, regardless of whether uniformity was the best production strategy in a given area. Significant parts of the arable land in the world are at high risk from diseases. These areas are climatically and edaphically diverse, with disease occurrence and severity being closely related to the ecology of the area. Yet, the same agricultural practices that work in low disease risk areas are commonplace in the high risk areas. Spaciogenic uniformity has been a major determining factor in several plant disease epidemics (Ullstrup, 1972; Dubin and Torres, 1981; Marshall, 1988).

A common thread woven through most disease epidemics is the crowded nature of genetically similar host plants. However, there is still a great practical need to cultivate large areas of similar plants in order to produce enough food. A possible answer to this problem lies in the use of diversity to manage disease. Viglizzo and Roberto (1989) studied the effects of diverse farming practices on 38 farms in Argentina, and found that diversification was a relevant stabilizing factor and was positively correlated with increased performance of agroecosystems. However, they cautioned against an excess of diversity that may impose additional management costs without extra benefits.

CROP RESIDUE MANAGEMENT

When an annual crop is harvested of its economic yield, what remains in the field is the crop residue. Sometimes the residue can be harvested and used for animal feed, animal bedding or other uses. Most of the time, however, the amount of residue on a field must be reduced in volume in order to plant the succeeding crop. Incorporation of the residue into the soil by ploughing or discing is the most common method of managing it. Incorporation into the soil brings about breakdown of the residue by soil microorganisms. The two main factors involved in crop residue management are the intensity of the management programme and the soil depth to which residue is incorporates. In some areas, tillage has been eliminated (no-till) or reduced (minimum-till, conservation tillage or ecofallow). The advantages of reduced tillage that leave crop residues at or near the soil surface are reduced soil erosion, improved soil physical condition, less tractor use and improved soil water conservation (Knowles et al., 1993).

As a disease control measure, the breakdown or removal of crop residues can decrease the amount of inoculum of those pathogens that can survive on plant debris. Thus, in general, reduced tillage tends to exacerbate problems of disease control. Some very important diseases are involved, including southern corn leaf blight (Ullstrup, 1972), *Septoria* (Harrower, 1974; Hewett, 1975; Garcia and Marshall, 1992) tan spot (Wright and Sutton, 1990), diseases of wheat, *Verticillium* wilt of cotton and bacterial blight of soybean (Daft and Leben, 1973). For example, *Septoria* leaf blotch has increased in frequency, which may be due to reduced tillage practices in many areas. If residues are buried, then the viability of *S. tritici* and *S. nodorum* spores is rapidly lost. However, if residues remain on the surface, the primary inoculum is increased for the subsequent wheat crop (Shaner and Buechley, 1995). Plant debris is also the primary habitat for the formation of the sexual, ascogenous stage of *S. tritici* (Garcia and Marshall, 1992).

In general, the incorporation or breakdown of crop residues by other means is designed to hasten the natural disappearance of a pathogen from the crop residues in the soil. Kommedahl and Todd (1991) has listed approximately 20 diseases that have been minimized by crop residue management. In some cases, tillage and residue management gave variable results over a period of years, but the practice generally had no effect on seedling diseases or pathogen inoculum densities (mainly *Rhizoctonia*, *Pythium* and *Fusarium*), although completely disposing of triticale residue by burning helped to reduce some cotton and soybean seedling disease (Sumner et al., 1995). Symptoms of soybean sudden death syndrome were increased under no-till conditions (Wrather et al., 1995). Double-crop soybeans after wheat with no-till reduced population levels of *Heterodera glycines*, the soybean cyst nematode, but yields were unaffected, and therefore the advantage may be over a long time period (Hershman and Bachi, 1995). In an 8-year study, no-till had a positive effect, or no effect, on soybean yields early in the study, but yields decreased later in the study due to weed pressure compared with conventional till (where weeds were controlled by cultivation). *H. glycines* populations were highly variable with respect to tillage treatment (Koenning et al., 1995).

In the Pacific northwest of the USA, reduced till or no-till for soil erosion has not been accepted because of lower yields (some of which is disease-related), particularly on wheat. An increase in take-all in wheat was found following wheat seeded directly into no-till soil conditions because more infested debris was present and because the inoculum source was ideally positioned close to the host for infection (Moore and Cook, 1984). In order to avoid water stress, which predisposes the plants to *Fusarium* root rot, fields with standing stubble should be worked with a chisel plough to improve water infiltration and reduce water stress (Cook, 1980).

SOIL FERTILITY MANAGEMENT

There are several steps which farmers may take to optimize the fertility level of a crop. In doing so, plant health is increased and the crop is less likely to suffer from the effects of disease. Some nutrients, if used in excess, in particular nitrogen, can cause plants to become too lush, perhaps leading to lodging or other problems. More often, plants do not have enough of a particular nutrient, or their nutrient status may be out of balance. Either way, the crop could be predisposed to infection by pathogens. There are few consistencies as to how nutrient imbalances affect disease, because the interactions are complex and vary with environment (Huber, 1991).

Nutrients

On wheat, both leaf rust and powdery mildew have shown a positive linear association with an increase in nitrogen, while additional phosphorus tended to bring about a decrease in both diseases (Boquet and Johnson, 1987). In a separate study, nitrogen was found to increase powdery mildew and *Septoria nodorum* blotch (Cox et al., 1989). Nitrogen must be kept at optimum levels in order to minimize the effects of soil-borne *Fusarium* on wheat and barley (Cook, 1980). In general, high rates of nitrogen were inclined to increased disease (Broscious et al., 1985; Harms et al., 1989; Karlen and Gooden, 1990; Knowles et al., 1995). Nitrogen form may also influence disease occurrence and severity. Huber (1991) listed over 200 plant diseases that are increased or decreased by nitrogen in some form, and an additional 98 diseases that were increased or decreased by the nitrate or ammonium forms of nitrogen. The timing of nitrogen application may also affect the crop's response to disease (Knowles et al., 1994). In The Netherlands, increases in fertilizer (mainly nitrogen) on winter wheat resulted in large increases in powdery mildew and yellow (stripe) rust severity, a moderate increase in snow mold, and no change in eyespot, *Fusarium* foot rot, leaf rust or speckled leaf blotch (Daamen et al., 1989).

In general, mineral fertilizers can be applied to soils in order to supplement the available nutrients. Disease severity may be minimized or reduced with proper plant nutrition by enhancing the plant's ability to defend itself. The exact minerals involved and their interaction with the environment, the crop and the pathogen differ in nearly every example. The availability of nutrients is probably as important as the quantity of nutrients present. The overall nutritional balance of the crop is more important than the presence or absence of any one nutrient. The form of nutrient is important in some cases and under some circumstances. In relation to integrated disease control, the impact of a proper balance of nutrients is greater on cultivars that are already tolerant or partially resistant to the disease in question (Huber, 1980).

Organic soil amendments

A cover crop of winter legumes may be used to provide nitrogen for a subsequent cash crop of cotton. In doing this, cotton seedling diseases were somewhat increased, but the resulting increase in cotton yield offset any decreases in yield from the seedling diseases (Rothrock et al., 1995). This method of increasing soil organic content by ploughing in a cover crop immediately prior to planting the main crop does help to bolster soil nitrogen, but rarely causes a direct decrease in a particular disease. The main effect of organic soil amendments is in the enhancement of biological control activities in the soil (Baker and Cook, 1974). However, cruciferous residues can suppress some soil-borne pathogens and root diseases owing to the high content of sulfur-containing compounds in the residues (Gamliel and Stapleton, 1993).

ERADICATION

Eradication in its purest sense would mean the removal of all traces of a pathogen, or an alternate host of a pathogen. Even though 100% eradication is unlikely to occur, a high enough level of eradication may be attainable for practical purposes. At a farm level, eradication can take the form of roguing parasitic plants, roguing infected hosts and solar heating of soil to eliminate pathogen propagules.

Roguing

A special example of eradication that helps on a small scale is the weeding of witchweed (*Striga* spp.) from infested fields in Africa. Removing and destroying the obligate, root-parasitic plant helps to decrease the spread of the pathogen by seed (Berner et al., 1995). Other parasitic higher plants, such as dwarf mistletoe (*Arceuthobium* spp.; Kuijt, 1955) and dodder (*Cuscuta* spp.), can also be controlled by roguing.

Roguing of infected host plants can help eradicate some diseases in perennial crops such as banana, cacao and sugarcane (Kommedahl and Todd, 1991). For those pathogens that require two distinct hosts for completion of their life cycle, it may be worthwhile to eradicate the non-economic alternate host, particularly if the asexual, repeating spore stage occurs on the non-economic host. Such is the case in blister rust (caused by *Cronartium ribicola*) of pine, where the removal of the non-economic host, *Ribes* spp., could result in an interruption of the pathogens' life cycle. However, eradication was effective only in areas where the disease presented a low to moderate hazard (Anderson, 1973). In the case of wheat stem rust, the repeating stage occurs on the primary host, wheat, and the sexual stage on barberry. Nevertheless, barberry eradication began in 1918 in many north-central states in the USA. Even though stem rust epidemics have occurred in the USA after barberry eradication, the roguing did serve to delay disease onset in the northern states, reduce the level of virulent inoculum in the spring, decrease the number of pathogenic races of *Puccinia graminis tritici*, and stabilize the pathogenic races present (Roelfs, 1982). In addition to physical removal, infected plants or trees can also be rogued by burning, as was the case in the eradication of citrus trees infected with citrus canker (caused by *Xanthomonas citri*; Stevens, 1915).

Solar heating of soil

The use of solar heat to pasteurize soil has been tried on a small scale, and is effective at controlling some soil-borne diseases. The practice has been used in about 38 countries to suppress soil-borne pests (Katan, 1981). The majority of successes with solar heating have been in climates with an abundance of hot, dry and cloudless days. However, the method was effective in reducing the incidence of southern blight of tomato in North Carolina (Ristaino et al., 1991). Effectiveness in reducing pathogen population levels of tomato diseases differed with the pathogen involved and the soil depth of the pathogenic propagules. Thus, solar heating resulted in a significant decrease in the density of *Phytophthora nicotianae* to depth of 25 cm and of *Pseudomonas solanacearum* to depth of 15 cm; two *Fusarium* spp. were reduced down to only 5 cm (Chellemi et al., 1994). In combination with organic amendments, solarization produced several antifungal compounds in the soil (Gamliel and Stapleton, 1993).

The method requires soils to be moist and to be covered with clear polyethylene for several days or weeks in order to trap heat produced by solar radiation, thereby increasing soil temperatures to lethal levels for pathogen propagules (Kommedahl and Todd, 1991). The method can also be effective in destroying weed seeds, but again, the depth of the seeds in the soil is critical to its effectiveness.

DISEASE AVOIDANCE

Although much of this topic would seem to be intuitive, it is often overlooked as a method for controlling diseases. It follows the same basic principle as has often been stated at field days in south Texas, 'The only way to guarantee you won't have a leaf rust problem on wheat, is to not plant wheat'. Although this is a satirical statement, avoidance is simply the process of staying away from potential disease problems or the evasion of other practices that may lead to disease situations. For example, field selection or topography within a field for a crop or a cropping system can help producers to avoid potential disease problems associated with soil-borne diseases, as in not planting wheat in low-lying areas in order to minimize infections by *Polymyxa graminis*, the vector of soil-borne mosaic virus (Bockus and Niblett, 1984). Disease avoidance strategies include planting time, disease-free planting material and seeding rate.

Planting time

Alterations in planting date can take advantage of unfavourable temperatures for pathogen infection and reproduction. For example, relatively cool soil temperatures are needed for spore germination of the wheat bunt pathogen, *Tilletia caries*. An early planting date in the late summer or early fall for winter wheat will avoid bunt infection because of the higher soil temperatures. However, in areas where barley yellow dwarf virus (BYDV) is a problem in winter cereals, early planting results in early exposure to aphid populations which can vector BYDV, thereby increasing the chances of severe yellow dwarf virus (Cisar et al., 1982; Irwin and Thresh, 1990; McGrath and Bale, 1990). In sugarbeet, an early planting date in the spring delays the onset of rhizomania, because soil temperatures are too low for infection by *Polymyxa betae*, the vector of beet necrotic yellow vein virus (Rush and Heidel, 1995). Planting date is important in avoiding *Fusarium* root rot in the Pacific northwest. If producers plant too early the young plants may suffer from water stress, which predisposes plants to *Fusarium* infection. However, if planting is too late, yield potential is typically reduced (Cook, 1980). On the other hand, for different disease in the same geographic area, delayed seeding helps reduce the severity of take-all (caused by *Gaeumannomyces graminis tritici*) in parts of Oregon (Taylor et al., 1983). This situation of different diseases reacting in opposite ways to a change in planting date underscores the need to understand disease cycles and how geography and climate affect crops and their potential pathogens. In order to avoid *Fusarium* in the soil, winter wheat must be planted at the optimum time (not too early or too late), spring wheat should be planted early and corn should be planted late, all different adjustments in planting date aimed at avoiding soil temperatures that favour pathogen progress (Dickson, 1923).

Altered planting dates are most effective when all the farmers within a region cooperate. Risk can be associated with planting at a time which is different from the optimum (most often determined by experience). For example, the fall planting date for winter wheat may be delayed to avoid a pathogen, but abundant rainfall during that time may delay planting too much, as soils dry much slower when soil temperatures are low. Delayed fall planting of winter wheat causes the plants to be less mature in spring than wheat planted earlier. Less mature plants are more susceptible to spring infections of wheat streak mosaic than are more mature plants (Hunger et al., 1992).

Flexibility in planting time is greater in some crops and in some areas than in others. In the temperate zone, crop distribution is defined by the frost-free period. The longer the time period a crop can reasonably avoid freezing temperatures, the more flexibility there is in planting. For example, in central Texas, winter wheat can typically be planted from September to December. Earlier plantings are normally used if the wheat is to serve as winter pasture for grazing animals, and later plantings are generally for grain production only. Yet in more northern areas, such as in Nebraska and the Dakotas, the window for winter wheat planting is quite narrow, because plants must be established prior to the onset of low temperatures.

Disease-free planting material

Some pathogens can be carried in or on the seed or other planted portions of a crop. Selection of plant material or seed which is free from pathogens can go a long way to mitigate potential disease problems. Clearly, when the option is available, a farmer will plant disease-free seed instead of seed bearing a pathogen. This is one of the basic reasons behind the production of certified seed and the existence of quarantine programmes. A topic typically not included in texts dealing with disease-free planting materials is cultivar selection (Mink, 1991). To me, this is the logical area where cultivar selection should be discussed. From the point of view of a farmer who wishes to minimize disease problems, the topic of disease-free planting materials would seem to be the most likely place to find information on disease-resistant cultivars. With most annual crops, farmers commonly have a choice between selecting a disease-resistant or a disease-susceptible cultivar (Harms et al., 1989; Marshall, 1989; Karlen and Gooden, 1990). Many interacting factors enter into the choice, but the practice should be considered to be an important cultural control of plant diseases.

Certification programmes have been developed for many perennial fruit crops, vegetatively propagated ornamentals, vegetable seeds and seeds of some field crops. Most states in the USA, as well as many other countries, have minimum tolerances for pathogens that planting materials must meet (Mink, 1991). Unfortunately, most places do not have minimum tolerances for genetic resistance to plant diseases, although lists of recommended cultivars are often available.

Seeding rate and spacing

Seeding rate can be used to increase or decrease the number of plants per unit area, which determines the spacing of plants within a field row, and indirectly affects the relative size of plants. Wheat and other small-grain cereals have different plant architecture at different seeding rates. At low rates, plants will produce more fertile secondary tillers than they will at high seeding rates, where more effort is put into production of primary tillers. Decreasing the seeding rate decreases the number of fertile heads, but increases the number of seeds per head. This does not necessarily directly affect yield, for example a 38% increase in the number of seeds per unit area had no effect on soft red winter wheat yield or disease outbreak in the Great Lakes region of USA (Beuerlein et al., 1989). In barley, seeding rates from 3/4 bushels per acre to 6 bushels per acre had no effect on yield. A combination of increased seeding rate and narrow row spacing is important for increasing grain yield (Marshall and Ohm, 1987). However, diseases may be influenced by seeding rate, particularly in the case of crop cultivar mixtures, because low seeding rates tend to produce large individual plants, which in turn can influence the amount of auto-versus allo-infection (Marshall, 1991). If a spring wheat crop in Utah is planted late, seeding rates need to be increased in order to decrease the incidence of stem rust and powdery mildew (Woodward, 1956). In general, powdery mildew is more severe at low seeding rates because of increased tillering (Broscious et al., 1985). Wider row spacing also helps to promote the progress of powdery mildew (Broscious et al., 1985). In general, plant density may affect how pathogens are distributed within a canopy, thereby influencing both the short- and long-distance spread of the disease. Plant density and growth habit also affect the relative humidity in the canopy and the rate at which moisture dries on leaves.

Indirect effects of cultural controls on plant resistance and biological control

Although the focus of this chapter has been on the direct effects that cultural controls have on plant diseases, some of the practices mentioned can influence a plant's response to infection, and most can influence biological control processes. Soils suppressive to *Phytophthora cinnamomi* root rot of avocado were high in organic content, with a good nutrient balance, owing to the incorporation of organic amendments and other nutrients (Broadbent and Baker, 1974). Nitrogen and other nutrients serve to keep the crop nutritionally balanced, thereby allowing the plants to defend themselves from pathogens.

Cropping sequence with fallow periods can allow certain biological control micro-organisms to proliferate and replace pathogenic micro-organisms in the soil environment.

CONCLUSIONS

The definition of cultural controls for plant diseases has been somewhat expanded here to include decision-making on the part of the farmer, particularly in regard to crop and cultivar selection. The term crop sequence has been promoted over crop rotation to emphasize the important aspect of time. Cultivar sequence within a cropping system is an overlooked, yet important, management tool for farmers. Time, as a component of disease control, allows the results of practices such as residue management to be fully expressed. The effectiveness and sustainability of cultural control practices will be measured over time. As was pointed out at the start of this chapter, matters of scale are important in cultural controls. If knowledgeable and forward-looking decisions concerning disease control can be made at the farm level, then regional, national and global concerns about a sustainable agriculture may be minimized.

REFERENCES

Anderson, R.L. (1973) *A Summary of White Pine Blister Rust Research in the Lake States.* North Central Forestry Experiment Station, USDA, St. Paul, MN.

Baker, K.F. and Cook, R.J. (1974) Biological Control of Plant Pathogens. Freeman, San Francisco, CA.

Belmar, S.B., Jones, R.K. and Starr, J.L. (1987) Influence of cropping rotation on inoculum density of *Rhizoctonia solani* and sheath blight incidence in rice. *Phytopathology* 77, 1138–1143.

Berner, D.K., Kling, J.G. and Singh, B.B. (1995) *Striga* research and control—A perspective from Africa. *Plant Disease* 79, 652–660.

Beuerlein, J.E., Oplinger, E.S. and Reicosky, D. (1989) Yield and yield components of winter wheat cultivars as influenced by management—a regional study. *Journal of Production Agriculture* 2, 257–261.

Black, M.C., Beute, M.K. and Leonard, K.J. (1984) Effects of monoculture with susceptible and resistant peanuts on the virulence of *Cylindrocladium crotalaraie. Phytopathology* 74, 945–950.

Bockus, W.W. and Niblett, C.L. (1984) A procedure to identify resistance to wheat soilborne mosaic in wheat seedlings. *Plant Disease* 68, 123–124.

Boquet, D.J. and Johnson, C.C. (1987) Fertilizer effects on yield, grain composition, and foliar disease of doublecrop soft red winter wheat. *Agronomy Journal* 79, 135–141.

Broadbent, P. and Baker, K.F. (1974) Behavior of *Phytophthora cinnamomi* in soils suppressive and conducive to root rot. *Australian Journal of Agricultural Research* 25, 121–137.

Broadfoot, W.C. (1934) Studies on foot and root rot of wheat. IV. Effect of crop rotation and cultural practice on the development of foot rot of wheat. *Canadian Journal of Research* 10, 95.

Broscious, S.C., Frank, J.A. and Frederick, J.R. (1985) Influence of winter wheat management practices on the severity of powdery mildew and *Septoria* blotch in Pennsylvania. *Phytopathology* 75, 538–542.

Brown, L.R. (1994) World grain yield drops. In Starke, L. (Ed.) *Vital Signs 1994.* W.W. Norton, New York.

Chellemi, D.O., Olson, S.M. and Mitchell, D.J. (1994) Effects of soil solarization and fumigation on survival of soilborne pathogens of tomato in northern Florida. *Plant Disease* 78, 1167–1172.

Chinn, S.H.F. (1976) *Cochliobolus sativus conidia* populations in soils following various cereal crops. *Phytopathology* 66, 1082–1084.

Cisar, G., Brown, C.M. and Jedlinski, H. (1982) Effect of fall or spring infection and sources of tolerance of barley yellow dwarf of winter wheat. *Crop Science* 22, 474–477.

Cook, R.J. (1980) Fusarium foot rot of wheat and its control in the Pacific Northwest. *Plant Disease* **64**, 1061–1066.

Cook, R.J. (1981) The influence of rotation crops on take-all decline phenomenon. *Phytopathology* **71**, 189–192.

Cook, R.J. (1986) Wheat management systems in the Pacific Northwest. *Plant Disease* **70**, 894–898.

Cox, W.J., Bergstrom, G.C., Reid, W.S., Sorrells, M.E. and Otis, D.J. (1989) Fungicide and nitrogen effects on winter wheat under low foliar disease severity. *Crop Science* **29**, 164–170.

Crookston, R.K., Kurle, J.E. and Lueschen, W.E. (1991) Rotational cropping sequence affects yield of corn and soybean. *Agronomy Journal* **83**, 108–113.

Daamen, R.A., Wijnands, F.G. and Vandervliet, G. (1989) Epidemics of diseases and pests of winter wheat at different levels of agrochemical input. *Journal Phytopathology* **125**, 305–319.

Daft, G.G. and Leben, C. (1973) Bacterial blight of soybeans: Field overwintered *Pseudomonas glycinea* as possible primary inoculum. *Plant Disease Reporter* **65**, 156–161.

Dickson, J.G. (1923) Influence of soil temperature and moisture on the development of seedling blight of wheat and corn caused by *Gibberella saubinettii*. *Journal of Agricultural Research* **23**, 837–870.

Dubin, H.J. and Torres, E. (1981) Causes and consequences of the 1976–77 wheat leaf rust epidemic in northwest Mexico. *Annual Review of Phytopathology* **19**, 41–49.

Dunleavy, J.M. and Weber, C.R. (1967) Control of brown stem rot of soybeans with corn–soybean rotations. *Phytopathology* **57**, 114–117.

Fiddian, W.E.H. (1973) The changing pattern of cereal growing. *Annals of Applied Biology* **75**, 123–149.

Foster, P. (1992) *The World Food Problem*. Lynne Rienner, Boulder, CO.

Fry, W.E. (1982) *Principles of Plant Disease Management*. Academic Press, New York.

Fussell, G.E. (1965) *Farming Technique from Prehistoric to Modern Times*. Pergamon Press, New York.

Gamliel, A. and Stapleton, J.J. (1993) Characterization of antifungal volatile compounds evolved from solarized soil amended with cabbage residues. *Phytopathology* **83**, 899–905.

Garcia, C. and Marshall, D. (1992) Observations on the ascogenous stage of *Septoria tritici* in Texas. *Mycology Research* **96**, 65–70.

Harlan, J.R. (1992) Crops and Man. 2nd edn. American Society of Agronomy, Madison, WI.

Harms, C.L., Beuerlein, J.E. and Oplinger, E.S. (1989) Effects of intensive and current recommended management systems on soft winter wheat in the US corn belt. *Journal of Production Agriculture* **2**, 325–332.

Harrower, K.M. (1974) Survival and regeneration of *Leptosphaeria nodorum* in wheat debris. *Transactions of the British Mycological Society* **63**, 527–533.

Hershman, D.E. and Bachi, P.R. (1995) Effect of wheat residue and tillage on *Heterodera glycines* and yield of doublecrop soybean in Kentucky. *Plant Disease* **79**, 631–633.

Hesseltine, W.B. (1962) *Third-Party Movements in the United States*. Van Nostrand & Co., New York.

Hewett, P.D. (1975) *Septoria nodorum* on seedlings and stubble of winter wheat. *Transactions of the British Mycological Society* **65**, 7–18.

Holmes, S.J.I. and Colhoun, J. (1975) Straw-borne inoculum of *Septoria nodorum* and *S. tritici* in relation to incidence of disease on wheat plants. *Plant Pathology* **24**, 63–66.

Huber, D.M. (1980) The role of mineral nutrition in defense. *Plant Disease* **5**, 381–406.

Huber, D.M. (1991) The use of fertilizers and organic amendments in the control of plant disease. In Pimentel, D. (Ed.) *CRC Handbook of Pest Management in Agriculture*. CRC Press, Boca Raton, FL.

Hunger, R.M., Sherwood, J.L., Evans, C.K. and Montana, J.R. (1992) Effects of planting dates and inoculation date on severity of wheat streak mosaic in hard winter wheat cultivars. *Plant Disease* **76**, 1056–1060.

Hunter, H. and Leake, H.M. (1933) *Recent Advances in Agricultural Plant Breeding*. P. Blakistons, Philadelphia, PA.

Irwin, M.E. and Thresh, J.M. (1990) Epidemiology of barley yellow dwarf: A study in ecological complexity. *Annual Review of Phytopathology* **28**, 393–424.

Jasny, N. (1944) *The Wheats of Classical Antiquity*. Johns Hopkins University Press, Baltimore,

MD.

Karlen, D.L. and Gooden, D.T. (1990) Intensive management practices for wheat in the south-eastern coastal plains. *Journal of Production Agriculture* **3**, 558–563.

Katan, J. (1981) Solar heating (solarization) of soil for control of soilborne pests. *Annual Review of Phytopathology* **19**, 211–236.

Kennedy, B.W. and Lambert, J.W. (1981) Influence of brown stem rot and cropping history on soybean performance. *Plant Disease* **65**, 896–897.

King, F.H. (1911) *Farmers of Forty Centuries*. Rodale Press, Emmaus, PA.

Knowles, T.C., Hipp, B.W., Graff, P.S. and Marshall, D.S. (1993) Nitrogen nutrition of rainfed winter wheat in tilled and no-till sorghum and wheat residues. *Agronomy Journal* **85**, 886–893.

Knowles, T.C., Hipp, B.W., Graff, P.S. and Marshall, D.S. (1994) Timing and rate of topdress nitrogen for rainfed winter wheat. *Journal of Production Agriculture* **7**, 216–220.

Knowles, T.C., Hipp, B.W., Marshall, D.S. and Sutton, R.L. (1995) Plant nutrition and fertilizer management for winter wheat production in the Blackland Praire. *Texas Agricultural Experiment Station Bulletin* No. 1725, 28 pp.

Koenning, S.R., Schmitt, D.P., Barker, K.R. and Gumpertz, M.L. (1995) Impact of crop rotation and tillage system on *Heterodera glycines* population density and soybean yield. *Plant Disease* **79**, 282–286.

Kollmorgen, J.F., Griffiths, J.B. and Walsgott, D.N. (1985) Effects of cropping sequences on saprophytic survival and carry-over of *Gaeumannomyces graminis var. tritici*. In Parker, C.A., Rovira, A.D., Moore, K.J. and Wong, P.T.W. (Eds) *Ecology and Management of Soilborne Plant Pathogens*. American Phytopathology Society, St. Paul, MN.

Kommedahl, T. and Todd, L.A. (1991) The environmental control of plant pathogens using eradication. In Pimentel, D. (Ed.) *CRC Handbook of Pest Management in Agriculture*. CRC Press, Boca Raton, FL.

Kuijt, J. (1955) Dwarf mistletoes. *Botanical Review* **21**, 569–628.

Marshall, D. (1988) Characteristics of the 1984–85 wheat leaf rust epidemic in central Texas. *Plant Disease* **72**, 239–241.

Marshall, D. (1989) National and international breeding programs and deployment of plant germplasm: New solutions or new problems. In Jeger, M.J. (Ed.) *Spatial Components of Plant Disease Epidemics*. Prentice-Hall, Orlando, FL.

Marshall, D. (1991) Crop diversity for plant pathogen control. In Pimentel, D. (Ed.) *CRC Handbook of Pest Management in Agriculture*. CRC Press, Boca Raton, FL.

Marshall, G.C. and Ohm, H.W. (1987) Yield response of 16 winter wheat cultivars to row spacing and seeding rate. *Agronomy Journal* **79**, 1027–1030.

McGrath, P.F. and Bale, J.S. (1990) The effects of sowing date and choice of insecticide on cereal aphids and barley yellow dwarf virus epidemiology in northern England. *Annals of Applied Biology* **117**, 31–43.

Melton, T.A. and Powell, N.T. (1991) Effects of two-year crop rotations and cultivar resistance on bacterial wilt in flu-cured tobacco. *Plant Disease* **75**, 695–698.

Mink, G.I. (1991) Control of plant diseases using disease-free stock. In Pimentel, D. (Ed.) *CRC Handbook of Pest Management in Agriculture*. CRC Press, Boca Raton, FL.

Moore, K.J. and Cook, R.J. (1984) Increased take-all of wheat with direct drilling in the Pacific Northwest. *Phytopathology* **74**, 1044–1049.

Nusbaum, C.J. and Ferris, H. (1973) The role of cropping systems in nematode population management. *Annual Review of Phytopathology* **11**, 423.

Piening, L.J. and Orr, D. (1988) Effects of crop rotation on common root rot of barley. *Canadian Journal of Plant Pathology* **10**, 61–65.

Power, J.F. and Follett, R.F. (1987) Monoculture. *Scientific American* 79–86.

Ristaino, J.B., Perry, K.B. and Lumsden, R.D. (1991) Effect of solarization and Gliocladium virens on clerotia of *Sclerotium rolfsii*, soil microbiota, and the incidence of southern blight of tomato. *Phytopathology* **81**, 1117–1124.

Roelfs, A.P. (1982) Effects of barberry eradication on stem rust in the United States. *Plant Disease* **66**, 177–181.

Rothrock, C.S., Kirkpatrick, T.L., Frans, R.E. and Scott, H.D. (1995) The influence of winter legume cover crops on soilborne plant pathogens and cotton seedling diseases. *Plant Disease* **79**,

167–171.

Rovira, A.D. and Venn, N.R. (1985) Effect of rotation and tillage on take-all and *Rhizoctonia* root rot in wheat. In Parker, C.A., Rovira, A.D., Moore, K.J. and Wong, P.T.W. (Eds) *Ecology and Management of Soilborne Plant Pathogens*. American Phytopathology Society, St. Paul, MN.

Rush, C.M. and Heidel, G.B. (1995) Furovirus diseases of sugar beets in the United States. *Plant Disease* **79**, 868–875.

Schippers, B., Geels, F.P., Hoekstra, O., Lamers, J.G., Maenhout, C.A.A.A. and Scholte, K. (1985) Yield depression in narrow rotations caused by unknown microbial factors and their suppression by selected Pseudomonads. In Parker, C.A, Rovira, A.D., Moore, K.J. and Wong, P.T.W. (Eds) *Ecology and Management of Soilborne Plant Pathogens*. American Phytopathology Society, St. Paul, MN.

Schippers, B., Bakker, A.W. and Bakker, P.A.H.M. (1987) Interactions of deleterious and beneficial rhizosphere microorganisms and the effect of cropping practices. *Annual Review of Phytopathology* **25**, 339–358.

Schmitthenner, A.F. and Van Doren, D.M. (1985) Integrated control of root rot of soybean caused by *Phytophthora megásperma* f. sp. *glycinea*. In Parker, C.A., Rovira, A.D., Moore, K.J. and Wong, P.T.W. (Eds) *Ecology and Management of Soilborne Plant Pathogens*. American Phytopathology Society, St. Paul, MN.

Shaner, G. and Buechley, G. (1995) Epidemiology of leaf blotch of soft red winter wheat caused by *Septoria tritici* and *Stagnospora nodorum*. Plant Disease **79**, 928–938.

Stevens, H.E. (1915) Citrus canker. *Florida Agricultural Experiment Station Bulletin* **128**, 1–20.

Sumner, D.R., Dowler, C.C., Johnson, A.W. and Baker, S.H. (1995) Conservation tillage and seedling diseases in cotton and soybean double-cropped with triticale. *Plant Disease* **79**, 372–375.

Taylor, R.G., Jackson, T.L., Powelson, R.L. and Christensen, N.W. (1983) Chloride, nitrogen form, lime, and planting date effects on take-all root rot of winter wheat. *Plant Disease* **67**, 1116–1120.

Tivy, J. (1990) *Agricultural Ecology*, Longman Scientific & Technical, Harlow.

Ullstrup, A.J. (1972) The impacts of the southern corn leaf blight epidemic of 1970–1971. *Annual Review of Phytopathology* **10**, 37–50.

Viglizzo, E.F. and Roberto, Z.E. (1989) Diversification, productivity and stability of agroecosystems in the semi-arid pampas of Argentina. *Agricultural Systems* **31**, 279–290.

Whiting, K.R. and Crookston, R.K. (1993) Host-specific pathogens do not account for the corn–soybean rotation effect. *Crop Science* **33**, 539–543.

Williams, L.E. and Schmitthenner, A.F. (1962) Effect of crop rotation on soil fungus populations. *Phytopathology* **52**, 241.

Woodward, R.W. (1956) The effect of rate and date of seedling of small grains on yields. *Agronomy Journal* 48: 160–162.

Wrather, J.A., Kendig, S.R., Anand, S.C., Niblack T.L. and Smith, G.S. (1995) Effects of tillage, cultivar, and planting date on percentage of soybean leaves with symptoms of sudden death syndrome. *Plant Disease* **79**, 560–562.

Wright, K.H. and Sutton, J.C. (1990) Inoculum of *Phyrenophora tritici-repentis* in relation to epidemics of tan spot of winter wheat in Ontario. *Canadian Journal of Plant Pathology* **12**, 149–157.

CHAPTER 12

Cultural Control for Weeds

K.S. Gill[1], M.A. Arshad[1] and J.R. Moyer[2]

[1] *Agriculture and Agri-Food Canada, Beaverlodge, Alberta, Canada*
[2] *Agriculture and Agri-Food Canada, Lethbridge, Alberta, Canada*

INTRODUCTION

A feature of natural disturbance is that the open patches are transient and unpredictable in both space and time, which renders adaptation by weeds difficult. On the other hand, modern crop husbandry aims to minimize the heterogeneity of agroecosystems to produce uniform growing conditions. Therefore, it creates vast areas of relative environmental homogeneity with a high level of predictability associated with land use, which allows adapted weed species to flourish well and interfere with the crop.

Weed control with herbicides is under scrutiny. They produce off-site effects such as surface- and ground-water contamination as well as health hazards. They often lead to the development of herbicide-resistant weed biotypes, generally do not provide effective control across the weed species and may not be economical in many cases. In addition to their effect on weeds and crops, changes in the nutrient content and growth pattern of plants caused by herbicides may alter the abundance of insects (Pimentel, 1971; Campbell, 1988). After an extensive literature search, Pimentel (1994) tabulated 31 instances of increased insect attack following various herbicide applications. In contrast, insect pest numbers were reduced on herbicide-stressed plants in some cases (Pimentel, 1971; Campbell, 1988). The reduced insect abundance in these instances was attributed to an increase in natural plant toxicants, such as cyanide and potassium nitrate, which were stimulated by the herbicide (Pimentel, 1971).

Many studies have indicated that minimization of soil contamination and disturbance by reducing the rate of herbicide use and the intensity of cultivation is a step toward sustainable agriculture that attains satisfactory crop production with as little impact on the environment as possible (Diebert, 1989; Moseley and Hagood, 1990; Glen and Anderson, 1993). The recent focus on non-chemical means of pest control has revived interest in the use of cultural weed control.

Various climatic, soil, biological and cultural management factors affect weed–crop interactions. Crop husbandry practices such as planting a crop, weed control, irrigation and fertilization are done in a manner to assist the crop to benefit most from the available resources. Cultural weed control methods include techniques which supplement crop competition in order to shift the balance in favour of the crop, or to present growing

Techniques for Reducing Pesticide Use. Edited by D. Pimentel
© 1997 John Wiley & Sons Ltd.

conditions unfavourable for weeds. For example, rotation of plant types and management techniques tend to make the existing weed flora on a site less well adapted, and thus cause less interference with crop growth. Cultural control exploits the principle of plant competition wherein the first plants to occupy an area have an advantage over the latecomers. Due to variation in climate, soil, crop, social and economic factors, it is unrealistic to apply one cultural weed control technology to all types of agriculture and weed complexes. However, the underlying principles regarding crop/weed interactions are applicable to a variety of situations which can be used to affect the weed community in several agroecosystems.

A number of excellent reviews have been published on various cultural control techniques (Zimdahl, 1980; Altieri and Liebman, 1988a; Worsham, 1989; Liebman and Dyck, 1993; Moyer et al., 1994). The objectives of this review are to examine the role of cultural practices in weed suppression, and suggest how these techniques can contribute to the development of weed management strategies which maintain weed infestation below economic thresholds, with minimal requirements for herbicides and tillage. To facilitate the readability of the text, common names of the plants and herbicides are used in the text and their corresponding scientific names are listed in Appendix A.

CROP COMPETITION

Fallow vs. cropping

Light is an important factor in the germination of weeds such as Canada thistle, dandelion and some grasses (Maguire and Overland, 1959; Wilson, 1979; Froud-Williams et al., 1983). Thus, the establishment of weeds can be inhibited by shading the soil surface with a vigorous crop.

The beneficial effect of crop competition in reducing weed growth is obvious in several situations. For example, on cropped lands of the Northern Great Plains, weeds can be effectively controlled until fall with a single herbicide application. However, if the land is summer-fallowed, weeds continue to emerge throughout the season and three to six tillage operations are required for their control (Foster and Lindwall, 1986; Donald and Nalewaja, 1990). Similarly, weed dry-matter production during the establishment of non-competitive forages is reduced in the presence of a cereal companion crop (Moyer, 1985), and some of the problem weeds such as dandelions increase only in overgrazed rangelands (Froud-Williams et al., 1983). According to Derksen et al. (1993), perennial weeds may have less potential to invade no-till fields in locations where crops require most of the growing season to mature, compared with locations where the climate is favourable for weed growth after harvest. In areas where there is a long period that is favourable for weed growth after harvest, short-term cover crops and their residues are used to suppress weed growth (Teasdale et al., 1991). The yield difference in soybeans between poor (less herbicide use) and good (more herbicide use) weed control in a double-cropped soybean–wheat system was 13%, while the yield difference for similar treatments in the single-crop soybeans was 57% (Wax et al., 1983). However, during the last few decades producers have depended largely on herbicides or intense cultivation for weed control and have overlooked important agronomic factors that influence weed growth (Zimdahl, 1990). From a weed control perspective, keeping a crop or some type of plant cover on the soil during the period when climatic conditions are favourable for plant growth is clearly a better option than leaving it bare. The above-ground part of the plants can be exploited for mulching, and the roots will add organic matter to the soil. Furthermore, planting arrangements, crop density, crop and cultivar types, seeding methods and time, as well as soil fertility, can be manipulated to increase crop competition and influence the degree of weed infestation and the weed control requirements.

Seeding time, arrangement and density

Changes in seeding date and cropping season result in reduced weed infestation owing to the accompanying agronomic and weed control measures and disruption of an established weed growth cycle. Germination or establishment of the crop before the weeds, even by few days, confers an advantage on the crop (Shurtleff and Coble, 1985). Weeds from spring-germinating seeds were found more frequently in spring-sown crops, while the weeds from fall-germinating seeds occurred more commonly in fall-seeded crops (Chancellor, 1985). Two years of an autumn-seeded grain reduced infestations of wild oats owing to the dense crop growth in spring (Thurston, 1962). Use of a relatively late-planted beans crop has been found to result in decreased weed-seed numbers in a subsequent sugarbeet crop (Dotzenko et al., 1969). Control of wild garlic in Sweden was associated with a change in planting date and the resultant change in time of tillage (Hakansson, 1982). In the mid-1970s a large reduction in the weed infestation of Tanzanian wheat fields was made by changing the seeding date from the beginning of the rainy season to the middle of the rainy season (Nielsen, 1982). Delaying planting to early June in Minnesota allowed early-germinating weeds to be controlled by pre-plant tillage, but it reduced the maximum yield potential by about 25% in corn and by about 10% in soybean (Gunsolus, 1990). However, Moyer et al., (1989) concluded that cultural practices such as seeder type, wheat variety or number of cultivations prior to seeding did not influence weed densities on the wheat farms in Tanzania. Johnson and Mullinex (1995) compared same-day tillage and seeding of peanut to a stale seedbed (deep tillage 6 weeks before seeding, with or without shallow tillage prior to seeding). The stale seedbed stimulated weed emergence. Weed populations were reduced and crop yield improved when a stale seedbed plus shallow tillage were used, but the weeds on the untreated stale seedbed were difficult to control after the peanut planting.

Narrower row spacing has been shown to assist weed suppression in many cases. Reducing the row spacing was found to decrease interference from weeds (Regnier and Janke, 1990). Narrow row spacings suppressed growth of weeds in snap beans because of earlier row closure (Teasdale and Frank, 1983). The same result was obtained in soybean, with less weed growth in 38- or 46-cm row spacing, than in 76-cm row spacing (Orwick and Schreiber, 1979). Burnside and Colville (1964) compared 25-, 51-, 76- and 102-cm row spacings and suggested that soybeans in narrow crop rows competed with weeds at an earlier stage of growth than those in wide rows, because they had better root distribution and complete shading of the soil surface earlier. These data were confirmed by Walker et al. (1984), who found that soybean yield was higher in 20-cm than in 40- or 80-cm row spacing when sicklepod was present for the whole season. Averaged across different seeding densities, green foxtail biomass was slightly reduced by narrowing safflower spacing from 22 to 11 cm in one out of two years, whereas its seed yield was unaffected in both years (Blackshaw, 1993). Gunsolus (1990) stated that narrow rows of corn and soybean allowed the crop canopy to close earlier, preventing some emerged weeds from developing. However, for mechanical weed control the row spacing should allow for inter-row cultivation.

Weed suppression by crops increases at higher seeding density in a given area. Shetty and Krantz (1980) reported a progressive decrease in weed biomass as the pigeonpea population was increased. When sorghum was introduced as an intercrop with pigeonpea, the total weed-suppressing effect of the two crops was almost four times that of the pigeonpea alone under low populations. Similarly, increased pea populations reduced weed biomass (Lawson, 1982). Seeding rate of commercial rice was inversely related to the panicle density and floral development of red rice, but did not affect its stand density (Dunand et al., 1984). Increasing safflower density from 0 to 1992 plants m^{-2} progressively decreased the green foxtail growth during the 1990 and 1991 growing seasons (Blackshaw, 1993). At the highest density of safflower, the green foxtail biomass was reduced by 66 and 72% and seed yield was reduced by 74 and 85% in 1990 and 1991, respectively. Increasing the seed rate from 50 to 150% of the recommended rate changed the weed control by the crop from 72 to 99% in pea and from 33 to 70% in lentil under favourable growing conditions (Boerboom and Young, 1995). The weed control in both crops under less favourable crop growing conditions was 21 and 39% under the 50 and 150% seeding rates, respectively. Higher sowing rates of wheat have been recommended as a control measure for annual ryegrass (Zimdahl, 1980). In other crops, increased seed rates hastened the formation of dense canopies and increased the crops' ability to compete for incoming photosynthetic radiation (Walker and Buchanan, 1982; Berkowitz, 1988), thus reducing the establishment of late weed flushes. The effect of both row spacing and seed rate have been investigated in some studies (Moyer et al., 1991). Dry matter of weeds decreased as row spacings decreased from 108 to 36 cm or the broadcast seeding rate of alfalfa increased from 0.33 to 3.0 kg ha^{-1}. They found significant herbicide × seeding interaction in one season, with essentially no benefit from herbicide application at the highest plant density and a larger benefit at the lowest plant density. However, this trend was absent in other cases, apparently owing to the presence of perennial weeds which competed successfully with dense stands of alfalfa. Accordingly, the use of higher seeding rates than those recommended was considered to be only a partial solution to the weed control problem.

It appears that manipulation of seeding time, row spacing and seed rate can be used to suppress weeds. Seeding time works via disruption of the established weed growth cycle and the associated agronomic and weed control measures. Reduced row space and increased seeding rates tend to hasten canopy development and thus enhance the crop's ability to compete for incoming photosynthetic radiation. The shading effect of a dense crop canopy slows the growth of existing weeds and prevent establishment of late weed flushes. Additionally, dense crop stands compete vigorously with weeds for soil nutrients and water (Berkowitz, 1988).

Crop and cultivar types

There are differences among crops in their ability to compete with weeds. The most competitive cereal was found to be winter rye, followed by winter barley and winter wheat (Zimdahl, 1990). In a study of 26 crops, wheat was the most competitive with weeds (Van Heemst, 1985). Therefore, there is a possibility of reducing weed numbers and biomass by selectively growing the most competitive crops. However, the choice of crop may be limited by economics, food requirements and resource availability.

Some studies have reported differences among cultivars within crops in their ability to compete with weeds (Walker and Buchanan, 1982; Berkowitz, 1988). 'Amosy' soybean was more competitive than 'Beesom' soybean, which had a lower emergence rate and was slower to emerge (Burnside, 1979). A semi-dwarf determinate soybean cultivar ('Riply') had a greater early-season effect on common cocklebur growth compared with an indeterminate late-maturing cultivar ('Douglas'), but the opposite was true during the late season (Jordan, 1992). Soybean competition reduced early-season weed density by 30–50%, depending on the cultivar (Shilling, et al., 1995). The cultivars that either grew taller or produced a more competitive crop canopy had a greater effect of sicklepod. For example, the centennial and dwarf 'Tracy M' cultivars caused a 30% reduction in early-season sicklepod biomass, while 'Sharkey' and 'Biloxi' reduced sicklepod growth by 40%. By late-season, sicklepod biomass reduction ranged from 18% ('Tracy M') to 55% ('Biloxi') and was directly related to soybean cultivar height. Studies with wheat have demonstrated differences among cultivars in their ability to compete with weeds (Appleby et al., 1976; Blackshaw et al., 1981; Wicks et al., 1986; Balyan et al., 1991). Semi-dwarf wheats have a yield advantage over normal-height wheats, but provide a more open canopy that permits weeds to compete more effectively with the crop for light and thus other resources (Zimdahl, 1990). Challaiah et al. (1986) reported that winter wheat yield reductions caused by downy brome varied as much as 21% depending on the cultivar grown. Wheat height was more highly correlated with increased competitiveness with downy brome than was canopy diameter or number of tillers. Downy brome caused 14–30% greater yield reductions in the semi-dwarf cultivars, 'Archer' and 'Norwin' than in the tall cultivars 'Norstar' and 'Redwin' (Blackshaw, 1994a), which was partly attributable to competition for light. At comparable densities, downy brome biomass differed among winter wheat cultivars in all years, mirroring the wheat biomass and yield data. Similarly, Koscelny et al. (1990) noted that wheat often produced more seed when grown with semi-dwarf than with tall cultivars of winter wheat in Oklahoma. In rice, plant height was the character most strongly correlated with increased weed competitiveness (Kwon et al., 1991; Garrity et al., 1992). Weeds were more effectively shaded by tall than by semi-dwarf rice cultivars.

Characteristic other than plant height of a cultivar have influenced its ability to compete with weeds. At Long Island, New York, owing to the rapid development of a dense canopy, the 'Green Mountain' cultivar of potato was found to reduce yellow nut sedge shoot dry mass by 89% and the number of tubers by 70% in comparison with the 'Katahdim' cultivar (Minotti, 1991). In some instances later-maturing cultivars of soybean were more competitive with weeds, while other research indicated no influence of soybean maturity group on weed growth (Callaway, 1993). No correlation was found between soybean cultivar maturity group and sicklepod growth when 20 soybean cultivars were compared (James et al., 1988). However, the two cultivars producing the least amount of shoot biomass when grown alone were affected less by interference from sicklepod than the cultivars producing the greater shoot mass.

Use of tall, later-maturing and dense-canopy cultivars appear to enhance the competitive ability of crops and reduce the negative effect of weeds. Therefore, the merits of developing short-statured varieties to attain greater yields should be weighed against greater potential yield reductions due to weeds. For sustainable agricultural systems the crop cultivars needs to be both high yielding and competitive with weeds.

Soil fertility and fertilization

Nutrient source, time and method of application changed the weed suppressing ability of crops. Weed growth in soybean plots that had received N each year for 4 years of maize production was higher than in plots with no residual N (Staniforth, 1962). The nitrogen-fixing ability of soybean gave it a competitive advantage over weeds in a nitrogen-deficient soil. Similarly, the use of crimson clover as an N source in maize suppressed lamb's-quarters growth in comparison to the use of synthetic fertilizer N by 80–91% at 23 days after planting, and by 42–59% at harvest, with 9–20% higher maize biomass at harvest in the legume treatment than in the fertilizer treatments (E. Dyck unpublished data, cited in Liebman and Dyck, 1993). However, N fertilization under weed-infested conditions increased weed growth and did not change or reduce crop yield in rice (Okafor and De Datta, 1976), sugar beet (Scott and Wilcockson, 1976), wheat (Carlson and Hill, 1985), faba bean (Patriquin et al., 1988) and barley/pea intercrops (Liebman, 1989). Rasmussen (1995) observed that broadcasting of fertilizer resulted in higher downy brome density and growth compared with banding, whereas wheat had greater growth and N uptake from the banded fertilizer. Addition of NPS increases both wheat and downy brome growth compared with N alone, but wheat appeared to be more responsive. Similarly, banding of N reduced the competitive effect of downy brome in winter wheat (Cochran et al., 1990). Fertilization of cassava or maize, planted alone, with NPK reduced weed dry matter by about 36 and 25% at 4 and 8 weeks after planting, respectively (Olasantan et al., 1994). In the same study, fertilization of intercropped plots reduced weed dry mass by 21–31% at 4 and 8 weeks after planting, respectively, as compared with unfertilized controls. Reductions in weed populations due to application of NPK and organic manure have been observed by Ramakrishnan and Kumar (1976) and Fawcett and Slife (1978). Split applications of N at planting and at ear development in a pot study increased maize yield and depressed growth of lamb's-quarters or wild mustard compared with a single application at planting (Larson and Hanway, 1977) owing to the slower growth of maize in the early season relative to weeds.

The timing of nutrient availability in soil to match crop requirements can reduce weed growth. This can be accomplished by time of application or by slow release of nutrients by organic matter. Fertilizer placement by banding can also give the crop a competitive advantage over weeds.

ALLELOPATHY

Allelopathy is defined as any negative or positive plant response to biochemicals produced by another plant (Rice, 1984). However, the term allelopathy generally refers to the detrimental effects of donor species on the germination, growth and development of the recipient. Allelochemicals may be exuded by living roots, leached from leaves, roots, stems, fruits or litter on the soil surface, released as gases, released from dead tissues in the soil, or released when micro-organisms degrade plant residues. Almost all cases of allelopathy seem to involve a complex of chemicals rather than one specific phytotoxin. The phytotoxins include alkaloids, benzoxazinones, cinnamic acid derivatives, coumarins, cyanogenic compounds, flavonoids, ployacetylenes, quinones and terpenes (Putman, 1994). These phytotoxins may inhibit seed germination, seedling establishment, growth of established plants or the beneficial activities of free-living as well as symbiotic micro-organisms (Klein and Miller, 1980). The allelopathic effect of crop residue compliments its soil and water conservation benefit.

A major challenge for the exploitation of allelopathy for weed control is to minimize its negative impacts on the crop. Although the mechanism is not well understood, there are several references in the literature indicating that plant residues reduce weed populations following harvest of a crop or the killing of a crop (Putman and Defrank, 1983; Shilling et al., 1985; Crutchfield et al., 1986). Experiments have demonstrated that cover crop residues left on the soil surface may have significant suppressive effects on weeds that cannot be attributed to either the physical presence of a mulch, e.g. shading and cooling of the soil, or to lack of tillage (Putman et al., 1983; Shilling et al., 1986; Worsham, 1989). Einheling and Rasmussen (1989) examined the allelopathic effects of grain sorghum, soybean and maize to control weeds in the subsequent year. Sorghum areas had consistently lower weed biomass in midsummer of the following year than did maize or soybean areas owing to the suppression of broadleaved weeds, while grass weeds remained unaffected.

Screening for allelopathic types in the germplasm of crops has been attempted by some researchers. Potential weed-suppressing types of oats (Fay and Duke, 1977), sunflowers (Leather, 1983), soybeans (Rose et al., 1983) and sorghum (Alsaadawi et al., 1986) have been observed. These indicate that allelopathic activity could possibly be exploited through breeding.

Allelopathy may be minimized by tilling the residue into the soil, but soil disturbance may compromise weed control by stimulating weed seed germination (Aldrich, 1984). However, certain species have also been identified as possessing allelopathic potential when their residue is mixed into soil. In a pot study, White et al. (1989) found residues of crimson clover and hairy vetch to decrease emergence and dry mass of pitted morning glory by 60–80% if they were incorporated into the soil. Residue left on the surface had less effect. Byodston and Al-Khatib (1994) observed reduced shepherds purse, kochia and green foxtail emergence and reduced dry weights of shepherds purse and kochia due to mixing of rapeseed leaves into a LaConner loamy sand in the glasshouse. Reduced emergence of all three weeds was observed by mixing soil with white mustard leaves. Rapeseed tissue incorporated into a Quincy loamy sand did not affect the emergence of shepherds purse, kochia, green foxtail and hairy nightshade, but reduced the emergence of puncturevine and the biomass of hairy nightshade, puncturevine and longspine sandbur. Mixing of white mustard tissue reduced emergence and biomass of all six weeds. Incorporation of rapeseed green manure into plots reduced weed biomass in the potato crop by 50–90%.

Despite considerable evidence of allelopathy, the effect is difficult to separate from plant competition or the physical effect of a mulch cover on weeds (Quasem and Hill, 1989; Banks, 1990).

CROPS FOR WEED SUPPRESSION

Cover crop, companion crop, nurse crop, live mulch, alley crop or smother crop are the names given to crops grown specifically for weed control, either during the fallow period or along with the crops grown for economic purposes. Relay cropping also uses the same principle. Cover crops are grown by many farmers to suppress weeds in vegetable and other non-competitive crops. In a field trial, Mucuna and sword bean grown as cover crops significantly decreased both weed density and dry mass when compared with maize alone (Gliessman, 1983). A Survey of 55 vegetable farms in the northeastern USA showed that over 90% of growers use cover crops regularly to maintain fertility, prevent erosion and/or control weeds, although the results of weed control with cover crops were mixed (Schonbeck, 1988). Farmers seed one or more cover crops, such as buckwheat, tartary buckwheat, winter rye, sudangrass, Japanese millet or mammoth red clover, in weedy fields for 1 year. Fast-growing crops such as buckwheat and oats are grown to suppress weeds during short fallow periods or between rows of widely spaced crops. Schonbeck et al. (1991) compared two plantings of buckwheat followed by winter rye (BW/R), sudangrass followed by winter rye (SG/R), mammoth red clover + oats (C+O) and Italian ryegrass (RG) cover crops for weed control from the spring of 1988 until the spring of 1989. Lettuce was grown in the summer of 1989. Some buckwheat plantings suppressed summer weeds effectively, but others failed because of drought. Sudangrass tolerated drought, but out-competed weeds only at the most fertile site. Clover established slowly, but suppressed weeds more effectively than rye during fall 1988 and spring 1989. Biomass of weeds + cover crop regrowth in lettuce was significantly lower after BW/R and SG/R than after C+O. Weed biomass excluding cover crop regrowth was in the order BW/R < SG/R < RG < C+O. A study by Putman et al. (1983) found that the residues of certain fall-planted cereal and grass cover crops resulted in significantly reduced dry mass of weeds the following summer as compared with unplanted controls. Rye, wheat and barley, which overwintered and were killed with non-persistent herbicide in spring, showed greater suppression of weeds than did the winter-killed crops of oats, grain sorghum and sorghum–sudangrass. The early season weed population (plants m^{-2}) under the residues of barley (11.5), corn (26.5), oats (7.5), rye (11), sorghum (28.7), sorghum × sudangrass (14.8) and wheat (11.9) was at least 75% lower in comparison to the population of 113.2 plants m^{-2} under the control. Common purslane and smooth crabgrass were relatively more sensitive to residues of sorghum and oats. Emergence of weed species germinating near the soil surface, including common lamb's-quarters, common purslane and redroot pigweed, was particularly inhibited by the residues of cereals. Among 14 crops tested for use as spring cover crops in Oklahoma, Nelson et al. (1991) found rye, barley, wheat and annual ryegrass to be most effective against weeds.

Highly competitive crops may be grown as smother crops during the fallow period or as a phase within a rotation. Fall rye and winter wheat were as effective as fall-applied 2,4-D plus spring-applied glyphosate/dicamba in controlling weeds from September to June (Moyer, 1995). Residues from fall rye and winter wheat reduced weed growth from June to September compared with bare fallow. A sweet clover cover crop effectively suppressed weed growth from September to June. Sweet clover was killed by harrowing in June. In September, weed dry matter (kochia, redroot pigweed, green foxtail, Russian thistle and dandelion) in sweet clover residue was <10% of that on bare fallow (Moyer, 1995). Smeda and Putman (1988) observed that 17 days after desiccation of the fall-planted barley, rye and wheat cover crops, the weed biomass was reduced by 80–90% in comparison to the bare-ground treatment. Seventy-one days after desiccation, only wheat and rye reduced weed biomass by 51–73%. The weed dry mass, 118 days after desiccation of lupine, wheat, rape, radish and oats, was between 300 and 1350 g m^{-2}, with 83–100% being grasses (de Almeida, 1985). Weed biomass in rye and triticale was 500 and 1100 g m^{-2}, respectively, with 64–77% being broadleaves. The residue biomass, 118 days after desiccation, was in the order oats < rye < rape < triticale < wheat < lupine. The degree of weed suppression by these crops also ranked in the same order.

Weed suppression by cover crops and tillage have been compared in some studies. Rye residue was found to suppress weeds in comparison to tillage (Robertson et al., 1976). Rye residue (rye seeded in fall and mowed in spring, with or without being killed by glyphosate) suppressed weeds for 4–8 weeks after transplanting of tomatoes, which was similar to the suppression from conventional tillage (fall ploughing, spring discing and spring harrowing) plus herbicides (trifuralin and metribuzin) (Masiunas et al., 1995). However, all treatments required supplemental weed management for commercially acceptable control. Senseman et al. (1988) compared a burndown (killed by herbicides) cover crop of hairy vetch to conventional tillage and a stale (1–2 months) seedbed in soybean plots. Three herbicide treatments (control, pre-emergence and pre-emergence plus post-emergence) were superimposed during the soybean growing season. In 1986, the control of pitted morning glory and prickly sida was excellent under all treatments except for the no-herbicide treatment. However, yields from vetch plots were lower than the conventional tilled plots, as wet conditions prevented planting into flowering vetch. On the other hand, vetch plot yields were similar to the conventional tillage and stale seedbed plots in 1987, as planting was timely. Satisfactory weed control was again obtained for all tillage and herbicide treatments except pitted morning glory, which was effectively controlled only by pre-emergence plus post-emergence herbicide treatment.

The use of nurse crops is common for seeding of pastures. Hall et al. (1995) compared an oat nurse crop, a pre-plant or post-emergence herbicide, and a check for weed suppression during establishment of band-seeded alfalfa, seeded in spring (April/May) or late summer (August/September). For spring-seeded alfalfa, treatment effects on alfalfa and weed dry matter during the seeding-year were sporadic, but net return was generally greatest when no weed control was used. In summer-seeded alfalfa, the treatments had limited impact on weed growth, alfalfa dry matter, forage quality or net return. they concluded that weed control practices were not economical and may actually reduce net return because of the additional cost. Lanini et al. (1988) reported reduced weed biomass and increased forage yield (alfalfa + oats) with increased planting rates (between 0 and 36 kg ha^{-1}) of an oat companion crop in establishing alfalfa. Weed growth during forage establishment has a minimal effect on forage yield in subsequent years (Moyer et al., 1995). In addition, the annual weeds that grow during establishment are almost completely suppressed by established forages in subsequent years.

Living mulches, usually low-growing legumes, are used to suppress weeds and conserve the soil in the open spaces under taller-growing crops. Crops are directly planted into the living perennial legume without tillage. Living mulches are usually chemically suppressed before or after crop seeding to avoid competition. Seeding of the crop is done with an appropriate no-tillage planter, which requires less energy and labour than conventional tillage and seeding. Lanini et al. (1989) compared subclover living mulch with cultivation, herbicide application and a control for weed control in the vegetable plots at Davis, Salinas and Riverside in California. In these trials subclover living mulches showed some potential to suppress weeds. When managed with a combination of mowing and herbicides, subclovers produced dense stands resistant to weed invasion. Moving alone was effective if subclover stands were dense and weed populations low. High maize yield in the no-tilled crop without a living mulch was obtained through the use of a high rate of N, while the live mulch produced a favourable maize yield at low N. From a pot experiment on undersowing winter wheat with black medic and persian clover, Hartl (1989) reported significantly reduced dry matter of weeds, especially the weeds still living at harvest. His field trial with white clover likewise showed a reduction in the dry matter of weeds. He concluded that undersowing wheat with clover efficiently restricts the negative consequences of late weed development.

Budelman (1988) showed that weed incidence can be considerably reduced with alley cropping. The weed reduction potential of alley cropping systems is due to the complimentary weed suppression functions of a mulch of tree prunings covering the ground and the shading effect of the trees. The extent of weed suppression depends mainly on the canopy development and shading characteristics of the trees, as well as on the rate of decomposition of pruned residues (Yamoah et al., 1986; Budelman, 1988). A reduction of up to 90% in weed biomass was observed by Jama et al. (1991) under horse tamarind (*Leucaena*) alley cropping in comparison with a crop-only control. Cocoa shadetree (*Gliricidia*) was found to have a detrimental effect on certain weed types and its mulch effectively suppressed weeds, especially dicot species (Inostrosa and Fournier, 1982). Rippin et al. (1994) reported that plant residues of *Erythrina poeppigiana* trees (1o t ha^{-1} dry matter, planted at 6 m × 3 m reduced weed biomass by 52%, while residues of the *Gliricidia* trees (12 t ha^{-1} dry matter, planted at 6 m × 0.5 m) reduced weed biomass by 28% in comparison with crop-only control. Weed suppression contributed to the higher maize grain yield of 3.8 t ha^{-1} under both alley croppings as opposed to the unmulched yield of 2.0 t ha^{-1}. Highly significant differences were found when comparing the relative incidence of monocot:dicot weeds in an *Erythrina* mulch (24% monocot:76% dicot) or *Erythrina* alley cropping (19% monocot:81% dicot) as opposed to the control (62% monocot:38% dicot). The *Gliricidia* alley crop showed some ability to reduce dicot weed biomass. *Erythrina* tended to have greater weed suppressing ability owing to the slower rate of decomposition of its mulch than that of *Gliricidia* mulch (Budelman, 1988; Haggar and Beer, 1993).

Weed-suppressive crops compete for light, soil moisture and nutrients with weeds, or may be allelopathic to weeds. There are reports of differential effects of crop residues on both weeds and subsequent crops (Putman et al., 1983). Small seeds, mainly annuals, may germinate but lack adequate seed reserves to emerge through the mulch layer. Perennial weeds may grow through or out from under the residue, as their seeds or roots generally have more stored energy reserves. Allelopathy may also inhibit emergence or early growth of small-seeded crops (Putman et al., 1983). Einhellig and Leather (1988) observed that generally the broadleaved weeds were more negatively affected by residue than the grassy weeds, and that the large-seeded crops (maize, cucumber, pea, snapbean) were less affected than the small-seeded crops (carrots, tomato, lettuce). In addition, weed-suppressive crops and their residues can affect nutrient availability through N fixation (legumes) and temporarily immobilize nutrients during decomposition (Schonbeck et al., 1993). Akobundu (1982) recommended that research should aim at ground-cover management that would replace weeds with plant species which are easier to manage than weeds. Research emphasis should include the evaluation of legumes for use as live and residue mulches, and cropping systems related to their use.

MULCHING

The addition of mulch to the soil surface affects the soil's physical, chemical and biological properties and processes, the transfer of energy and matter between the soil and the atmosphere, and crop and weed growth, which may influence the need for weed control. Organic mulches may consist of plant by-products such as crop residues, rice hulls, peanut shells, grass clippings, wood dust or chips and bark from trees, waste products such as newsprint, and crushed rock or small stones. Bark mulches are reported to be more effective than straw mulches, as bark also releases tannins and phenols which may reduce weed growth (Ashworth and Harrison, 1983). Synthetic mulches include woven and spun-bound sheets of polyethylene (plastic), polypropylene or polyester. Elmore and Tafoya (1993) reported that the degree of weed control by a mulch is likely to depend on the mulch material, placement, depth and mulch maintenance. These mulches can be considered as barrier-type mulches, since the degree of weed control depends principally on their ability to block light and prevent seed germination. They can also present a physical barrier restricting the emergence of certain weed seedlings (Facelli and Pickett, 1991). In addition to acting as a barrier, herbicides may leach more rapidly owing to greater water infiltration under mulch, or they may be tied up by the residue, which will lessen weed control. However, corn residue (1100 vs. 6200 kg ha^{-1}) was not found to hinder weed control with herbicides (atrazine or alachlor) in Iowa (Erbach and Lovely, 1975). In most instances the benefits of residues more than compensate for a reduction in weed control due to reduced penetration of soil-applied herbicides (Fawcett, 1987; Mills and Witt, 1989).

There is considerable evidence in the literature indicating a reduced degree of weed establishment due to the presence of residue mulch, and a number of studies have indicated that herbicides or tillage can be substituted by mulches in certain cases. Residues from oats and ryegrass are used successfully in Brazil for weed control (de Almeida, 1990) and to reduce herbicide usage in conservation tillage systems (Peeten, 1990). Rye foliage mulch (2500 kg ha^{-1}) reduced sicklepod biomass in soybean plots, but its combination with no tillage offered no advantage over either of these alone (Shilling et al., 1995). Residues from a range of winter crops (black oats, rapeseed, oats, wheat, turnipseed, barley, rue, common vetch, flax, chikling vetch, triticale, serradela, black oats + common vetch, and rye) were found to provide selective control of arrowleaf sida, marmelad grass and common blackjack in soybeans (Roman, 1990). The degree of control at 40 days after planting of soybeans varied from 0 to 100%. Herbicide treatment was not required for weed control before seeding soybeans when most effective crop residues were present. In a 2-year field trial with onion and garlic planted in autumn, 3 t ha^{-1} of wheat straw mulch was found to be as effective as herbicide application for weed control (Duranti and Cuocolo, 1989). It seemed to be more appropriate for onion than for garlic, because with low crop density a relatively uniform distribution of the mulching layer was possible. Amongst the ploughing, ripping, rotovation, no-till and no-till plus straw mulch (5 t ha^{-1}) treatments, the lowest number of weeds as observed under the last treatment during both seasons of growing maize in Zambia (Gill and Lungu, 1988). Apparently the presence of mulch on the soil surface reduced weed establishment. Similarly, residue mulch, early inter-cultivation with a hoe and pre-emergence herbicide reduced weed mass and enhanced maize yield in Zambia (Gill et al., 1992). However, the scope of pre-emergence herbicide and inter-cultivation is limited owing to resource constraints on small-scale farmers. Therefore, residue mulch plus minimum tillage, to remove any weeds present at seeding, was considered to be an appropriate alternative to the conventional labour-intensive system, eliminating the need for pre-seeding land preparation and an early weeding with hand hoes. Birzins (1983) tested fresh coniferous sawdust and chip mulches (5 cm thickness), white (2 mm thick) and black (4 mm thick) polyethylene plastics, glyphosate and simazine herbicides, and an untreated check for weed suppression in a 1.2 m × 1.2 m area around recently planted interior spruce ramets between 8 July and 3 September 1981. Black and white plastics provided the most and least effective weed control, respectively. Glyphosate killed the weeds present on contact, but new weeds appeared within a month. Simazine provided excellent control for the duration of the study except for the regrowth of perennials from roots not removed by pre-application hand-weeding. Sawdust and chip mulches provided fair weed control, with some perennials growing through the mulch. Pine bark and bitumenized felt mulches degraded after only one growing season, and thereafter provided ineffective weed control (Davis, 1984). The use of mulch is recommended for suppression of weeds, thereby reducing weed pressure and herbicide use for growing crops in advancing countries (Akobundu, 1982).

The thickness and fineness of the mulch layer have been found to affect weed control by mulches. Finely pulverized mulches may be better able to exclude light from the soil than coarser mulches, although blown-in seed often establish better in finer material as it is a good growth medium (Campbell-Loyd, 1986). Gartner (1978) obtained better weed control with 10 cm than with 5 cm mulch of bark and wood chips. Billeaud and Zajicek (1989) found that pinebark nuggets resulted in a reduced weed population when compared with a pinebark mulch or an untreated control. Control of the field-planted purple nutsedge was significantly better with woven polypropylene fabric or 15 cm organic mulch than with 5 cm and 10 cm organic mulch. Applying 5, 10 or 15 cm organic mulch above the fabric mulch did not improve weed control relative to the fabric alone. Weed populations were observed to decrease as winter wheat straw mulch levels increased, and thus herbicide rates could be reduced with increasing amounts of straw (Crutchfield et al., 1986).

The effectiveness of organic and synthetic mulches have been compared with each other as well as with other weed-control measures. Owing to the reported adverse effects of plastic on plant growth (Robinson, 1986; Dilatush, 1988), landscape fabrics were tested and considered as a viable weed control alternative to plastic mulches because of their porosity (Appleton and Derr, 1987). Ashworth and Harrison (1983) compared six synthetic and two organic mulches over a 4-month period in a field study. Heavy-duty green plastic, black polyethylene and woven black polypropylene provided the best weed control. Shredded hardwood bark (5 cm deep) provided slightly less control, and clear polyethylene, straw mulch and the untreated control showed the poorest weed control. Derr and Appleton (1989) compared six woven polypropylene landscape fabrics with black plastic and pre-emergence herbicides for weed control. Large crabgrass shoots and roots and yellow nutsedge shoots penetrated all the polypropylene fabrics and developed into large plants. On the other hand, black polyethylene controlled crabgrass when seeds were planted above or below the plastic, and controlled yellow nutsedge bulbs planted below the plastic. Black plastic and herbicide plus mulch provided equal or greater control of large crabgrass than the landscape fabrics. More time was required to hand-weed landscape fabrics covered with mulch than uncovered fabrics. Lennartsson (1990) compared the killing of a mown grass pasture with carpet, cardboard, black polythene, a 20 cm layer of hay or rotovation. The abundance of weeds in the first season after killing was greater in rotovated plots than in those killed by mulching. Mulching with black polythene, plantex (a non-woven, water-permeable, 100% polypropylene) or, plantex with a 3–5 cm layer of leaf-mould on top were able to control weeds successfully in dwarf french beans, while leaf mould alone partially controlled some weeds. In strawberries, only black polythene and plantex with leaf-mould on top provided effective weed control, as plantex alone was not sufficiently opaque to prevent weed growth. Similarly, the growth of bermudagrass and johnsongrass was suppressed more by spun-bound than by meshed fabrics, and all mulches gave only partial control of yellow nutsedge (Martin et al., 1991a). McLean et al. (1987) noted mixed results using two organic mulches, one herbicide and one polyester fabric. Powell et al. (1987) reported that a longleaf straw plus a fabric barrier was superior for weed control when compared with longleaf straw or pine bark mulch alone. Potter (1988) compared several types of sheet mulch materials, including clear, white and black polyethylene (0.125 and 0.038 mm), a permeable woven polythene and several other permeable materials. Weed control was best in the heavier grade black polythene, and tree growth was related to the degree of weed control. Treatments using clear, white or thin black polythene allowed enough light transmission to permit weed growth, and weeds grew through the permeable materials. According to Derr and Appleton (1989), landscape fabrics have both positive and negative attributes. Landscape fabrics can effectively reduce annual broadleaf and grassy weeds, but most are far less effective against perennial weeds, which are capable of growing through several inches of mulch, and then thrive in the absence of annual-weed competition. A desirable weed control alternative would appear to be a UV-resistant fabric, which would not transmit light.

Synthetic mulches have been used for weed control through soil solarization, a technique that uses the entrapped solar energy to kill weeds at a level nearly equal to a fumigant, and which can be a viable alternative to herbicide use in many cases. To solarize an area, pre-moistened soil is covered during summer with clear polyethylene to trap long-wave radiation. On sunny days the soil temperature can rise to $68°C$. It is thought that because heat accumulates slowly, dormant weed seeds may be induced to germinate, and then die from heat or disease. The principles of soil solarization are discussed by Katan (1981). According to Rubin and Benjamin (1984), the following mechanisms may be involved in the weed-control process using solarization: direct thermal killing of germinating or even dormant seeds; thermal breaking of dormancy followed by thermal killing; thermally induced changes in CO_2/O_2 ethylene and other volatiles which are involved in seed dormancy release followed by thermal killing; the direct effect of high temperatures interacting with toxic volatiles released from decomposing organic matter or seed metabolism; indirect effects via microbial attack of seeds weakened by sublethal temperatures. The susceptibility level of 40 weeds is given by Bell (1993).

Soil solarization for 98 days (14 June to 10 September 1985) resulted in increased yield of collard green and in a 91% reduction of weeds during 1986, and was more effective than pre-emergence application of DCPA herbicide to non-solarized plots (Stevens et al., 1990). There were significant negative effects on the marketable yield and root growth of collard green and soil microflora from the DCPA application to solarized soil. Studies in Israel (Horowitz et al., 1983; Rubin and Benjamin, 1983) showed that many weeds can be controlled in solarization periods of 30 days. However, a short solarization period was found only partially to control nutsedge in Israel (Rubin and Benjamin, 1983) and stimulated its emergence in Mississippi (Egley, 1983). Wild mustard, redroot pigweed and jimsonweed emergence from seeds was significantly reduced by heating the soil for 30 min to 40, 50 and $70°C$, respectively (Rubin and Benjamin, 1984). Sweet clover, however, was not affected even at $90°C$. Rhizomes of johnsongrass and bermudagrass, when heated in

Table 12.1   Reduction in weed dry mass (%) by maize/cassava intercrop in comparison with single crops at 4 and 8 weeks after planting (WAP), as influenced by fertilizer application

	Unfertilized		Fertilized	
Compared with:	4 WAP	8 WAP	4 WAP	8 WAP
Sole cassava	37	15	50	40
Sole maize	15	20	37	45

Source: modified from Olasantan et al., 1994

soil for 30 min, were killed at 40°C, while purple nutsedge tubers were significantly affected only when heated to 60°C or higher. Generally, the longer the soil solarization period, the more effective the weed control (Katan, 1981; Horowitz et al., 1983). Solarization has been observed to work in the San Joaquin, Imperial and Watsonville Valleys of California, and nearly as well in northern states such as Idaho, to control a number of weeds (Bell, 1993).

INTERCROPPING

Intercropping has been reported to suppress weeds in many cases. In pigeonpea-based intercropping systems at ICRISAT, intercropping with sorghum reduced weed growth by 10–75% compared with pigeonpea alone (Rao and Shetty, 1976). Shetty and Rao (1981) observed that in a pearlmillet/groundnut intercrop system, an arrangement of one pearlmillet row for every three groundnut rows resulted in optimal weed suppression and crop yield. Adding Italian ryegrass or red clover to barley or faba beans was found to decrease perennial ryegrass growth from both rhizome fragments (Dyke and Barnard, 1976) and seeds (Williams, 1972). Olasantan et al. (1994) reported that intercropping cassava and maize is an effective weed-control practice in traditional farming (Table 12.1). Weed suppression from 45 main plus smother crop type intercropping systems is summarized in Table 12.2 (modified from Liebman and Dyck, 1993). Although intercrops were not always superior to the single crops in weed suppression, the weed biomass accumulation by intercrops was generally lower than that of at least one of the component species grown as a single crop.

Despite considerable evidence of weed suppression by intercropping, many studies have shown that intercrops do not necessarily lead to reduction of weed biomass below those observed in plots of the component crops. Ayeni et al. (1984) have shown that three cropping patterns (maize alone, cowpea alone and an intercrop of maize and cowpea) had no effect on weed growth for the first 6 weeks of the cropping season. Alternate rows of cowpea in sorghum did not reduce purple witchweed (*Striga*, a parasitic weed) density, as planting cowpea rows 80 cm from sorghum rows provided little cover at the base of the sorghum (Carsky et al., 1994), but planting the cowpea and sorghum in the same row, and in the same or alternating hills, reduced the density and numbers of witchweed per sorghum plant, which means that any spatial arrangement that increases cowpea ground cover at the base of the sorghum can reduce *Striga* emergence. Production of mature *Striga* capsules decreased with increasing cowpea ground cover. Earlier, Carson (1989) found a positive relationship between the temperature at 10 cm soil depth under groundnut and *Striga* density. Reduction in temperature by 2°C reduced *Striga* density at

Table 12.2   Weed suppression by 45 main crop(s) plus a smother-crop-type intercropping system. See Liebman and Dyck (1993) for original references on different systems

---

Intercrops showing stronger than main crop(s) weed suppression effect
  Alfalfa + barley or oats
  Bahiagrass + subterranean clover or arrowleaf clover
  Barley + Italian ryegrass or red clover
  Bermudagrass + subterranean clover or arrowleaf clover
  Cassava + bean or *Desmodium heterophylum* (Wild.) DC.
  Cassava/plantain + *Desmodium ovalifollium* Wall. ex Merr or *D. ovalifolium/Inga edulis* Mart.
  Faba bean + Italian ryegrass or red clover
  Maize + Italian ryegrass or perennial ryegrass or red clover or hairy vetch or subterranean
    clover or soybean or *Arachis repens Handro* or *Centrosema pubescens* Benth or
    Psophocarpus palustris Desv.
  Maize/cassava + cowpea or egusi melon or peanut or sweet potato
  Pigeonpea + urdbean (= blackgram) or mungbean or soybean or cowpea or sorghum
  Rice + blackgram
  Sainfoin (*Onobrychis viciifolia* Scop.) + barley
  Sorghum + cowpea or mungbean or peanut or soybean
  Wheat + black medic (*Medicago lupulina* L.) or Persian clover (*Trifolium resupinatum* L.) or
    white clover

Intercrops showing weaker than main crop(s) weed suppression effect
  Maize + soybean or *Desmodium trifolium* (L.) DC. or *Indigofera spicata* Forsk. or *Axonopus*
    compressus (SW) Beauv.

Intercrops showing variable results among experiments for the weed suppression effect
  Barley/oats + red clover or alfalfa
  Rice + *Azollo pinnata*

Source: modified from Liebman and Dyck, 1993

sorghum harvest to 60–70%. Results from 24 intercropping systems in which all the component crops were considered as main crops are summarized in Table 12.3 (modified from Liebman and Dyck, 1993). The success of weed control through intercropping is variable, as none of the species is sown for its weed-suppressing ability.

Large differences in weed suppression ability have been noted among the crop species intercropped with barley and faba bean (Williams, 1972; Dyke and Barnard, 1976), bean, cassava and maize (Soria et al., 1975; Fleck et al., 1984), cassava/maize mixtures (Unamma et al., 1986), pigeonpea (Ali, 1988), and sorghum (Abraham and Singh, 1984). In general, intercrops that include species with rapid early growth and dense canopy formation over the ground surface are found to be most weed-suppressive.

Intercrops may suppress the growth of weeds through greater preemptive use of resources, or give higher yields than single crops without superior suppression of weed growth. Negative correlations between weed growth and intercrop yield advantage are reported from experiments with a variety of crop combinations (Shetty and Rao, 1981; Abraham and Singh, 1984; Ali, 1988), which indicates that intercrops suppress weed growth through greater preemptive use of resources (Igbozurike, 1971; Hart, 1980; Mohlar and Liebman, 1987). Similarly, as LER (land equivalent ratio × the sum of I_i/S_i, where I_i and S_i are the yields per unit area of the ith component crop under intercropping and single cropping conditions, respectively) values of the intercrops increased the weed biomass decreased, both in terms of the absolute value and as a percentage of the amount

Table 12.3   Weed suppression by 24 intercropping systems in which all component crops are considered as main crops. See Liebman and Dyck (1993) for original references on different systems

Effect of combined crops stronger than single crops of each component
Bean + cassava or sunflower
Maize + mungbean or sweet potato or peanut or soybean or bean or sunflower or cassava
Maize + bean + cassavà
Pigeonpea + soybean or sorghum
Effect of combined crops intermediate between single crops of each component
Barley + pea
Flax + oats or wheat
Maize + peanut or soybean or bean
Maize + bean + cassava
pearlmillet + peanut
Pigeonpea + soybean or sorghum
Effect of combined crops weaker than single crops of each component
Maize + cowpea or cassava

Source: modified from Liebman and Dyck, 1993

produced in the corresponding single crops (Soria et al., 1975). On the other hand, data is available indicating no consistent relationship between weed growth and intercrop yield advantage (Bantilan et al., 1974). Selective negative allelopathy by some intercrops on weed growth was indicated by Gliessman (1983). In addition to crop yield, the ability of intercrops to compete with weeds depends upon the type, number, proportion, density and spatial arrangement of component crops, selected cultivars and fertility, and the moisture status of the soil (Moody and Shetty, 1981; Liebman, 1988).

Shifts in species composition of weed flora resulting from intercropping has been the subject of a few studies. The shift was found to be influenced by the weeds present in the single crops (Shetty and Rao, 1981; Janiya and Moody, 1984) and by the competitive dominance of the crop or crops planted, but not on the amount of difference between the crops (Mohlar and Liebman, 1987).

While many studies have concentrated on weed suppression by intercrops relative to single crops, attempts have been made to compare weed control by intercrops with other weed-control techniques. Unamma et al. (1986) compared maize/cassava, cowpea/maize/cassava, and egusi melon/maize/cassava with the use of two hand-weedings or herbicides in Nigeria. Even though the intercrops failed to control weeds as effectively as hoeing or herbicides late in the growing season, crop yields were not depressed by weed competition since there was adequate weed suppression in the critical period of 4–8 weeks after planting. In each of 2 years, the unweeded cowpea/maize/cassava or melon/maize/cassava intercrop systems produced as much cassava and maize as a weed-free monoculture of each crop. In contrast, across the 2 years in the maize/cassava system, uncontrolled weeds reduced cassava yield by 49% and maize yield by 53%. Economic returns for the different weed control systems were in the order intercrops > herbicides > hand-weeding. Addition experiments in Nigeria have demonstrated that intercropping with egusi melon is an effective labour-saving method of weed control in yam or yam/maize/cassava (Akobundu, 1980), and plantain (Obiefuna, 1989) production systems. Sengupta et al. (1985) demonstrated that rice + blackgram intercropping effectively suppressed weed growth and eliminated one-handweeding. Ali (1988) reported that the weed growth under an unweeded pigeonpea/mungbean intercrop was 22–38% lower than in a single crop of unweeded pigeonpea, and seed yield of intercrops was similar to that of a weeded sole crop of pigeonpea. Zuofa and Tariah (1992) compared three hoe-weedings (at 3, 6 and 9 weeks after planting), altrazine + metolachlor, intercrops (sweet potato, cowpea, groundnut or melon) plus a weeding at 3 weeks after planting for the early (March–July) and late (September–November) season maize crops. Herbicide in both seasons and three hoe-weedings late in the season controlled weeds more effectively than the intercrops. Weed control by intercrops and three weedings early in the season was similar at 4 weeks after planting, whereas at 6 weeks the sweet potato and groundnut intercrops provided better weed control than three weedings. Intercrops were recommended because of acceptable weed control, additional nutritional produce, little additional cash input, and ease of adoption.

In general, reduced weed growth and greater crop yields may be achieved through intercropping than through sole cropping. Intercrops that include species with rapid early growth and dense canopy formation to cover the soil surface are the most weed-suppressive. Greater weed suppression may result from larger quantities of resources being captured by crops and smaller quantities being captured by weeds, or increased allelopathic suppression of weeds by intercrops. Crop yield advantages over sole crops, even when the intercrops fail to suppress weed growth below levels obtained from the component sole crops, suggests that intercropping may result in the use of resources by the crops which cannot be exploited by weeds, increased resource conversion efficiency by crops, or shifts in crop biomass allocation. Understanding how component crops and weeds respond to the manipulation of environmental factors in intercropping should benefit the design of weed-suppressive and agronomically productive cropping systems.

CROP ROTATION

Crop rotation involves growing different crops in a systematic and recurring sequence on the same land, as opposed to growing a single crop continuously in the same field. The weed flora under a single crop tends to be dominated by the weeds which are highly adapted to the growing conditions maintained for the crop (Froud-Williams, 1988). Such weed flora are difficult to control without killing the crop or placing the land under fallow. For example, weed infestation in the Broadbank long-term continuous wheat at Rothamsted made fallowing necessary for 1 year in every 5, until herbicides were introduced in 1954 (Jenkinson, 1991). Continuous wheat under rainfed conditions in Saskatchewan suffered two crop failures in 13 years as a result of green foxtail infestation (Austenson et al., 1970). Green foxtail and wild oat densities increased in continuous spring cereal production (Hume et al., 1991). In continuous spring wheat, annual grasses become an increasing problem (Moyer et al., 1989; Donald and Nalewaja, 1990).

The above-mentioned observations on weed infestations in systems utilizing single crops, whether short or long term, strongly suggest that rotation among diverse crops and the resultant environments may prohibit the domination of weed flora by certain species and minimize weed interference with crop growth. The recent focus on non-chemical means of pest control has revived interest in the use of crop rotation (Walker and Buchanan, 1982; Cook, 1986; Gunsolus, 1990). Traditional crop rotations include legumes for N source, legume or grass sod for maintenance of organic matter, and one or two main crops. A crop rotation can make use of fallow (tilled or chemical), green

Table 12.4   Examples of crop rotations used for weed suppression in California without use of herbicides

Rotation	Comments
Winter cereals/summer crops	Used primarily for annual weed seed reduction. Relay-planting large-seeded crops (corn, beans or corn) into winter cereal or after shearing off all alfalfa below the crown has been useful. Seeding of large-seeded crops deep into pre-irrigated soil allows the crop to germinate but prevents small-seeded weeds from establishing
Flooded rice/upland crops	For field bindweed, johnsongrass, bermudagrass, knapweed and other perennial weeds. Use of dry-environment crops such as safflower, sunflower and dry beans aids in control of weeds in a rice crop. Has worked for control of many, but not all, perennial weeds
Winter cereal/summer fallow/row crop	For johnsongrass, bermudagrass and purple nutsedge control. Works best on sandy loam soils and where summer rains are rare. Tillage about five times at 10–15-day intervals allows desiccation of perennial plant parts
Cereals/safflower or garlic/row crop	For field bindweed in fairly dry soils. Use blade 45–60 cm deep on a dry soil in fall, prior to spring-planted row crops. Blading delays bindweed emergence for about a year and aids control of weed seedlings in row crops

Source: modified from Kempen et al., 1991

manures, cover crops and sequential cropping systems (two or more crops grown in sequence in the same field in a single year). Growing conditions in which each crop differs from its predecessor in terms of the nature and timing of agronomic practices serve to disrupt the weed growth cycle and prevent domination of the weed flora by only a few species (Froud-Williams, 1988; Liebman and Janke, 1990). Results of 29 test crop × rotation combinations in which herbicides were not used were summarized by Liebman and Dyck (1993). Compared with a single-crop system, the number of weeds in rotations was lower in 21 cases, the same in five cases, and more in one case. The weed-seed density in crop rotations was lower in nine cases and similar in three cases to that in continuous cropping with one of the component crops. Kempen et al. (1991) summarized the rotation techniques used for weed reduction in California, with and without herbicides (Tables 12.4 and 12.5). They concluded that the weed management programs which include rotation provide many options to the growers, and crop rotation results in better weed control than continuous cropping with a single crop.

Control of some problem weeds has been reported to be much better when crop rotation is used. According to Blackshaw (1994b), crop rotation is a key component of an improved management system for control of downy brome. Inclusion of fallow or spring canola in rotation with winter wheat suppressed downy brome densities during 6 years to less than 55 and 100 plants m^{-2}, respectively. In continuous winter wheat, yields decreased as downy brome densities increased progressively over the years, indicating that continuous winter wheat production will not be viable in regions where downy brome is prevalent unless effective herbicides are developed. In a second study, fewer weeds were found in winter wheat–fallow, winter wheat–lentil and winter wheat–canola

Table 12.5   Examples of the crop rotations used for weed suppression in California which utilize integrated weed management (IWM) techniques

Rotation	Comments
Rotation using fallow for weed control	
Crop/fallow	Fallow to control johnsongrass, bermudagrass, field bindweed and knapweed. Use tillage to break up root systems after crop harvest. Increase weed vigour with irrigation in late summer and maybe before herbicide application. Spray foliage at the appropriate stage, 1–6 weeks before frost
Rotations that take advantage of herbicides and crop characteristics	
Spinach/summer carrots or dry beans	Takes advantage of herbicides and crop characteristics
Winter cereal/dry beans/tomatoes	Herbicides are used in the cereal. Tillage follows irrigation after harvest. Dry beans are planted into dry mulch and herbicides are used to control weeds. This reduces the population in the subsequent spring tomato crop
August carrots/April cotton	Use herbicides, which also control cereals, to avoid volunteers. Cotton benefits from N carryover. A similar rotation can be used for many other crops
Spring potatoes/November onions or garlic	Pre-emergence tillage allows weed control at potato emergence. Potatoes, being a competitive crop, out-compete subsequent emerging weeds. Fall-planted onion or garlic shortens the fallow period. Herbicides used during both crops allow weed management at a reasonable cost

Source: modified from Kempen et al., 1991

rotations than in continuous winter wheat, and a dense infestation of downy brome developed in the continuous winter wheat rotation (Blackshaw et al., 1994). Schreiber (1992) reported that the use of corn/soybean or soybean/wheat/corn rotation reduced the giant foxtail seed count as well as the plant population relative to that with continuous corn under three tillage systems (conventional, chisel and no-till). The foxtail population increase due to a reduction in tillage intensity was significantly less when crops were rotated. Economic analysis of these results indicated that weed management levels above the minimum were not warranted when crops were rotated, because of equal corn yields at all herbicide levels (Martin et al., 1991b).

The importance of including different crop types for weed control within rotations is illustrated by many studies. Zemanek et al. (1985) compared winter wheat/spring wheat/spring barley/pea or faba bean to winter wheat/sugarbeet/spring barley/red clover or faba bean rotations with and without herbicides. After three cycles, the untreated second rotation contained a 30% lower weed density than did the first rotation, which showed that inclusion of the cleaning crop (using a very high level of weed control) of sugarbeet reduced weed density. Yield reductions due to the absence of herbicide occurred only in the rotation without sugarbeet. Initial infestation of wild oats, estimated at > 2000 seeds m^{-2}, was reduced to 2 seeds m^{-2} in a wheat crop following 3 years of a sod crop, perennial ryegrass or white clover (Wilson and Phipps, 1985). However, after 6 years the wild oat population increased to 22 seeds m^{-2}. Dense planting of barley as found to suppress the growth of quackgrass and wild oats (Evans et al., 1991). A traditional cure for land infested with wild oats appears to be a rotation including a grass or sod crop. The undisturbed soil conditions of the perennial crops suppress seedling emergence, although they may also increase the longevity of weed seeds in the soil (Roberts and Feast, 1973).

Weed population shifts associated with changing crops are due to complex and dynamic processes, which are not easily predicted. Crop rotation results in different growing conditions every season, requiring a different weed-management approach. Even if most of the cultural practices do not change, the weed population certainly will. For example, in irrigated transplanted rice, where water management can control grassy weeds, floating aquatic weeds can pose serious problems. Effective weed control can be facilitated by selection of crop sequences that use competitive crops to control particular weeds and allow for the application of selective herbicides to control specific weeds (Campbell et al., 1990). The cumulative effect of crop rotation is to reduce the population of weeds that are adapted to any one specific crop–herbicide combination (Froud-Williams, 1988; Derksen et al., 1994). One example is the severe infestation of *Bromus* spp. where winter wheat is grown continuously (Zentner et al., 1988; Peeper and Wiese, 1990), because it has a life cycle similar to that of winter wheat. A rotation with a spring-seeded crop is a very effective method of controlling *Bromus* spp. However, formation of a dense canopy in the early spring by winter wheat inhibits the germination and growth of several annual weeds (Peters, 1990). In addition to crop-related effects on weeds, it is possible to apply more effective herbicides for annual and perennial grass control in broadleaved crops such as canola, flax or soybeans than in wheat, and for broadleaved weed control in wheat than in flax or canola. Bermudagrass, a problem in no-till systems in Uruguay, is more effectively controlled in sunflower than in wheat (Gim'enez et al., 1991). Pawlowski and Wesolowski (1980) studied weed biomass in rotations containing 50, 75 or 100% cereal crops. Weed biomass was 143% higher and grain yield 55% lower in a 100% cereal rotation (spring wheat/winter wheat/oats/winter heat) than in a 50% cereal rotation (sugarbeet/winter wheat/mixed legumes/winter wheat). Increasing cereals from 50 to 75% had little effect on weed infestation or yield when sugarbeet was retained in the rotation. However, when mixed legumes were utilized in a 75% cereal rotation, weed biomass was 26% higher and yield 44% lower as compared with 50% cereal rotation.

The success of rotation systems for weed control appears to be based on the use of crop sequences that employ varying patterns of resource competition, allelopathic interference, soil disturbance and mechanical damage to provide an unstable and frequently inhospitable environment that prevents the proliferation of particular weed species.

TILLAGE

Soil disturbance has been found to simulate breaking of seed dormancy (Chancellor, 1985; Forcella and Lindstrom, 1988) and to reduce weed species diversity (Cardina et al., 1991; Gill and Arshad, 1995). While stimulation of weed emergence by tillage may need increased weed management efforts in the short run, repeated tillage can also be a useful tool to reduce weed seeds in the tilled layer (Warnes and Anderson, 1984; Chancellor, 1985).

Advances in tillage-related crop production techniques have occurred by replacing mechanical tillage with chemical and cultural methods for weed control. Replacement of tillage with herbicides offers some significant advantages, such as better control of some weeds and improved soil moisture levels and soil conservation, with reduced labour, fuel, lubricant and machinery costs (Lindwall, 1980; Brandt, 1989). Potential disadvantages include the heavy reliance on herbicides to control weeds, with their possible negative impact upon health and the environment, and reduced control of certain weed species. The subject of weed control under various tillage systems has been reviewed in a number of publications (Wiese, 1985; Donald, 1990; Moyer et al., 1994). We will concentrate on the use of mechanical tillage for reducing herbicide use.

Tillage and fallow management

Weed control during a 12–20-month fallow period requires 4–8 tillage operations or 4–5 herbicide applications (Lindwall and Anderson, 1981; Fenster and Wicks, 1982). Intensive and often excessive tillage pulverizes the soil and exposes it to the sun, leading to increased oxidation and increased soil erosion, which results in reduced soil organic matter (Arshad et al., 1990). Excessive tillage also causes the soil tilth to deteriorate and increases salinization (Larney et al., 1994). Studies in Australia and western US on combinations of tillage and herbicides have shown potential for reducing the cost of conservation fallow while maintaining many of its benefits (Fenster and Wicks, 1982; Leys et al., 1990; Smika, 1990). Conventional cultivation during the fallow year with 168-cm sweeps controlled most spring-germinating weeds but did not adequately control overwintered flixweed or downy brome (Blackshaw and Lindwall, 1995). Treatments involving a combination of herbicides and tillage gave the best control of all weed species.

Tillage and seedbed preparation

Several researchers have compared the use of mechanical and chemical techniques for seedbed preparation and related weed control. Weed emergence stimulated by tillage allowed mechanical control to partially replace herbicides (Forcella et al., 1993). Downy brome was controlled better with conventional tillage than with glyphosate-based zero tillage in most cases during 6 years of continuous winter wheat, winter wheat–fallow and winter wheat–spring canola cropping systems (Blackshaw, 1994b). However, the level of control was not sufficient to permit continuous winter wheat production. Vogel (1994) observed that the four reduced tillage systems quickly developed a severe perennial weed problem, whereas conventional mouldboard ploughing, when done immediately after crop harvest, gave the best control. On the other hand, during an 18-year period in a winter wheat–sorghum–fallow rotation, no-tillage plots treated with atrazine after winter wheat had higher crop yields and amount of residue on the soil surface and lower weed yields compared with the tilled plots (Wicks et al., 1988). The use of no-tillage or the presence of rye mulch usually reduced sicklepod growth, but the combination offered no advantage over either factor alone (Shilling et al., 1995). The weed populations (Table 12.6) under conventional (one fall and two spring cultivations), reduced (one spring cultivation) and zero (spring herbicide application) tillage did not show a consistent trend in favour of any tillage system and were found to vary more due to the year than the tillage system (Arshad et al., 1994, 1995). The yields of spring barley, canola and wheat under the reduced tillage were generally higher compared with both the conventional and zero tillage systems, and this system was considered to be agronomically as well as environmentally desirable.

Table 12.6   Total population density (Plants m^{-2} of 10 most common weeds under the three tillage systems during 3 years of continuous barley, canola and wheat

Year	Conventional tillage	Reduced tillage	Zero tillage
Barley			
1989	22.7	37.1	42.0
1990	4.0	8.2	10.2
1991	96.8	136.5	159.8
Canola			
1989	21.1	41.1	48.8
1990	12.5	9.2	15.8
1991	220.2	239.2	340.5
Wheat			
1989	16.2	15.3	42.2
1990	9.0	3.4	6.9
1991	184.1	111.7	98.2

Source: modified from Arshad et al. 1994 and Arshad et al. 1995.

Because reduced tillage systems are usually accompanied by increased reliance on herbicides for weed control, the specific weeds favoured in these systems depend a great deal on the specific herbicides (Froud-Williams et al., 1981) and the weed flora may shift toward a higher frequency of perennial weeds (Gill and Arshad, 1995). However, the predictions that conservation tillage systems would lead to a decrease in annual weeds and an increase in perennial weeds were not realised in the study by Blackshaw et al. (1994). Dandelion and perennial sowthistle densities did not increase dramatically in reduced-tillage treatments, and a mixed response to the tillage treatments was noted with annual weeds. Flixweed. field pennycress, wild buckwheat and common lamb's-quarters were often less prevalent in zero tillage, but downy brome, redroot pigweed and Russian thistle increased in minimum and zero tillage systems. Gill and Arshad (1995) observed that the relative contributions to the number of species observed and the population density of the weed flora were greater for broadleaf species under conventional tillage, and for the perennial species under the zero tillage system, with reduced tillage showing results between the two (Table 12.7). The tillage effect on the frequency, density and relative abundance of weed populations was found to be species-specific, and there was no consistent increase in the weed population during a 5-year period under any of the three tillage systems. In a review of weed management in north and south America, Moyer et al. (1994) have shown that some weed species are likely to increase whereas others are likely to decrease with a change from conventional to conservation tillage systems, and have also indicated that the change is greatly influenced by the climate, cropping system and agronomic practices used.

There are claims that herbicide use has increased in conservation compared with conventional tillage. This view is supported by economic analyses of the wheat–fallow cropping system (Zentner et al., 1988), a continuous wheat cropping system (Malhi et al., 1988) and the conservation tillage systems in Argentina (Hensen and Zeljkovich, 1982). However, there are a number of reports which are contrary to the above observations. In a review of pest management in conservation tillage systems, Fawcett (1987) found either

Table 12.7 Five-year (1989–1993) mean of the number of species and population density of broadleaf and perennial weeds under conventional (CT), reduced (RT) and zero (ZT) tillage systems, averaged across barley, canola and wheat crops

Species types	Number of species			Population density (plants m^{-2})		
	CT	RT	ZT	CT	RT	ZT
Broadleaf (total = 15 species)	3.9	4.6	4.6	72	96	117
Perennial (total = 6 species)	0.5	0.9	1.7	0.8	2.1	7.7

Source: modified from Gill and Arshad, 1995.

no significant increase or small increases in herbicide use with conservation tillage. Calculations by Zentner et al. (1988) indicated that there was little increase in the cost of herbicide inputs for continuous winter wheat in conservation compared with conventional tillage systems. In conservation tillage techniques used in southern Brazil, herbicide inputs are similar to those in conventional tillage, except for one additional herbicide application for weed control at seeding (Peeten, 1990).

There is no consistent trend that herbicide application prior to seeding is more effective than cultivation in controlling weeds prior to seeding, or in reducing weed germination in the following crop. Herbicide application prior to seeding can be replaced with one or two operations with tillage implements such as the wide-blade cultivator, or a single pass with an air-seeder on a wide-blade cultivator, and leave sufficient plant residue cover to prevent soil erosion (Moyer et al., 1994).

Tillage in growing crops

Post-planting tillage can kill weeds and damage the crop. Post-seeding hoeing, harrowing and between-row cultivation have been tested for weed control in various crops. Gunsolus (1990) stated that mechanical weed control in corn and soybean is directly related to the timeliness of the operation. The rotary hoe was found to be effective on weeds that have germinated but not yet emerged, except for weeds that germinate from deeper than 5 cm, in no-till fields and in fields with > 20% residue cover. Inter-row cultivation was most effective for weeds which were up to 10–15 cm tall. Wilson et al. (1993) reported that crop harrowing in the fall was effective in reducing the densities of some broadleaf weed species in winter wheat without any yield reduction. Harrowing in spring had little effect if the weeds or crops were well rooted and the ground was too compact for the tines to penetrate. Research in soybean demonstrated that rotary hoeing provided 30–70% weed suppression (Buhler et al., 1992). Similarly, weed control was improved by inter-row cultivation under the ridge tillage and no-till systems (Forcella and Lindstrom, 1988; Buhler, 1992). Kirkland (1995) tested several harrowing frequencies to control wild oats in spring wheat. Two passes with diamond-tooth harrows at the 1–2-leaf stage of wild oats reduced panicle density and fresh weight in 2 of 3 years, whereas multiple harrowing reduced wheat clums, fresh weight and yield in 2 of 3 years. In corn, one pass with a harrow was as effective as herbicide treatment in controlling green foxtail, but did not control wild oats, which can germinate and emerge even from 8 cm below the soil surface (Moyer and Dryden, 1979). The preceding observations indicate that inter-row crop cultivation may improve weed control. However, it has been also indicated that an acceptable level of weed control cannot be achieved within crop cultivation alone (Snipes et al., 1984; Holm and Foster, 1990; Snipes and Mueller, 1992).

Some researchers have compared tillage or herbicide application to their combinations for weed control in growing crops. Gebhardt (1981) reported that neither herbicide nor cultivation alone were able to maintain optimal weed control and soybean yield, whereas satisfactory results were obtained when both were employed in various combinations. Hamill et al. (1995) compared combinations of metribuzin rates and cultivation for soybean production in southeastern Ontario. One cultivation combined with a herbicide application at half the recommended rate not only provided effective weed control, but also maintained reliable economic return, whereas neither of them alone gave satisfactory results. The success of a given weed management system appeared to depend on the environment. Weed control was more critical in years of below-normal crop growth conditions than in years of favourable growth conditions. Similarly, combining reduced rates of herbicides with inter-row cultivation resulted in crop yields which were generally similar to those obtained with the full rate of herbicide (Mulder and Doll, 1993; Buhler et al., 1994). Rotary hoeing supplemented the weed control in soybean provided by row cultivation (Buhler et al., 1992) and herbicides (Peters et al., 1959). Inter-cultivation of herbicide-treated conventional tillage plots significantly reduced weed dry matter production in cotton plots (Yadava et al., 1984). In the absence of herbicide application, one dry hoeing plus three inter-cultivations were required to reduce the weed dry matter production to a similar level to that obtained when herbicides were used. However, post-planting tillage in pea and lentil did not affect weed control or crop yield compared with post-planting herbicide weed control (Boerboom and Young, 1995). According to Gebhardt (1981) and Eadie et al. (1992), if a satisfactory crop yield is maintained under reduced herbicide use and cultivation, the gross margin (total income minus cost) will increase as a result of reduced expenditure on herbicide and/or cultivation. In the case of suboptimal weed control and crop yield, the financial return may be offset by reduced management costs. Integrating post-seeding tillage into the system may also prevent the development of herbicide-resistant weed populations, reduce dependence on herbicides, and delay or prevent an increase in perennial weed species often associated with conservation tillage systems.

It appears that combinations of herbicides and tillage may reduce the costs of weed control, be useful in managing or preventing the development of herbicide-resistant weeds, and offer growers flexibility in their management programs.

COMBINATIONS OF CULTURAL PRACTICES

Simultaneous alterations of more than one crop management practice have been found to improve the overall weed control. Dunand et al. (1984) stated that combining an early date of planting, a high rate of rice seeding and permanent flooding from seeding to drainage at harvest limited red rice infestation and maximized commercial rice production. Singh and Ghosh (1992) showed that timely sowing (last week of June), two weedings (15 and 30 days after sowing), recommended fertilizer application (40 kg N, 13 kg P, 17 kg k ha^{-1}) and optimum land preparation (two ploughings, one at 15 days before and one at sowing) are essential for reducing the weed infestation and maximizing the productivity of rain-fed upland rice. Seeding rate, line sowing, row direction and plant protection alone had no significant effect on weeds or on crop productivity within the farmers' normal production practices for rice. The responses were clearly influenced by better weed suppression by the more vigorous crop resulting from a number of management factors. Their results also emphasized the need for improved varieties and organic manuring to obtain high rice yields, although without any improvement in weed suppression.

Tillage systems and crop rotations could be integrated to accelerate the decline of problem weeds in the seedbank (Ball, 1992). Minimum tillage to leave weed seeds near the surface, coupled with effective crop rotation/herbicide regimes, could accelerate the depletion of problem weed species in the seedbank. Farmers in Brazil are including more cover crops (crushed but not harvested) for weed suppression in their cropping sequences (Peeten, 1990). Herbicide use and cost is reduced by about 50% when cover crops are used in conservation tillage systems. Corn or soybean are seeded directly into the cover crop residue without a soil-applied herbicide. Post-emergence herbicides are used to control weeds if they emerge through the cover crops. Successful conservation tillage systems usually involve cropping sequences of three or more crop types, in which the ground is covered with a crop during most of the period conducive for weed growth (Moyer et al., 1994). The amount and cost of herbicides used is then similar to that for herbicides used in conventional tillage systems.

Altieri and Liebman (1988b) listed non-chemical weed-control practices and the corresponding ecological principles. Most of these techniques, such as mulching, solarization by mulching, cultivation, crop cultivars, seeding time, seeding rate, row spacing, cover or smother crops, crop rotation and intercropping, come under cultural control methods.

In organic agriculture, weeds are controlled by combinations of non-chemical methods that do not otherwise harm the soil or crop, and low levels of weeds or certain species of weeds are tolerated (Patriquin, 1988). These methods of weed control include tillage, seeding time, intercropping, crop rotation, grazing by animals, mulching, composting and soil fertility management.

CONCLUSIONS AND RECOMMENDATIONS

Cultural weed control practices are an important component of weed management, and are especially useful in crops where weed-control options are limited. They can provide substantial weed-control benefits, although these should not be considered as a panacea for all weed problems.

Intensive cropping systems help to reduce the severity of weed infestation because of vigorous crops, more shading of the soil surface and frequent weed-control measures. Their weed-control effectiveness is increased by the use of cultural techniques such as crop rotation, intercropping, targeting of inputs to the crop, timeliness of planting and establishment of a uniform crop using appropriate plant densities and arrangements.

Despite spectacular successes with herbicides, it is increasingly felt that total reliance on herbicides to control all weeds is undesirable and may be economically unfeasible in the long run, although herbicides may be necessary to meet some weed-control requirements. Therefore, combinations of various weed-control techniques is the most feasible approach, especially for conservation production systems, since no single method should be considered or relied on by itself as a remedy for weed management. Each control method will provide some measure of weed inhibition, and the sum total of the methods used should result in dependable and adequate control of the weed flora present in a given situation. By combining appropriate cultural, mechanical, biological, ecological and chemical methods, an environment should be created that is detrimental to weeds and favourable to the crop, which might need a lesser degree of both mechanical tillage and herbicides. This approach to controlling weeds will require greater management skill from the grower, but it should also provide more effective, dependable and economic weed control. Herbicide resistance and the herbicide-tolerant species are less likely to develop in systems that combine as many effective weed-control strategies as possible. Adoption of conservation tillage systems may also be more successful if growers plan crop management carefully and use herbicides and tillage judiciously.

There is a need to understand the underlying mechanisms through which diversification of cropping systems in time and space enhances agroecosystem performance, including weed management. This will facilitate the integration of cultural practices for weed suppression into agronomic practices and utilize resources most effectively.

Researchers need to continue to develop economical and effective integrated control systems for the numerous agroecosystems which exist worldwide. More information is needed concerning the effects of cultural practices on factors that influence production, dormancy and longevity of seeds, as well as weed seedling emergence, mortality mechanisms for weed plants and seeds, relative resource consumption by crops and weeds, and crop–weed allelopathic interactions. How do slow-release nutrients differ from quick-release synthetic sources in terms of their effects on crop–weed interactions? Isolation and improvement of crop rotation, intercropping and weed-suppressive crop elements or combinations of elements may be especially important for weed control. The identification and breeding of crop cultivars that are both agronomically productive and weed-suppressive should be a priority. there must be continued attention to the study of both weed population dynamics and crop–weed interactions as influenced by management practices. Addressing these areas of research can result in significant advances in the design and improvement of effective weed-suppressive practices with minimal negative effects.

REFERENCES

Abraham, C.T. and Singh, S.P.; (1984) Weed management in sorghum–legume intercropping systems. *Journal of Agricultural Science, Cambridge* **103**, 103–115.

Akobundu, I.O. (1980) Weed control strategies for multiple cropping systems of the humid and subhumid tropics. In Akobundu, I.O. (Ed.) *Weeds and their Control in the Humid and Subhumid Tropics.* IITA, Ibadan, pp. 80–101.

Akobundu, I.O. (1982) The role of conservation tillage in weed management in the advancing countries. In *Improving Weed Management.* Proceedings of the FAO/IWSS Expert Consultation on Improving Weed Management in Developing Countries. Plant Production and Protection Paper No. 44, FAO, Rome pp. 23–38.

Aldrich, R.J. (1984) *Weed-Crop Ecology: Principles in Weed Management.* Breton, North Scituate, MA.

Ali, M. (1988) Weed suppressing ability and productivity of short duration legumes intercropped with pigeonpea under rainfed conditions. *Tropical Pest Management* **34**, 384–387.

Alsaadawi, I.S., Al-uqaili, J.K., Alrubeaa, A.J. and Al-Hadithy, S.M. (1986) Allelopathic suppression of weeds and nitrification by selected cultivars of *Sorghum bicolor* (L.) Moench. *Journal of Chemical Ecology* **12**, 200–206.

Altieri, M.A. and Liebman, M. (1988a) *Weed Management in Agroecosystems: Ecological Approaches.* CRC Press, Boca Raton, FL. 354 pp.

Altieri, M.A. and Liebman, M. (1988b) Weed management: Ecological guidelines. In Altieri, M.A.

and Liebman, M. (Eds) *Weed Management in Agroecosystems: Ecological Approaches*. CRC Press, Boca Raton, FL, pp. 331–337.

Appleby, A.P., Olsen, P.D. and Colbert, D.R. (1976) Winter wheat yield reduction from interference by Italian ryegrass. *Agronomy Journal* **68**, 463–466.

Appleton, B.L. and Derr, J.F. (1987) The wide word of geotextiles. *Ground Maintenance* **23**, 42–48.

Arshad, M.A., Schintzer, M., Angers, D.A. and Ripmeester, J.A. (1990) Effects of till vs. no-till on the quality of soil organic matter. *Soil Biology and Biochemistry* **22**, 595–599.

Arshad, M.A., Gill, K.S. and Coy, G.R. (1994) Wheat yield and weed population as influenced by three tillage systems on a clay soil in a temperate continental climate. *Soil Tillage Research* **28**, 227–238.

Arshad, M.A., Gill, K.S. and Coy, G.R. (1995) Barley, canola, and weed growth with decreasing tillage in a cold, semiarid climate. *Agronomy Journal* **87**, 49–55.

Ashworth, S. and Harrison, H. (1983) Evaluation of mulches for use in the home garden. *HortScience* **18**, 180–182.

Austenson, H.M., Wenhardt, A. and White, W.J. (1970) Effect of summer-fallowing and rotation on yield of wheat, barley and flax. *Canadian Journal of Plant Science* **50**, 659–666.

Ayeni, A.O., Duke, W.B. and Akobundu, I.O. (1984) Weed interference in maize, cowpea and maize/cowpea intercrop in a subhumid tropical environment. I. Influence of cropping season. *Weed Research* **24**, 269–279.

Ball, D.A. (1992) Weed seedbank response to tillage, herbicides, and crop rotation sequence. *Weed Science* **40**, 654–659.

Balyan, R.S., Malik, R.K., Panwar, R.S. and Singh, S. (1991) Competitive ability of winter wheat cultivars with wild oat (*Avena ludoviciana*). *Weed Science* **39**, 154–158.

Banks, P.A. (1990) Southeast. In Donald, W.W. (Ed.) *Systems of Weed Control in Wheat in North America*. Weed Science Society of America, Champaign, IL, pp. 182–190.

Bantilan, R.T., Palada, M. and Harwood, R.R. (1974) Integrated weed management. I. Key factors affecting weed crop balance. *Philippine Weed Science Bulletin* **1**, 14–36.

Bell, C.E. (1993) Soil solarization: Weed control using solar energy. *Proceedings of the California Weed Conference*, Fremont, CA Vol. 45, pp. 4–7.

Berkowitz, A.R. (1988) Competition for resources in weed–crop mixtures. In *Weed Management in Agroecosystems: Ecological approaches*. Altieri, M.A. and Liebman, M. (Eds) CRC Press, Boca Raton, FL, pp. 89–119.

Billeaud, L.A. and Zajicek, J.M. (1989) Influences of mulches on weed control, soil pH, soil nitrogen content, and growth of *Ligustrum japonicum*. *Journal of Environmental Horticulture* **7**, 155–157.

Birzins, P.J. (1983) Evaluation of six weed control treatments in an interior spruce seed orchard. *Tree Planters' Notes* **34**(1), 10–12.

Blackshaw, R.E. (1993) Safflower (*Carthamus tinctorius*) density and row spacing effects on competition with green foxtail (*Setaria viridis*). *Weed Science* **41**, 403–408.

Blackshaw, R.E. (1994a) Differential competitive ability of winter wheat cultivars against downy brome. *Agronomy Journal* **86**, 649–654.

Blackshaw, R.E. (1994b) Rotation affects downy brome (*Bromus tectorum*) in winter wheat (*Triticum aestivum*). *Weed Technology* **8**, 728–732.

Blackshaw, R.E. and Lindwall, C.W. (1995) Management systems for conservation fallow on the southern Canadian prairies. *Canadian Journal of Soil Science* **75**, 93–99.

Blackshaw, R.E., Stobbe, E.H. and Sturko, A.R.W. (1981) Effect of seeding dates and densities of green foxtail (*Setaria virdis*) on the growth and productivity of spring wheat (*Triticum aestivum*). *Weed Science* **29**, 212–217.

Blackshaw, R.E., Larney, F.O., Lindwall, C.W. and Kozub, G.C. (1994) Crop rotation and tillage effects on weed populations on the semi-arid Canadian prairies. *Weed Technology* **8**, 231–237.

Boerboom, C.M. and Young, F.L. (1995) Effect of postplant tillage and crop density on broadleaf weed control in dry pea (*Pisum sativum*) and lentil (*Lens culinaris*). *Weed Technology* **9**, 99–106.

Brandt, S.A. (1989) Zerotill vs. conventional tillage with two rotations: Crop production over the last 10 years *Proceedings Soils and Crops Workshop*, University of Saskatchewan, Saskatoon, SK, Canada pp. 330–338.

Budelman, A. (1988) The performance of the leaf mulches of *Leucaena leucocephala*, *Fleminga macrophylla* and *Gliricidia sepium* in weed control. *Agroforestry Systems* **6**, 137–145.

Buhler, D.D. (1992) Population dynamics and control of annual weeds in corn (*Zea mays*) as influenced by tillage systems. *Weed Science* **40**, 241–248.

Buhler, D.D., Gunsolus, J.G. and Ralston, D.F. (1992) Integrated weed management techniques to reduce herbicide inputs in soybean. *Agronomy Journal* **84**, 973–978.

Buhler, D.D., Doll, J.D., Proost, R.T. and Visocky, M.R. (1994) Interrow cultivation to reduce herbicide use in corn following alfalfa without tillage. *Agronomy Journal* **86**, 66–72.

Burnside, O.C. (1979) Soybean (*Glycine max*) growth as affected by weed removal, cultivar, and row spacing. *Weed Science* **27**, 562–565.

Burnside, O.C. and Colville, W.L. (1964) Soybean and weed yields as affected by irrigation, row spacing, tillage, and amiben. *Weeds* **12**, 109–112.

Byodston, R.A. and Al-Khatib, K. (1994) Brassica green manure crops suppress weeds. *Western Society Weed Science Society Proceedings* **47**, 24–27.

Callaway, M.B. (1993) A compendium of crop varietal tolerance to weeds. *Journal of Sustainable Agriculture* **7**, 169–180.

Campbell, B.C. (1988) The effects of plant growth regulators and herbicides on host plant quality to insects. In Heinrichs, E.A. (Ed.) *Plant Stress–Insect Interactions*. Wiley-Interscience, New York, pp. 206–247.

Campbell, C.A., Zentner, R.P., Janzen, H.H. and Bowren, K.E. (1990) *Crop Rotation Studies on the Canadian Prairies*. Publication 1841/E, Research Branch Agriculture Canada, Ottawa, 13 pp.

Campbell-Loyd, R. (1986) Mulches are often misunderstood and misused. *Landscape Design* **163**, 75.

Cardina, J., Reginer, E. and Harrison, K. (1991) Long-term effects on seed banks in three Ohio soils. *Weed Science* **39**, 186–194.

Carlson, H.L. and Hill, J.E. (1985) Wild oat (*Avena fatua*) competition with spring wheat: Effects of nitrogen fertilization. *Weed Science* **34**, 29–33.

Carsky, R.J., Singh, L. and Ndikawa, R. (1994) Suppression of *Striga hermonthica* on sorghum using a cowpea intercrop. *Experimental Agriculture* **30**, 349–358.

Carson, A.G. (1989) Effect of intercropping sorghum and groundnuts on density of *Striga hermonthica* in The Gambia. *Tropical Pest Management* **35**, 130–132.

Challaiah, R.E., Burnside, O.C., Wicks, G.A. and Johnson, V.A. (1986) Competition between winter wheat (*Triticum aestivum*) cultivars and downy brome (*Bromus tectorum*). *Weed Science* **34**, 689–693.

Chancellor, R.J. (1985) Changes in the weed flora of an arable field cultivated for 20 years. *Journal of Applied Ecology* **22**, 491–501.

Cochran, V.L., Morrow, L.A. and Schirman, R. D. (1990) The effect of N placement on grass weeds and winter wheat in three tillage systems. *Soil Tillage Research* **18**, 347–355.

Cook, R.J. (1986) Interrelationships of plant health and the sustainability of agriculture with special reference to plant diseases. *American Journal of Alternative Agriculture* **1**, 19–24.

Crutchfield, D.A., Wicks, G.A. and Burnside, O.C. (1986) Effect of winter wheat (*Triticum aestivum*) straw mulch level on weed control. *Weed Science* **34**, 110–114.

Davis, R.J. (1984) Weed control for amenity trees on man-made sites. *Aspects of Applied Biology* **5**, 55–64.

de Almeida, F.S. (1985) Effect of some winter crop mulches on soil weed infestation. In *Proceedsings of the 1985 British Crop Protection Conference on Weeds*. Croydon, UK Vol. 2, pp. 621–630.

de Almeida, F.S. (1990) Alolpatia. *Ciencia Hoje* **11**, 39–45.

Derksen, D.A., Lafond, G.P., Swanton, C.J., Thomas, A.G. and Loepky, H.A. (1993) The impact of agronomic practices on weed communities: Tillage systems. *Weed Science* **41**, 409–417.

Derksen, D.A., Thomas, A.G., Lafond, G.P., Swanton, C.J. and Loeppky, H.A. (1994) The influence of agronomic practices on weed communities: Fallow within tillage systems. *Weed Science* **42**, 184–194.

Derr, J.F. and Appleton, B.L. (1989) Weed control with landscape fabrics. *Journal of Environmental Horticulture* **7**, 129–133.

Diebert, E.J. (1989) Soybean cultivar response to reduced tillage in northern dryland areas. *Agronomy Journal* **81**, 672–676.

Dilatush, T.R. (1988) More on black plastic mulch for shrubs. *Hortideas* **5**, 3–4.

Donald, W.W. (Ed.) (1990) *Systems of Weed Control in Wheat in North America*. Monograph Series of the Weed Science Society of America, No. 6. Weed Science Society of America, Champaign, IL, 488 pp.

Donald, W.W. and Nalewaja, J.D. (1990) Northern Great Plains. In Donald, W.W. (Ed.) *Systems of Weed Control in North America*. Weed Science Society of America, Champaign, ILL, pp. 90–126.

Dotzenko, A.D., Ozkan, M. and Stover, K.R. (1969) Influence of cropping sequence, nitrogen fertilizer and herbicides on weed seed populations in sugar beet fields. *Agronomy Journal* 61, 34–37.

Dunand, R., Baker, J. and Dilly, Jr., R. (1984) Cultural practices for red rice suppression. *Annual Progress Report, Louisiana Agricultural Experiment Station* 76, 158–164.

Duranti, A. and Cuocolo, L. (1989) Chemical weed control and mulching in onion (*Allium cepa* L.) and garlic (*Allium sativum* L.) *Advances in Horticultural Science* 3, 7–12.

Dyke, G.V. and Barnard, A.J. (1976) Suppression of couchgrass by Italian ryegrass and broad red clover undersown in barley and field beans. *Journal of Agricultural Science, Cambridge* 87, 123–126.

Eadie, A.G,., Swanton, C.J., Shaw, J.E. and Anderson, G.W. (1992) Banded herbicide application and cultivation in a modified no-till corn (*Zea mays*) system. *Weed Technology* 6, 535–542.

Egley, G.H. (1983) Weed seed and seedling reductions by soil solarization with transparent polyethylene sheets. *Weed Science* 31, 404–409.

Einheling, F.A. and Leather, G.R. (1988) Potentials for exploiting allelopathy to enhance crop production. *Journal of Chemical Ecology* 14, 1829–1844.

Einheling, F.A. and Rasmussen, J.A. (1989) Prior cropping with grain sorghum inhibits weeds. *Journal of Chemical Ecology* 15, 951–960.

Elmore, C.L. and Tafoya, S.M. (1993) Water savings and weed control with mulches and plastics. *Proceedings of the California Weed Conference*, Fremont, CA Vol. 45, pp. 147–154.

Erbach, D.C. and Lovely, W.G. (1975) Effects of plant residue on herbicide performance in no-tillage corn. *Weed Science* 23, 512–515.

Evans, R.M., Thill, D.C., Tapia, L., Shafii, B. and Lish, J.M. (1991) Wild oat (*Avena fatua*) and spring barley (*Hordeum vulgare*) density affect spring barley grain yield. *Weed Technology* 5, 33–39.

Facelli, J.M. and Pickett, S.T. (1991) Plant litter: Its dynamics and effects on plant community structure. *Botanical Review* 57, 2–32.

Fawcett, R.S. (1987) Overview of pest management for conservation tillage systems. In Logan, T.J., Davidson, J.M., Baker, J.L. and Overcash, M.R. (Eds) *Effect of Conservation Tillage on Groundwater Quality: Nitrates and Pesticides*. Lewis, Boca Raton, FL. pp. 19–37.

Fawcett, R.S. and Slife, F.W. (1978) Effects of field application of nitrate on weed seed germination and dormancy, *Weed Science* 26, 594–596.

Fay, P.K. and Duke, W.B. (1977) An assessment of allelopathic potential in *Avena* germ plasm. *Weed Science* 25, 224–228.

Fenster, C.R. and Wicks, G.A. (1982) Fallow systems for winter wheat in western Nebraska. *Agronomy Journal* 74, 9–13.

Fleck, N.G., Machado, C.M.N. and de Souza, R.S. (1984) Eficiencia da consosrciacao de culturas no conrole de plantas daninhas. *Pesquisa Agropecuaria Brasileira* 19, 591–598.

Forcella, F. and Lindstrom, M.J. (1988) Weed seed populations in ridge and conventional tillage. *Weed Science* 36, 500–503.

Forcella, F., Eradat-Oskoui, K. and Wagner, S.W. (1993) Applications of weed seedbank ecology to low-input crop management. *Ecological Applications* 3, 74–83.

Foster, R.K. and Lindwall, C.W. (1986) Minimum tillage and wheat production in western Canada. In Slinkard, A.E. and Fowler, D.B. (Ed.) *Wheat Production in Canada—A Review*. University of Saskatchewan, Saskatoon, pp. 354–366.

Froud-Williams, R.J. (1988) Changes in weed flora with different tillage and agronomic management systems. In Altieri, M.A. and Liebman, M. (Eds) *Weed Management in Agroecosystems: Ecological approaches*. CRC Press, Boca Raton, FL, pp. 213–236.

Froud-Williams, R.J., Chancellor, R.J. and Drennan, D.S.H. (1981) Potential changes in weed floras associated with reduced-cultivation systems for cereal production in temperate regions. *Weed Research* 21, 99–109.

Froud-Williams, R.J., Drennan, D.S.H. and Chancellor, R.J. (1983) Influence of cultivation regime on weed floras of arable cropping systems. *Journal of Applied Ecology* 20, 187–197.

Garrity, D.P., Movillon, M. and Moody, K. (1992) Differential weed suppression ability in upland rice cultivars. *Agronomy Journal* **84**, 586–591.

Gartner, J.B. (1978) Using bark and wood chips as a mulch for shrubs and evergreens. *American Nurseryman* **147**, 9, 53, 55.

Beghardt, M.R. (1981) Pre-emergence herbicides and cultivations for soybeans (*Glycine max*). *Weed Science* **29**, 165–168.

Gill, K.S. and Arshad, M.A. (1995) Weed flora in the early growth period of spring crops under conventional, reduced, and zero tillage systems on a clay soil in northern Alberta, Canada. *Soil Tillage Research* **33**, 65–79.

Gill, K.S. and Lungu, O.I.M. (1988) Evaluation of tillage systems for *Zea mays* grown on a compacted Oxic Paleustalf. In Unger, P.W., Sneed, T.V., Jordan, W.R. and Jensen, R. (Eds) *Challenges in Dryland Agriculture—A Global Perspective*. Proceedings of the International Conference on Dryland Farming, 15–19 August 1988, Amarillo Bushland, TX, pp. 559–561.

Gill, K.S., Arshad, M.A., Chivunda, B.K., Phiri, B. and Gumbo, M. (1992) Influence of residue mulch, tillage and cultural practices on weed mass and corn yield from three field experiments. *Soil Tillage Research* **24**, 211–223.

Gim'enez, A., Sawchik, J., Cibils, R. and Augsburger, H. (1991) *Control de Gramilla en un Sistema con Labranza Reducida*. Jornada Cultvos de verano, Instituto Nacional de Investigaci'on Agropecuaria, La Estanzuela, Colonia, Uruguay, pp. 85–92.

Glen, S. and Anderson, N.G. (1993) Hemp dogbane (*Apocynum cannabinum*) and wild blackberry (*Rubus allegheniensis*) control in no-tillage corn (*Zea mays*). *Weed Technology* **7**, 47–51.

Gliessman, S.R. (1983) Allelopathic interactions in crop weed mixtures: Applications for weed management. *Journal of Chemical Ecology* **9**, 991–999.

Gunsolus, J.L. (1990) Mechanical and cultural weed control in corn and soybeans. *American Journal of Alternative Agriculture* **5**, 114–119.

Haggar, J.P. and Beer, J.W. (1993) Effect on maize growth on the interaction between increased nitrogen availability and competition with trees in alley cropping. *Agroforestry Systems* **21**, 239–249.

Hakansson, S. (1982) Multiplication, growth, and persistence of perennial weeds. In Holzner, W. and Numata, M. (Eds) *Biology and Ecology of Weeds*. W. Junk, The Hague, pp. 123–135.

Hall, M.H., Curran, W.S., Werner, E.L. and Marshall, L.E. (1995) Evaluation of weed control practices during spring and summer alfalfa establishment. *Journal of Production Agriculture* **8**, 360–365.

Hamill, A.S., Zhang, J. and Swanton, C. (1995) Reducing herbicide use for weed control in soybean (*Glycine max*) grown in two soil types in southeastern Ontario. *Canadian Journal of Plant Science* **75**, 283–292.

Hart, R.D. (1980) A natural ecosystem analog approach to the design of a successful crop system for tropical forest environments. *Biotropica* **12** (Suppl.), 73–82.

Hartl, W. (1989) Influence of undersown clovers on weeds and on the yield of winter wheat in organic farming. *Agriculture, Ecosystems and Environment* **27**, 389–396.

Hensen, O.M. and Zeljkovich, V.J. (1982) Investigacion en labranza reducida en Argentina. In Caballero, H.D. and Diaz, R. (Eds) *Semiario labranza reducida en el conosur*. Instituto Interamericano de Cooperation para la Agricultera/Centro de Investigaciones Agricolas Alberto Boerger, Colonia, Uruguay, pp. 1–5.

Holm, F.A. and Foster, R.K. (1990) *Non-Chemical Weed Control. Transition to Organic Agriculture*. Conference Division of Extension, University of Saskatchewan, Saskatoon, SK, pp. 12–141.

Horowitz, M., Regev, Y. and Herzlinger, G. (1983) Solarization for weed control. *Weed Science* **31**, 170–179.

Hume, L., Tessier, S. and Dyck, F.B. (1991) Tillage and rotation influences on weed communities in wheat (*Triticum aestivum*) in south-western Saskatchewan. *Canadian Journal of Plant Science* **71**, 783–789.

Igbozurike, M.V. (1971) Ecological balance in tropical agriculture. *Geographical Reviews* **61**, 519–527.

Inostrosa, S.I. and Fournier, O. (1982) Effecto alelopatico de *Gliricidia sepium*. *Revista de Biologia Tropical (CR)* **30**, 35–39.

Jama, B., Betahun, A. and Ngugi, D.N. (1991) Shading effects of alley cropped *Leucaena leuco-*

cephala on weed biomass and maize yield at Mtwapa, Coast Province, Kenya. *Agroforestry Systems* **13**, 1–11.

James, K.L., Banks, P.A. and Karnok, K.J. (1988) Interference of soybean, *Glycine max*, cultivars with sicklepod, *Cassia obtusifolia*. *Weed Technology* **2**, 404–409.

Janiya, J.D. and Moody, K. (1984) Use of *Azolla* to suppress weeds in transplanted rice. *Tropical Pest Management* **30**, 1–6.

Jenkinson, D.S. (1991) The Rothamsted long-term experiments: Are they still of use. *Agronomy Journal* **83**, 2–10.

Johnson, III, W.C. and Mullinix Jr., B.G. (1995) Weed management in peanut using stale seedbed techniques. *Weed Science* **43**, 293–297.

Jordan, N. (1992) Differential interference between soybean (*Glycine max*) varieties and common cocklebur (*Xanthium struarium*): A path analysis. *Weed Science* **40**, 614–620.

Katan, J. (1981) Solar heating (solarization) of soil for control of soilborne pests. *Annual Review of Phytopathology* **19**, 211–236.

Kempen, H.M., Agamalian, H., Elmore, C. and Cudney, D. (1991) Rotation techniques for control of weeds *Proceedings of the California Weed Conference*, Fremont, CA Vol. 43, pp. 70–77.

Kirkland, K.J. (1995) Frequency of post-emergence harrowing affects wild oat control and spring wheat yield. *Canadian Journal of Plant Science* **75**, 163–165.

Klein, R.R. and Miller, D.A. (1980) Allelopathy and its role in agriculture. *Communications in Soil Science and Plant Analysis* **11**, 43–56.

Koscelny, J.A., Peeper, T.F., Solie, J.B. and Solomon, S.G. (1990) Effect of wheat (*Triticum aestivum*) row spacing, seeding rate, and cultivar on yield loss from cheat (*Bromus secalinus* L.). *Weed Technology* **4**, 487–492.

Kwon, S.L., Smith Jr., R.J. and Talbert, R.E. (1991) Interference of red rice (*Oryza sativa*) densities in rice (*O. sativa*). *Weed Science* **39**, 169–174.

Lanini, W.T., Orloff, S.B., Vargas, R.N., Marble, V.L., Orr, J. and Grattan, S.R. (1988) The effect of an oat companion crop on weeds in seedling alfalfa. *Proceedings of the California Weed Conference*, Fremont, CA Vol. 40, pp. 79–89.

Lanini, W.T., Pittenger, D.R., Graves, W.L., Munoz, F. and Agamalian, H.S. (1989) Subclovers as living mulches for managing weeds in vegetables. *California Agriculture* November–December, 25–27.

Larney, F.J., Lindwall, C.W., Izaurralde, R.C. and Moulin, A.P. (1994) Tillage systems for water conservation on the Canadian prairies. In Carter, M.R. (Ed.) *Conservation Tillage in Temperate Agroecosystems*. CRC Press, Boca Raton, FL, pp. 305–328.

Larson, W.E. and Hanway, J.J. (1977) Corn production. In Sprague, G.F. (Ed.) *Corn and Corn Improvement*. American Society of Agronomy, Madison, WI, pp. 625–669.

Lawson, H.M. (1982) Competition between weeds and vining peas grown at a range of population densities: Effects on the crop. *Weed Research* **22**, 27–38.

Leather, G.R. (1983) Sunflowers (*Helianthus annus*) are allelopathic to weeds. *Weed Science* **31**, 37–42.

Lennartsson, E.K.M. (1990) The use of surface mulches to clear grass pasture and control weeds in organic horticultural systems. *British Crop Protection Council Monographs* **45**, 187–192.

Leys, A.R., Armor, R.L., Barnett, A.G. and Plater, B. (1990) Evaluation of herbicides for control of summer-growing weeds on fallows in southern-eastern Australia. *Australian Journal of Experimental Agriculture* **30**, 271–279.

Liebman, M. (1988) Ecological suppression of weeds in intercropping systems: A review. In Altieri, M.A. and Liebman, M. (Eds) *Weed Management in Agroecosystems: Ecological Approaches*. CRC Press, Boca Raton, FL, pp. 197–212.

Liebman, M. (1989) Effects of nitrogen fertilizer, irrigation, and crop genotype on canopy relations and yields of an intercrop/weed mixture. *Field Crops Research* **22**, 83–100.

Liebman, M. and Dyck, E. (1993) Crop rotation and intercropping strategies for weed management. *Ecological Applications* **3**, 92–122.

Liebman, M. and Janke, R. (1990) Sustainable weed management practices. In Francis, C.A, Flora, C.B. and King, L.D. (Eds) *Sustainable Agriculture in Temperate Zones*. Wiley, New York, pp. 111–143.

Lindwall, C.W. (1980) Zero and minimum tillage, *Agrologist* **9**, 910.

Lindwall, C.W. and Anderson, D.T. (1981) Agronomic evaluation of minimum tillage systems for summer fallow in southern Alberta. *Canadian Journal of Plant Science* **61**, 247–253.

Maguire, J.D. and Overland, A. (1959) *Laboratory Germination of Seeds of Weedy and Native Plants.* Circular 349, Washington Agricultural Experiment Station, Pullman, 15 pp.

Malhi, S.S., Mumey, G., O'Sullivan, P.A. and Harker, K.N. (1988) An economic comparison of barley production under zero and conventional tillage. *Soil Tillage Research* **11**, 159–166.

Martin, C.A., Ponder, H.G. and Gilliam, C.H. (1991a) Evaluation of landscape fabrics in suppressing growth of weed species *Journal of Environmental Horticulture* **9**, 38–40.

Martin, M.A., Schreiber, M.M., Riepe, J.R. and Bahr, J.R. (1991b) The economics of alternate tillage systems, crop rotations, and herbicide use on three representative east-central corn belt farms. *Weed Science* **39**, 299–307.

Masiunas, J.B., Weston, L.A. and Weller, S.C. (1995) The impact of cover crops on weed population in a tomato cropping system. *Weed Science* **43**, 318–323.

McLean, M., Kobyashi, K. and deFrank, J. (1987) Weed control with various mulching and herbicide methods in a new lime orchard. *HortScience* **22**, 151.

Mills, J.A. and Witt, W.W. (1989) Effect of tillage systems on the efficacy and phytotoxicity of imazaquin and impazethapyr in soybean (*Glycine max*). *Weed Science* **37**, 233–238.

Minotti, P.L. (1991) Role of crop interference in limiting losses from weeds. In Pimentel, D. (Ed.) *CRC Handbook of Pest Management in Agriculture II.* 2nd edn. CRC Press, Boca Raton, FL, pp. 358–368.

Mohlar, C.L. and Liebman, M. (1987) Weed productivity and composition in sole crops and intercrops of barley and field pea. *Journal of Applied Ecology* **24**, 685–699.

Moody, C.L. and Shetty, S.V.R. (1981) Weed management in intercropping systems. In *Proceedings of the International Workshop on Intercropping,* 10–13 January 1979, Hyderabad. International Crops Research Institute for the Semi-arid Tropics (ICRISAT), Patencheru, India pp. 229–237.

Moseley, C.M. and Hagood Jr., E.S. (1990) Reducing herbicide inputs when establishing no-till soybeans (*Glycine max*). *Weed Technology* **4**, 14–19.

Moyer, J.R. (1985) Effect of weed control and a companion crop on alfalfa and sainfoin establishment, yields and nutrient composition. *Canadian Journal of Plant Science* **65**, 107–116.

Moyer, J.R. (1995) Effect of cover crops on weed populations. *Proceedings of the 42 Annual Meeting of the Canadian Pest Management Society,* 11 July 1995, Ottawa, Ontario, p. 11.

Moyer, J.R. and Dryden, R.D. (1979) Wild oats, green foxtail, and broad-leaved weeds: Control and effect on corn yield at Brandon, Manitoba, *Canadian Journal of Plant Science* **59**, 383–389.

Moyer, J.R., Owenya, Z.J. and Kibuwa, S.P. (1989) Weed population and agronomic practices at wheat farms on the Hanang plains in Tanzania. *Tropical Pest Management* **35**, 26–29.

Moyer, J.R., Richards, K.W. and Schaalje, G.B. (1991) Effect of plant density and herbicide application on alfalfa seed and weed yields. *Canadian Journal of Plant Science* **71**, 481–489.

Moyer, J.R., Roman, E.S., Lindwall, C.W. and Blackshaw, R.E. (1994) Weed management in conservation tillage systems for wheat production in north and south America. *Crop Protection* **13**, 243–259.

Moyer, J.R., Cole, D.E., Maurice, D.C. and Darwent, A.L. (1995) Companion crop, herbicide and weed effects on establishment and yields of alfalfa–bromegrass mixture. *Canadian Journal of Plant Science* **75**, 121–127.

Mulder, T.A. and Doll, J.D. (1993) Integrating reduced herbicide use with mechanical weeding in corn (*Zea mays*). *Weed Technology* **7**, 382–389.

Nelson, W.A., Kahn, B.A. and Roberts, B.W. (1991) Screening cover crops for use in conservation tillage systems for vegetables following spring plowing. *HortScience* **26**, 860–862.

Nielsen, J.G. (1982) *History of the Tanzania–Canada Wheat Programme 1967–1982.* Report to the Department of Supply and Services, Government of Canada, Ottawa.

Obiefuna, J.C. (1989) Biological weed control in plantains (*Musa* AAB) with egusi melon (*Colocynthis citrulus* L.). *Biological Agriculture and Horticulture* **6**, 221–227.

Okafor, L.I. and De Datta, S.K. (1976) Competition between upland rice and purple nutsedge for nitrogen, moisture, and light. *Weed Science* **24**, 43–46.

Olasantan, F.O., Lucas, E.O. and Ezumah, H.C. (1994) Effects of intercropping and fertilizer application on weed control and performance of cassava and maize. *Field Crops Research* **39**, 63–69.

Orwick, P.L. and Schreiber, M.M. (1979) Interference of redroot pigweed (*Amaranthus retroflexus*) and robust foxtail (*Setaria viridis* var. *robusta-alba* or var. *robusta-perpurea*) in soybeans (*Glycine max*). *Weed Science* **27**, 665–674.

Patriquin, D.G. (1988) Weed control in organic farming systems. In Altieri, M.A. and Liebman, M. (Eds) *Weed Management in Agroecosystems: Ecological Approaches.* CRC Press, Boca Raton, FL, pp. 303–317.

Patriquin, D.G., Baines, D., Lewis, J. and MacDougall, A. (1988) Aphid infestation of faba beans on an organic farm in relation to weeds, intercrops, and added nitrogen. *Agriculture, Ecosystems and Environment* **20**, 279–288.

Pawlowski, F. and Wesolowski, M. (1980) Plonowanie i zachwaszczenie roslin w plodozmianach o roznym udziale zboz na glebie lessowej. *Zeszyty Naukowe Akademii Rolniczo-Technicznej w Olsztynie, Rolnictwo* **29**, 91–100.

Peeper, T.F. and Wiese, A.F. (1990) Southern great Plains. In Donald, W.W. (Ed.) *Systems of Weed Control in Wheat in North America.* Weed Science Society of America, Champaign, ILL, pp. 158–181.

Peeten, H. (1990) Fifteen years of conservation tillage in Paran'a. International Workshop on Conservation Tillage Systems, Canadian International Development Agency/Empresa Brasileira de Pesquisa Agropecu'aria–Centro Nacional de Pesquisa Agropecu'aria, Passo Fundo, Rio Grande do Sul, pp. 28–31.

Peters, E.J. (1990) The midwest and northeast. In Donald, W.W. (Ed.) *Systems of Weed Control in Wheat in North America.* Weed Science Society of America, Champaign, ILL, pp. 191–199.

Peters, E.J., Klingman, D.L. and Larson, R.E. (1959) Rotary hoeing in combination with herbicides and other cultivations for weed control in soybeans. *Weeds* **7**, 449–458.

Pimentel, D. (1971) *Ecological Effects of Pesticides on Non-Target Species.* US Government Printing Office, Washington, DC.

Pimentel, D. (1994) Insect population responses to environmental stress and pollutants. *Environmental Review* **2**, 1–15.

Potter, C.J. (1988) An evaluation of weed control techniques for three establishment. *Aspects of Applied Biology* **16**, 337–346.

Powell, M.A., Bilderback, T.E. and Skroch, W.A. (1987) Landscape mulch evaluation. *Proceedings of the Southern Nurserymens Association Research Conference* Nashville, TN Vol. 32, pp. 345–346.

Putman, A.R. (1994) Phytotoxicity of plant residues. In Unger, P.W. (Ed.) *Managing Agricultural Residues.* Lewis, Boca Raton, FL, pp. 285–314.

Putman, A.R. and DeFrank, J. (1983) Use of phytotoxic plant residues for selective weed control. *Crop Protection* **2**, 173–181.

Putman, A.R., DeFrank, J. and Barnes, J.P. (1983) Exploitation of allelopathy for weed control in annual and perennial cropping systems. *Journal of Chemical Ecology* **9**, 1001–1010.

Quasem, J.R. and Hill, T.A. (1989) On difficulties with allelopathy methodology. *Weed Research* **29**, 345–347.

Ramakrishnan, P.S. and Kumar, R. (1976) Adaptive responses of an alkaline soil population of *Cynodon dactylon* L. to N–P–K nutrition. *Journal of Ecology* **64**, 187–193.

Rao, M.R. and Shetty, S.V.R. (1976) Some biological aspects of intercropping systems on crop–weed balance. *Indian Journal of Weed Science* **8**, 32–34.

Rasmussen, P.E. (1995) Effects of fertilizer and stubble burning on downy brome competition in winter wheat. *Communications in Soil Science and Plant Analysis* **26**, 951–960.

Regnier, E.E. and Janke, R.R. (1990) Evolving strategies for managing weeds. In Edwards, C.A., Lal, R., Madden, P., Miller, R.H. and House, G. (Eds) *Sustainable Agriculture Systems.* Soil and Water Conservation Society, Ankeny, IA, pp. 174–202.

Rice, E.L. (1984) *Allelopathy.* 2nd edn. Academic Press, New York, 422 pp.

Rippin, M., Haggar, J.P., Kass, D. and Kopke, U. (1994) Alley cropping and mulching with *Erythrina poeppigiana* (Walp.) O.F. Cook and *Gliricidia sepium* (Jacq.) walp.: Effects on maize/weed competition. *Agroforestry Systems* **25**, 119–134.

Roberts, H.A. and Feast, P.M. (1973) Emergence and longevity of seeds of annual weeds in cultivated and undisturbed soil. *Journal of Applied Ecology* **10**, 133–143.

Robertson, W.K., Lundy, H.W., Prine, G.M. and Currey, W.L. (1976) Planting corn in sod and small grain residue with minimum tillage. *Agronomy Journal* **68**, 271–274.

Robinson, D.W. (1986) Mulches and herbicides in ornamental plantings. *HortScience* **23**, 547–552.

Roman, E.S. (1990) Effect of cover crops on the development of weeds. International Workshop on Conservation Tillage Systems. Canadian International Development Agency/Empresa Brasileira de Pesquisa Agropecu'aria–Centro Nacional de Pesquisa Agropecu'aria, Passo Fundo, Rio Grande do Sul, pp. 258–262.

Rose, S.J., Burnside, O.C., Specht, J.E., and Swisher, B.A. (1984) Competition and allelopathy between soybeans and weeds. Agron. J. 76: 523–528.

Rubin, B. and Benjamin, A. (1983) Solar heating of the soil: Effect on weed control and on soil-incorporated herbicides. *Weed Science* **31**, 819–825.

Rubin, R. and Benjamin, A. (1984) Solar heating of the soil: Involvement of environmental factors in the weed control process. *Weed Science* **32**, 138–142.

Schonbeck, M. (1988) Cover cropping and green manuring on small farms in New England and New York: An information survey. Research Report No. 10, New Alchemy Institute, East Palmouth, MA. Cited in Schonbeck, M., Browne, J., Deziel, G. and DeGregorio, R. (1991) *Biological Agriculture and Horticulture* **8**, 123–143.

Schonbeck, M., Browne, J., Deziel, G. and DeGregorio, R. (1991) Comparison of weed biomass and flora in four cover crops and a subsequent lettuce crop on three New England organic farms. *Biological Agriculture and Horticulture* **8**, 123–143.

Schonbeck, M., Herbert, S., DeGregorio, R., Mangam, F., Guillard, K., Sideman, E., Herbst, J. and Jaye, R. (1993) Cover cropping systems for brassicas in northeastern United States. I. Cover crop and vegetable yields, nutrients and soil conditions. *Journal of Sustainable Agriculture* **3**, 105–132.

Schreiber, M.M. (1992) Influence of tillage, crop rotation, and weed management on giant foxtail (*Setaria faberi*) population dynamics and corn yield. *Weed Science* **40**, 645–653.

Scott, R.K. and Wilcockson, S.J. (1976) Weed biology and growth of sugar beet. *Annals of Applied Biology* **83**, 331–335.

Sengupta, K., Bhattacharya, K.K. and Chatterjee, B.N. (1985) Intercropping upland rice with blackgram (*Vigna mungo*). *Journal of Agricultural Science, Cambridge* **104**, 217–221.

Senseman, S.A., Bozsa, R.C. and Oliver, L.R. (1988) Influence of hairy vetch on weed control and soybean production. *Southern Weed Science Society Proceedings* **41**, 48.

Shetty, S.V.R. and Krantz, B.A. (1980) Weed research at ICRISAT (International Crops Research Institute for the Semi-Arid Tropics). *Weed Science* **28**, 451–454.

Shetty, S.V.R. and Rao, A.N. 1981. Weed management studies in sorghum/pigeonpea and pearlmillet/groundnut intercrop systems—some observations. In *Proceedings of the International Workshop on Intercropping.* 10–13 January 1979, Hyderabad. ICRISAT, Patencheru, India, pp. 238–248.

Shilling, D.G., Liebel, R.A. and Worshman, A.D. (1985) Rye (*Secale cereal*) and wheat (*Triticum aestivum* L.) mulch: The suppression of certain broadleaf weeds and the isolation and identification of phytotoxins. In Thompson, A.C. (Ed.) *The Chemistry of Allelopathy. Biochemical Interactions among Plants.* American Chemical Society, Washington, DC, pp. 243–271.

Shilling, D.G., Worsham, A.D. and Danehower, D.A. (1986) Influences of mulch, tillage, and diphenamid on weed control, yield, and quality in no-till flue-cured tobacco (*Nicotiana tabacum*). *Weed Science* **34**, 738–744.

Shilling, D.G., Brecke, B.J., Hiebsch, C. and MacDonald, G. (1995) Effect of soybean (*Glycine max*) cultivar, tillage, and rye (*Secale cereale*) mulch on sicklepod (*Senna obtusifolia*). *Weed Technology* **9**, 339–342.

Shurtleff, J.L. and Coble, H.D. (1985) The interaction of soybean (*Glycine max*) and five weed species in the greenhouse. *Weed Science* **33**, 669–672.

Singh, R.S. and Ghosh, D.C. (1992) Effect of cultural practices on weed management in rainfed upland rice. *Tropical Pest Management* **38**, 119–121.

Smeda, R.J. and Putman, A.R. (1988) Cover crop suppression of weeds and influence on strawberry yields. *HortScience* **23**, 132–134

Smika, D.E. (1990) Fallow management practices for wheat production in the Central Great Plains. *Agronomy Journal* **82**, 319–323.

Snipes, C.E. and Mueller, T.C. (1992) Cotton (*Gossypium hirsutum*) yield response to mechanical and chemical weed control systems. *Weed Science* **40**, 249–254.

Snipes, C.E., Walker, R.H., Whitwell, T., Buchanan, G.A., McGuire, J.A. and Martin, N.R. (1984) Efficacy and economics of weed control methods in cotton (*Gossypium hirsutum*). *Weed Science* **32**, 95–100.

Soria, J., Bajan, R., Pinchinat, A.M., Paez, G., Mateo, N., Moreno, R., Fargas, J. and Forsythe, W. (1975) Investigacion sobre sistemas de produccion agricola para el pequeno agricultor del tropico. *Turrialba* **25**, 283–293.

Staniforth, D.W. (1962) Responses of soybean varieties to weed competition. *Agronomy Journal* **54**, 11–13.

Stevens, C., Khan, V.A., Okoronkwo, T., Tang, A.Y., Wilson, M.A. and Lu, J. (1990) Soil solarization and dacthal: Influence on weeds, growth, and root microflora of Collards. *HortScience* **25**, 1260–1262.

Teasdale, J.R. and Frank, J.R. (1983) Effect of row spacing on weed competition with snap beans (*Phaseolus vulgaris* L.). *Weed Science* **31**, 81–85.

Teasdale, J.R., Beste, C.E. and Potts, W.E. (1991) Response of weeds to tillage and cover crop residue. *Weed Science* **39**, 195–199.

Thurston, J.M. (1962) The effect of competition from cereal crops on the germination and growth of *Avena fatua* in a naturally infested field. *Weed Research* **2**, 192–207.

Unamma, R.P.A., Ene, L.S.O., Odurukwe, S.O. and Enyinnia, T. (1986) Integrated weed management for cassava intercropped with maize. *Weed Research* **26**, 9–17.

Van Heemst, H.D.J. (1985) The influence of weed competition on crop yield. *Agricultural Systems* **18**, 81–93.

Vogel, H. (1994) Weeds in single-crop conservation farming in Zimbabwe. *Soil Tillage Research* **31**, 169–185.

Walker, R.H. and Buchanan, G.A. (1982) Crop manipulation in integrated weed management systems. *Weed Science* **30** (Suppl.), 17–24.

Walker, R.H., Patterson, M.G., Hauser, E., Isenhour, D.J., Todd, J.W. and Buchanan, G.A. (1984) Effects of insecticide, weed-free period, and row spacing on soybean (*Glycine max*) and sicklepod (*Casia obtusifolia*) growth. *Weed Science* **32**, 702–706.

Warnes, D.D. and Anderson, R.N. (1984) Decline of wild mustard (*Brassica kaber*) seeds in soil under various cultural and chemical practices. *Weed Science* **32**, 214–217.

Wax, L.M., Kogan, M., Kuhlman, D.E., Lim, S.M., Schoper, J.B., Burkhardt, L.F. and McKibben, G.E. (1983) Effect of tillage, crop rotation, and level of pest management on crop yields and pest populations. *Update 83. 48th Annual Report*. Dixon Springs Agricultural Centre, Urbana–Champaign, IL, pp. 35–40.

White, R.H., Worsham, A.D. and Blum, U. (1989) Allelopathic potential of legume debris and aqueous extracts. *Weed Science* **37**, 674–679.

Wicks, G.A., Ramsel, R.E., Nordquist, P.T., Schmidt, J.W. and Challaiah, R.E. (1986) Impact of wheat cultivars on establishment and suppression of summer annual weeds. *Agronomy Journal* **78**, 59–62.

Wicks, G.A., Smika, D.E. and Hergert, G.W. (1988) Long term effect on no-tillage in a winter wheat (*Triticum aestivum*)–sorghum (*Sorghum bicolor*)–fallow rotation. *Weed Science* **36**, 384–393.

Wiese, A.F. (Ed.) (1985) *Weed Control in Limited-Tillage Systems*. Weed Science Society of America, Monograph No. 2, Champaign, IL, 297 pp.

Williams, E.D. (1972) Growth of *Agropyron repens* seedlings in cereals and field beans. *Proceedings of the 11th British Weed Control Conference*, Nottingham, UK pp. 32–37.

Wilson, B.J. and Phipps, P.A. (1985) A long-term experiment of tillage, rotation and herbicide use for the control of *A. fatua* in cereals. *Proceedings of the 1985 British Crop Protection Conference—Weeds*, Croydon, UK pp. 693–700.

Wilson, B.J., Wright, K.J. and Butler, R.C. (1993) The effect of different frequencies of harrowing in the autumn or spring on winter wheat and on the control of *Stellaria media* (L.) vill., *Gallium aparine* L. and *Brassica napus* L. *Weed Research* **33**, 501–506.

Wilson, R.G. Jr. (1979) Germination and seedling development of Canada thistle (*Cirsium arvense*). *Weed Science* **27**, 146–151.

Worsham, A.D. (1989) Current and potential techniques using allelopathy as an aid in weed management. In Chou, C.H. and Waller, G.R. (Eds) *Photochemical Ecology: Allelochemicals, Mycotoxins and Insect Pherones, and Allomones*. Institute of Botany, Academia Sinica, Taipei,

Taiwan, Monograph Series No. 9, pp. 275–291.

Yadava, B.D., Pahuja, S.S., Nandwal, A.S. and Hooda, I.S. (1984) Cultural and chemical control of weeds in cotton (*Gossypium hirsutum* L.). *Indian Journal of Agricultural Research* **18**, 215–220.

Yamoah, C.F., Agboola, A.A. and Mulongoy, K. (1986) Decomposition, nitrogen release and weed control by pruning of selected alley-cropping shrubs. *Agroforestry Systems* **4**, 239–246.

Zemanek, J., Mikulka, J., Ludva, L. and Ludvova, A. (1985) The effect of long-term application of herbicides on weed infestation and crop yields at the research station Hnevceves, *Annals of the Research Institute for Crop Production Prague–Ruzyne* **23**, 99–118.

Zentner, R.P., Lindwall, C.W. and Carefoot, J.M. (1988) Economics of rotations and tillage systems for winter wheat production in southern Alberta. *Canadian Farm Economics* **22**, 3–13.

Zimdahl, R.L. (1980) *Weed-Crop Competition: A Review.* International Plant Protection Centre, Oregon State University, Corvallis, OR, 196 pp.

Zimdahl, R.L. (1990) The effect of weeds on wheat. In Donald, W.W. (Ed.) *Systems of Weed Control in Wheat in North America.* Weed Science Society of America, Champaign, ILL, pp. 11–32.

Zuofa, K. and Tariah, N.M. (1992) Effects of weed control methods on maize yields and net income of small-holder farmers, Nigeria. *Tropical Agriculture (Trinidad)* **69**, 167–170.

APPENDIX A

Scientific names of the plants and herbicides arranged alphabetically by their common names

Common name	Scientific name
Plants	
Alfalfa	*Medicago sativa* L.
Annual ryegrass or Italian ryegrass	*Lolium multiflorum* Lam.
Arrowleaf clover	*Trifolium vesiculosum* Savi
Arrowleaf sida	*Sida rhombifloria* L.
Axonopus	*Axonopus compressus* (SW) Beauv.
Azolla	*Azolla pinnata* R. Br.
Bahiagrass	*Paspalum notatum* Fluegge.
Barley	*Hordeum vulgare* L.
Bean or snap bean	*Phaseolus vulgaris* L.
Bermudagrass	*Cynodon dactylon* (L.) Pers.
Black oats	*Avena strigosa* Schreb.
Black medic	*Medicago lupulina* L.
Blackgram or urdbean	*Vigno mungo* (L.) Hepper
Brome	*Bromus* spp.
Buckwheat	*Fagopyron esculentum* Moench
Canada Thistle	*Cirsium arvense* (L.) Scop.
Canola	*Brassica campestris* L.
Carrot	*Daucus carota* var. *sativusa* L.
Cassava	*Manihot esculenta* Crantz
Centro	*Centrosema pubescens* Benth.
Cheat	*Bromus secalinus* L.
Chickling vetch	*Lathyrus sativus* L.
Clover	*Trifolium* spp.
Cocoa shadetree	*Gliricidia sepium* (Jacq.)
Collard green	*Brassica oleracea acephala* L.
Common blackjack	Bidens pilosa L.
Common lamb's-quarters	*Chenopodium album* L.
Common purslane	*Portulaca oleracea* L.
Common vetch	*Vicia sativa* L.
Corn or maize	*Zea mays* L.

Appendix A (*cont.*)

Common name	Scientific name
Cotton	*Gossypium hiysutum* L.
Cowpea	*Vigna unguiculata* (L.) Walp.
Crabgrass	*Digitaria* spp.
Crimson clover	*Trifolium incarnatum* L.
Cucumber	*Cucumis sativa* L.
Dandelion	*Taraxacum officinale* Weber in Wiggwer
Desmodium	*Desmodium heterophyllum* (Willd.) DC.
	Desmodium ovalifolium Wall. ex Merr.
	Desmodium triflorum (L.) DC.
Downy brome	*Bromus tectorum* L.
Egusi melon	*Citrullus lanatus* (Thunb.) Matsum. & Nakai = *Colocynthis citrullus* L.
Faba bean	*Vicia faba* L.
Field bindweed	*Convolvulus arvensis* L.
Field pennycress	*Thlaspi arvense* L.
Flax	*Linum usitatissimum* L.
Flixweed	*Descurainia sophia* (L.) Webb.
Flower morning glory	*Jacquemontia tamnifolia*
French beans	*Phaseolus vulgaris* L. ssp. *nanus* (L.)
Garlic	*Allium sativum* L.
Giant foxtail	*Setaria faberii* (L.) Beauv.
Green foxtail	*Setaria viridis* (L.) Beauv.
Hairy vetch	*Vicia villosa* Roth
Hairy nightshade	*Solanum sarachoides* Sendt.
Horse tamarind	*Leucaena leucocephela*
Indigofera	*Indigofera spicata* Forsk.
Japanese millet	*Ethinochloa crtusgalli* var. *frumentacea* Roxb. W.F. Wight
Jimsonweed	*Datura stramonium*
Johnsongrass	*Sorghum halepense*
Knapweed	*Centaurea* spp.
Kochia	*Kochia scoparia* (L.) Schrad.
Large crabgrass	*Diginaria sanguinalis* (L.) Scoop.
Lentil	*Lens culinaris* Medik.
Lettuce	*Lactuca sativa* L.
Longspine sandbur	*Cenchrus panciflorus* Benth.
Lupine	*Lupinus albus* L.
Marmeladgrass	*Brachiaria plantaginea* (L.) Hitchc.
Mucuna	*Stizolobium deeringianum* Bort = *Mucuna pruriens* (L.) DC. var. *utilis* (Wight) Burck.
Mungbean	*Vigna radiata* (L.) R. Wilczek
Nutsedge	*Cyperus* spp.
Oat	*Avena sativa* L.
Onion	*Allium cepa*
Pea	*Pisum sativum* L.
Peanut or groundnut	*Arachis hypogaea* L.
Pearl millet	*Pennisetum americanum* (L.) K. Schum.
Perennial ryegrass	*Lolium perenne* L.
Perennial sowthistle	*Sonchus arvensis* L.
Persian clover	*Trifolium resupinatum* L.
Pigeonpea	*Cajanus cajun* (L.) Huth
Pigweed	*Amaranthus* spp.

Appendix A   (*cont.*)

Common name	Scientific name
Pitted morning glory	*Ipomoea lacunosa* L.
Plantain	*Musa paradisiaca* L. (AAB group)
Potato	*Solanum tuberosum* L.
Prickly sida	*Sida spinosa* L.
Psopho	*Psophocarpus palustris* Desv.
Puncturevine	*Tribulus terrestris* L.
Purple witchweed	*Striga hermonthrica* (Del.) Benth.
Purple nutsedge	*Cyprus rotundus* L.
Quackgrass	*Agropyron repens*
Radish	*Raphanus sativus*
Rape	*Brassica napus*
Red rice	*Oryza rufipogon* Griff.
Red clover	*Trifolium pratense* L.
Redrood pigweed	*Amaranthus retroflexus* L.
Rice	*Oryza sativa* L.
Russian thistle	*Salsala pertifer* A. Nels. (= S. Kali L. var. *tennuifolia* Tausch.)
Rye (winter)	*Secale cereale* L.
Safflower	*Carthamus lanatus* L.
Sainfoin	*Onobrychis viciifolia* Scop.
Seradella	*Ornithopus sativus* Brot.
Sheperd's-purse	*Capsella bursa-pastoris* (L.) Medic.
Sicklepod	*Cassia obtusifolia*
Smooth crabgrass	*Digitaria ischaemum* (Schreb.) Muhl.
Sorghum	*Sorghum bicolor* (L.) Moench
Sorghum-sudangrass	*Sorghum arundinaceum* (Desv.) Stapf var. *sundanense* (Stapf) Hitchc.
Soybean	*Glycine max* (L.) Merr.
Strawberries	*Fragaria grandiflora* Ehrh.
Subterranean clover	*Trifolium subterraneum* L.
Sudangrass	*Soerghum sudanense* (Piper) Staff.
Sugar beet	*Beta vulgaris* L.
Sunflower	*Helianthus annuus* L.
Sweet potato	*Ipomoea batatas* (L.) Lam.
Sweet clover	*Melilotus* spp.
Sweet clover	*Melilotus sulcatus* Desf.
Sword bean	*Canavalia ensiformis* (L.) DC.
Tartary buckwheat	*Fagopyrum tartaricum* L. Gearth
Tomato	*Lycopersicon lycopersicum* (L.) Karst. ex Farw.
Triticale	*Triticale triticosecale* Wittmack
Turnip	*Brassica rapa* L.
Wheat	*Triticum aestivum* L.
White clover	*Trifolium repens* L.
White mustard	*Brassica hirta* Moench
Wild mustard	*Brassica kaber* (Dc.) L.C. Wheeler = *Sinapis arvensis* L.
Wild garlic	*Allium vineale* L.
Wild buckwheat	*Polygenum convolvulus* L.
Wild oat	*Avena fatua* L.
Wild garlic	*Allium vineale* L.
Wild groundnut	*Arachis repens* Handro
Yam	*Dioscorea* spp.

Appendix A   (*cont.*)

Common name	Scientific name
Yellow nutsedge	*Cyperus esculentus* L.
Herbicides	
Alachlor	2-Chloro-N-(2,6-diethylphenyl)-N-(methoxy-methyl) acetamide
2,4-D	(2,4-Dichlorophenoxy) acetic acid
Atrazine	6-Chloro-N-ethyl-N'-(1-methylethyl)-1,3,5-triazine-2,4-diamine
DCPA	Dimethyl 2,3,5,6-tetrachloro-1,4-benzenedicarboxylate
Dicamba	3,6-Dichloro-2-methoxybenzoic acid
Glyphosate	N-(phosphonomethyl) glycine
Metolachlor	2-Chloro-N-(2-ethyl-6-methylphenyl)-N-(2-methoxy-1-methylethyl) acetamide
Metribuzin	4-Amino-6-(1,1-dimethylethyl)-3-(methylthio)-1,2,4-triazin-5(4H)-one
Simazine	6-Chloro-N,N'-diethyl-1,3,5-triazine-2,4-diamine
Trifluralin	2,6-Dinitro-N,N-dipropyl-4-(trifluoromethyl) benzenamine

CHAPTER 13

IPM Practices for Reducing Pesticide Use in US Field Crops

David J. Horn

The Ohio State University, Columbus, OH, USA

INTRODUCTION

Approximately 26 million kg of insecticides are applied annually to major field crops in the USA (Gianessi and Anderson, 1995), which constitute 78% of USA cropland (USDA, 1995a,b). Most insecticide applications to field crops are applied as a water-based or ultra-low-volume (ULV) spray from aircraft or ground equipment, and a smaller amount is applied as granules to the soil surface or mixed with the soil. Regardless of the method of application, some environmental contamination is inevitable, whether through spray drift or runoff of soil particles to which insecticides adsorb. Reductions in insecticide use on field crops would therefore have valuable positive environmental and economic effects (Whittaker et al., 1992). If insecticide use were limited to situations when it is really necessary, and if amounts of insecticide could be focused more selectively onto the target organisms, a goal of 50% reduction in insecticide usage on field crops might be reached within 10 years. This chapter investigates current patterns of insecticide use in insect pest management on field crops, and explores the potential for a 50% reduction in that use on major field crops through the use of improved application techniques and timing, alternative control systems and more effective integrated pest management (IPM). Together, corn, cotton, wheat, alfalfa, soybeans and sorghum constitute over 90% of field-crop hectareage in the USA, and a reduction of insecticide use on these six crops alone would have a significant impact on overall insecticide use. As well as the foregoing major crops, pasture and hay also have a significant hectareage in the USA, although current insecticide application (80 000 kg annually) is negligible. However, pasture is often located so close to crops such as corn, alfalfa and cotton that there is a risk of spray drift, and a reduction of insecticide use on major field crops would reduce the hazards of drift if pasture is nearby.

A decade ago, Pimentel and Levitan (1986) estimated that 93% of row crops and 10% of forage crops received pesticides, although insecticides comprised a small proportion of this amount. (Most pesticide applied to field crops is herbicide, especially to corn, soybeans and cotton under conventional tillage). One quarter of *all* insecticide used in agriculture in 1986 was applied to cotton and corn. By 1993 (the most recent year for

Techniques for Reducing Pesticide Use. Edited by D. Pimentel
© 1997 John Wiley & Sons Ltd.

Table 13.1   Insecticides use on major field crops 1986–1993. Figures are approximate because reports are summarizations from various sources

Crop	Area (ha × 10³)	Area treated (%)	Insecticide used (kg × 10⁶)		% change 1986–1993
			1986*	1993**	
Alfalfa	9 711	9	1	2	+ 100
Corn	31 262	29	14	12	− 15
Cotton	4 448	65	8	9	+ 11
Rice	1 252	18†	0.2	0.2	0
Sorghum	4 783	16	1.1	1.2	+ 10
Soybeans	23 366	1	5.0	0.5	− 90
Tobacco	314	85	1.6	1.6	0
Wheat	24 963	2	1.0	1.0	0
Other grains	4 899	3	0.1	<0.1	− 10
Pasture	251 314	<0.0001	Negligible		

* Pimentel et al. (1991).
** Gianessi and Anderson (1995).
† The figures for insecticide applied to rice for mosquito control were unavailable, but they probably increase this figure.

which reliable figures are available) there had been some (16%) reduction in insecticide use on major field crops (Table 13.1), mainly soybeans, corn and other grains. This reduction has been accomplished partly by application of some of the strategies and techniques discussed in this review, and is an encouraging trend in efforts to curb unnecessary insecticide applications. (Amounts of insecticide applied to cotton, sorghum and (especially) alfalfa have increased from 1985 to 1995). This chapter stresses insecticide reduction on the major field crops of corn, alfalfa, soybeans, cotton and wheat, although the principles apply to other field crops, as well as horticultural and other crop production systems. This is not intended to be an exhaustive review, but significant examples are emphasized to illustrate current and proposed strategies for IPM. More detailed information is available from the references to this chapter.

CONCEPTS CENTRAL TO IPM FOR FIELD CROPS

Economic injury level and economic threshold

An understanding of economic injury levels (EILs) (Stern et al., 1959; Headley, 1972) is central to an appreciation of pest management on field crops. The economic injury level is defined most simply as the amount of insect damage that can be tolerated without a loss of economic yield. Normally the direct costs of control are included in calculating economic injury levels (Headley, 1972, 1975), although the environmental impacts and social costs (from wildlife poisoning to lawsuits) resulting from insecticide use or misuse have not been considered in determining these levels. (Economic injury levels would be higher if these 'externalities' were included). Adoption of economic injury levels has moved IPM away from so-called 'preventive' (or prophylactic) insecticide treatments and toward the use of insecticides 'as needed' (prescription) only when economic injury is anticipated.

Because many non-insecticide insect management techniques (such as biological control) have a delayed impact, during which pest populations may continue to increase, an 'economic threshold' (or 'action level') is often designated, which is that a lower population density should occur earlier than the economic injury level. The economic threshold is the point at which action is taken to prevent later injury. In the application of broad-spectrum chemical insecticides, the economic threshold and economic injury level are very nearly identical because the impact of such chemicals is almost instantaneous. However, biological controls such as *Bacillus thuringiensis* or insect viruses must be ingested, and then replicate internally, while the insects continue to feed and grow. Thus, to use biological control effectively, the economic threshold must be set at a lower level and earlier than the EIL. The economic threshold is usually expressed either as a direct measurement of insect pest populations (e.g. corn borer larvae per stem (Straum, 1983), or an indirect indicator of insect impact (e.g. corn rootworm infestation ratings from 1 to 6 depending on the severity of infestation (Hill and Peters, 1971), or percentage of alfalfa tips damaged by weevil larvae).

Field crops offer favourable prospects for reduction of insecticide use because generally they have relatively high economic injury levels, meaning that low densities of many pests can be tolerated. This is because field crops often have a low value per hectare (when compared with such high-value crops as fruits, vegetables and flowers), and in most cases (e.g. field corn, soybeans, cotton, alfalfa) the crop is not destined for direct human consumption, so that the cosmetic standards so important in fresh market and processed foods are secondary considerations. (Tobacco is an exception because the crop is intended for human use). Again, toleration of low population densities of pests means that biological and cultural management techniques can be used effectively, and pest population densities can be monitored adequately in relation to economic injury levels.

Pest complexes and key pests

Most field crops, like other agricultural crops, are attacked by several, sometimes many, species of phytophagous insects and mites. It is not unusual to find 20 different species capable of causing economically significant damage to a single crop. This melange of pestiferous insects attacking a crop at once is termed a pest complex. Most often, there is one (or two) pest species within the pest complex that are particularly troublesome. Such pests are present in potentially damaging numbers consistently, season after season. They may cause damage at lower population densities than do other members of the complex. They are the pests around which most management programs are designed, and are often the primary targets for insecticide applications. These so-called 'key pests' include such insects as the European corn borer and the corn rootworm complex on field corn, the boll weevil and *Heliothis* species on cotton, the alfalfa weevil and potato leafhopper on alfalfa, and the Hessian fly, greenbug and recently the Russian wheat aphid on wheat. Effective non-insecticide management of the key pest(s) is essential for a significant reduction of insecticide applications to field crops. Historically, extensive use of chemical insecticides to control key pests on field crops (and other crops) has led to the creation of additional 'secondary' pests (e.g. spider mites on cotton and soybeans) by eliminating the activities of arthropod natural enemies that normally keep these secondary pests below damaging levels.

PAST AND CURRENT IMPACT OF INSECTICIDE USE ON MAJOR FIELD CROPS

Corn maize

The key pests of field corn in most areas of the USA are the European corn borer, *Ostrinia nubilalis*, and three beetles (*Diabrotica* spp.) whose larvae constitute the corn rootworm complex. The European corn borer was introduced accidentally into the USA in 1917 and by the 1940s had become established in most of eastern and central USA and in adjacent areas of Canada. Its larvae burrow into the stems of many wild herbaceous plants other than corn, and also infest potatoes, peppers and other cultivated plants (in which they are occasional pests). The corn rootworm complex is native to North America. Their larvae attack roots and weaken the plants, although corn plants have a high degree of physiological tolerance, producing additional roots in response to rootworm feeding. Weakened plants lean (lodge) and are difficult to harvest with machinery.

Very little insecticide was used on field corn prior to the 1940s (USDA, 1954). As organic insecticides became available and conventional tillage and mechanical harvesting were widely utilized, insecticides were routinely applied, especially to manage the European corn borer and the rootworm complex. Conventional corn production under clean cultivation continues to involve broad-spectrum chemical application for rootworm management, usually as a granular formulation applied as a preventive treatment at planting, when it is easiest to apply. The western and southern corn rootworms have shown increasing resistance to many soil insecticides (Levine and Oloumi-Sadeghi, 1991). Soil insecticides applied against corn rootworms significantly reduce populations of predatory ground beetles (Brust et al., 1986), with consequent reduced predation on rootworms and cutworms. Levine and Oloumi-Sadeghi (1991) have recommended increased efforts in checking the developing crop to detect potentially damaging corn rootworm infestations, although accurate measurement of rootworm infestations is difficult and time-consuming. The efficiency of soil insecticides can be enhanced by the addition of a larval attractant, although Hibbard et al. (1995) found that the efficacy of the insecticide is not consistent under these conditions. European corn borer management involves routine checks, with insecticides being applied only when 75% of the young plants show injury to leaf whorls. This assessment is aided by computer simulations based on rises in temperature. Mature corn is more resistant to the European corn borer, and the economic threshold is higher for the second generation. Pest management checking and monitoring programs have also been implemented for other field corn pests: Periera and Hellman (1993), for example directly counted the numbers of early-instar fall armyworm larvae per plants. The hectareage of corn produced with some sort of minimum tillage has increased steadily in recent years, with a consequent reduction in rootworm populations (Levine and Oloumi-Sadeghi, 1991).

A recent development is the insertion of a gene from *Bacillus thuringiensis* var. *kurstaki* into the genome of corn (Armstrong et al., 1995). This gene codes for the delta-endotoxin that leads to death in Lepidoptera, including the European corn borer. Transgenic plants are highly resistant to attack by the borer. Widespread acceptance of these varieties would result in a reduction of the need for insecticide applications to corn. However, if a large proportion of the crop is genetically engineered and widely planted, we might expect evolution of resistance by the European corn borer and other Lepidoptera, as has been documented for the Hessian fly and greenbug to overcoming antibiosis in wheat (Maxwell and Jennings, 1980). Wholesale adoption of transgenically resistant corn (or any other crop) should therefore be viewed with caution owing to the potential for insect resistance (Boulter, 1993; Armstrong et al., 1995).

Wheat

Wheat insect management presents an outstanding example of what can be accomplished through emphasis on IPM without significant insecticide application. Only 2% of the total USA hectareage was treated with insecticide in 1992. Wheat possesses a rather 'cooperative' pest complex, with no routinely devastating key pest, although unpredictable outbreaks of several species warrant ongoing monitoring of the crop. For most insect pests of wheat, EILs are rather high and average population densities are normally rather low. The Hessian fly (*Myaeitolia destructor*) is a key pest; its larvae feed within the stem and weaken it to the point that it will not support the ripening grain. The adult Hessian fly has a short flight season in late summer, and it is recommended that wheat planting simply be delayed until the end of the flight period of adults. A 'fly-free date' for planting wheat is widely promulgated by IPM advisers, and this adjustment of planting date has practically eliminated the need to control Hessian fly chemically. Planting of resistant varieties has helped in management of the Hessian fly, wheat stem sawfly and greenbug; wheat represents an outstanding classical example of the application of host-plant resistance to insect pest management. In some cases, effective use of resistant varieties requires monitoring. For example, solid-stemmed wheat varieties that resist wheat stem sawfly are only advantageous in outbreak years of the sawfly; otherwise the yield is lower than that of susceptible hollow-stemmed varieties (Weiss and Morrill, 1992). This requires monitoring of the sawfly population. Planting a mixture of hollow- and solid-stemmed varieties is also a viable management option (Weiss et al., 1990). Tillage and rotation (Hatchett et al., 1987) have also done much to improve overall management of the pest complex of wheat without resorting to insecticides.

The advent of a newly introduced pest, for example the Russian wheat aphid, may initially result in increased insecticide input. The Russian wheat aphid entered the USA in the mid-1980s, and is particularly difficult to manage biologically owing to its tendency to curl the stem, affording protection from predators and parasitoids (Morrison et al., 1991). There is evidence to indicate that a complex of natural enemies in Eurasia may provide partial biological control of the Russian wheat aphid there (Hopper et al., 1995), and classical biological control should be pursued. Additional research into the development of EILs in relation to statistically accurate sampling is necessary (Elberson and Johnson, 1995). Hill et al. (1993) demonstrated that judicious monitoring could result in halving chlorpyrifos applications for Russian wheat aphid with no loss of yield. Wheat varieties that resist Russian wheat aphid are also under development.

Alfalfa

In the USA in 1994, nearly 10 million hectares were planted to alfalfa; only corn, wheat and soybeans exceeded this amount. When well watered and well fertilized, most varieties of alfalfa display tolerance to phytophagous insects. Nearly all alfalfa is destined to be eaten by livestock, so that EILs are generally rather high, and low insect numbers do not normally constitute an economically important threat.

The development of alfalfa pest management is an instructive example of application of IPM principles that might result in reduced pesticide application. While management of alfalfa is largely a function of which species of the pest complex are present, a key pest in the alfalfa ecosystem has been the alfalfa weevil, *Hypera postica*, introduced into North America from Eurasia in the early part of this century.

Until 1962, damage was prevented by application of a granular cyclodiene insecticide (heptachlor) to alfalfa stubble in autumn, which provided inexpensive season-long protection by killing adult weevils at the outset of oviposition. Use of heptachlor on alfalfa was terminated after small amounts of heptachlor epoxide residue were detected in milk and meat. Since then, shorter-residual organophosphate or carbamate insecticides sprayed directly onto larvae in the plant tips reduce damage to below the EIL, but they are expensive and the timing is critical, with great care being necessary to avoid excessive insecticide residue on the crop at harvest. Application equipment must be very precisely calibrated, and the economic threshold of '75% tip damage' is difficult for growers to measure and interpret.

Overall management of alfalfa insects depends upon the integration of a variety of techniques. A biological control program has resulted in the establishment of 12 species of exotic parasitic wasps, with some success in reducing weevil densities (Kingsley et al., 1993). The naturally occurring fungus *Zoophthora phytonomi* is the key factor in reducing alfalfa weevil numbers in much of Ontario (Nordin et al., 1983). Several varieties of alfalfa which are resistant to the weevil are commercially available. The alfalfa ecosystem is subject to periodic disturbance by harvesting, and management of alfalfa weevil by adjustment of harvesting date often destroys many larvae before significant plant damage occurs.

Managing alfalfa insects is aided by the application of model-based simulations. Ruesink et al. (1980) developed a coupled insect–plant model to which hymenopteran parasitism was added. A temperature-driven computer model accurately predicts weekly weevil populations and plant development, and economic thresholds and treatment decisions vary as functions of these parameters. Censuses of weevil life-stages are input into a computer model, producing grower advisories that recommend insecticides only when necessary, i.e. when the model predicts a high probability that the EIL will be exceeded. These insecticide treatments are timed for application when natural enemies are least likely to be active in the field.

In eastern North America, the potato leafhopper (*Empoasca fabae*) is also a key pest of alfalfa. The potato leafhopper sucks sap from a wide variety of agricultural and non-crop plants, and becomes abundant on alfalfa in mid-summer, causing yellowing, stunting and loss of yield in the second and subsequent annual crops. Its damage is more insidious than that done by the weevil, for the yield of protein is reduced significantly before outward yellowing becomes visible. Biological controls are generally ineffective, despite efforts to locate suitable parasitoids. The increase in insecticide use on alfalfa since 1986 (Table 13.1) is a function of increasing concern about potato leafhopper damage. IPM programs similar to those for the alfalfa weevil have been implemented for potato leafhopper (Flinn et al., 1986), but accurate assessment of leafhopper populations in relation to economic injury remains a problem.

Where alfalfa grows throughout the year it can be harvested up to nine or ten times annually, and the pest complex in these areas differs from that of eastern and mid-western North America. The key pest is the Egyptian alfalfa weevil, *Hypera bruneipennis*, and a temperature-dependent model very similar to that used for the alfalfa weevil in the eastern USA is used to predict future densities of the Egyptian alfalfa weevil. Optimal management depends on estimating post-aestivation migration rates of adults returning to the fields, and again insecticides are used only when necessary. Avoidance of broad-spectrum insecticides further conserves natural enemies, which are usually abundant and diverse enough to control most other pests. These additional, occasional pests include several species of aphids and the alfalfa caterpillar (*Colias eurytheme*), while the potato leafhopper is rare and negligible (as a pest). Because the crop and its pests are present throughout the year, strip-harvesting is a viable option. Segments of each field are harvested alternately, so that there is always a standing crop to serve as a refuge for predators and parasitoids, including lacewings, lady beetles, damsel bugs and wasps. The microbial insecticide *Bacillus thuringiensis*, although expensive, can be applied as needed against outbreaks of the alfalfa caterpillar, thereby conserving natural enemy populations.

Cotton

Cotton production often involves significant insecticide input; in 1992, 20 million kg was applied to USA cotton. The key pests are the boll weevil (*Anthonomus grandis*), pink bollworm (*Pectinophora gossypiella*), *Heliothis* spp., *Lygus* bugs, spider mites, thrips and whiteflies, all depending on location. Generally the boll weevil is the key pest in southeastern and south central USA, whereas the pink bollworm is more of a problem on irrigated cotton in the southwestern states. In the mid-1980s, 40% of all agricultural insecticide applied in the USA was used on cotton (Pimentel and Levitan, 1986), and application cost up to $700 ha^{-1} (Reynolds et al., 1982). The boll weevil accounts for up to 94% of the insecticide applied to cotton (Slosser et al., 1994). Insecticides directed against the boll weevil have resulted in outbreaks of secondary pests, notably the *Heliothis* complex, which feeds on a wide variety of other crop plants (corn, soybeans, tobacco, etc.). Aphids, thrips and whiteflies have all become secondary pests of cotton, and the boll weevil, pink bollworm and *Heliothis* have all become resistant to insecticides in large portions of their geographic ranges.

Insecticide use on cotton can be significantly reduced through the use of integrated management techniques. Current cotton insect management practices continue a recent trend toward reduced insecticide use and applications on an as-needed basis (Frisbie et al., 1989). Regular checking has resulted in the reduction of insecticide treatments to cotton from as many as 18 annual applications to as few as two with no apparent loss of yield. Pest management checking in cotton increased by 15% from 1982 to 1989 (Ferguson, 1990). Smith et al. (1974) found that Mississippi cotton growers who monitored pest damage spent an average of $39.93 ha^{-1}, nearly half the insecticide costs of growers who used a preventive programme of weekly sprays. Adoption of monitoring in Texas reduced insecticide input from 6.14 kg ha^{-1} in 1976 to 1.68 kg ha^{-1} by 1982 (Frisbie and Adkisson, 1985) and to less than 1 kg ha^{-1} by 1995. Pest management scouts in Texas sample once or twice weekly, and recommend spraying when 15–25% of squares or bolls show feeding punctures, or when there are 40 fleahoppers per 100 terminals (Wilson et al., 1989). Arkansas scouts use an economic threshold of 1–1.5 punctured squares per foot-row (Cate, 1985). Planting date is considered when deciding whether or not to apply insecticide to cotton; Slosser et al. (1994) showed that the proportion of squares damaged by boll weevil was unaffected by insecticide use on some varieties of late-planted cotton. Pheromone traps were used on 34% of cotton acreage in 1989 to monitor boll weevil numbers and avoid unnecessary insecticide treatments (Ferguson, 1990). In California, pink bollworm densities are monitored via pheromone traps baited with hexalure; 3.5–4 bollworms per trap constitutes a population capable of damage, and the grower is advised to initiate spraying (Pfadt, 1985).

Coupled plant–insect models for the cotton ecosystem have been devised despite the challenge of unpredictability in plant growth. A temperature-dependent model developed in Texas shows that small changes in the plant are enough to reduce boll weevil densities to below the EIL (Curry et al., 1980). Gutierrez et al. (1977) and Wang et al. (1977) developed similar models for the pink bollworm in California, and other models currently exist for the boll weevil, several Lepidoptera, including pink bollworm and *Heliothis*, and for spider mites and *Lygus* bugs (Frisbie et al., 1989). 'COMAX', a comprehensive cotton management system, includes a weather-driven plant simulation model providing recommendations on irrigation, fertilization, pesticides and harvest (Lemmon, 1986). Growers using COMAX may save up to $38 ha^{-1} in production costs, a large proportion of which is insecticide cost. 'TEXCIM' is a similar management model currently used by cotton producers in Texas.

To control cotton pests biologically, three European parasitoid species have been released in the USA against the tarnished plant bug, and although their impact to date has been limited, there is promise for an increase in the future (Sterling et al., 1989). Reduced insecticide use on cotton enhances the prospects for effective biological control using parasitoids and predators. Many parasitoids and predators are known (Clausen, 1978), including specialists that feed on aphids, leafhoppers, whiteflies and spider mites, all of which are under partial biological control in untreated cotton (and other) fields (Sterling et al., 1989). Nuclear polyhedrosis virus of *Heliothis* is effective when application is timed to coincide with egg hatching.

Resistant varieties are under development; 'frego bract' cotton has a single gene that results in resistance to boll weevil, and pubescent and red varieties also resist boll weevils to some extent (Gallun and Khush, 1980). Varieties high in gossypiol concentration resist both *Lygus* and *Heliothis* spp. As in corn, the gene from *Bacillus thuringiensis* coding for delta-endotoxin has been incorporated into cotton plants (El-Zik and Thaxton, 1989; Fischhoff, 1992); similar work is underway to develop varieties resistant to boll weevil.

Other techniques to reduce insecticide use in cotton include uniform planting dates after peak boll weevil emergence in spring, and growing and early harvesting of short-season varieties to reduce infestations of boll weevil and pink bollworm (Masud et al., 1981). Frisbie and Adkisson (1985) reported that planting early-maturing varieties resulted in savings of up to $117 ha^{-1} in insecticide use. Additional control may be achieved by clean tillage of crop residues which destroys weevil-infested bolls (along with other pest insects) late in the season, reducing the over-wintering weevil population. Regional management, with cooperation among growers, has shown promise in the management of bollworms on cotton by the destruction of wild host plants (Knipling and Stadelbacher, 1983). Delayed, uniform spring planting results in avoiding boll weevil damage in much the same way as does the 'fly-free date' against Hessian flies in wheat; boll weevils simply emerge too early to develop a significant infestation (Rummel and Carroll, 1983). Area-wide eradication of the boll weevil through chemo-sterilization of males has been attempted using pheromone traps to lure male weevils to bait laced with a potent chemo-sterilant. Insecticide-treated trap crops are also used, along with pheromone trapping. Initial results have been encouraging on the periphery of the weevil's range (North and South Carolina), resulting in enhanced profit to cotton producers due to savings in insecticide costs (Carlson et al., 1989). Insecticides might be used to reduce weevil populations to levels at which the input of sterile males might be effective.

Adoption of management techniques such as these has resulted in 90% reduction in insecticide applied to cotton in Texas from 1966 to 1986 (Pimentel et al., 1991). Potentially, cotton is an outstanding example of achieving pesticide reduction through integrated management using monitoring, biological control, host-plant resistance, destruction of crop residues, adjustment of planting and harvest dates, and careful attention to model-based crop advisories to ensure that insecticide is applied only when needed and absolutely necessary.

MANAGEMENT PRACTICES FOR THE FUTURE

Several approaches, used alone or in concert, show promise for reducing insecticide use not only on field crops, but widely by IPM of many agricultural crops. A 50% insecticide reduction might readily by achieved through wider adoption of the following practices. These are discussed in what I believe to be decreasing order of potential significance for reducing insecticide use.

Improved scouting and monitoring, together with application of sophisticated and accurate crop simulations, to predict economic injury levels accurately

By monitoring key pests in the complex, insecticide application moves from 'preventive; to 'as-needed'. Insecticides should be used only as necessary, on an as-needed basis, much as prescription medicines are used. Field crops, with their relatively high economic injury levels, afford the luxury of such flexibility. The development of economic thresholds combined with statistically accurate and cost-effective sampling techniques allows the monitoring of insect populations with respect to economic damage. For instance, sequential sampling for European corn borers provides accurate timing for insecticide applications on corn (Sorenson et al., 1995). The sampling plan is modifiable according to the value and purpose of the crop; economic thresholds for the corn borer are likely to be lower on corn grown for seed rather than for feed or processing (Slosser, 1993), with the consequent use of more insecticide but timed more precisely. Significant progress is also being made on developing economic thresholds for corn rootworms (Mulock et al., 1995).

Application of insecticide at reduced rates and with more efficiency

An estimated 1% of insecticide reaches the intended target and actually kills an insect, and often the efficiency is even less than this (Crosby, 1973). Improved design and calibration of delivery equipment, together with improved formulations such as ultra-low volumes and efficient stickers and spreaders, would ensure more effective delivery of insecticides, and lower rates could be applied. Reduced insecticide rates may mean that the same area of cropland is treated, but with greater efficiency, lower cost and reduced hazard. Willson (personal communication, 1995) showed an equal kill of rootworms with more than 50% reduction of standard granular soil insecticides. Insecticide companies are reluctant to adopt these practices because of the increased uncertainty of control. It is important to monitor the spectrum of insecticide resistance in the target population. Kanga et al. (1995) present an example of field monitoring for boll weevil insecticide resistance. Application of ultra-low-volume sprays from aircraft resulted in 50% reduction on cotton (Akesson and Yates, 1984; Pimentel and Levitan, 1986). Combining attractant and insecticide has resulted in 99% reduction of rootworm insecticide use (Paul, 1989).

Improved crop rotations, including modified tillage practices, and efficient handling of crop residue

The benefits of reduced tillage to soil conservation in many field crops have been evident for a generation. Reduced tillage often results in ecosystem diversification, sometimes including the increased activity of natural enemies. The combined biotic impact of parasitoids and predators, coupled with the reduction in populations of phytophagous insects that may accompany vegetational diversification (Root, 1973; Horn, 1981, 1988), often contributes to reducing field crop pests below the EIL. Diversification may not improve insect pest management in every case; for example, Tonhasca (1994) found that intercropping soybeans with maize had no discernible impact on rates of predation by generalist herbivores, including some soybean pests, although for 3 years none of the pests of soybean exceeded economic levels. Corn and soybean rotation reduces the impact of the corn rootworm, and most cornfields in rotation need not receive a soil insecticide at planting (Levine and Oloumi-Sadeghi, 1991), although a legume cover crop may actually increase corn rootworm injury (Buntin et al., 1994), and tillage without control of grassy weeds (probably by a herbicide) might increase populations of stalk borer (Levine, 1993). If reduced tillage is combined with planting corn varieties resistant to European corn borer, insecticides can be reduced by 80% (Lockeretz et al., 1981). Corn–oat–meadow rotations in Iowa resulted in high yields with reduced input of all chemicals, including insecticides (Chase and Duffy, 1991). The impact of generalist predators can be increased through changes in cropping practices; for instance, predation of Mexican bean beetle increased in maize dicultures rather than bean monocultures (Coll and Bottrell, 1995). Brust and King (1994) found that reduced inputs of insecticides combined with crop rotation yielded higher predator populations and reduced pest numbers, including corn rootworms. In some of their plots, rootworm numbers in insecticide-treated plots were no different from those in untreated plots, probably because of effective predation. Brust and King (1994) recommend that the most effective corn management strategy is to reduce insecticide inputs and practice multi-year rotation. Populations of southern corn rootworms are reduced, and predation is increased, in no-till compared with conventional corn production (Brust and House, 1990). The same held true for cutworms (Turnock et al., 1993). Soil compaction due to agricultural equipment has been found to reduce survivorship and movement of corn rootworm larvae (Ellsbury et al., 1994), preventing infestation of corn adjacent to field previously planted to corn and thus reducing the need for soil insecticides. Alfalfa strips interplanted with cotton harbour large numbers of *Lygus* bugs and predators, and serve both as a trap crop for *Lygus* and a refuge and source for predators that help control cotton pests (Godfrey and Leigh, 1994). Alteration of planting date is the classic technique for avoiding damage due to Hessian fly and other pests. Adjustment of planting dates in cotton can result in 10% insecticide reduction (Pimentel and Shoemaker, 1974). A shift to dryland varieties in cotton could reduce insecticide use by 97% (Shaunak et al., 1982).

Improved crop varieties that resist insect attack more readily, including release of bioengineered varieties

Incorporation of the gene producing the delta endotoxin of *Bacillus thuringiensis* into crop plants has great potential for reducing insecticide use (Lambert and Peferoen, 1992). A bioengineered crop plant is an effective means of delivering an insecticidal agent while avoiding environmental contamination, because the insecticide is essentially contained within the plant and a much higher proportion reaches the target pest (Thacker, 1994). Corn containing the delta-endotoxin has shown high resistance to European corn borer in field trials (Armstrong et al., 1995) and is now commercially available. Bioengineered cotton should be available soon for more effective management of Lepidoptera. *Bacillus thuringiensis* varieties are also pathogenic in Coleoptera, opening the potential for management of boll weevil, corn rootworms and Mexican bean beetle. Wholesale reliance on bioengineered plant resistance should be viewed with caution, however, as evolution of resistance by the insects is a distinct possibility. (A few Lepidoptera have developed resistance to *Bacillus thuringiensis*, although none of the pests of field crops has yet done so.)

Conventional approaches to plant breeding should likewise be continued in an effort to produce additional resistance to pests. The outstanding successes with wheat breeding illustrate the utility of this approach. In particular, efforts in plant breeding should be applied to soybeans; an example of this application is the relationship between trichome density and the resistance of soybeans to whiteflies. Plants with more erect hairs apparently interfered with whitefly feeding (Lambert et al., 1995).

Biological control

Biological control of field crop pests can be enhanced by manipulations such as minimum tillage, rotations and other cropping practices that preserve parasitoids and predators. A renewal of effort is warranted to review some of the apparent 'failures' in the past in classical biological control programmes on field crops. Numerous exotic parasitoids and predators of field crop pests were reviewed and some were introduced without successful establishment during the early 20th century (Clausen, 1978). Many of these modest efforts at biological control were abandoned with the advent of synthetic organic insecticides. Such pests might be targets for additional biological control efforts, as both the theory and the practice of biocontrol have changed and improved in the past 60 years. Also, new parasitoids or predators may become available for release with political changes in eastern Europe and Central Asia, allowing for more open exploration and subsequent importation of biocontrol agents. If these efforts are focused on key pests, a substantial reduction in insecticide use can result.

REFERENCES

Akesson, N.B. and Yates, W.E. (1984) Physical parameters affecting aircraft spray application. In Harvey, Y.W. and Garner, J. (Eds) *Chemical and Biological Controls in Forestry*. American Chemical Society Series 238, Washington, DC.

Armstrong, C.L. and 23 co-authors (1995) Field evaluation of European corn borer control in progeny of 173 transgenic corn events expressing an insecticidal protein from *Bacillus thuringiensis*. *Crop Science* **35**, 550–557.

Boulter, D. (1993) Insect pest control by copying nature using genetically engineered crops. *Phytochemistry* **34**, 1453–1466.

Brust, G.E. and House, G.J. (1990) Effects of soil moisture, no-tillage and predators on southern corn rootworm (*Diabrotica undecimpunctata howardi*) survival in corn agroecosystems. *Agriculture Ecosystems and Environment* **31**, 199–216.

Brust, G.E. and King, L.R. (1994) Effects of crop rotation and reduced chemical inputs on pests and predators in maize agroecosystems. *Agriculture Ecosystems and Environment* **48**, 77–89.

Brust, G.E., Stinner, B.R. and McCartney, D.A. (1986) Predator activity and predation in corn agroecosystems. *Environmental Entomology* **15**, 1017–1020.

Buntin, G.D., All, J.N., McCracken, D.V. and Hargrove, W.L. (1994) Cover crop and nitrogen fertility effects on southern corn rootworm (Coleoptera: Curculionidae) damage to corn. *Journal of Economic Entomology* **87**, 1683–1688.

Carlson, G.A., Sappie, G. and Hamming, M. (1989) *Economic Returns to Boll Weevil Eradication*. Economic Research Service, USDA, AER 621. Washington, DC.

Cate, J.R. (1985) Cotton: Status and current limitations to biological control in Texas and Arkansas. In Hoy, M.A. and Herzog, D.C. (Eds) *Biological Control in Agricultural IPM Systems*. Academic Press, New York.

Chase, C. and Duffy, M. (1991) An economic comparison of conventional and reduced-chemical farming systems in Iowa. *American Journal of Alternative Agriculture* **6**, 168–173.

Clausen, C.P. (Eds) (1978) *Introduced Parasites and Predators of Arthropod Pests and Weeds: A World Review*. USDA Agricultural Handbook No. 480. Washington, DC.

Coll, M. and Bottrell, D.G. (1995) Predator-prey association in mono- and dicultures: Effect of maize and bean vegetation. *Agriculture Ecosystems and Environment* **54**, 115–125.

Crosby, D.G. (1973) The fate of pesticides in the environment. *Annual Review of Plant Physiology* **24**, 467–492.

Curry, G.L., Sharpe, P.J., DiMichele, D.W. and Cate, J.R. (1980) Towards a management model of the cotton–boll weevil ecosystem. *Journal of Environmental Management* **11**, 187–223.

Elberson, L.R. and Johnson, J.B. (1995) Population trends and comparison of sampling methods for *Diuraphis noxia* and other cereal aphids (Homoptera: Aphididae) in northern Idaho. *Environmental Entomology* **24**, 538–549.

Ellsbury, M.M., Schumacher, T.E., Gustin, R.D. and Woodson, W.D. (1994) Soil compaction effect on corn rootworm populations in maize artificially infested with eggs of western corn rootworm (Coleoptera: Chrysomelidae). *Environmental Entomology* **23**, 943–948.

El-Zik, K.M. and Thaxton, P.M. (1989) Genetic improvement for resistance to pests and stresses in cotton. In Frisbie, R.E., El-Zik, K.M. and Wilson, L.T. (Eds) *Integrated Pest Management Systems and Cotton Production*. Wiley, New York.

Ferguson, W.L. (1990) Cotton pest management practices. In *USDA Agricultural Resources Inputs Situation and Outlook*, pp. 24–27. Washington, DC.

Fischhoff, D.A. (1992) Management of lepidopteran pests with insect-resistant cotton. Beltwide Cotton Conference, pp. 751–753. Texas A&M Univ. College Station, TX.

Flinn, P.W., Taylor, R.A.J. and Hower, A.A. (1986) Predictive model for the population dynamics of the potato leafhopper, *Empoasca fabae*, on alfalfa. *Environmental Entomology* **15**, 898–904.

Frisbie, R.E. and Adkisson, P.L. (1985) IPM: Definitions and current status in US agriculture. In Hoy, M.A. and Herzog, D.C. (Eds) *Biological Control in Agricultural IPM Systems*. Academic Press, New York.

Frisbie, R.E., El-Zik, K.M. and Wilson, L.T. (Eds) (1989) *Integrated Pest Management Systems and Cotton Production*. Wiley, New York.

Gallum, R.L. and Khush, G.S. (1980) Genetic factors affecting expression and stability of resistance.

In Maxwell, F.G. and Jennings, P.R. (Eds) *Breeding Plants Resistant to Insects.* Wiley, New York.

Gianessi, L.P. and Anderson, J.E. (1995) *Pesticide Use in US Crop Production.* National Summary Report, National Center of Food and Agriculture Policy, Washington, DC.

Godfrey, L.D. and Leigh, T.F. (1994) Alfalfa harvest strategy effect on *Lygus* bug and insect predator population density: Implications for use as trap crop in cotton. *Environmental Entomology* **23**, 1106–1118.

Gutierrez, A.P., Leigh, T.F., Wang, Y. and Cave, R.D. (1977) An analysis of cotton production in California: *Lygus hesperis* injury—an evaluation. *Canadian Entomology* **109**, 1375–1386.

Hatchett, J.H., Starks, K.J. and Webster, J.A. (1987) Insect and mite pests of wheat. In Heyne, E.G. (Ed.) *Wheat and Winter Wheat Improvement.* American Society of Agronomy, Madison, WI.

Headley, J.C. (1972) Economics of agricultural pest control. *Annual Review of Entomology* **17**, 273–286.

Headley, J.C. (1975) The economics of pest management. In Metcalf, R.L. and Luckmann, W.L. (Eds) *Introduction to Insect Pest Management.* Wiley, New York.

Hibbard, B.E., Peniss, F.B., Pilcher, S.D., Schroeder, M.E., Jewett, D.K. and Bjostad, L.B. (1995) Germinating corn extracts and 6-methoxy-2-benzoxazolinone: Western corn rootworm (Coleoptera: Curculionidae) larval attractants evaluated with soil insecticides. *Journal of Economic Entomology* **88**, 716–724.

Hill, B.E., Butts, R.A. and Schaalje, G.B. (1993) Reduced rates of foliar insecticides for control of Russian wheat aphid in western Canada. *Journal of Economic Entomology* **86**, 1259–1265.

Hill, T.M. and Peters, D.C. (1971) A method of evaluating post-plant insecticide treatments for control of western corn rootworm larvae. *Journal of Economic Entomology* **64**, 764–765.

Hopper, K.R., Aidara, R., Agret, S., Cabal, J., Coutinot, D., Dabire, R., Lesieux, C., Kirk, G., Reichert, S., Tronchetti, F. and Vidal, J. (1995) Natural enemy impact on the abundance of *Diuraphis noxia* (Homoptera: Aphididae) in wheat in southern France. *Environmental Entomology* **24**, 402–408.

Horn, D.J. (1981) Effect of weedy backgrounds on colonization of collards by green peach aphid, *Myzus persicae*, and its major predators. *Environmental Entomology* **10**, 285–289.

Horn, D.J. (1988) Parasitism of cabbage aphid (*Brevicoryne brassicae*) and green peach aphid (*Myzus persicae*) (Homoptera: Aphidae) on collards in relation to weed management. *Environmental Entomology* **17**, 354–358.

Kanga, L.H.B., Plapp Jr., F.W., Wall, M.L., Karner, M.A., Huffman, R.L., Fuchs, T.W., Elzen, G.W. and Martinez-Carrillo, J.L. (1995) Monitoring tolerance to insecticides in boll weevil populations (Coleoptera: Curculionidae) from Texas, Arkansas, Oklahoma, Mississippi and Mexico. *Journal of Economic Entomology* **88**, 198–204.

Kingsley, P.C., Bryan, M.D., Day, W.H., Burger, T.L., Dysart, R.J. Schwalbe, C.P. (1993) Alfalfa weevil (Coleoptera: Curculionidae) biological control: Spreading the benefits. *Environmental Entomology* **22**, 1234–1250.

Knipling, E.F. and Stadelbacher, E.A. (1983) The rationale for areawide management of *Heliothis* populations. *Bulletin of the Entomological Society of America* **29**(4), 29–37.

Lambert, A.L., McPherson, R.M. and Espelie, K.E. (1995) Soybean host-plant resistance mechanisms that alter abundance of whiteflies (Homoptera: Aleyrodidae). *Environmental Entomology* **24**, 1381–1386.

Lambert, B. and Peferoen, M. (1992) Insecticidal promise of *Bacillus thuringiensis*. *BioScience* **42**, 112–121.

Lemmon, H. (1986) Comax: An expert system for cotton crop management. *Science* **233**, 29–33.

Levine, E. (1993) Effect of tillage practices and weed management on survival of stalk borer (Lepidoptera: Noctuidae) eggs and larvae. *Journal of Economic Entomology* **86**, 1924–1928.

Levine, E. and Oloumi-Sadeghi, H. (1991) Management of diabroticine rootworms in corn. *Annual Review of Entomology* **36**, 229–255.

Lockeretz, W., Shearer, G. and Kohl, D.H. (1981) Organic farming in the corn belt. *Science* **211**, 540–547.

Masud, S.M., Lacewell, R.D., Taylor, C.R., Benedict, J.H. and Lippke, L.A. (1981) Economic impact of integrated pest management strategies for cotton production in the coastal bend region of Texas. *Southern Journal of Agricultural Economics* 13(2:)47–52.

Maxwell, F.G. and Jennings, P.R. (Eds) (1980) *Breeding Plants Resistant to Insects.* Wiley, New

York.

Morrison, P.L., Brooks, G., Hein, G., Johnson, G., Massey, W., McBride, D., Peairs, F., Schulz, J.T. and Legg, D. (1991) *Economic Impact of the Russian Wheat Aphid in the Western United States: 1989–1990.* Great Plains Agricultural Council Publication No. 139.

Mulock, B.S., Ellis, C.R. and Whitfield, G.H. (1995) Evaluation of an oviposition tap for monitoring egg populations of *Diabrotica* spp. (Coleoptera: Chrysomelidae) in field corn. *Canadian Entomology* **127**, 839–849.

Nordin, G.L., Brown, G.C. and Millstein, J.A. (1983) Epizootic phenology of *Erynia* disease of the alfalfa weevil, *Hypera postica*, in central Kentucky. *Environmental Entomology* **12**, 1350–1355.

Paul, J. (1989) Getting tricky with rootworms. *Agrichemical Age* **33**(3), 6.

Pereira, C. and Hellman, J.L. (1993) Economic injury levels for *Spodoptera frugiperda* on silage corn in Maryland. *Journal of Economic Entomology* **86**, 1266–1270.

Pfadt, R.E. (Ed.) (1985) *Fundamentals of Applied Entomology.* 4th edn. Macmillan, New York.

Pimentel, D. and Levitan, L. (1986) Pesticides: Amounts applied and amounts reaching pests. *BioScience* **36**, 86–91.

Pimentel, D. and Shoemaker, C.A. (1974) An economic and land-use model for reducing insecticides on cotton and corn. *Environmental Entomology* **3**, 10–21.

Pimentel, D., McLaughlin, L., Zepp, A., Lakitan, B., Kraus, T., Kleinman, P., Vancini, F., Roach, J., Graap, E., Keeton, W.S. and Selig, G. (1991) Environmental and economic impacts of reducing US agricultural pesticide use. In Pimentel, D. (Ed.) *CRC Handbook of Pest Management in Agriculture.* Vol. I. 2nd edn. CRC Press, Boca Raton, FL, pp. 679–718.

Reynolds, H.T., Adkisson, P.L., Smith, R.F. and Frisbie, R.F. (1982) Cotton insect pest management. In Metcalf, R.L. and Luckmann, W.H. (Eds) *Introduction to Insect Pest Management.* Wiley, New York.

Root, R.B. (1973) Organization of a plant–arthropod association in simple and diverse habitats: The fauna of collards. *Ecological Monographs* **4**, 95–124.

Ruesink, W.G., Shoemaker, C.A., Gutierrez, A.P. and Fick, G.W. (1980) The systems approach to research and decision making for alfalfa pest control. In Huffaker, C.B. (Ed.) *New Technology of Pest Control.* Wiley, New York.

Rummel, D.R. and Carroll, S.C. (1983) Winter survival and effective emergence of boll weevil cohorts entering winter habitat at different times. *Southwest Entomology* **8**, 101–106.

Shaunak, R.K., Lacewell, R.D. and Norman, J. (1982) *Economic Implications of Alternative Cotton Production Strategies in the Lower Rio Grande Valley of Texas, 1923–1978.* Texas Agricultural Experimental Station Bulletin 1420.

Slosser, J.E. (1993) Influence of planting date and insecticide treatment on insect pest abundance and damage in dryland cotton. *Journal of Economic Entomology* **86**, 1213–1222.

Slosser, J.E., Bordovsky, D.G. and Bevers, S.J. (1994) Damage and costs associated with insect management options in irrigated cotton. *Journal of Economic Entomology* **87**, 436–455.

Smith, R.F., Adkisson, P.L., Huffaker, C.B. and Newsom, L.D. (1974) Progress achieved in the implementation of integrated control projects in the USA and tropical countries. *EPPO Bulletin* **4**, 221–239.

Sorenson, C.E., VanDuyn, J.W., Kennedy, G.G., Bradley Jr., J.R., Eckel, C.S. and Fernandez, G.C.J. (1995) Evaluation of a sequential egg mass sampling system for predicting second generation damage by European corn borer (Lepidoptera: Pyralidae) in field corn in North Carolina. *Journal of Economic Entomology* **88**, 1316–1323.

Sterling, W.L., El-Zik, K.M. and Wilson, L.T. (1989) Biological control of pest populations. In Frisbie, R.E., El Zik, K.M. and Wilson, L.T. (Eds) *Integrated Pest Management Systems and Cotton Production.* Wiley, New York.

Stern, V.M., Smith, R.F., van den Bosch, R. and Hagen, K.S. (1959) The integration of chemical and biological control of the spotted alfalfa aphid: The integrated control concept. *Hilgardia* **29**, 81–101.

Straub, R.W. (1983) Minimization of insecticide treatment for first generation European corn borer (Lepidoptera: Pyralidae) control in sweet corn. *Journal of Economic Entomology* **76**, 345–348.

Thacker, J.R.M. (1994) Transgenic crop plants and pest control. *Science Progress* **77**, 207–219.

Tonhasca, A. Jr. (1994) Response of soybean herbivores to two agronomic practices increasing agroecosystem diversity. *Agricultural Ecosystems and Environment* **48**, 57–65.

Turnock, W.J., Timlick, B. and Planiswany, P. (1993) Species abundance of cutworms (Noctuidae) and their parasitoids in conservation and conventional tillage fields. *Agriculture Ecosystems and Environmental* **45**, 213–227.

USDA (1954) *Losses in agriculture*. Agricultural Research Services Report 20–1.

USDA. *Agricultural Resources, Inputs Situation and Outlook Report*. Economic Research Services, Washington, DC.

USDA (1995a) *Agricul;tural Chemical Usage. 1994 Field Crops Summary*. USDA National Agricultural Statistics Survey, Economic Research Services.

USDA (1995b) *Agricultural Statistics*. US Government Printing Office, Washington, DC.

Wang, Y., Gutierrez, A.P., Oster, G. and Daxl, R. (1977) A population model for plant growth and development: Coupling cotton–herbivore interactions. *Canadian Entomology* **109**, 1359–1374.

Weiss, M.J. and Morill, W.L. (1992) Wheat stem sawfly (Hymenoptera: Cephidae) revisited. *American Entomology* **38**, 241–245.

Weiss, M.J., Riveland, N.R., Reitz, L.L. and Olson, R.C. (1990) Influence of resistant and susceptible cultivar blends of hard red spring wheat on wheat stem sawfly (Hymenoptera: Cephidae) damage and wheat quality parameters. *Journal of Economic Entomology* **83**, 255–259.

Whittaker, G., Lin, B.-H. and Vasavada, U. (1992) Restricting pesticide use: The impact on profitability by farm size. *USDA Agricultural Resources Inputs Situation and Outlook*, pp. 43–46.

Wilson, L.T., Sterling, W.L., Rummel, D.R. and DeVay, J.E. (1989) Quantitative sampling principles in cotton IPM. In Frisbie, R.E., El-Zik, K.M. and Wilson, L.T. (Eds) *Integrated Pest Management Systems and Cotton Production*. Wiley, New York.

CHAPTER 14

IPM Practices for Reducing Fungicide Use in Field Crops

K.L. Bailey

Agriculture and Agri-Food Canada, Saskatoon, Saskatchewan, Canada

INTRODUCTION

A strong reliance has developed recently on the use of chemicals (i.e. pesticides, fertilizers, growth regulators) to solve many farm production problems. By the early 1980s, the consequences of ever-increasing chemical use emerged with concerns about contamination of the environment, long-term health effects on producers and consumers, and the need for producers to reduce production costs (Lasley et al., 1990). The concept of sustainable agriculture, low-input farming systems and integrated pest management (IPM) were promoted because they were associated with conservation of natural resources, minimization of environmental impact, benefits to human health, contributions to world food stability, enhancement of long-term productivity and growth of rural communities. This chapter discusses IPM practices that may aid in the reduction of fungicide use in field crops. The aspects covered include perceptions and realities of IPM use, the components and tools used to implement IPM, and concluding remarks on ways to put IPM into practice.

IPM: PERCEPTIONS AND REALITIES

Definition of IPM

Integrated pest management (IPM) is a dynamic system that coordinates all known control methods to achieve an acceptable level of pest control. The basic concept sounds simple and intuitive, but the process of integrating the components is complex. IPM requires: (i) planning for the prevention or reduction of pest problems, (ii) learning pest identification, (iii) monitoring of crops, pests and local environments, (iv) determining damage thresholds that delimit acceptable levels of crop quality and yield, (v) coordinating genetic, cultural, physical, biological, behavioural and chemical control methods to optimize and synergize their effects, and (vi) evaluating and adjusting IPM to strive for improved control.

    For IPM to be successful, it must continually evolve and integrate perceptions of

Techniques for Reducing Pesticide Use. Edited by D. Pimentel

Table 14.1   Examples of average grain yield losses caused by fungal diseases on field crops grown
in Canada and the United States from 1992 to 1994

Crop	Diseases controlled	Test locations	Loss (%)	Source*
Wheat	Septoria leaf blotch, tan spot, leaf rust, powdery mildew	Indiana, New York, Kansas, Pennsylvania, Kentucky, South Carolina, Washington, Ohio, North Dakota	16	F&N Tests 48: 223–229 and 231, 233, 235, 237 F&N Tests 49: 202, 212
	Common bunt	Kansas, Montana	22	F&N Tests 48: 314, 320, 321
	Common root rot	Oregon	8	F&N Tests 48 340
Barley	Net blotch	PEI	6	PMRR 1994: 136
	Smuts	North Dakota, Idaho, Montana	15	F&N Tests 48: 290, 292, 294, 295
Corn	Leaf spots	Indiana, Virginia	32	F&N Tests 48: 217, 218
Canola	Sclerotinia stem rot	North Dakota, Alberta	16	F&N Tests 48: 25
				PMRR 1994: 104
Field pea	Ascochyta blight	Manitoba, Alberta	14	PMRR 1994: 119–121
	Powdery mildew	Manitoba	14	PMRR 1994: 122
Soybeans	Leaf spots	Tennessee	11	F&N Tests 48: 286
	Root rots	Kansas	6	F&N Tests 48: 308, 309
Dry beans	Leaf spots	North Dakota	7	F&N Tests 48: 251
	Rust	North Dakota	28	F&N Tests 48: 252, 253
	Sclerotinia	North Dakota	17	F&N Tests 48: 254
	Root rots	New York, Indiana	10	F&N Tests 48: 297, 298
Peanuts	Sclerotinia	Georgia	15	F&N Tests 48: 260–264
	Leaf spots	Georgia, Texas, Alabama	42	F&N Tests 48: 266, 267, 269 F&N Tests 49: 257
	Root rots	Texas, Virginia	22	F&N Tests 48: 304, 305

*Summarized from reports published in: Fungicide and Nematicide (F & N) Tests, American
Phytopathological Society, USA, and the Pest Management Research Report (PMRR), 1994, Agriculture
Canada, Ottawa.

acceptance from the public at large (urban and rural communities) as well as the
farming/production community. It requires an understanding of the food production
system, both biologically and socio-economically, and the willingness to move away from
reliance on only one or two control options. The objective is to achieve a balanced and
complementary approach to pest management.

Field crop losses and fungicide use in Canada and the United States

In North America, fungal diseases can cause substantial losses in both the yield and
quality of field crops. Small to moderate losses can have a large impact on national
economics because the total land-base planted to field crops is substantial. For example,
wheat is grown on approximately 13.7 million ha in Canada and produces 26.2 million
tonnes of grain (Anonymous, 1994). Significant grain shortages resulted when epidemics
of wheat stem rust caused losses averaging 2.7 million tonnes in 1916, 1927, 1935 and 1954
(Martens et al., 1984).

Table 14.2   Estimated amounts of fungicides used on field crops, fruits, and vegetables in Ontario

Crop	Area grown (thousand hectares)	Quantity of fungicide (tonnes)
Corn	1020	3.7
Small grains	908	0.5
Soybeans	285	1.7
Dry beans	67	0.9
Tobacco	43	4.2
Hay	1153	0.2
Grapes	10	43.2
Apples	8	101.0
Potatoes	18	87.3
Tomatoes	11	63.8

Adapted from Gibson, 1984.

Losses from specific diseases are difficult to estimate, since more than one disease may be present on a single plant and the complex of diseases may be different among years and regions. Credibility of estimates become an issue when the sum of the losses from several diseases assessed independently totals more than 100%. Appraisals derived from fungicide trials may overestimate actual losses encountered in an average field, but probably underestimate the impact of a major epidemic. Despite these difficulties, estimates of loss provide a general understanding of the importance of diseases on a crop. Estimated average losses due to fungal pathogens in field crops in North America in 1992–1994 are presented in Table 14.1, and in these years, fungi caused about 15% loss in seed yield. In individual reports, the losses ranged from less than 1% to greater than 50% depending on the crop, disease, location and year. For example, the disease incidence of common bunt of wheat was 61% in Kansas, but less than 3% in Montana.

In 1980 the value of fungicide sales in the world was US$2.57 billion; Canada accounted for only 1% of this total and the United States accounted for 7% (Gibson, 1984). By 1991, global fungicide sales reached US$5.03 billion (Powell and Jutsum, 1993). The market size in the United States was US$2.8 billion, and 31% of the fungicides sold were applied to small grain crops, 14% on rice, 2% on cotton and 2% on maize. The greatest proportion of fungicides (39%) was used on fruit, vegetables and vines.

In Canada, 9% of agricultural pesticides purchased in 1977 were fungicides (Gibson, 1984), but by 1989, only 3.3% of all pest control products sold were fungicides (Billett, 1992). It is estimated that fungicide seed treatments are applied annually to 10% of the cereal acreage and 80% of canola acreage in the Canadian prairies. The quantity of fungicides used on field crops in Ontario is small compared with those used on fruit and vegetable crops (Table 14.2). Climatic conditions, favourable for specific diseases, determine the quantity of fungicides applied yearly for all crops, but approximately US$30 million is sold annually in Canada.

Reducing pesticide use

IPM can lead to a reduction in the amount of pesticide used on field crops. Burrows (1983) tested this hypothesis on cotton grown in California. Consultants offered a service to growers on a fee-per-acre basis by giving specific field recommendations on strategic

Table 14.3   Percentage of farmers using selected practices to reduce pesticide use in Iowa

Practice	Not used	Limited use	Moderate use	Heavy use
Crop rotation	2	11	43	44
Soil testing	6	20	47	27
Cultivation	2	15	60	23
Sow legumes	11	33	39	17
IPM	41	34	22	3
Examine fields for pests	10	29	46	15
Employ scouting service	85	10	4	1
Non-conventional products	78	17	4	1

Adapted from Lasley et al., 1990.

spraying schedules, applications of biological control agents, availability of resistant plant varieties, and changing cultural practices. By using the IPM recommendations to achieve an acceptable level of pest control, pesticide use on cotton decreased by 31–47%.

Before IPM practices are adopted by farmers, they must feel that there is a need to reduce their dependency on chemicals. Changes in farm management practices may result from decisions based on economic viewpoints, such as the need to reduce costly inputs, or from personal viewpoints influenced by environmental or health risks.

Lasley et al. (1990) conducted a survey in Iowa to determine what practices farmers were using to reduce pesticide use and to investigate their opinions on the need for lower input costs in their farming practice. Crop rotation, soil testing and mechanical cultivation were the most widely used practices to reduce chemical use (Table 14.3). Most farmers believed that agriculture relies too heavily on chemicals for pest control, but this did not translate into a shift towards using lower inputs. Farmers also believed that lower-input farming would decrease health and safety concerns. There was no relationship to age or education in their opinions, but operators of larger farms used more practices to reduce chemical use although they had less supportive opinions on low-input farming. The study concluded that more effort is needed to gain farmers' acceptance of existing recommended practices to reduce pesticide use.

Some producers may not readily adopt IPM because they believe that pesticide use will be less risky and reduce the probability of crop failure. However, if it is perceived that reducing the use of chemical pesticides will result in less profitability, then it is unlikely that IPM practices will be adopted. If, in fact, lower profits do result, then that is a measure of the 'incentives' required to motivate more farmers to adopt IPM. There is little information on costs related to reducing chemical use and employing alternative strategies, but some studies do show that IPM may be more profitable than using chemical control alone.

Pimentel et al. (1991) estimated that fungicide use on cotton and tobacco in the United States could be reduced by two-thirds through IPM techniques. The cost per hectare of alternative disease control measures for cotton and tobacco was US$10 as compared with fungicide treatment costs of US$17 for cotton and US$30 for tobacco. Fungicide use on other field crops, such as rice, peanuts and sugarbeets, could be reduced by approximately one-third. The cost per hectare of substituting IPM practices in place of fungicides only was US$3 vs. US$121 for rice, US$10 vs. US$60 for peanuts, and US$5 vs. US$21 for sugarbeet. In all cases, the cost of employing alternative strategies was dramatically less than chemical control.

However, the impact of reducing fungicide use on wheat grown in the United States was relatively small. Without the use of fungicides, reductions in grain yield ranged from 2 to 4% in the Central and Northern Plains, and from 6 to 7% in the Southern Plains and southwest (Knutson et al., 1990). Removing fungicides had little effect on wheat production costs; the total economic cost per bushel increased by 2% when fungicides were eliminated from production and disease control was maintained primarily by crop rotation.

Whittaker et al. (1992) designed a simulation model to predict declines in profits by restricting the dollars spent on pesticides used on farms producing corn, soybeans and wheat in the Lake-States corn belt production area in the United States. In this study, the term pesticide included herbicides, insecticides, fungicides, nematicides, defoliants, fumigants, growth regulators, frost protectants and biological control agents. Data on production costs and expenditures were collected from 226 farms. The median profit margin averaged US$83 acre^{-1} for small farms, US$147 acre^{-1} for medium farms, and US$125 acre^{-1} for large farms. The average pesticide expenditure was US$21 acre^{-1}. The model showed that small farms could reduce their pesticide expenditure to US$10 acre^{-1} before significantly decreasing their profits, while the threshold for pesticide reduction was US$16 acre^{-1} and US$22 acre^{-1} for medium and large farms, respectively. This study indicates that producers could reduce pesticide use while not affecting their profit margin.

COMPONENTS OF IPM

Fungicide application technology

Disease control by fungicide application can be an integral component of IPM, but the emphasis should be on the appropriate and correct use of these chemicals. Fungicide use could be reduced by lowering rates and frequency of applications. A study in Sweden showed that the general public perceived that chemical manufacturers recommended pesticide rates for best performance rather than rates based on varying thresholds that would prevent significant yield losses (Bellinder et al., 1994). Adequate control of sclerotinia stem rot of canola was obtained with lower doses of benomyl and lower water rates than normally recommended (Morrall, 1993; Morrall et al., 1989). Gaudet et al. (1994) demonstrated that the threshold rate of efficacy of carbathiin seed treatments for control of bunt in winter wheat was 0.76 g kg^{-1} for early fall seeding, but that a higher rate of 1.01 g kg^{-1} was need for control with late fall seeding. Adjusting rates to maximize performance for locations and environments may improve disease control while using less chemical.

Combining chemicals to create synergists and the use of adjuvants may also reduce fungicide inputs and rates of application (Gressel, 1993). For example, the adjuvants Bond, Soydex and Kinetic added to a fungicide used to control stem rot in peanuts improved yields by 16%, 12% and 8% compared with the fungicide used alone (Seebold and Backman, 1994). Kharbanda (1992) found that adjuvants used with fungicide formulations of propiconazole and prochloraz significantly increased the number of healthy plants and plot yields as compared with fungicides applied alone for the control of blackleg on canola.

The timing of applications to correspond to disease severity would reduce fungicide use and increase its efficacy. Weather and local environmental parameters will strongly influence the severity and spread of many fungal pathogens. McCabe and Gallagher (1993) demonstrated that in years when disease was low (in their examples *Septoria tritici* Roberge in Desmaz on the flag leaf of wheat was less than 10% and *Erysiphe graminis* D.C. ex Merat f.sp. *tritici* Em. Marchal was less than 5%), the yield response to fungicide inputs was low. In these cases, the net margins increased by 5% when fungicides were omitted. In years when *S. tritici* severity approached 30% and powdery mildew 15%, then spraying the fields for disease control resulted in a 20% yield increase and a net margin increase of 15% over the unsprayed fields.

New nozzle types, penetration of crop canopy, improved sprayer precision, changing droplet size and using shrouded booms are technological improvements that can be further developed to reduce fungicide use. These techniques will also minimize the effects on non-target organisms. However, the benefits derived from these technologies will depend upon proper application. For example, shrouded booms and shields can reduce pesticide drift by up to 85% under moderate wind conditions (Ford, 1986), but shields may also provide a false sense of security, leading to less responsible pesticide application (e.g. under higher wind conditions when the shields lose their effectiveness) (Wolf et al., 1993). Another study of 422 farm sprayers showed that 52% had faulty nozzles and 26% had problems with the delivery and pump systems (Bellinder et al., 1994). Billett (1992) estimated that a producer could save about US$1.00 ha^{-1} on pesticide costs by using the correct sprayer application, maintaining nozzles, adjusting boom height, nozzle angle and travel speed, preventing misses and overlaps in the travel path, and spraying under optimal weather conditions. The problem with new technology is getting the end-user to adopt it and use it correctly.

Cultivar resistance

Development of plant varieties with resistance to disease is regarded as being an economically and environmentally sound method to limit crop losses. Additional inputs, such as fungicides, may help increase yield but could also result in a loss in profit. Paxton et al. (1992) studied 10 wheat varieties under conventional and more intensive management regimes involving additional nitrogen fertilizer plus two applications of the foliar fungicides Dithane and Bayleton. Relative variety performance was the same under both management regimes. Over all varieties, yields with intensive management were 13% higher than those with conventional management. However, the costs with intensive management were also higher by 36%, resulting in an economic loss. In this case, the use of genetic resistance alone was more economical to control fungal leaf spot diseases than the additional use of fungicides.

Breeding for resistance relies upon genetic variation occurring in the host/pathogen relationship. Genes for resistance may originate and be preserved in both wild and cultivated crop species. Jana and Bailey (1995) showed that *in situ* conservation of wild barley and cultivated landraces from the Middle East preserved genes for resistance that were effective towards Canadian isolates of *Cochliobolus sativus* (Ito & Kurib.) Drechsl. ex Dastur and *Pyrenophora teres* (Died.) Drechsl. Traditional breeding methods have transferred resistance genes among cultivated species, and also from wild species to cultivated ones (Heaton and Klisiewicz, 1981; Conner et al., 1988; Bailey et al., 1993, 1995).

Development of genetically modified plants should increase the possibility of introducing new genes for resistance into plant varieties. This should also allow the modification of pathways involved in resistance. For example, using transgenic tobacco plants with suppressed phenylalanine ammonia-lyase, Maher et al. (1994) showed that plants were more susceptible to infection by *Cercospora nicotianae* Ell. & Ev. Suppression of this enzyme reduced the quantity of the pathway end-product, chlorogenic acid, in the leaf tissues. This product was not formed in response to fungal infection, but was a pre-formed protectant. Breeding programs using either traditional recurrent selection methods or gene transfer can incorporate partial protection by pre-formed phenylpropanoid products, which may contribute to more durable resistance.

Tobacco and canola plants exhibited enhanced resistance to *Rhizoctonia solani* Kühn after insertion of a bean chitanase gene regulated by a cauliflower mosaic virus 35S promoter (Broglie et al., 1991). These transgenic plants had a 23–44-fold increase in chitanase activity in the leaves as compared with control plants. After infection with the fungus, seedling mortality of tobacco was 53% in the control and 23% in the transgenics; mortality of canola seedlings was 78% in the control and 43% in the transgenic plants. This provides a new control strategy for protection against damping off and could become a method replacing chemical seed treatment.

Breeding for qualitatively inherited traits often involves specific genes for resistance, but reliance on the gene-for-gene relationship requires constant monitoring of the pathogen populations to detect early changes in the occurrence and prevalence of different virulence genes (Gaudet and Puchalski, 1989; Martens et al., 1989; Tekauz, 1990; Chong and Seaman, 1993). The effectiveness of resistance genes can decrease when changes occur in the pathogen population. For example, most Canadian barley cultivars possessed the T-gene that gave barley resistance to all the prevalent races of wheat stem rust. In the late 1980s, Race QCC, which was virulent to the T-gene, became the most common isolate and all barley cultivars became susceptible (Harder and Dunsmore, 1991, 1993). Pyramiding multiple resistance genes into one cultivar and selection for quantitatively inherited resistance may help offset the devastating epidemics that can result from the breakdown of host resistance.

In Britain, cultivar mixtures of cereal grains have been studied to evaluate their effectiveness in delaying the breakdown of host resistance. The use of cultivar mixtures may also be useful in some IPM programs to reduce disease problems when the varietal purity of the plant product is not strictly needed, such as when producers grow their own livestock feed. Cultivar mixtures of three wheat varieties possessing different types of genetic resistance delayed disease development of powdery mildew, leaf rust and stripe rust by 50% (Manthey and Fehrmann, 1993). These mixtures had less disease and higher yields (ca. 5–6%) than the controls, which consisted of the combined average of pure stands of the cultivars used in the mixture. Profit was calculated from the gross minus the cost of inputs (i.e. cost of seed, fertilizer, pesticides, machine costs, interest charges and insurance), with the mixture having the additional expense (US$7 ha^{-1}) of mixing the seed. The profit margin was greater in cultivar mixtures, resulting in an economic increase of 8% for the best-performing mixture as compared with the corresponding pure stands.

Cultural practices

Cultural practices have evolved in an effort to attain more affordable returns from crop production. Adopting such practices has been a learning process, as each change has some cost:benefit ratio. It takes considerable time to evaluate all the parameters that are altered with each change in cultural practice.

Tillage

Tillage results in numerous physical changes to the soil environment (i.e. water potential, aeration, compaction, porosity and temperature). In turn, these changes have an impact on the survival and activity of the pathogen, host susceptibility and the prevalence of other soil micro-organisms (Rothrock, 1992). The response of a pathogen to tillage will depend on the host–pathogen combination and the influence of environment. Changes in tillage practices often alter disease levels in a crop, and the severity of the problem may either increase, decrease or remain unchanged with reduced tillage. In field beans grown in Ontario, deep tillage reduced root damage caused by *R. solani, Fusarium solani* (Mart.) Appel. & Wr. f.sp. *phaseoli* (Burk.) Synd. & Hans and *Pythium ultimum* Trow and increased yield (Tan and Tu, 1995). In the Canadian prairies, zero tillage reduced common root rot of cereals and increased yield (Bailey et al., 1992; Bailey and Duczek, 1996). Until there is a better understanding of the biological and physical factors influencing diseases, it will be difficult to predict each pathogen's response to tillage for every environment. Also, even though tillage may reduce disease, the cost involved should be considered. Excessive tillage destroys soil organic matter, upsets the balance of soil microbial populations, increases soil erosion risks and consumes large amounts of energy.

Concerns have been raised that reduced tillage systems may increase fungicide use because the crop residue left on the soil surface acts as an overwintering substrate for many fungal pathogens. Leaf spot diseases of cereals grown on the Canadian prairies have been shown to increase under reduced tillage systems (Bailey et al., 1992), but the severity of the infection was more dependent on the local weather conditions than on tillage practice. After 8 years in a tillage × rotation study, yield losses due to the foliar diseases were 7% greater under zero tillage compared with conventional tillage (Bailey and Duczek, 1996). In drier environments, diseases of wheat have not dramatically increased due to the presence of more crop residue, and so fungicide use in these areas should not increase with reduced tillage. Fawcett (1987) noted that in the United States the primary methods of disease control for field crops is genetic resistance and crop rotation. Disease control by foliar fungicides is not common, and only 1% of corn, 1% of soybeans and 2% of wheat received fungicide treatments in 1982.

Even though crop residue on the soil surface acts as a source of primary inoculum within a field, secondary inoculum coming from an outside source may create epidemics in fields, if weather conditions are conducive, regardless of the tillage practice used. Jenkyn et al. (1995) found that net blotch on winter barley was more severe in the autumn when straw residue from the previous crop was left on the surface. By spring, there were no consistent differences between treatments that removed the residue by burning and ploughing or those that left the residue by non-inversive minimum tillage. Similarly, Prew et al. (1995) observed no effect of straw disposal methods on septoria leaf spot severity on winter wheat. Fungicide applications to the plots increased yield, but there was no interaction with the straw disposal method. Crop rotation and genetic resistance used in conjunction with reduced tillage should not result in increased fungicide use compared with conventional tillage.

Rotation

Cook and Veseth (1991, p. 70) state that 'Crop rotation is possibly the oldest method of pest and disease control, and it needs to be rediscovered in this age of monoculture and intensive or specialized cropping.' It gives an opportunity to manage soil- and residue-borne pathogens with natural enemies and residue decomposition. The rotations available to producers are determined by the type and number of crops that are suitable for production and those that also give a satisfactory economic return. Some rotations carry more risk than others. For example, growing wheat, canola and peas in rotation will lessen diseases on wheat but increase the risk of infection by sclerotinia stem rot on peas and canola. Rotation among small grain crops such as wheat, barley and corn pose a very high risk where *Fusarium graminearum* Schwabe is prevalent, as this pathogen affects all these crops.

Diverse crop rotations reduce disease risk, whereas monoculture cropping systems increase the risk of severe pest problems with an accompanying loss in yield (Sutton and Vyn, 1990; Bailey et al., 1992; Bailey and Duczek, 1996). If unrelated crops are grown in succession, then the risk of diseases carrying over is lessened. Crop rotation is a more important factor affecting disease severity than tillage. With continuously grown corn, *Fusarium* spp. can survive and increase very quickly regardless of tillage practice (Skoglund and Brown, 1988). Monoculture can also lead to selection of strains of pathogenic fungi which are more virulent to that crop (Conner and Atkinson, 1989; El-Nashaar and Stack, 1989).

Also known to occur with monoculture is the phenomenon of disease-suppressive soils, such as take-all decline of cereals. With suppressive soils, a soil-borne pathogen may become established and increase in the initial years of cropping, but it then diminishes with continued culture of the crop (Cook, 1982; Bruehl, 1987). The decline effect is lost when a different crop is introduced into the production cycle. It may take up to 7 years before the effects of take-all decline reduce disease severity to tolerable levels, and during this time severe losses in yield may be expected. Manipulating crop rotations to take advantage of take-all decline does not provide a satisfactory method for management of this disease. However, the identification of the micro-organisms responsible for the phenomenon has led to the development of biological control agents for more successful disease control.

Crop rotation reduces the level of initial inoculum in a field. Pedersen and Hughes (1992) found that a 2-year interval between wheat crops could delay epidemics caused by the septoria disease complex by 9 days. Climate influenced the length of the interval required to reduce disease such that in years that were not favourable for disease development, only 1 year between wheat crops was needed to lower disease severity significantly. However, consecutive plantings of similar crops, such as barley and wheat, could lead to increased disease severity because *Septoria* spp. originating on wheat can overwinter and sporulate on barley residue, thus preventing a decrease in the level of initial inoculum (L.J. Duczek and K.L. Bailey, Agriculture and Agri-Food Canada, Saskatoon, unpublished data 1994–95).

Table 14.4   Effect of controlling diseases with soil fumigation on wheat yields under different crop rotations

Rotation	Yield increase in disease-free check (%)
Continuous wheat	70
Wheat–lentils	22
Wheat–barley–lentils	7

Adapted from Cook and Veseth, 1991.

The largest impact of crop rotation on yield can be demonstrated in higher rainfall areas or under irrigation, where disease pressure is greater. Using soil fumigation to create diseased and disease-free plots, Cook and Veseth (1991) showed that the yield losses in wheat due to disease were higher with continuous wheat rotations than with 2- or 3-year rotations with other cereals and pulse crops (Table 14.4). In the short term, crop rotation may not return the largest profits to the producers if lower-value crops need to be grown, but in the long term it will be sustainable because yield will be maintained.

Other practices

Disease incidence and severity can be influenced by other agronomic practices such as seeding date, seeding depth, row spacing, herbicide use, irrigation and nutrition. The effect of a single practice on disease levels is not usually sufficient to give adequate disease control, but the degree of control may be cumulative when compounded with other practices. The incidence of leaf stripe in barley in western Canada can be reduced by 45% with later seeding dates (i.e. last week of May), but owing to location and seasonal variability, seed-treatment fungicides should also be used to provide reliable protection (Tekauz et al., 1985). Seeding date had no effect on the germination of sclerotia of *Sclerotinia sclerotiorum* (Lib.) de Bary (Teo et al., 1989), but yield losses in sunflowers were minimized with plant densities of $26–49 \times 10^3$ plants ha^{-1} and a plant spacing of 36 cm (Hoes and Huang, 1985). The disease was limited to infections by direct contact with sclerotia, and the spacing prevented plant-to-plant spread of the disease by mycelia growing in the soil.

IPM recommendations should select those practices that can maximize agronomic production while minimizing disease development. However, the recommendations need to be considered in conjunction with other biological or epidemiological conditions. Teo et al. (1988) found that damping off caused by rhizoctonia root rot of canola could be partially controlled by later seeding dates if the isolates were predominantly the AG 2-1 type, but not if they were the AG 4 type. If both anastomosis groups are present in a field, then seed treatment is recommended.

Herbicide applications can influence the biology of the pathogens. In some cases the chemicals may have a direct effect on the pathogen, and in other cases it may be an effect of the method of application. Tinline and Hunter (1982) observed that the intensity of common root rot of wheat increased in field plots that incorporated triallate or trifluralin as pre-emergent herbicides. Upon further investigation, common root rot intensity also increased in plots that used the incorporation practice without the herbicide. Therefore, the application method, which involved the physical disturbance of the soil, and not the chemical was responsible for the increased intensity of the disease.

The herbicides glyphosate, triallate and trifluralin directly affected sporulation of the blackleg pathogen *Leptosphaeria maculans* (Desm.) Ces. & de Not. on canola (Petrie, 1995). Stubble infected with the virulent strain of *L. maculans* was collected in the spring, dipped into herbicide suspensions, and incubated outdoors until the fall when sporulation was estimated. Glyphosate completely inhibited spore production, triallate reduced spore production by 50%, and trifluralin increased spore production by 128%. Petrie (1995) suggested that this work should be evaluated in the field to determine if reduced tillage systems which use glyphosate for weed control would also control blackleg and allow for shortened rotations between canola crops.

Nutrition has been used as a means of disease control for many years, although it is frequently not recognized as a primary component of control. Nutrients can be supplied either directly or indirectly by altering solubilities with microbial activity. Copper-deficient wheat plants are more susceptible to take-all and have lower yields than plants with higher copper levels (Brennan, 1991). Copper levels can be supplemented by the incorporation of copper sulfate into the soil. The fungal organism *Penicillium bilaji* Chalabuda was developed into a commercial fungal inoculant (marketed by Philom Bios Inc., Saskatoon, SK, Canada) that acidifies the micro-environment to solubilize rock phosphate and improve phosphorus availability and uptake by plants. One study in North Dakota showed that direct additions of phosphorus to soil increased common root rot severity on spring wheat, whereas the seed inoculant had no effect on common root rot (Goos et al., 1994). Both the inoculant and direct additions of phosphate fertilizers increased grain yields by 3–7% above the untreated control.

Tactics such as crop sequence, organic amendments, liming for pH adjustment, tillage and irrigation can influence disease levels through nutritional interactions (Engelhard, 1989). Crop production efficiency may be improved, but data on nutritional effects on plant disease are often conflicting. For example, in some years, the addition of chloride fertilizers to soil has been associated with reduced disease severity of common root rot on cereals and increased yields, but inconsistent results have prevented this practice from being recommended for disease control (Shefelbine et al., 1986; Timm et al., 1986; Goos et al., 1989; Tinline et al., 1993). Timm et al. (1986) suggested that chloride either directly interferes with NO_3^- uptake or inhibits nitrification. More research is needed to document the interactions between nutrient release and uptake with plant disease, and also to clarify the mechanisms of action. The development of tools such as decision support systems which integrate information on nutrition should help incorporate this data into IPM systems and help to synthesize the information available.

Biological control agents

In an ecologically balanced crop production system, the use of microbial agents or their by-products should provide an alternative method of disease control for both foliar and soil-borne pathogens. The development of microbial pest control agents is estimated to take 3 years and cost US$5 million, as compared with 8–12 years and US$40–80 million for chemical pesticides (Woodhead et al., 1990). In actuality, it has taken much longer to develop microbial agents into marketable products. The commercial success of biofungicides has been limited, although sales in 1991 reached approximately US$1 million in the United States (Powell and Jutsum, 1993). Success has been limited because most research has focused on the development of biocontrol agents to replace chemicals, and has ignored opportunities based on biological balance and interactions with current farming practices. Chemical control of diseases is not always completely effective, and may require multiple applications. Yet, the expectation for biological control products often exceeds our expectations for chemicals. Deacon and Berry (1993) review the application of concepts on how biocontrol agents should be used and managed in field settings.

Biological control agents are often desirable because of their specificity towards the targeted pathogen. Organisms that colonize residues and interact with plant pathogens through competition, predation or antagonism may assist in lowering disease incidence and severity under reduced tillage farming practices (Pfender et al., 1993), but to achieve a reasonable level of control over a wide range of environments, less specificity may be beneficial. In Brazil, *Trichoderma harzianum* Rifai can colonize both wheat and soybean residues and reduce the incidence of overwintering pathogens on both crops (Fernandez, 1992a,b).

Principles discovered in plant breeding could be adopted in the development of biological control agents. For example, multiple gene resistance often gives crops greater and more durable protection over a wide range of environments than single-gene resistance. Pierson and Weller (1994) used mixtures of several fluorescent *Pseudomonas* strains that suppressed take-all of wheat in greenhouse and field trials. Application of these mixtures resulted in 20% increased yield compared with untreated controls, but the individual strains had no significant effect on yield. The effectiveness of strain combinations was not the same at all locations, but further research will be done to identify 10–15 core strains that can be combined for maximum effect depending on cultivar, soil type, tillage level and disease severity. Isolates of organisms that are introduced for disease control may react differentially with the other genera and species in the rhizosphere or phylloplane. Co-inoculation with several isolates, each suppressing different pathogens, may be needed to enhance the efficacy of the biocontrol agent.

As an alternative to applying living organisms on crops, agrochemical companies are also developing natural product chemicals from secondary metabolites produced by microbial agents. There are many hurdles to the commercialization of these products, but advantages such as target specificity and novel mechanisms of action may outweigh disadvantages such as the cost of synthesis and poor stability of the product. Morton and Nyfeler (1993) believe that understanding how natural products are formed and function will contribute to the development of new principles for crop protection. Natural products should also fit into IPM since they are generally degraded in nature, but there is no guarantee that they will always be environmentally friendly.

Monsanto isolated arginidienne from a cell-free filtrate of fermented *Bacillus megaterium* to inhibit *Phytophthora megasperma* Drechsl., which causes root rot of soybeans (Stonard et al., 1994). Ciba-Geigy developed phenylpyrrole fungicides from a secondary metabolite (pyrrolnitrin) isolated from *Pseudomonas pyrrocinia* (Morton and Nyfeler, 1993). This is available under the trade name of Fenpiclonil™ as a cereal seed treatment in Europe and has broad-spectrum activity against a number of phytopathogenic fungi.

Another way to use natural products would be to screen for activity inducing local or systemic acquired resistance. Chemical interactions that induce resistance could be used in conjunction with resistant cultivars to improve the level and increase the durability of resistance. Field trials with isonicotinic acids, which reduce lesion size, showed that the compounds must be applied prior to infection as a protectant, and that they worked more effectively on monocot than dicot hosts (Morton and Nyfeler, 1993). Obligate pathogens (i.e. powdery and downy mildews) were more easily controlled than the facultative necrotrophs such as *Rhizoctonia solani and Septoria* spp. However, the phytotoxicity of these products when applied as a foliar spray has precluded further commercial development.

TOOLS ASSISTING WITH THE INCORPORATION OF IPM STRATEGIES

Disease forecasting and monitoring

Monitoring and forecasting are useful tools for disease management, especially if disease outbreaks could be predicted so that timely and effective controls can be implemented. EPINFORM is a computerized program designed to model the progress and yield-loss due to stripe rust and *Septoria nodorum* Berk. blotch on wheat in Montana using information on disease cycles and local climatological data (Caristi et al., 1987). Knowing the epidemiological factors that increase disease severity, the relationships between severity and damage, and developing tools for rapid assessment of severity should allow disease control components to be used more effectively. Shaw and Royle (1986) calculated that at least one fungicide spray could be saved by using disease forecasting to manage *Septoria tritici* on winter wheat in Britain. They concluded that further savings could be attained if there was a better understanding of the epidemiological interaction between *S. tritici* and *S. nodorum*, with the prediction of actual disease levels.

In Germany, economic thresholds have been established for control of some fungal pathogens in wheat; foliar fungicides are recommended when *S. tritici* on leaves reaches 50% severity, and when *S. nodorum* on the heads reaches 12% severity (Habermeyer and Hoffmann, 1994). In the Canadian prairies, a decision guide was developed recommending that a one-time application of propiconazole on spring wheat should increase yield by 6–11% if leaf spot incidence (a pathogen complex of *S. tritici, S. nodorum* and *Pyrenophora tritici-repentis* (Died.) Drechsl.) was greater than 50% on the flag leaf, or on flag-1, or flag-2 leaves at the time between flag leaf emergence and early–mid-flowering growth stages; multiple or later applications of the fungicide did not result in higher yield (L.J. Duczek, Agriculture and Agri-Food Canada, Saskatoon, Saskatchewan, personal communication 1996). Hershman et al. (1994) found that after growth stage 55, the best indicator to predict whether fungicides should be applied to control *Septoria* on wheat in Kentucky was if the severity was greater than 25% on the flag-2 leaf.

Some forecasting tools have been developed that educate the producer by showing them how to identify and assess the level of risk in fields. One example is with sclerotinia stem rot of canola in western Canada. A simple agar-plate test kit allows producers to determine ascospore infestation of the petals (Morrall et al., 1991; Turkington et al., 1991a,b; Turkington and Morrall, 1993). Monitoring of individual fields may be done from early bloom through to late bloom, and percentage infestation at each sample time indicates a new risk level for disease. The kit includes a guide with colour plates to teach growers how to distinguish the pathogen from other fungi growing on the plates. The guide also gives instructions for producers to assess the risk level in a field and the potential economic value of control by calculating if, when and how much fungicide should be applied. During a 4-year test-marketing period, the kit retailed for about US$30 and the number of kits sold rose from 330 in 1991 to 1100 in 1994 (Morrall and Thomson, 1995). However, this only represents about 1.5% of the acreage of canola in western Canada at risk from infection. The kits were primarily used by highly motivated growers and agri-service companies.

Most research on monitoring and disease forecasting has been directed at crops in regions where the primary method of control has been achieved with fungicides, and often for crops with high cash value. These tools should have a much broader application. Monitoring and disease forecasting should also enhance application of innundative biological control agents or biofungicides, improve crop rotation planning and risk assessment, and help preserve the integrity of crop resistance.

Site-specific farming

Site-specific farming uses information about field conditions to determine appropriate application levels of fertilizers and pesticides, that are then applied at variable rates to reduce costs and environmental harm. Disease management could benefit from site-specific farming by: (i) spot-treating sites with fungicides or microbial agents where disease epidemics are initiated, (ii) allowing for dose-adjusted responses, (iii) selecting more resistant cultivars at sites with higher disease potential, (iv) adjusting the tillage level in parts of fields to reduce high-residue areas that initiate disease foci, (v) adjusting fertility levels to give better disease control and (vi) predicting risk assessment levels.

Stevenson and van Kessel (1996) found that leaf spot severity of wheat was not uniform across a field and was influenced by the topography. High-water-catchment footslopes (depressions at the bottom of a slope) had lower leaf spot severity than low-water-catchment footslopes, or the shoulders of these slopes in both a wheat–wheat and a pea–wheat rotation. Root disease was not affected by topography. The difference in seed yield between the high- and low-catchment footslopes was 27% in the wheat–wheat rotation and 18% in the wheat–pea rotation. Analysis showed that the reduction in yield was caused largely by leaf spot disease. In a concurrent study on uniformly level ground, leaf spot and root rot disease severities were similar under both rotations, but wheat yield was increased by 44% in the more diverse rotation. These results also indicate that small-plot research may not truly reflect the development of disease in large non-uniform plots or the interaction of disease with micro-environments.

Site-specific farming requires technology such as a geographical information system for storing information on spatially defined characteristics, and a global positioning system to pinpoint locations in a field. Applicators, sensors and developmental databases are being tested to assess the feasibility of making this technology available to growers (Becker and Senft, 1992; Schueller, 1992; Hayes et al., 1994; Schueller and Wang, 1994). Based on systems developed for site-specific applications for fertilizer, increased returns for corn and wheat ranged from US$4 to US$32 ha^{-1}. The expected purchase price of an on-farm system will be approximately US$10 000.

Decision support systems

Decision support systems (DSS) should become an integral part of intensive IPM systems to provide tools for synthesis and analysis of relevant information on numerous aspects of crop production. The databases required for computerized interactions should be derived from discipline-based research programs, and concentrate on decision areas perceived to be difficult and important by the end-users of the systems. DSS should assist farmers in reducing pesticide use through economy of application (dose and frequency) and by integrating other control procedures that will be equally effective as chemical control.

Kearney (1992) suggests that for a DSS to be useful the following requirements should be met: (i) a balance should be struck between what is perceived to be useful information in recommendations without overwhelming the user with excessive tactical and strategic information; (ii) the system must deal with relevant problems encountered by end users; (iii) the benefits of using the system must outweigh the costs of using the system; (iv) the system must be easily accessible for short-term problems requiring immediate solutions and also for long-term planning. Greer et al. (1994) observed that farmers are reluctant to adopt DSS because the systems are too complex, and they do not want to use unfamiliar terminology and logic. Modifications to DSS and complemented by better education and training on DSS should increase the acceptability of these systems.

Many DSS have been developed for farming situations that require intensive management (e.g. orchards, Atkins et al., 1992), or for rapidly changing biological systems (e.g. insect forecasting and management; Knight and Cammell, 1994; Longstaff and Cornish, 1994). There are few DSS that incorporate recommendations for fungal disease control for many different types of field crops. Those that do have concentrated on foliar diseases of grain crops, and are used primarily to make decisions on the benefits of spraying fungicides. Few systems are broad enough to include recommendations on disease control practices other than chemical control. Three expert systems have been developed and tested for foliar disease control in wheat in various countries, and one for disease control in corn.

EPIPRE is a computerized disease management system for wheat, providing recommendations on timing of field monitoring, when to use fungicides, and choice of fungicides (Zadoks, 1981; Rabbinge and Rijsdijk, 1983). The program was tested from 1981 to 1984 to evaluate the economic benefits of EPIPRE recommendations to farmers in The Netherlands as compared with three control situations (i.e. no disease control, general extension service recommendations, and an intensive routine spraying schedule) (Reinink, 1986). The results showed that EPIPRE was more restrictive in pesticide use than the general recommendations given by an experienced extension agent, but did not result in a large reduction in pesticide use. Comparing EPIPRE to an intensively managed farm system showed that the latter had slightly higher yields (0% in 1982, 10% in 1983 and 1% in 1984) but more than double the pesticide costs. Surprisingly, the ultimate benefit derived from EPIPRE was not from its decision support system, but from its information system. The information educated farmers in disease awareness and in monitoring techniques. It influenced general crop protection practices by preventing the introduction of high-input routine spraying practices by demonstrating to growers and extension agents that these intensive practices were not necessary.

The Wheat Disease Control Advisory (WDCA) was developed in Israel for the management of *Septoria tritici* blotch, leaf rust and yellow rust on wheat grown in semi-arid conditions (Shtienberg et al., 1990). Under these conditions the severity of the disease fluctuates between fields and years resulting in erratic benefits from chemical control. The knowledge base for WDCA was developed from data accumulated in field trials and included studies on disease thresholds, optimal spray periods and the epidemiology of the diseases. The action thresholds indicating when to start spraying were 50% disease severity on the 4th leaf from the top of the plant for *Septoria* control, and 0.5 pustules per tiller for rust control; spray programs started later would not suppress the disease. The

Table 14.5   Harvested yields and net profit in field experiments conducted in order to evaluate the performance of the Wheat Disease Control Advisory system (WDCA)

WDCA prototype	No. of fields	Yield (t ha^{-1})		Difference (%)	Net profit* (US$ ha^{-1})
		Common practice	WDCA recommendation		
Evaluating WDCA recommendation to spray					
Researchers	23	5.11	5.69	11	58.60
		(0.78)**	(0.79)		(53.70)
Growers	58	4.43	5.29	19	122.30
		(1.03)	(1.00)		(99.50)
Evaluating WDCA recommendation not to spray					
Researchers	10	4.84	5.17	7	16.10
		(0.89)	(0.90)		(41.40)

*In plots managed according to WDCA recommendation relative to plots managed according to the common practice. The cost of one application was 40 and 50 US$ ha^{-1}, and the value of harvested grains was 170 and 210 US$ ton^{-1} in 1985–87 and 1988, respectively.
**Standard deviation in parentheses.
Adapted from Shtienberg et al., 1990.

threshold to discontinue chemical control was based on plant growth stage; applications after the late milk stage resulted in reduced yields and net income. Special attention was paid to transferring the system to growers so that it could become widely adopted and fully tested. The system was independently assessed by researchers and then by growers (without assistance from scientists or extension personnel). The recommendations of WDCA were carried out on one half of a field and the local practices were used on the other half. The WDCA recommendations resulted in higher yields and net profits (Table 14.5). If WDCA had been used commercially on only one half of the wheat growing area, the average increase in net profit on a medium-sized wheat farm of 300 ha would have been US$36 690, and more than US$3 million nationwide.

In Germany, Habermeyer and Hoffmann (1994) introduced the IPS WEIZEN-MODELL, a plant protection decision model for fungal diseases of wheat. The model used epidemiological thresholds to determine the timing of fungicide applications. At 20 of 25 sites tested over 2 years, grain yields were similar between the prophylactic fungicide treatments (i.e. 3–7 applications) and the treatments recommended by IPS WEIZEN-MODELL (i.e. 1–2 applications). Therefore, the model saved production costs for producers because fewer fungicide applications were required to achieve the same level of disease control.

Heinemann et al. (1992) developed an integrated pest management expert system called MAIZE for field corn production in Pennsylvania. It is a very extensive system, and includes specific modules for insect identification and management, disease diagnosis and management, variety selection, herbicide selection, and grain drying and storage. It was designed to assist crop management advisors, consultants, county extension agents, agri-business personnel and educators in making within-season crop production recommendations and developing crop production plans. The production module is divided into management, diagnosis and agronomic considerations, and pest evaluation, which is further divided into four submodules: problem diagnosis, pre-season planning, within-season management and monitoring procedures. The submodules can be accessed direc-

Table 14.6 Model illustrating four databases and interactive criteria for developing a decision support system to manage fungal diseases in field crops

A. Diagnosis	B. Pest management	C. Economics	D. Agronomics
DATABASE NETWORKS FOR DISEASE MANAGEMENT			
Agents	**Urgency**	**Control**	**Regional data**
Insect	Within season	Fungicide options	Variety
Disease	Pre-season planning	Alternative options	Fertility
Nutrient deficiency			Tillage
Chemical injury	**Management options**	**Returns**	Rotation
Environmental injury	*Within season*	Calculate cost of	Phenology
	Forecasting,	treatment	Cultural practice
Probable agents	monitoring	Re-calculate with	Soil type
Lists	Fertility	different values	Other chemicals used
Pictorial	Biological control		
	agents		
Background information	Fungicides		
General information	Sanitation, burial		
Distribution	Control of vector,		
Life cycle	volunteers and		
Damage, losses	alternate hosts		
Symptoms			
General control tactics	*Pre-season planning*		
Monitoring tools	Variety		
	Fertility		
Risk assessment	Tillage		
Historical risk	Rotation		
Immediate risk	Biological control		
Economic thresholds	agents		
Prediction period	Sanitation		
Damage expectation	Fungicides, seed		
Risk level and rationale	treatments		
	Seed quality		
	Seeding depth		
	Vector control		
	Forecasting,		
	monitoring		
	Final recommendation		
	Best options		
	Rationale		
INTERACTIVE NETWORKS FOR DISEASE MANAGEMENT			
Select region, location, variety	*Select agent, crop, region, location, variety*	*Select agent, crop, region*	*Select crop, region, practice*
Areas affected in field	*Growth stage*	*Projected yield*	
Crop growth stage	*Plant parts affected*	*Cost of treatment*	
Symptoms	*Cropping history*	*Cost of applying treatment*	
Field cropping history	*Rotation plans*	*Monetary loss due to disease*	
Weather conditions	*Agronomic practices*		
Planting date	*Disease severity*	*Monetary loss in yield due to application of treatment*	
Plant density	*Within-season weather*		
Seeding depth	*Projected long-term weather*		
Fertilizer applied	*Potential loss based on severity*		
Pesticides applied			
Presence of insect vectors	*Use of monitoring techniques*		

tly, and the diagnoses and recommendations are developed from a network that asks the user questions until sufficient information has been collected to make a recommendation. Recommendations can be obtained for the following diseases of corn: stalk rots, seedling blight, ear rot, northern corn leaf blight, southern corn leaf blight, bacterial wilt, bacterial leaf blight, eyespot, *Helminthosporium* leaf spot, anthracnose, corn rust, common smut and grey leaf spot.

More work needs to be done on developing DSS for field crops and for managing fungal diseases in these crops. Table 14.6 is a model illustrating the types of databases and interactive criteria needed to develop such a DSS. Interacting with the end user and resourcing among several databases will allow decisions to be made based on complex interactions. Much of the information is available already, but there are some areas where research would need to be conducted to complete the information base. Investigations on developing economic thresholds for diseases, simple monitoring techniques to be used on farms for risk assessment, and identifying the economic impact of non-chemical control practices should make the system more relevant and appealing to the end user. To remain relevant, the system would also need to evolve because of new information and changes in technology that occur over time.

RECOMMENDATIONS FOR THE ADOPTION OF IPM PRACTICES BY PRODUCERS

Opportunities presently exist to reduce fungicide use on field crops and adopt alternative methods of disease control. Although the quantity of fungicides used on field crops is relatively small compared with that used on other crops, the amount applied could be reduced by a least one-third through improvements in fungicide application technology and disease monitoring tools. In the absence of fungicide control, the use of genetic resistance, crop rotation and other cultural practices to manage disease problems should be able to improve field crop yields by 5–15%. The overall profits will increase by reducing expenses (i.e. the cost of fungicides and reduced applications) and by increasing net income (i.e. increased yields). Therefore adopting IPM practices would be beneficial to producers.

Although the knowledge and opportunity exists, the use of IPM in field crops is still very limited. Research shows that new technologies are being developed for each component of IPM, and new tools are being developed to improve adoption at the farm level. However, a problem remains in getting IPM strategies adopted into practice at the farm level. There are several issues governed by factors relating to production, economic and social aspects that inhibit this process.

One issue relates to attitudes on acceptable levels of plant disease. Disease eradication or elimination is not necessary for most situations. Some disease on crops should be tolerated, but at levels that maintain economic and aesthetic acceptability. Consensus on accepting this concept is difficult, especially for all the numerous interacting groups in the agri-food community.

There should be more research on alternative control strategies for disease management. Developing new resistant cultivars requires continual evaluation and development to keep ahead of changing pathogen populations. More research is needed before biological control agents and decision support systems can be used as common practices over different environments. The work needs to be divided between the development of new methods and the strengthening of traditional management strategies, with multi-disciplinary research aimed at understanding the interactions among all components.

To get IPM adopted by producers, it must be demonstrated that the practices have value and applicability. Detailed cost–benefit analyses of alternative practices demonstrate the economic value of changing to IPM systems. Data collected in support of IPM often lack economic details on pesticide prices and the comparative value of non-pesticide control measures. Individual farm demonstrations help to show the practicality of the production practices and instill enthusiasm for promoting successful technologies. Regional demonstrations help to show that the technologies can be used successfully by more than one individual. Successful adoption of IPM will involve an education process that will take some considerable time.

In the United States, suggestions have been made to introduce incentives for farmers who practice sustainable agriculture, and to prohibit the unnecessary use of pesticides (Williams, 1992). These incentives include imposing fees and licensing for both users and manufacturers of pesticides for use on particular crops in some geographic areas and for use with 'high-risk' pesticides. The fee structure would be based on the frequency of pesticides used beyond what is necessary for minimum crop-quality needs. Lichtenberg (1992) suggested that efficient solutions to pesticide use may be achievable through the transmission of information to users and consumers on pesticide efficacy and associated health risks compared with alternative methods, and through taxation on pesticides. However, he noted there are few databases that allow for estimations of the economics of pesticide use versus those of alternative practices.

In Sweden, the government mandated a pesticide reduction program in 1985 to achieve a 50% reduction by 1990, and this was met by reducing the use of insecticides by 64%, of herbicides by 54%, and of fungicides by 2% (Bellinder et al., 1994). There is much to be learned from the implementation of this model. It was based on risk reduction (i.e. cost–benefit risk assessments for all compounds) and use reduction by employing other economically equivalent control measures. Pesticides were used only if no alternative was available, and adjusted to rates which would achieve acceptable (not maximum) control. The key factors that allowed this goal to be met were the high degree of cooperation between grower groups, scientists, environmentalists and policy makers, and the fact that each group kept the economic viability of the growers as their primary criterion.

CONCLUSIONS

Adopting a balanced complement of IPM strategies should improve crop productivity and returns as compared with strategies where only fungicides are used in field crops. Some individual producers may greatly benefit from adopting IPM strategies as compared with intensive fungicide management strategies. Overall, reductions in fungicide use will be small because application of fungicides on field crops is low relative to other crops.

For IPM to be truly effective, there must be a process by which the component strategies can be integrated at the farm level. The most effective IPM strategy should be developed for the individual producer to deal with the particular problems at that specific location. This means identifying key problems and locations, assessing all known variables, simulating with known variables to predict various opportunities, and interpreting a complex situation into a practical, affordable solution. The development of any IPM strategy will require significant interdisciplinary cooperation among scientists, extension officers, producers, simulation modellers, economists and consumers, but inevitably, all tools that are developed must become readily available and be used easily by the grower.

IPM programs that will return reasonable profits while maintaining aesthetic crop appeal also requires a willingness by the producer to learn complex management skills. This process can only be accomplished in a communicative, cooperative environment. Ways to advance the adoption of IPM is through public (urban, rural and on-farm) education, the use of old fashioned extension agrology, and small- to large-scale demonstrations. The responsibility to improve agriculture with IPM does not rest solely with producers and researchers, but with the entire community, as everyone is affected by the food production industry.

REFERENCES

Anonymous (1994) *Canadian Grains Industry Statistical Handbook, 1993.* Canada Grains Council, Winnipeg, Manitoba, 267 pp.

Atkins, T.A., Laurenson, M.R., Mills, T.M. and Ogilvie, D.K. (1992) Orchard 2000: Towards a decision support system for New Zealand's orchard industries. *Acta Horticulturae* **313**, 173–182.

Bailey, K.L. and Duczek, L.J. (1996) Managing cereal diseases under reduced tillage. *Canadian Journal of Plant Pathology* **18** 159–167.

Bailey, K.L., Mortensen, K. and Lafond, G.P. (1992) Effects of tillage systems and crop rotations on root and foliar diseases of wheat, flax, and peas in Saskatchewan. *Canadian Journal of Plant Science* **72**, 583–591.

Bailey, K.L., Harding, H. and Knott, D.R. (1993) Transfer to bread wheat of resistance to common root rot [*Cochliobolus sativus*] identified in *Triticum timopheevii* and *Aegilops ovata. Canadian Journal of Plant Pathology* **15**, 211–219.

Bailey, K.L., Hucl, P. and Harding. H. (1995) Three spring wheat germplasm lines (839–984, 841–2, and 839–1076) with resistance to common root rot. *Canadian Journal of Plant Science* **75**, 695–696.

Becker, H. and Senft, D. (1992) Satellites key to new farming aids. *Agricultural Research* **40**, 4–8.

Bellinder, R.R., Gummesson, G. and Karlsson, C. (1994) Percentage-driven government mandates for pesticide reduction: The Swedish model. *Weed Technology* **8**, 350–359.

Billett, D. (1992) Importance of application technology on the farm. In Holm, F.A. (Ed.) *Proceedings of Appli-Tech'92: Agricultural Chemical Application Technology for the 90s.* 17–19 June, Regina Saskatchewan. Extension Division, University of Saskatchewan, Saskatoon, pp. 3–6.

Brennan, R.F. (1991) Effect of copper application on take-all severity and grain yield of wheat in field experiments near Esperance, Western Australia. *Australian Journal of Experimental Agriculture* **31**, 255–258.

Broglie, K., Chet, I., Holliday, M., Cressman, R., Biddle, P., Knowlton, S., Mauvais, C.J. and Broglie, R. (1991) Transgenic plants with enhanced resistance to the fungal pathogen *Rhizoctonia solani. Science* **254**, 1194–1197.

Bruehl, G.W. (1987) *Soilborne Plant Pathogens.* MacMillan, New York, 368 pp.

Burrows, T.M. (1983) Pesticide demand and integrated pest management: A limited dependent variable analysis. *American Journal of Agricultural Economics* **65**, 806–810.

Caristi, J., Scharen, A.L., Sharp, E.L. and Sands, D.C. (1987) Development and preliminary testing of EPINFORM, an expert system for predicting wheat disease epidemics. *Plant Disease* **71**, 1147–1150.

Chong, J. and Seaman, W.L. (1993) Distribution and virulence of *Puccinia coronata* f.sp. *avenae* in Canada in 1991. *Canadian Journal of Plant Pathology* **15**, 41–45.

Conner, R.L. and Atkinson, T.G. (1989) Influence of continuous cropping on severity of common root rot in wheat and barley. *Canadian Journal of Plant Pathology* **11**, 127–132.

Conner, R.L., MacDonald, M.D. and Whelan, E.D.P. (1988) Evaluation of take-all resistance in wheat-alien amphiploid and chromosome substitution lines. *Genome* **30**, 597–602.

Cook, R.J. (1982) Use of pathogen-suppressive soils for disease control. In Schneider, R.W. (Ed.) *Suppressive Soils and Plant Disease*. APS Press, Minnesota, pp. 51–65.

Cook, R.J. and Veseth, R.J. (1991) *Wheat Health Management*. APS Press, Minnesota, 152 pp.

Deacon, J.W. and Berry, L.A. (1993) Biocontrol of soil-borne plant pathogens: Concepts and their application. *Pesticide Science* **37**, 417–426.

El-Nashaar, H.M. and Stack, R.W. (1989) Effect of long-term continuous cropping of spring wheat on aggressiveness of *Cochliobolus sativus*. *Canadian Journal of Plant Science* **69**, 395–400.

Engelhard, A.W. (1989) *Management of Diseases with Macro- and Microelements*. APS Press, Minnesota, 217 pp.

Fawcett, R.S. (1987) Overview of pest management for conservation tillage systems. In Logan, T.J., Davidson, J.M., Baker, J.L. and Overcash, M.R. (Eds) *Effects of Conservation Tillage on Groundwater Quality: Nitrates and Pesticides*. Lewis, Chelsea, MI, pp. 19–36.

Fernandez, M.R. (1992a) The effect of *Trichoderma harzianum* on fungal pathogens infesting soybean residues. *Soil Biology and Biochemistry* **24**, 1027–1029.

Fernandez, M.R. (1992b) The effect of *Trichoderma harzianum* on fungal pathogens infesting wheat and black oat straw. *Soil Biology and Biochemistry* **24**, 1031–1034.

Ford, R.J. (1986) Field trials of a method for reducing drift from agricultural sprayers. *Canadian Agricultural Engineering* **28**, 81–83.

Gaudet, D.A. and Puchalski, B.J. (1989) Races of common bunt (*Tilletia caries* and *T. foetida*) of wheat in western Canada. *Canadian Journal of Plant Pathology* **11**, 415–418.

Gaudet, D.A., Puchalski, B.J. and Entz, T. (1994) Effects of seeding date and cultivar susceptibility on effectiveness of carbathiin for control of common bunt (*Tilletia tritici* and *T. laevis*) in winter wheat in southern Alberta. *Canadian Journal of Plant Pathology* **16**, 304–310.

Gibson, K.M. (1984) A conceptual framework for estimating the impact of Canadian pesticide regulations on the decision to introduce new pesticide entities. M.Sc. Thesis, University of Manitoba, Winnipeg, 151 pp.

Goos, R.J., Johnson, B.E. and Stack, R.W. (1989) Effect of potassium chloride, imazalil, and method of imazalil application on barley infected with common root rot. *Canadian Journal of Plant Science* **69**, 437–444.

Goos, R.J., Johnson, B.E. and Stack, R.W. (1994) *Penicillium bilaji* and phosphorus fertilization effects on the growth, development, yield and common root rot severity of spring wheat. *Fertilizer Research* **39**, 97–103.

Greer, J.E., Falk, S., Greer. K.J. and Bentham, M. (1994) Explaining and justifying recommendations in an agriculture decision support system. *Computers and Electronics in Agriculture* **11**, 195–214.

Gressel, J. (1993) Synergizing pesticides to reduce use rates. In Duke, S.O., Menn, J.J. and Plimmer, J.R. (Eds) *Pest Control with Enhanced Environmental Safety*. American Chemical Society, Washington, DC, pp. 48–61.

Habermeyer, J. and Hoffmann, G.M. (1994) Strategie und Realisation der Einführung des PflanzenschutzEntscheidungsmodelles ('IPS WEIZENMODELL') gegen Pilzkrankheiten an Weizen in die landwirtschaftliche Praxis. *Zeitschrift für Pflanzenkrankheiten und Pflanzenschutz* **101**, 617–633.

Harder, D.E. and Dunsmore, K.M. (1991) Incidence and virulence of *Puccinia graminis* f.sp. *tritici* on wheat and barley in Canada in 1990. *Canadian Journal of Plant Pathology* **13**, 361–364.

Harder, D.E. and Dunsmore, K.M. (1993) Incidence and virulence of *Puccinia graminis* f.sp. *tritici* on wheat and barley in Canada in 1991. *Canadian Journal of Plant Pathology* **15**, 37–40.

Hayes, J.C., Overton, A. and Price, J.W. (1994) Feasibility of site-specific nutrient and pesticide applications. In Campbell, K.L., Graham, W.D. and 'Del' Bottcher, A.B. (Eds) *Environmentally Sound Agriculture*. Proceedings of the 2nd Conference, 20–22 April, Orlando, FL. American Society of Agricultural Engineers, St. Joseph, MI, USA. pp. 62–68.

Heaton, T.C. and Klisiewicz, J.M. (1981) A disease-resistant safflower alloploid from *Carthamus tinctorius* L. × *C. lanatus* L. *Canadian Journal of Plant Science* **61**, 219–224.

Heinemann, P.H., Calvin, D.D, Ayers, J., Carson, J.M., Curran, W.S., Eby, V., Hartzler, R.L., Kelley, J.G.W., McClure, J., Roth, G. and Tollefson, J. (1992) MAIZE: A decision support system for management of field corn. *Applied Engineering in Agriculture* **8**(3), 407–414.

Hershman, D.E., Perkins, D.M., Morgan, D.C. and Bachi, P.R. (1994) Performance of growth stage and threshold-based foliar fungicide treatments for the soft red winter wheat cultivar Clark, 1993. *Fungicide and Nematicide Tests* **49**, 212.

Hoes, J.A. and Huang, H.C. (1985) Effect of between-row and within-row spacings on development of sclerotinia wilt and yield of sunflower. *Canadian Journal of Plant Pathology* **7**, 98–102.

Jana, S. and Bailey, K.L. (1995) Responses to wild and cultivated barley from West Asia to net blotch and spot blotch. *Crop Science* **35**, 242–246.

Jenkyn, J.F., Gutteridge, R.J. and Todd, A.D. (1995) Effects of incorporating straw, using different cultivation systems, and of burning it, on diseases of winter barley. *Journal of Agricultural Science, Cambridge* **124**, 195–204.

Kearney, M. (1992) User requirements of computer models in decision support systems in orchards. *Acta Horticulturae* **313**, 165–171.

Kharbanda, P.D. (1992) Performance of fungicides to control blackleg of canola. *Canadian Journal of Plant Pathology* **14**, 169–176.

Knight, J.D. and Cammell, M.E. (1994) A decision support system for forecasting infestations of the black bean aphid, *Aphis fabae* Scop., on spring-sown field beans, *Vicia faba. Computers and Electronics in Agriculture* **10**, 269–279.

Knutson, R.D,. Smith, E.G., Miller, J.W., Taylor, C.R. and Penson, J.B. (1990) *Impacts of Chemical Reduction on Wheat Yields and Costs.* AFPC Policy Working Paper 90-8, Texas A&M University, College Station, TX, 46 pp.

Lasley, P., Duffy, M., Kettner, K. and Chase, C. (1990) Factors affecting farmers' use of practices to reduce commercial fertilizers and pesticides. *Journal of Soil and Water Conservation* **45**, 132–136.

Lichtenberg, E. (1992) Alternative approaches to pesticide regulation. *Northeastern Journal of Agricultural Research Economics* **21**, 83–97.

Longstaff, B.C. and Cornish, P. (1994) Pestman: A decision support system for pest management in the Australian central grain-handling system. *AI Applications,* **8**, 13–23.

Maher, E.A., Bate, N.J., Ni, W., Elkind, Y., Dixon, R.A. and Lamb, C.J. (1994) Increased disease susceptibility of transgenic tobacco plants with suppressed levels of preformed phenylpropanoid products. *Proceedings of the National Academy of Science USA* **91**, 7802–7806.

Manthey, R. and Fehrmann, H. (1993) Effect of cultivar mixtures in wheat on fungal diseases, yield, and profitability. *Crop Protection* **12**, 63–68.

Martens, J.W., Seaman, W.L. and Atkinson, T.G. (1984) *Diseases of Field Crops in Canada.* Canadian Phytopathological Society, Ottawa, Canada, 160 pp.

Martens, J.W., Dunsmore, K.W. and Harder, D.E. (1989) Incidence and virulence of *Puccinia graminis* in Canada on wheat and barley in 1988. *Canadian Journal of Plant Pathology* **11**, 424–430.

McCabe, T. and Gallagher, E.J. (1993) Winter wheat production systems: Effect of reduced inputs on grain yield, quality and economic return. *Aspects of Applied Biology* **36**, 251–256.

Morrall, R.A.A. (1993) Rationalizing the control of sclerotinia stem rot of spring canola with benomyl in western Canada. In *Proceedings of the 6th International Congress of Plant Pathology,* 28 July–6 August, Montreal, p. 50, (abstract).

Morrall, R.A.A. and Thomson, J.R. (1995) Four year's experience in western Canada with commercial petal testing to forecast sclerotinia. In *Proceedings of the 9th International Rapeseed Congress.* 4–7 July, Cambridge, pp. 1013–1015.

Morrall, R.A.A., Rogers, R.B. and Rude, S.V. (1989) Improved techniques of controlling sclerotinia stem rot of canola (oilseed rape) with fungicides in western Canada. *Mededelingen vande Faculteit Landbouwwetenschappen, Rijksuniversiteit Gent* **54/2b**, 643–649.

Morrall, R.A.A., Turkington, T.K., Kaminski, D.A., Thomson, J.R., Gugel, R.K. and Rude, S.V. (1991) Forecasting sclerotinia stem rot of spring rapeseed by petal testing. In *Proceedings of the 8th International Rapeseed Congress.* 9–11 July, Saskatoon. GCIRC, pp. 483–488.

Morton, H.V. and Nyfeler, R. (1993) Utilizing derivatives of microbial metabolites and plant defenses to control diseases. In Duke, S.O., Menn, J.J. and Plimmer, J.R. (Eds) *Pest Control with Enhanced Environmental Safety.* American Chemical Society, Washington, DC. pp. 316–322.

Paxton, K.W., Hallmark, W.B., Harrison, S.A., Hutchinson, R.L., Boquet, D.J., Moore, S.H., Rabb, J.L., Habetz, R.J. and Colyer, P.D. (1992) An economic analysis of management practices for wheat production in Louisiana. *Louisiana Agriculture* **35**, 13–15.

Pedersen, E.A. and Hughes, G.R. (1992) The effect of crop rotation on development of the septoria disease complex on spring wheat in Saskatchewan. *Canadian Journal of Plant Pathology* **14**, 152–158.

Petrie, G.A. (1995) Effects of chemicals on ascospore production by *Leptosphaeria maculans* on blackleg-infected canola stubble in Saskatchewan. *Canadian Plant Disease Survey* **75**, 45–50.

Pfender, W.F., Zhang, W. and Nus, A. (1993) Biological control to reduce inoculum of the tan spot pathogen *Pyrenophora tritici-repentis* in surface-borne residues of wheat fields. *Phytopathology* **83**, 371–375.

Pierson, E. and Weller, D.M. (1994) Use of mixtures of fluorescent pseudomonads to suppress take-all and improve the growth of wheat. *Phytopathology* **84**, 940–947.

Pimentel, D., McLaughlin, L., Zepp, A., Lakitan, B., Kraus, T., Kleineman, P., Vancini, F., Roach, W.J., Graap, E., Keeton, W.S. and Selig, G. (1991) Environmental and economic impacts of reducing US agricultural pesticide use. In Pimentel, D. (Ed.) *CRC Handbook of Pest Management in Agriculture.* Vol. 1. 2nd edn. CRC Press, Boca Raton, FL, pp. 679–718.

Powell, K.A. and Jutsum, A.R. (1993) Technical and commercial aspects of biocontrol products. *Pesticide Science* **37**, 315–321.

Prew, R.D., Ashby, J.E., Bacon, E.T.G., Christian, D.G., Gutteridge, R.J., Jenkyn, J.F., Powell, W. and Todd, A.D. (1995) Effects of incorporating or burning straw, and of different cultivation systems, on winter wheat grown on two soil types, 1985–91. *Journal of Agricultural Science, Cambridge* **124**, 173–194.

Rabbinge, R. and Rijsdijk, F.H. (1983) EPIPRE: A disease and pest management system for winter wheat, taking account of micrometeorological factors. EPPO Bulletin **13**, 297–305.

Reinink, K. (1986) Experimental verification and development of EPIPRE, a supervised disease and pest management system for wheat. *Netherlands Journal of Plant Pathology* **92**, 3–14.

Rothrock, C.S. (1992) Tillage systems and plant disease. *Soil Science* **154**, 308–315.

Schueller, J.K. (1992) A review and integrating analysis of spatially variable control of crop production *Fertilizer Research* **33**, 1–34.

Schueller, J.K. and Wang, Min-Wen (1994) Spatially variable fertilizer and pesticide application with GPS and DGPS. *Computers and Electronics in Agriculture* **11** 69–83.

Seebold, K.W. and Backman, P.A. (1994) Effects of adjuvants and application methods on fungicide performance in peanut. 1993. *Fungicide and Nematicide Tests* **49**, 261.

Shaw, M.W. and Royle, D.J. (1986) Saving septoria sprays: The use of disease forecasts. In *Proceedings of the British Crop Protection Conference: Pests and Diseases.* Brighton, pp. 1193–1200.

Shefelbine, P.A., Mathre, D.E. and Carlson, G. (1986) Effects of chloride fertilizer and systemic fungicide seed treatments on common root rot of barley. *Plant Disease* **70**, 639–642.

Shtienberg, D., Dinoor, A. and Marani, A. (1990) Wheat Disease Control Advisory, a decision support system for management of foliar diseases of wheat in Israel. *Canadian Journal of Plant Pathology* **12**, 195–203..

Skoglund, L.G. and Brown, W.M. (1988) Effects of tillage regimes and herbicides on *Fusarium* species associated with corn stalk rot. *Canadian Journal of Plant Pathology* **10**, 332–338.

Stevenson, F.C. and van Kessel, C. (1996) A landscape-scale assessment of the nitrogen and non-nitrogen benefits of peas in a crop rotation. *Soil Science Society American Journal* **60**, In Press.

Stonard, R.J., Ayer, S.W., Kotyk, J.J., Letendre, L.J., McGary, C.I., Nickson, T.E., LeVan, N. and Lavrik, P.B. (1994) Microbial secondary metabolites as a source of agrochemicals. In Hedin, P.A., Menn, J.J. and Hollingworth, R.M. (Eds) *Natural and Engineered Pest Management Agents.* American Chemical Society, Washington, DC, pp. 25–36.

Sutton, J.C. and Vyn, T.J. (1990) Crop sequences and tillage practices in relation to diseases of winter wheat in Ontario. *Canadian Journal of Plant Pathology* **12**, 358–368.

Tan, C.S. and Tu, J.C. (1995) Tillage effect on root rot severity, growth and yield of beans. *Canadian Journal of Plant Science* **75**, 183–186.

Tekauz, A. (1990) Characterization and distribution of pathogenic variation in *Pyrenophora teres* f. *teres* and *P. teres* f. *maculata* from western Canada. *Canadian Journal of Plant Pathology* **12**, 141–148.

Tekauz, A., Harper, F.R. and Davidson, J.G.N. (1985) Effect of seeding and seed treatment fungicides on infection of barley by *Pyrenophora graminea. Canadian Journal of Plant Pathology*

7, 408–416.

Teo, B.K., Yitbarek, S.M., Verma, P.R. and Morrall, R.A.A. (1988) Influence of soil moisture, seeding date, and *Rhizoctonia solani* isolates (AG2-1 and AG4) on disease incidence and yield in canola. *Canadian Journal of Plant Pathology* **10**, 151–158.

Teo, B.K., Morrall, R.A.A. and Verma, P.R. (1989) Influence of soil moisture, seeding date, and canola cultivars (Tobin and Westar) on the germination and rotting of sclerotia of *Sclerotinia sclerotiorum. Canadian Journal of Plant Pathology* **11**, 393–399.

Timm, C.A., Goos, R.J., Johnson, B.E., Sobolik, F.J. and Stack, R.W. (1986) Effect of potassium fertilizers on malting barley infected with common root rot. *Agronomy Journal* **78**, 197–200.

Tinline, R.D. and Hunter, J.H. (1982) Herbicides and common root rot of wheat in Saskatchewan. *Canadian Journal of Plant Pathology* **4**, 341–348.

Tinline, RD., Ukrainetz, H. and Spurr, D.T. (1993) Effect of fertilizers and of liming acid soil on common root rot in wheat, and of chloride on the disease in wheat and barley. *Canadian Journal of Plant Pathology* **15**, 65–73.

Turkington, T.K. and Morrall, R.A.A. (1993) Use of petal infestation to forecast sclerotinia stem rot of canola: The influence of inoculum variation over the flowering period and canopy density. *Phytopathology* **83**, 682–689.

Turkington, T.K., Morrall, R.A.A. and Gugel, R.K. (1991a) Use of petal infestation to forecast sclerotinia stem rot of canola: Evaluation of early bloom sampling, 1985–90. *Canadian Journal of Plant Pathology* **13**, 50–59.

Turkington, T.K., Morrall, R.A.A. and Rude, S.V. (1991b) Use of petal infestation to forecast sclerotinia stem rot of canola: The impact of diurnal and weather-related inoculum fluctuations. *Canadian Journal of Plant Pathology* **13**, 347–355.

Whittaker, G., Lin, B.H. and Vasavada, U. (1992) A look of pesticide reduction and profits. *Agricultural Outlook* **188** (August), 23–25.

Williams, M.E. (1992) Pesticides: The potential for change. *EPA Journal* **18**(2), 15–16.

Wolf, T.M., Grover, R., Wallace, K., Shewchuk, S.R. and Maybank, J. (1993) Effect of protective shields on drift and deposition characteristics of field sprayers. *Canadian Journal of Plant Science* **73**, 1261–1273.

Woodhead, S.H., O'Leary, A.L., O'Leary, D.J. and Rabatin, S.C. (1990) Discovery, development, and registration of a biocontrol agent from an industrial perspective. *Canadian Journal of Plant Pathology* **12**, 328–331.

Zadoks, J.C. (1981) EPIPRE: A disease and pest management system for winter wheat developed in The Netherlands. *EPPO Bulletin* **11**, 365–369.

CHAPTER 15

IPM Practices for Reducing Insecticide Use in US Fruit Crops

Frank G. Zalom

Statewide IPM Project and Department of Entomology, University of California, Davis, CA, USA

INTRODUCTION

About 40 species of fruit are grown commercially in the United States, on plants as diverse as trees (e.g. apples, peaches, citrus), vines (e.g. grapes, raspberries), bushes (e.g. blueberries, cranberries) and a few non-perennials (e.g. strawberries). Some are grown for the fresh market, but a significant proportion are processed as canned, frozen or juiced products. Post-harvest storage and handling is important in both fresh and processed fruit production. Fruit crops are grown in virtually every state, with commercial production units ranging from small market gardens to large corporate farms that may produce a few thousand hectares of a particular crop. Each of these crops, growing regions and production units is influenced by its own unique production concerns such as climate, soils, water availability and quality, pest complexes, marketing and economics.

Insects are one of the major production concerns for fruit growers because they can damage plants directly as well as damage the marketable product, reducing yield and quality. Because most fruit are grown on perennial plants, growers must be concerned about the direct damage insects will cause to trees or vines which could weaken or kill them. Unlike annual production systems, perennial fruit crops take several years to reach production, and can remain productive for many years. Therefore, growers have an investment in each tree or vine on their farms, and the productivity of those trees or vines will influence their value over time. Fruit producers must also meet demands for high quality, blemish-free fruit for both fresh and processed markets. In some cases, insect damage standards are established which must be met by growers in order for them to market their crops. If these standards are not met, the grower must sort damaged fruit, which is typically quite expensive, or face losing the ability to go to market with the fruit. In other cases, incentives are given by marketers to growers who can reduce insect damage below established levels.

Interestingly, organophosphates continue to be the most widely used category of insecticides for fruit production. The widespread use of synthetic organic insecticides has

---

Techniques for Reducing Pesticide Use. Edited by D. Pimentel

led to a number of issues which are causing the general public, regulatory agencies and growers themselves to seek ways of managing insects with reduced use of these materials or with alternative practices.

Recently, food safety concerns as they relate to pesticide residues, whether real or perceived, has had a profound affect on the attitudes of growers and those marketing fruit products. The risk to producers of a single pesticide residue being detected which might be perceived as harmful (e.g. daminozide in apples) is great. Residue concerns are also significant for processors who may be storing products for a considerable period of time, and for whom brand recognition is possible. Products with residues of materials that are within tolerance levels and generally considered safe today may prove to be a liability if consumer perceptions change. Both growers and those bringing fruit to market would prefer the presence of minimal residues if it is possible to do so while maintaining fruit quality.

Farm worker exposure is an especially serious concern in labour-intensive crops such as fruits, where workers are directly involved in performing many cultural practices where insecticides have been applied. In some crops, lengthened re-entry intervals and protection standards implemented to protect workers from exposure to residues have limited the practical use of the materials affected.

Insect resistance to some of the most widely used insecticides in fruit production is increasingly common. Coincidentally, the pesticide re-registration process is substantially reducing the number of available chemical alternatives, and some insecticides are being lost owing to other regulatory issues. A decreasing number of insecticide choices intensifies the likelihood of genetic resistance, as fewer materials become used more intensively.

The mortality of some protected, non-target species has been attributed to the use of certain insecticides in orchards. Being perennial plantations, a diverse fauna often becomes associated with most fruit production systems, including vertebrates such as rodents and birds. Although many of these vertebrates are pest species, they are part of a food chain that attracts predators such as raptors into the orchards where they can be exposed to potentially lethal doses of insecticides.

In spite of the relatively high value of most fruit crops and quality concerns faced by fruit growers, fruit systems provide some of the best examples of integrated pest management (IPM) development and adoption.

The concept of IPM originated in the late 1950s with the goal of more effectively controlling pest problems while reducing the undesirable side effects of pesticides. One of the first citations for 'integrated control' in the literature was in walnuts, and concerned the integration of pesticides with parasites of the walnut aphid, *Chromaphis juglandicola* (Kaltenbach) (Michelbacher and Bacon, 1952). Later, in a review article on insect control for deciduous fruit, Barnes (1959) emphasized the need for an integrated approach. Formal tree fruit IPM efforts began in 1972 when the Huffaker Project, funded by the National Science Foundation, and the Smith–Lever IPM Project, funded through the USDA Cooperative Extension Service, provided states with an opportunity to develop new knowledge and to transfer knowledge to growers. Under this effort, pilot IPM projects in apples were initiated in Michigan, New York, Pennsylvania and Washington (Whalon and Croft, 1984). Prior to this time, many states had already developed considerable IPM knowledge for a number of tree fruit and nut crops through regular programmatic efforts.

Progress in research was marked by interdisciplinary projects which facilitated interactions of the various crop protection disciplines. Progress in extension was marked by the preparation of manuals and fact sheets, and by reports of significant reductions in pesticide use as the result of IPM implementation projects (e.g. Rajotte et al., 1987).

Recently, there has been increased recognition that more biologically intensive approaches are needed to minimize the pest impact. Although biologically intensive IPM methods have been studied for many years, the results of this research have not been as widely accepted as the application of decision support methodology.

INTEGRATED PEST MANAGEMENT TACTICS

IPM systems for fruit crops require a diversity of pest control tools or tactics which are compatible with one another and with other production practices. The most widely used IPM tactics are those which provide decision support such as monitoring methods, including traps and phenology models. The application of these decision tools often results in both a reduction in pesticide use and more efficient pest control. Biologically intensive IPM tactics available to, and used by, fruit growers for managing insects include biological controls, resistant varieties, semiochemicals and cultural controls. Biological controls including parasites and predators have been widely used in fruit production, and perennial cropping systems, which include most of the fruit crops, are especially good targets. Insect pathogens, including viruses, fungi, bacteria and entomophagous nematodes, are also used for specific pests to prevent disruption by conventional pesticides. With a few notable exceptions, host-plant resistance using conventional breeding methods has not been as widely attempted in perennial fruit crop systems as in annual cropping systems, but variation in susceptibility to insects among cultivators of a particular species has been noted. Transgenic gene insertion to provide insect resistance is now being attempted in a few fruit crops. Mating disruption with pheromones has been used successfully for some fruit pests, and is becoming more widely available to growers. Canopy management and pruning are used to promote advantageous tree structure. Sanitation and removal of wild hosts are used to reduce sources of pests. Vegetation management, including the use of cover crops, has been promoted to enhance the establishment and efficacy of biocontrol agents.

Efforts aimed at the commercialization and adoption of these tactics have encountered significant technical, regulatory, economic and institutional barriers, which must be overcome if biologically intensive methods are to become common in commercial agriculture. Because they are often information-intensive, crop consultants will play an increasingly important role in the implementation process.

DECISION SUPPORT

Lack of practical monitoring procedures or use of those procedures often results in an excessive use of pesticides and poor timing of applications. In many instances (e.g. National Research Council, 1989), pesticides used for controlling a given pest have been reduced by 40% without affecting fruit quality or yield simply by using quantitative monitoring procedures in combination with realistic control action thresholds. The USDA Extension IPM pilot projects conducted on apples and pears during the 1970s demonstrated that monitoring also reduced the costs of pesticides to growers, and increased net returns (e.g. Barnett et al., 1978; Rajotte et al., 1987).

Monitoring techniques which incorporate the use of bait or pheromone traps and phenology models are simple tools which have become widely used for monitoring various orchard insects. They are useful in providing information on the mobile adult stages of insects. When used in conjunction with phenological models, they can be used to predict development of the target species and ultimately to time pesticide applications. A 1987 survey of California pest control advisors (Flint and Klonsky, 1989) indicated that a substantial number used pheromone traps or bait traps, degree-day calculations to predict insect phenology, and University of California guidelines to make treatment decisions for various orchard pests. The codling moth, *Cydia pomonella* (L.), was the most widely monitored of the insects mentioned. Attractant traps which use colour and shape are being used successfully in many apple growing regions to monitor the phenology of the apple maggot, *Rhagoletis pomonella* Walsh (e.g. Moericke et al., 1975; Prokopy and Owens, 1976; AliNiazee et al., 1987; Jones and Davis, 1989). Attractant baited traps are used to sample the western cherry fruit fly, *Rhagoletis indifferens* Curren, and to time pesticide applications reduced pesticide use by 20–100% (AliNiazee, 1978).

Pheromone traps and lures are commercially available for a number of other insect pests of fruit, including the American plum borer, *Euzophera semifuneralis* (Walker), amorbia, *Amorbia cuneana* (Busck), the apple ermine moth, *Yponomeuta malinellus* Zeller, the cranberry girdler, *Chrysotuechia topiaria* (Zeller), the filbertworm, *Cydia latiferreana* (Walsingham), the fruit tree leafroller, *Archips argrospila* (Walker), the grape berry moth, *Endopiza viteana* Clemens, the lesser appleworm, *Grapholita prunivora* (Walsh), the lesser peachtree borer, *Synanthedon pictipes* (Grote and Robinson), the obliquebanded leafroller, *Choristoneura rosaceana* (Harris), the omnivorous leafroller, *Platynota stultana* Walsingham, the omnivorous looper, *Sabulodes aegrotata* (Guenee), the orange tortrix, *Argyrotaenia citrana* (Fernald), the spotted tentiform leafminer, *Phyllonorcycter blancardella* (F.), and the tufted apple budmoth, *Platynota idaeusalis* (Walker).

Some traps are designed to capture insects without using semiochemical attractants, relying instead on some other behavioural response such as the target insect's visual stimulus to colour. Recently a modified pyramidal type of trap called the Tedders trap was developed which captures the plum curculio, *Conotrachelus nenuphar* Herbst, by capitalizing on knowledge of the insect's migratory behaviour. As no practical monitoring method has been available for this key eastern US fruit pest, the trap could have significant impact when widely used to evaluate phenology. The trap's efficiency could also be enhanced if orchard tree volatiles are added (e.g. Butkowich and Prokopy, 1993).

Direct sampling of pests and beneficial species by counting numbers per leaf or other appropriate unit is a more accurate way of assessing abundance, and permits the application of treatment thresholds. Pests that are relatively immobile, such as scale insects and aphids, and those that are typically found on leaves, such as mites, thrips, immature whiteflies and leafhopper nymphs, are especially good candidates for direct sampling. Pests or beneficials can be counted directly on the leaves, or dislodged onto another surface. Direct sampling has a distinct disadvantage in that it can be time-consuming, requiring more labour than a crop consultant or field scout can economically afford.

The development and implementation of binomial, presence–absence sampling programs has helped simplify direct sampling for some pests. Examples where both pests and beneficials are sampled in this manner include spider mites, *Tetranychus* spp., and the western orchard predatory mite, *Galandromus* (= *Metaseiulus* or *Typhlodromus*) *occiden-*

talis (Nesbitt), on almonds in California (Wilson et al., 1984; Zalom et al., 1984) and tart cherry in Utah (Jones, 1990), and citrus red mite, *Panonychus citri* (McGregor), and the predatory mite, *Euseius tularensis* Congdon, on citrus (Zalom et al., 1985). Approximately 70.6 and 52.4% of the pest control advisors working on almonds and citrus, respectively, in California have tried these presence–absence sampling programs (Flint and Klonsky, 1989).

Although insect monitoring procedures are usually used to make pesticide-use decisions, it can be argued that monitoring pests, or preferably pests, beneficial species and crop status in combination, is the fundamental practice separating minimum IPM use from non-use. Use of more biologically intensive methods in combination with monitoring to reduce pesticide use still further constituted more intensive IPM systems.

BIOLOGICAL CONTROLS

Perhaps the most famous example of classical biological control was the 1888 introduction of the vedalia beetle, *Rodolia cardinalis* (Mulsant), into California to control the cottonycushion scale, *Iceryi purchasi* Maskell. The scale was introduced into California about 1868, and soon devastated the State's citrus industry. An eradication program conducted at the time, which included the use of fumigation with hydrocyanic acid and burning infested trees, ultimately failed. The predaceous vedalia beetle, collected by Albert Koebele in Australia and imported to southern California, proved an immediate success. In 1 year, shipments of oranges from Los Angeles County tripled, returning millions of dollars to affected growers from an initial investment of about $2000.

Other biological control successes followed in citrus. Many were focused on work that was done in the Fillmore Citrus Protective District, a group of public and private organizations and individuals organized in 1922 to practice area-wide pest management on 1620 citrus hectares near Fillmore, California (Graebner et al., 1984). Although formed to share the costs of fumigation and spraying for California red scale, *Aonidiella aurantii* (Maskell), across growers in the District, the area-wide pest management approach also proved suited for managing the release of biological control agents. In 1926, the district began the construction of an insectary to rear a predator of the citrophilus mealybug, *Pseudococcus calceolariae* (Maskell). The mealybug had been introduced into California in 1913, and could not be controlled by fumigation. The mealybug destroyer, *Cryptolaemus montrouzieri* Mulsant, was introduced by University of California scientist Harry Smith, and it proved effective for control. Unfortunately, it did not overwinter well in the citrus groves, so it had to be released annually. Two parasites of *P. calceolariae*, *Coccophagus gurneyi* Compere and *Tetracnemus pretiosus* Timberlake were introduced from Australia by Harold Compere, another University of California scientist, in 1929, and were reared in the Fillmore insectary for release in District groves. By 1930, the mealybug was under complete biological control (DeBach, 1974).

In 1937, *Metaphycus helvolus* Compere, a parasite of the black scale, *Saissetia oleae* (Oliver), was introduced into California, and proved very successful when it was introduced into the Fillmore citrus groves. By 1944, the release of *M. helvolus* was considered so effective and important that a second insectary was constructed by the District to increase its production (Graebner et al., 1984). The use of *M. helvolus* to control black scale eliminated much of the fumigation that had been done in the district, but also resulted in increasing California red scale incidence.

Organophosphate insecticides introduced after World War II exacerbated the problem with California red scale, and dramatically increased eradication costs for District growers, who eventually decided to attempt the regulation of California red scale with releases of the imported parasite *Aphytis melinus* DeBach, which proved to be an excellent biological control agent in this region (Graebner et al., 1984). Its activity is now complemented by releases of other imported parasites, *Encarsia perniciosi* Tower and *Comperiella bifasciata* Howard, and the predatory coccinellids *Lindorus lophanthae* (Blaisdell) and *Chilocorus stigma* (Say).

Annual pest costs for growers of the Fillmore Citrus Protective District during the 10-year period of 1971 to 1980 were $28.75 ha^{-1} versus $145.00 ha^{-1} for those Ventura County growers not in the District (Graebner, 1982).

A recent example of classical biological control was the control of the filbert aphid, *Myzocallis coryli* (Goetze), which was introduced into Oregon from Europe by introducing a French strain of a parasite, *Trioxys pallidus* Haliday (Messing and AliNiazee, 1989). Filbert aphid had become a serious pest requiring two applications of organophosphate insecticides each year for control. Insecticides applied for this pest have declined substantially since the introduction of the parasitoid.

Another French strain of *T. pallidus* was introduced into California in 1959 to control the walnut aphid (Schlinger et al., 1960). It became established primarily in the mild coastal areas of southern California, but did not succeed in the hotter and drier areas of the San Joaquin Valley. In 1968, researchers collected and introduced a strain of *T. pallidus* from Iran which proved very successful in these hotter regions (van den Bosch et al., 1970). If not disturbed by the use of broad-spectrum insecticides to control the codling moth and the walnut husk fly, *Rhagoletis completa* Cresson, the parasite is capable of reducing walnut aphid populations to very low levels (van den Bosch et al., 1979).

T. pallidus has recently been introduced into New Mexico as well to control the pecan aphid complex (Ellington, 1991), in the Mesilla Valley. This introduction is part of a four-element biological control program that together with terminating pesticide use, initiating the mass release of aphid predators, and using ground cover to improve beneficial habitat, has eliminated insecticide sprays for pecan aphids which had previously received 6 sprays per year (Ellington et al., 1995).

CONSERVATION AND AUGMENTATION

Conserving natural enemies by avoiding disruptive sprays has become an essential practice in several orchard systems. Prior to the 1940s, spider mites were considered sporadic pests in most perennial fruit systems. Since the widespread use of broad-spectrum pesticides, spider mites have become annual pests in virtually all fruit-growing regions of North America. In the Pacific Northwest, *Tetranychus mcdanieli* McGregor is a primary spider mite pest of apples. Beginning in the mid-1960s, an effective mite management approach, based on the conservation of the predator mite *G. occidentalis* using selective insecticides for control of orchard pests, was developed and implemented. The strategy was based on detailed observations taken by Hoyt (1969) which indicated that by carefully using organophosphate cover sprays and encouraging an alternate prey, the apple rust mite *Aculus schlechtendali* (Nalepa), *G. occidentalis* could provide control of the spider mites. Implementation of this program reduced the average mite control cost for Washington State growers from $24 ha^{-1} in 1967 to $8–12 ha^{-1} in 1985 (Croft, 1990).

Similarly, conservation of the phytoseiid complex of *G. occidentalis*, *Typhlodromus pyri* Scheuten and *Typhlodromus arboreus* Chant in Oregon pear and apple orchards through the use of selective insecticides for major pests, including pheromone mating disruption for the codling moth, has been demonstrated to keep the yellow spider mite, *Eotetranychus carpini borealis* (Ewing), under natural biological control (Croft and AliNiazee, 1983; Moffitt and Westigard, 1984).

Pyrethroid insecticides have been shown to be highly disruptive in orchards by killing phytoseiids, and their use should be avoided (e.g. Croft and Hoyt, 1978; AliNiazee, 1984). There is some indication that the residual effects of pyrethroids applied to phytoseiids in orchards will persist into the subsequent growing season (Bentley et al., 1987).

Several biological control agents are commercially available for augmentative or inundative release (Hunter, 1994). Of these, the most widely used agents include the predaceous mites, particularly *Phytoseiulus persimilis* Athias-Henriot, *G. occidentalis* and *Amblyseius* (= *Neosieulus*) *californicus* (McGregor), which are released for the control of various spider mite species. *P. persimilis* was first introduced into North America in the mid-1960s (Oatman, 1965), and has become permanently established along the south and central coasts of California following releases into strawberry fields. *P. persimilis* is a highly mobile predator (Oatman, 1970), but it typically does not move into or across a field until it exhausts spider mite populations where is has been feeding. Oatman et al. (1976) indicated that *P. persimilis* must be released at rates of 5 or 10 per plant in order to be effective, but such releases would be uneconomical. Currently, releases are being made at lower rates, and are targeted to early detection of spider mite infestations. Occasionally, releases of *P. persimilis* are combined with releases of *A. californicus* and *G. occidentalis*, which have a greater ability to persist at low prey densities.

Inundative releases of the parasitoid *Trichogramma platneri* against the codling moth are being attempted commercially on California walnuts and apples in combination with orchard sanitation and pheromone mating disruption. First attempted on these crops by Flanders (1926), control of codling moth by parasites alone has not been commercially viable even at levels of parasitism which exceed 75% (Dolphin et al., 1972). However, integrated into a program of multiple tactics, the parasites may provide sufficient additional control to eliminate conventional pesticides for most of the season.

Integrating tactics was a key element of one program where the experiences of California coastal citrus growers were introduced and adapted for the San Joaquin Valley in 13 demonstration orchards. Under this program (Haney et al., 1992), California red scale was managed with releases of *A. melinus* and selective use of low rates of chlorpyriphos or narrow-range oil as needed. Citrus thrips, *Scirtothrips citri*, were managed with selective treatments of sabadilla, a botanical insecticide, applied as determined by treatment thresholds. Lepidoptera were managed with *Bacillus thuringiensis* Berliner and narrow-range oil treatments as needed. Insecticide treatments declined by 35% in these orchards, with a saving to growers averaging $25.20 ha^{-1}.

The release of laboratory-selected strains of natural enemies chosen for resistance to disruptive insecticides commonly used in the orchards allows the selected natural enemies to persist even when the disruptive materials are applied for control of other key pests. A laboratory-selected strain of the predator mite *G. occidentalis*, resistant to carbaryl, organophosphates and sulfur, is now used successfully to manage spider mites in California almond orchards (Hoy et al., 1984). Extensive research was conducted into economical mass-rearing (Hoy et al., 1982), sampling (Wilson et al., 1984; Zalom et al., 1984) and applications of selective acaricides at lower-than-label rates to help to reduce the spider mite population to predator:mite ratios which were in favour of the predator mites, allowed its use to be integrated with other almond orchard practices. Analysis of the implementation of the *G. occidentalis*-based integrated mite management program on California almonds proved that it was extraordinarily cost-effective (Headley and Hoy, 1987). Approximately 65% of the 170 000 ha of almonds produced in California now utilize this program, at an annual savings of over $21 million, primarily by reducing acaricide costs (mostly propargite). The annual return on the initial research investment has been between 500 and 600%.

With few exceptions, studies which document the effectiveness of most commercially available predators and parasites released in commercial situations are rare. Research to improve the rearing and release of these organisms, which would reduce costs and improve the predictability of controlling target pests, is necessary in order for them to become more widely accepted by growers.

MICROBIAL AGENTS

Research on the use of microbial agents specifically for insect control has been ongoing since the 1950s (e.g. Steinhaus, 1954), with identified agents including bacteria, fungi, viruses and entomophagous nematodes. Except for the entomophagous nematodes, commercial production and use of microbial agents are regulated in the same way as pesticides. Few microbial agents have been registered for use on fruit crops, and of those, the primary agent used has been *B. thuringiensis*. Recently, codling moth granulosis virus was registered, and has been used by organic growers to a limited extent (Caprile et al., 1994). Microbial agents or their toxins are primarily delivered like conventional pesticides, and are applied to a crop or plant as needed. While generally regarded as non-disruptive to non-target species, microbial agents have not been widely used because they are not perceived as being as effective as conventional pesticides, probably because of their relatively short residual activity and narrow-spectrum action against pest species.

A recent example where better timing of *B. thuringiensis* applications have led to widespread commercial use has been on California almonds, peaches and prunes for controlling the peach twig borer, *Anarsia lineatella* Zeller, where it is intended to replace dormant-season applications of organophosphates. The dormant-season application of an oil with an organophosphate insecticide (usually diazinon or methidathion, and parathion when it was available) had been the recommended control for peach twig borer for over 20 years, and was considered a good IPM practice because it was less disruptive to natural enemies than in-season sprays. Virtually all producers of stone fruits and almonds applied the dormant-season spray. The use of organophosphate insecticides in dormant-season sprays became controversial when diazinon, the most widely used material, was implicated in killing the red tailed hawk, a treaty-protected raptor, and was being found in the rivers that drain both the San Joaquin and Sacramento Valleys. Applications of *B. thuringiensis* during bloom at the 'popcorn' stage of almonds or the 'red bud' stage of peaches, and again 'petal fall' provide acceptable control of the peach twig borer (Barnett et al., 1993; Roltsch et al., 1995). It is estimated that about 15% of California's 220 000 ha of peaches, prunes and almonds now utilize this approach.

HOST-PLANT RESISTANCE

Host-plant resistance, often mentioned with biological control as one of the cornerstones of IPM (e.g. Huffaker, 1985), has not been a priority for varietal screening of perennial fruit crops. Conventional breeding programs only occasionally address disease resistance, and more rarely arthropod resistance. One important case of host-plant resistance in perennial fruit crops has been resistant grape rootstocks. The grape phylloxera *Daktulasphaira vitifoliae* (Fitch) is an aphid-like insect that attacks vinifera grape roots, causing stunted growth and vine death. The insect is native to the Mississippi Valley and the southeastern US, and devastated vineyards in Europe and California following its introduction into those areas in the mid-1800s. There are no effective chemical controls for phylloxera, although grape growers must make certain that roots of grape transplants and the vineyard soils are uninfested. Many resistant rootstock varieties have been selected to suit particular growing regions and conditions (Kido et al., 1982). Most of these are only moderately resistant to phylloxera attack, but are able to tolerate infestations. Recently, there is evidence that a new strain of grape phylloxera has developed which has overcome resistance to the A × R No. 1 rootstock (DeBenedictis and Granett, 1993), increasing the need to develop new rootstocks and a better understanding of the mechanisms which result in resistance or tolerance.

In some crops, variation in cultivar susceptibility to insect attack is known, but has not been the result of conscious selection for that characteristic. For example, some cultivars of almonds have consistently greater damage from navel orangeworm (University of California, 1985). Soft shell hardness and seal, and time and duration of hardness all influence susceptibility. Soft shell cultivars with a poor seal are more vulnerable than hard shell cultivars with tight seals. Later season maturation or a more lengthy hullsplit period compounds the problem. Unfortunately, cultivars are selected primarily for market factors, and the most widely grown and preferred cultivar 'Nonpareil' is also one of the most susceptible.

Breeding for insect resistance using transgenic gene insertion methods has been successfully accomplished for apples and walnuts using a gene which produces the toxins from *B. thuringiensis* (Dandekar et al., 1994). This technology holds great promise for controlling key pests in crop plants, potentially reducing the use of conventional pesticides. However, there is concern that selection from constant exposure of resident pest populations in orchards could accelerate the development of pest resistance to *B. thuringiensis* (Rissler and Mellon, 1993). Resistance management strategies should be developed and implemented (e.g. Alstad and Andow, 1995) before resistant perennial fruits containing the *B. thuringiensis* gene are used commercially in order to avoid the loss of *B. thuringiensis* for producers who have existing investments in susceptible orchards.

SEMIOCHEMICALS

Research on the use of semiochemicals to disrupt insect communication has led to new possibilities for manipulating insect behaviour. The most common approach has been to use pheromones, which are substances produced by insects that have a specific effect on members of their own species (Bartell, 1977). The most common behaviours affected are mating, aggregation, feeding, oviposition, alarm and defence.

The most common use of pheromones is in monitoring for decision support, and there are many examples of their importance in pest management for fruit growers. Of the possible approaches for controlling pests with semiochemicals, mating disruption has received the most attention. Mating disruption relies on the concept that releasing enough sex pheromone into the atmosphere will disrupt the orientation of male insects to females, thereby reducing the number of matings within a population and ultimately leading to pest suppression. In the US, mating disruption has become commercially viable for control of the oriental fruit moth, *Grapholita molesta* (Busck), in California (Weakley et al., 1987; Rice and Kirsch, 1990), the grape berry moth in New York (Dennehy et al., 1990) and the peachtree borer, *Synanthedon exitosa*, in the southeastern US (Snow, 1990). Pheromone mating disruption of the oriental fruit moth in California peaches in used on over 4000 ha annually, and has increased from 600 ha in 1987, the first year following its initial EPA registration.

Recently, some large-scale commercial demonstrations of mating disruption for the codling moth have been attempted on apples and pears in California, Oregon and Washington (e.g. Bentley et al., 1994; Brunner, 1994). In 1995 about 1200 ha were treated in five area-wide codling moth control sites. The area-wide approach is expected to be more effective than smaller individual sites because border effects would be reduced, mitigating the observation that damage is twice as great along borders as in the interior of orchards, and movement of natural enemies into untreated orchards is expected to increase. To date, acceptable control of this pest has been most predictable when mating disruption is combined with some other method to reduce initial populations.

The use of mating disruption for fruit pests has been limited by product registration, product cost, variations in dispenser technology and the resulting efficacy for different pests, and the cost of applying dispensers. Pheromones of relatively few insects are registered for use as a control method. Although specificity is considered to be one advantage of using pheromones, if insecticide sprays must be applied for other key pests of a particular fruit production system the benefits of using pheromones for control of a single pest might be lost. The cost of the pheromones and their application are significant determinants for use by growers. Insects such as the oriental fruit moth require multiple insecticide applications for control, making the cost of a pheromone control programme competitive with a conventional insecticide approach. In addition, the cost of applying pheromone dispensers to orchard systems can be greater than the cost of the pheromone dispensers themselves when they must be placed by hand in the tops of trees.

The benefits of mating disruption when its use can substitute for conventional insecticides includes conservation of biological control agents. Westigard and Moffitt (1984) noted that populations of beneficial organisms increased substantially in pear orchards where pheromone mating disruption was tested for control of the codling moth, and that fruit damaged by the pear psylla, a serious pest that is disrupted by organophosphate sprays for control of the codling moth, was only about 15% of that observed in insecticide-treated plots.

In mass-trapping, pheromones are intended to attract a segment of a population to traps for the purpose of removing those individuals from the population. Results have been mixed in agricultural systems because of the expense of implementing such systems, and failure to reduce damage to acceptable levels. Madsen et al. (1976) reported some success in controlling the codling moth in an isolated apple orchard with a low initial population by installing and servicing 10 pheromone traps per hectare.

Volatiles present in almond oil attract the navel orangeworm to the almond nut for oviposition, and this knowledge has been used in developing egg traps, which are widely used by California pest control advisors to monitor pest phenology (Rice et al., 1976). When the almond oil volatiles were sprayed on tree trunks, successful oviposition disruption resulted as progeny of eggs laid there did not survive (Phelan and Baker, 1987). This product holds promise for control, but because of its limited market may never receive registration and become commercially available.

Attracting insects to sources of insecticide, or 'attracticide' (Haynes and Baker, 1988), has been implemented commercially for a few insects, but no products have been registered for this purpose for key fruit pests.

CULTURAL CONTROLS

A number of horticultural production practices can serve to reduce arthropod pressure and increase biological diversity in fruit production systems. Pruning and canopy management are used to promote advantageous tree or vine structure, or to remove infested or injured plant parts. Sanitation and removal of wild hosts are used to reduce sources of pests. Vegetation management, including the use of cover crops and trap crops, have been promoted to enhance the establishment and efficacy of biocontrol agents. Planting location, fertility management and water management can also be important insect management considerations.

PRUNING

Cultural controls such as pruning are often compatible with other control tactics, and when combined they enhance the level of control compared with using either tactic alone. For example, many parasites have been introduced into California since 1891 for control of black scale on olives, but these parasites have generally only proved satisfactory in controlling the scale in coastal regions, while control in the central valley has been disappointing (Daane et al., 1991). Daane and Caltagirone (1989) showed that cultural practices used in olive orchards can affect the microclimate, which in turn can influence black scale development and survival Closed tree canopies and irrigating with high-volume sprinklers lower the orchard temperature and raise humidity, creating an environment conducive to black scale outbreaks. Pruning strategies can directly increase summer mortality and canopy air circulation, influencing scale development patterns to favour parasite establishment. Low-volume sprinklers and drip irrigation reduces orchard humidity, creating a less conducive environment for the scale. A monitoring program (Daane and Caltagirone, 1989) based upon adult scale counts, and the knowledge that scale fecundity is influenced by temperature, canopy structure and type of irrigation, permits growers to decide on appropriate management techniques, including parasite release and pruning.

Pruning citrus skirts so that leaves and branches do not touch the ground provides a physical barrier to the movement of the Fuller's rose beetle, *Pantomorus cervinus*, a flightless beetle which must crawl to the fruit for oviposition (Haney and Morse, 1988). The beetle was not previously considered an economic pest, but it now appears on a pest exclusion list in Japan, so growers exporting fruit to Japan must control Fuller's rose beetle in their groves. Skirt-pruning is a fairly expensive, but non-disruptive, method of control, which also reduces ant and snail damage.

CANOPY MANAGEMENT

Leaf removal around grape bunches to increase air flow in and around berry clusters has been used successfully to reduce the incidence of bunch rot caused by the fungus *Botrytis cinerea*. It has also been shown to affect the severity of powdery mildew (Gubler et al., 1987), and the colour and ripening of red and purple varieties, which improves grape quality. This practice has been adopted by at least 20% of California grape growers (Pence and Grieshop, 1990), and its use is estimated to be as high as 50% in the North Coast wine grape region. Leafhoppers and spider mites are also concentrated on the leaves, which are removed in the spring, and an incidental benefit of the practice is a significant reduction in their abundance. In addition, pesticides applied for control of insects associated with the bunches can more easily penetrate the altered canopy, which increases efficacy and reduces the need for repeated applications. Pence and Grieshop (1990) estimate that fungicides applied for bunch rot control and insecticides applied for leafhopper control have been reduced by 50% in vineyards by using leaf removal.

SANITATION

Insects can survive and reproduce in crop and non-crop host-plant materials in and around orchards, vineyards and fields. Removing these refugia can substantially reduce pest populations for the subsequent season.

Complete control of post-first-generation codling moths was obtained in a Massachusetts orchard over several successive growing seasons by removing all apple and pear trees within 100 m of the orchard perimeter (Prokopy et al., 1991). Potential problems with this approach are the lack of dilution of orchard populations with immigrant moths, possibly leading to increased pesticide resistance in the now isolated orchard population, and a reduction in the pool of natural enemies, which are no longer available on the untreated host trees that were removed (Prokopy, 1994). Similar concerns for disrupting natural enemy sources have been expressed when modifying surrounding habitats for pear psylla control (Gut et al., 1988; Van den Baan and Croft, 1990).

Sanitation was a key element of the successful almond IPM program implemented in California during the 1980s. The navel orangeworm became a key pest of California almonds in the 1960s, and by the late 1970s, growers had become dependent on azinphos-methyl and carbaryl for control. Damage by navel orangeworm increased dramatically because of poorly timed treatments and lack of natural controls. Secondary pest outbreaks of mites increased, requiring additional pesticide treatments. University of California and USDA research indicated that navel orangeworm populations could be substantially reduced by removing the old 'mummy' nuts remaining on the trees, which served as overwintering hosts for navel orangeworm larvae, and destroying the nuts once on the ground (e.g. Engle and Barnes, 1983). Another important IPM technique for navel orangeworm control is the removal of nuts before the start of the third-generation moth flight. If harvest is delayed, benefits derived from sanitation and any insecticide sprays applied will be seriously reduced. Sanitation and early harvest were adopted by over 80% of almond growers by 1985, while nut damage and insecticide use for navel orangeworm control declined by over 40% (Klonsky et al., 1990). In spite of increased production costs due to winter sanitation (about the cost of one in-season insecticide plus miticide treatment), the return to California almond growers from reduced damage and insecticide use has been in excess of $12 million annually.

Sanitation can be used after harvest to cull damaged fruit, but this approach is labour-intensive. It is typically more economical to control pests in the field. One interesting control approach for the California red scale has been a high-pressure washer, which removes scales from the fruit in the packing house. The washer was developed in South Africa when insecticide resistance to organophosphates in California red scale became serious (Honiball et al., 1979). The device allows field management strategies to focus on prevention of tree injury rather than cosmetic injury (Bedford, 1990), which increases treatment thresholds substantially. The washer has been installed in perhaps eight California and 15 Florida packing houses (J. Morse, personal communication 1995), and also effectively removes sooty moulds growing on fruit from honeydew secretions of black scale, citricola scale and whiteflies.

VEGETATION MANAGEMENT

Selective diversification of floor vegetation has been shown to increase natural enemy abundance within fruit crop systems, but few experiments have shown success in controlling key pests. Altieri and Schmidt (1986) reported that the half of a California apple orchard which had a planted cover crop of vetch and bean plants had fewer leafhoppers and more predatory coccinelid beetles, parasites and spiders than the other half of the orchard which was clean-cultivated. Bugg et al. (1991) showed that vetch and cereal rye planted in a Georgia pecan orchard increased the abundance of predatory coccinelids on both the ground cover and the orchard trees. Bugg and Dutcher (1993) observed that warm-season cover crops harboured alternate prey for syrphid flies, coccinelids and various predatory wasps which can be predators of pecan aphids and other pecan orchard pests. Phytoseiid mite populations were enhanced on vetch and other legumes used as cover crops in Massachusetts (e.g. Coli and Ciurlino, 1990).

Selection and management of cover crops, especially those which remain in the orchards during the summer, must be done in order to prevent induction of other pest problems. Some fruit or nut pests, such as the southern green stink bug, *Nezara viridula* (L.) (Dutcher and Todd, 1983), the tufted apple bud moth (Knight and Hull, 1988), the two-spotted spider mite, *Tetranychus urticae* Koch (e.g. Tedders et al., 1984; Meagher and Meyer, 1990), the green peach aphid, *Myzus persicae* (Sulzer) (Tamaki, 1975), the pavement ant, *Tetramorium caespitum*, the southern fire ant, *Solenopsis xyloni* (Barnett et al., 1989), the tarnished plant bug, *Lygus lineolaris* (Palisot de Beauvois) (Johnson, 1988), and a leafhopper vector of X-disease, *Paraphlepsius irroratus* (Larsen and Whalon, 1987) may be harboured in certain cover crops, and can become a problem if they disperse onto the host crop. Tillage, mowing or herbicide use during the season can also cause pests such as spider mites to migrate onto the trees (Flexner et al., 1991).

Resident vegetation can also harbour beneficial insects, much as do planted cover crops. Tamaki (1972) reported that *Geocoris bullatus* was abundant on peach orchard floor vegetation in Washington State, and appeared to be important in regulating green peach aphid populations. The western orchard predatory mite, *G. occidentalis*, increased on resident vegetation in Utah, controlling spider mites when sufficient populations occurred (Alston, 1994). Bugg and Waddington (1994) provide an excellent review of common orchard floor weeds and the beneficial insects they support.

IMPROVED APPLICATION TECHNOLOGY

Significant pesticide use reduction could be achieved, even without employing biologically based methods or changing current horticultural practices, by improving application technologies for orchard insect control. Hall (1985) estimated that 60% of applications with conventional air-blast sprayers were not placing the desired amounts of
pesticide on peach trees. Byers and Lyons (1985) estimated that about 35% of pesticides
applied by air-blast sprayers do not reach the target and are lost to the environment.
Seiber (1988) showed that a modified air-blast sprayer could reduce pesticide use by
almost 40% while maintaining effective control.

Spray dilution can be a factor in reducing pesticide rates. Haney et al. (1992) estimated
the efficiency for dilute sprays in citrus groves to be 56%, while the efficiency for
low-volume treatments was 73%. Reduced dilution also results in lower application and
energy costs.

Changing the pattern of spraying within orchards can reduce pesticide use. For
example, Prokopy et al. (1978) demonstrated that spraying alternate row middles in apple
orchards as opposed to every row middle reduced insecticide and acaricide use by 50%
without increasing damage or yields. An economic analysis of New Jersey apple growers
spraying alternate middles indicated that growers utilizing this technique saved
$9.90–17.06 ha^{-1} (Rossi et al., 1983), primarily from a reduction in the amount of
insecticides applied.

IPM CERTIFICATION

An IPM education and certification program was initiated in Massachusetts in 1989, in
response to growers' requests for recognition of their IPM practices and requests for
additional IPM education (Hollingsworth and Coli, 1992). Guidelines were developed for
several fruit crops, including apples, strawberries and cranberries, as a list of best
management practices identified by university research and extension specialists, and
private consultants. The guidelines were then sent to state commodity organizations for
review and comment. The program has been tested for several seasons by participating
growers under the soil conservation Service's ICM cost-sharing program. Over 80% of
consumers and almost 70% of the processors, retailers and wholesalers surveyed supported 'IPM-grown' labelling (Paschall et al., 1992). Seventy-five percent of growers
indicated that they would enroll in an IPM certification program (Hollingsworth et al.,
1992). Clearly, there is interest in IPM labelling of fruits as indicated by this experience,
but full implementation and acceptance in the marketplace remain to be tested.

USE OF IPM TACTICS IN FRUIT CROPS

Estimating the use of IPM in fruit crops is difficult largely because of the difficulty in
defining IPM, and the lack of a complete and static IPM system for any given crop or
production region. As discussed, IPM encompasses a wide range of practices which
individually can be thought of as IPM tactics. Perhaps the most fundamental IPM tactic
is field monitoring. IPM tactics also include the use of a number of alternative practices
such as biological and microbial controls, host-plant resistance, semiochemicals, selective

Table 15.1   Use of pesticides, IPM and professional scouting on bearing hectares of five fruit crops

Crop and State	No pesticides	No IPM	Level of IPM use			Professional scouting
			Low	Medium	High	
Apples						
Arizona	22	53	0	2	24	47
California	26	32	0	18	24	53
Michigan	0	46	2	27	25	66
New York	0	41	1	20	39	66
North Carolina	0	88	0	1	11	27
Oregon	0	81	0	1	18	28
Pennsylvania	0	37	1	13	49	74
South Carolina	0	72	0	0	28	66
Virginia	1	32	1	3	64	75
Washington	0	68	2	12	18	44
Grapes						
California	7	35	6	11	40	73
Michigan	0	66	3	29	2	46
New York	1	80	1	6	12	32
Oregon	20	77	0	1	2	6
Washington	7	57	7	8	21	42
Oranges						
Arizona	6	74	20	0	0	40
California	7	18	5	18	51	85
Florida	0	40	17	26	16	72
Peaches						
California	3	49	2	21	24	81
Michigan	0	48	9	25	18	65
New York	0	43	3	35	19	73
North Carolina	0	98	0	1	1	3
Pennsylvania	0	44	2	22	33	74
South Carolina	0	99	0	1	0	27
Virginia	0	52	2	3	43	56
Washington	4	64	14	7	11	42
Pears						
California	7	28	6	14	46	68
New York	0	41	3	40	16	71
Washington	0	64	3	12	21	57

Source: 1991 survey by the USDA Economic Research Service (USDA, 1994).

chemicals and cultural controls such as canopy management, sanitation and vegetation management.

Perhaps the most comprehensive survey of IPM use on US fruit crops was conducted by the USDA Economic Research Service, National Agricultural Statistics Service, in 1991 (USDA, 1994). The survey targeted 30 crops in 13 states, and accounted for most of the US hectares of major fruit and nut crops. The results of the survey showed that half of the US fruit and nut hectares was under some degree of IPM, which included the use of professional pest monitoring. Nearly 90% of the hectares under IPM were also managed using some additional alternative pest management practice. Table 15.1 presents the results for five of the crops in the NASS study. The states in the survey contained the following proportions of the total US hectares of each crop: apples, 82%; grapes, 99%; oranges, 100%; peaches, 79%; pears, 95%.

Most production hectares were treated with pesticides for all of these crops. Interestingly, over 20% of the apple hectares in Arizona and California reported no pesticide applications, and 20% of the grape hectares in Oregon were not treated. About 8% of the total fruit and nut hectares in the US did not receive a pesticide application, ranging from no untreated hectares in some crops and states to a high of 40% of untreated California avocado hectares. No IPM was used on over 80% of North Carolina and Oregon apples, on over 80% of New York grapes, and on over 90% of North and South Carolina peaches. Nationally, growers of 42% of the fruit and nut hectares did not use any IPM.

The use of professional monitoring varied across crops and states, with the highest uses overall in oranges and pears, and the lowest use in grapes. The highest uses of monitoring on any crop in any state were 90% on Florida grapefruit, 85% on California oranges and lemons, and also on Florida tangerines, 81% on California peaches, and 80% on California sweet cherries. Only 3% of North Carolina peach hectares were professionally monitored, as were 6% of Oregon grape hectares. Nationally, 65% of hectares were professionally monitored across all fruit and nut crops.

In the survey, the grower's IPM approach was characterized as low, medium or high depending on the number of IPM tactics employed. A low IPM approach was defined as the use of professional monitoring and economic thresholds to determine pesticide application decisions, with no additional alternative practices being used. Growers with a medium or high IPM approach used one, two, or three or more additional alternative practices such as biological controls, resistant varieties, pheromones, pruning or canopy management. Nationally, 6% of fruit and nut hectares were managed using a low IPM approach, while 17% and 27% were managed with a medium or high approach, respectively. Overall, half of the fruit and nut hectares were managed using some level of IPM.

Pest monitoring is not only fundamental to an IPM program, but appears to be related to the use of additional alternative practices as well. Figure 15.1 presents a regression analysis of professional monitoring and IPM use data reported in the Economic Research Service survey for apples, grapes, oranges, peaches and pears (USDA, 1994). Professional monitoring by fruit and nut growers was significantly related ($Y = -9.981 + 0.861x$, $R^2 = 0.801$, $n = 29$, $P < 0.05$) to the use of one or more alternative practices. It is likely that growers who use professional monitoring services are better informed, and are therefore more willing to utilize alternative practices which might otherwise seem risky. Monitoring information, like using pesticides preventatively, can be in important risk-management tool for growers.

Of the alternative pest management practices reported in the Economic Research Service Survey, field sanitation (60%) and pruning or canopy management (47%) had the highest levels of use, probably because these practices are also necessary for horticultural reasons. Planting locations were only used by 11% of growers, again probably because of horticultural factors. The relatively low use of trap crops (9%) and beneficials (19%) probably reflects the costs and risk perception associated with these practices. Table 15.2 details the use of the practices on five fruit crops and across states.

Kovach and Tette (1988) surveyed New York apple growers to determine the amount and extent of IPM use. In their study, low IPM users were defined as those who used at least one passive monitoring device (e.g. pheromone traps, leaf wetness sensors or sticky

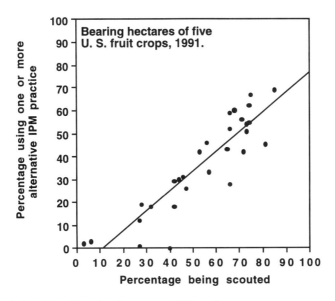

Figure 15.1   Relationship of bearing hectares of US apples, grapes, oranges, peaches and pears being scouted to the percentage of hectares being grown using one or more alternative pest management practices

traps) or one special application method (e.g. alternate row or perimeter spraying). High users were those who hired scouts to monitor pests, and who used passive monitoring devices and special spray methods. Their results indicated that over 80% of New York apple growers used some IPM tactic. Those growers who were high IPM users reduced insecticide use by 30%, acaricide use by 47% and fungicide use by 10%. These growers also saved on average $38.32 ha^{-1} for each year of the period 1976–1986 without significantly affecting fruit quality.

ESTIMATED PESTICIDE REDUCTION IN SELECTED US FRUIT CROPS BY USING IPM

It is difficult to estimate the total amount of pesticides applied to US fruit and nut crops or the number of hectares treated because of a lack of national use reporting. D. Pimentel (personal communication 1995) estimates the total annual use of insecticides on US fruit crops to be about 25 million kg. Since 1990, the State of California has required total reporting of pesticide use, so more accurate data are available on which to base estimates.

As pesticide use reductions of 40% often occur when pesticide use decisions are based on the use of monitoring and pest thresholds, significant reductions would be possible if these practices were universally applied to US fruit crops. Table 15.2 estimates the amount of insecticides and acaricides that would be applied to five US and California fruit crops if all hectares were professionally monitored and thresholds used compared to estimated current use. The estimated decreases in insecticide and acaricide use ranged from 16% in US orange production and 10.4% in California orange production to 25% in US apple production and 22.6% in California apple production.

More efficient pesticide application techniques and equipment could further reduce the

Table 15.2 Production and pesticide use in US and California apples, grapes, oranges, peaches and pears, with estimated use if all bearing hectares were grown with scouting and monitoring, and with scouting and monitoring with completely efficient pesticide applications

Corn	Production (ha × 10³)		% using and acaricides (kg × 10³)		Total use with scouting and thresholds		Total use with scouting and thresholds (ha × 10³)		thresholds and efficient application (ha × 10³)	
	US	CA	US	CA**	US	CA	US	CA	US	CA
Apples	167	14	7000***	53	45	42	5250	41.0	3150	24.6
Grapes	293	262	2000†	671	55	57	1640	555.6	984	333.4
Oranges	276	74	8500††	362	60	74	7140	324.4	4284	194.6
Peaches	37	24	700***	117	45	47	546	92.2	328	55.3
Pears	30	10	1000***	30	40	66	760	25.9	456	15.6

*   California Agricultural Statistics Service, 1994.
*** Pimentel et al., 1991 (1978 data).
†   Pimentel et al., 1991.
††   Haydu, 1981.
**   California Department of Pesticide Regulation, 1992.

amount of pesticides applied in orchard crops. Estimated additional decreases (Table 15.2) in insecticide and acaricide use resulting from efficient application technology averaged 31.5% for US production and 33.2% for California production.

Professional monitoring and more efficient application equipment might be expected to increase grower production costs, but studies cited earlier indicate that savings occur as the result of fewer pesticides being applied, and in many cases increased yield or quality. Monitoring and more efficient application technologies do not necessarily require growers to implement biologically based alternatives, but the use of these practices are more likely to be attempted when professional monitoring is applied (Figure 15.1).

Of all insecticide and acaricide treatments applied to California apples, about 49.9% are for control of the codling moth, and another 19.9% are acaricides for control of spider mites (California Department of Pesticide Regulation, 1990). If the implementation of pheromone mating disruption for control of codling moth were successful, many of the sprays applied for this pest could be eliminated. It is recognized that this technology must be combined with some other tactics to achieve acceptable control, so it is possible that not all insecticide sprays would be eliminated unless mass releases of parasitoids, use of microbial pesticides during a portion of the year, sanitation or cultural approaches would prove feasible on a large scale. Elimination of pesticides applied for control of the codling moth would decrease the amount of acaricide applied for control of spider mites, which are often induced by codling moth sprays. A similar potential for use reduction occurs in pears, where about 65.2% of the insecticides and acaricides applied are for control of the codling moth or for induced pests. The grower costs and associated risks of applying mating disruption on a large scale remain to be determined, but the potential for pesticide use reduction is significant.

On California grapes, about half of the total kilograms of pesticide applied is cryolite, a naturally occurring material which is organically acceptable in the mined form. Cryolite is applied for control of the omnivorous leafroller and several other Lepidoptera at rates of approximately $7\,kg\,ha^{-1}$. Alternatives to cryolite could significantly reduce the amount of insecticides applied to grapes. About 33 400 kg of carbofuran is applied annually (California Department of Pesticide Regulation, 1990), primarily to control infestations of grape phylloxera. If infested vineyards were replaced with tolerant grape rootstocks, about 5% of insecticides used on grapes would be eliminated. About 8% of the total uses identified in Table 15.2 are acaricides. Many of these acaricide treatments could be reduced through better water and fertility management, and the use of less disruptive controls for key pests.

Morse and Klonsky (1994) present an excellent summary of insecticides and acaricides used on California citrus, with potential implications, given the use of IPM tactics for specific pests. Of all insecticides and acaricides applied to California citrus, 54% were applied to control scale insects. As discussed, augmentative releases of parasites for California red scale and black scale have been effective in some coastal growing areas, and to some extent in the San Joaquin Valley, where there remains uncertainty about the reliability of such releases. High-pressure post-harvest washers are being tested commercially, and may prove to be efficient in removing post-harvest concerns about scales and sooty moulds on fruit. The widespread implementation of biological control for scale insects, possibly used in concert with narrow-range oils and the installation of post-harvest washers in packing houses, could have a significant impact on the total amount of insecticide applied to citrus in California and probably in other US citrus growing areas.

About 6.2% of insecticides were applied for various Lepidoptera, primarily citrus cut-worm and fruittree leafroller. Of the 47 600 kg applied for their control, approximately 4.1% of the sprays were *B. thuringiensis*. Increased use of *B. thuringiensis* for control of Lepidopterans on citrus is possible. About 4.4% of sprays were applied for spider mites, and the use of alternative methods for target insect pests could reduce the sprays needed for spider mite infestations resulting from disruption.

About 54.4% of insecticides applied to California peaches are for control of the oriental fruit moth (California Department of Pesticide Regulation, 1990). Pheromone mating disruption is being used commercially for this pest, but the tactic has not been universally accepted because of cost, potential risk of damage and lack of application on an area-wide basis, which limits its effectiveness to some extent. Improvement of the technology and area-wide use would limit insecticide treatment for the oriental fruit moth. Another 37.9% are applied primarily as dormant-season sprays for control of the peach twig borer. Bloom-time treatments with *B. thuringiensis* for control of the peach twig borer would eliminate many of the conventional pesticide treatments, although the kilograms of insecticides applied would not change, as the rates of *B. thuringiensis* used are approximately the same as those of conventional materials applied during the dormant season.

CONCLUSIONS

There are many challenges to the development and implementation of IPM on fruit crops, but there exists an excellent groundwork in the scientific literature and in experiences with successful field implementation. A joint effort by the US Environmental Protection Agency and the US Department of Agriculture was begun in 1991 to analyze the potential for IPM adoption in US agriculture, and momentum was built up to bring agencies, institutions, organizations and individual to consider the issue. Constraints to IPM implementation (Zalom and Fry, 1992) and resolutions (Sorenson, 1994) were identified by participants at the National Integrated Pest Management Forum which was held in Washington, DC, and the processes leading up to that meeting. A Fruit Action Team assembled as part of the forum (Tette and Jacobson, 1992) identified a lack of incentives to develop and adopt IPM methods, a lack of resources directed toward a biologically based IPM, economic pressures, risks and incentives to take a short-term research approach at academic institutions, the lack of trained IPM consultants, market demands for blemish-free produce, lack of information on the practical use of biologically based methods, and pesticide resistance as non-regulatory constraints. Regulatory constraints included unreasonable restrictions on testing procedures and plot size, the old pesticide-type registration mentality applied to new pesticide chemistry, regulation of genetically modified organisms, conflicting state and federal requirements on pesticide use, the economics of registration, commodity grading standards, lack of access to regulators, and lack of an accelerated process allowing the testing and registration of biologically based products. Some of these regulatory constraints have already been modified since the forum.

Future IPM systems which are developed in fruit crops must be economical, yet friendly to the environment and less risky to workers and consumers. History shows that there are risks when introducing new IPM tactics, and it is important that we consider ways of mitigating the risks—of pest resistance, and of damage to the environment and to human health—so that these new innovations can provide a more lasting and predictable approach to the management of fruit pests.

ACKNOWLEDGEMENTS

Many of the citations mentioned in this chapter were suggested as being especially relevant by several pest management specialists of fruit crops, including Martin Barnes, Jay Brunner, Robert Bugg, William Coli, Elizabeth Grafton-Cardwell, Daniel Horton, Marjorie Hoy, Joseph Kovach, Themis Michailides, Joseph Morse, Ronald Prokopy, James Stapleton and James Tette. I am grateful for their assistance in identifying relevant recent examples of successful fruit-crop IPM programs.

REFERENCES

AliNiazee, M.T. (1978) The western cherry fruit fly. 3. Developing a management program by utilizing attractive traps as monitoring devices. *Candian Entomology* 110, 1133–1139.

AliNiazee, M.T. (1984) Effect of two synthetic pyrethroids on the predatory mite, *Typhlodromus arboreus*, in the apple orchards of Western Oregon. In Griffiths, D.A. and Bowman, C.E. (Eds) *Acarology* Vol. VI. Wiley Interscience, New York, pp. 655–658.

AliNiazee, M.T., Mohammed, A.B. and Booth. S.R. (1987) Apple maggot response to traps in an unsprayed orchard in Oregon. *Journal of Economic Entomology* 80, 1143–1148.

Alstad, D.N. and Andow, D.A. (1995) Managing the evolution of resistance to transgenic plants. *Science* 268, 1894–1896.

Alston, D.G. (1994) Effect of apple orchard floor vegetation on density and dispersal of phytophagous and predaceous mites in Utah. *Agriculture, Ecosystems and Environment* 50, 73–84.

Altieri, M.A. and Schmidt, L.L. (1986) Cover crops affect insect and spider populations in apple orchards. *California Agriculture* 40(1), 15–17.

Barnes, M.M. (1959) Deciduous fruit insects and their control. *Annual Review of Entomology* 4, 343–362.

Barnett, W.W., Davis, C.S. and Rowe, G.A. (1978) Minimizing pear pest control costs through integrated pest management. *California Agriculture* 32(2), 12–13.

Barnett, W.W., Edstrom, J.P., Coviello, R.L. and Zalom, F.G. (1993) Insect pathogen 'Bt' controls peach twig borer on fruits and almonds. *California Agriculture* 47(5), 4–6.

Barnett, W.W., Hendricks, L.C., Asai, W.K., Elkins, R.B., Boquist, D. and Elmore, C.L. (1989) Management of navel orangeworm and ants. *California Agriculture* 43(4), 21–22.

Bartell, R.J. (1977) Behavioural responses of Lepidoptera pheromones. In Shorey, H.H. (Ed.) *Chemical Control of Insect Behavior: Theory and Application.* Wiley & Sons, New York, pp. 201–213.

Bedford, E.C.G. (1990) Mechanical control: High-pressure rinsing of fruit. In Rosen, D. (Ed.) *The Armored Scale Insects, Their Biology and Control.* Vol. B. Elsevier, Amsterdam, pp. 507–513.

Bentley, W.J., Zalom, F.G., Barnett, W.W. and Sanderson, J.P. (1987) Population densities of *Tetranychus* spp. (Acari: Tetranychidae) after treatment with insecticides for *Amyelois transitella* (Lepidoptera: Pyralidae). *Journal of Economic Entomology* 80, 193–200.

Bentley, W.J., Sherrill, L.B. and McLaughlin, A. (1994) Mating disruption of codling moth has mixed results. *California Agriculture* 48(6), 45–48.

Brunner, J.F. (1994) Integrated pest management in tree fruit crops. *Food Review International* 10, 135–157.

Bugg, R.L. and Dutcher, J.D. (1993) *Sesbania exaltata* (Rafinesque-Schmaltz), a warm-season cover crop in pecan orchards: Effects on aphidophagous Coccinelidae and pecan aphids. *Biology of Agriculture and Horticulture* 6, 123–148.

Bugg, R.L. and Waddington, C. (1994) Using cover crops to manage arthropod pests of orchards: A review. *Agriculture, Ecosystems and Environment* 50, 11–28.

Bugg, R.L., Dutcher, J.D. and McNeill, R.J. (1991) Cool-season cover crops in the pecan orchard understory: Effects on Coccinelidae and pecan aphids. *Biological Control* **1**, 8–15.

Butkowich, S.L. and Prokopy, R.J. (1993) The effect of short-range host odor stimuli on host fruit finding and feeding behavior of plum curculio adults (Coleoptera: Curculionidae). *Journal of Chemical Ecology* **19**, 825–835.

Byers, R.E. and Lyons, C.G. (1985) Effect of chemical deposits from spraying adjacent rows on efficacy of peach bloom thinners. *HortScience* **20**, 1076–1078.

California Agricultural Statistics Service (1994) *California Agricultural Statistics 1993*. California Department of Food and Agriculture, Sacramento, CA.

California Department of Pesticide Regulation (1992) *Annual Pesticide Use Report 1990*. California Department of Pesticide Regulation, Sacramento, CA.

Caprile, J., Klonsky, K., Mills, N., McDougall, S., Micke, W. and Van Steenwyk, R. (1994) Insect damage limits yield, profits of organic apples. *California Agriculture* **48**(6), 21–28.

Coli, W.M. and Ciurlino, R. (1990) Interaction of weeds and apple pests. *Proceedings of the New England Fruit Meetings* **96**, 52–58.

Croft, B.A. (1990) *Arthropod Biological Control Agents and Pesticides*. Wiley, New York.

Croft, B.A. and AliNiazee, M.T. (1983) Differential resistance to insecticides in *Typhlodromus arboreus* Chant and associated phytoseiid mites in apples in the Willamette Valley, Oregon. *Environmental Entomology* **12**, 1420–1423.

Croft, B.A. and Hoyt, S.C. (1978) Considerations for the use of pyrethroid insecticides for deciduous fruit pest control in the USA. *Environmental Entomology* **7**, 627–630.

Daane, K.M. and Caltagirone, L.E. (1989) Biological control in California olive orchards. Cultural practices affect biological control of black scale. *California Agriculture* **43**(1), 9–11.

Daane, K.M., Barzman, M.S., Kennett, C.E. and Caltagirone, L.E. (1991) Parasitoids of black scale in California. Establishment of *Prococcophagus probo* Annecke & Mynhardt, and *Coccophagus rusti* Compere (Hymenoptera: Aphelinidae) in olive orchards. *Pan-Pacific Entomology* **67**, 99–106.

Dandekar, A.M., McGranahan, G.H., Vail, P.V. and Uratsu, S.L. (1994) Low levels of expression of wild type *Bacillus thuringiensis* var. *kurstaki* cryia (C) sequences in transgenic walnut somatic embryos. *Plant Science* **96**, 151–162.

DeBach, P. (1974) *Biological Control by Natural Enemies*. Cambridge University Press, Cambridge.

DeBenedictis, J.A. and Granett, J. (1993) Laboratory evaluation of grape roots as hosts of California grape phylloxera biotypes. *American Journal of Enology and Viticulture* **44**, 285–291.

Dennehy, T.J., Roelofs, W.L., Taschenberg, E.F. and Taft, T.N. (1990) Mating disruption for control of grape berry moth in New York vineyards. In Ridgway, R.L., Silverstein, R.M. and Inscoe, M.N. (Eds) *Behavior-Modifying Chemicals for Insect Management*. Marcel Dekker, New York, pp. 223–240.

Dolphin, R.E., Cleveland, M.L., Mouzin, L.E. and Morrison, R.K. (1972) Releases of *Trichogramma minutum* and *T. cacoeciae* in an apple orchard and the effects on populations of codling moth. *Environmental Entomology* **1**, 481–484.

Dutcher, J.D. and Todd, J.W. (1983) Hemipteran kernel damage of pecan. In Payne, J.A. (Ed.) *Pecan Pest Management—Are we There*? Miscellaneous Publications of the Entomology Society of America, Vol. 13(2), pp. 1–11.

Ellington, J.J. (1991) Biological Control of pecan nut case bearer in the Mesilla Valley. *Proceedings of the Western Pecan Growers Association* **25**, 25–31.

Ellington, J.J., Carrillo, T.D., LaRock, D., Richman, D.B., Lewis, B.E. and Abd El-Salam, A.H. (1995) Biological control of pecan insects in New Mexico. *HortTechnology* **5**(3), 230–233.

Engle, C. and Barnes, M.M. (1983) Cultural control of navel orangeworm in almond orchards. *California Agriculture* **37**(9–10), 19.

Flanders, S.E. (1926) The mass production of *Trichogramma minutum* Riley and observations on the natural and artificial parasitism of the codling moth egg. *Proceedings of the International Congress on Entomology* **4**, 110–130.

Flexner, J.L., Westigard, P.H., Gonzalves, P. and Hitton, R. (1991) The effect of groundcover and herbicide treatment on twospotted spidermite density and dispersal in Southern Oregon pear orchards. *Entomol. Exp. Appl.* **60**: 111–121.

Flint, M.L. and Klonsky, K. (1989) IPM information delivery to pest control advisors. *California Agriculture* **43**(1), 18–20.

Graebner, L. (1982) An economic history of the Fillmore Citrus Protective District. Ph.D. Dissertation, University of California, Riverside, CA.

Graebner, L., Moreno, D.S. and Baritelle, J.L. (1984) The Fillmore Citrus Protective District: A success story in integrated pest management. *Bulletin of the Entomological Society of America* **30**(4), 27–33.

Gubler, W.D., Marois, J.J., Bledsoe, A.M. and Bettiga, L.J. (1987) Control of *Botrytis* bunch rot of grape with canopy management. *Plant Disease* **71**, 599–601.

Gut, L.J., Westigard, P.H. and Liss, W.J. (1988) Arthropod colonization and community development on young pear trees in southern Oregon. *Melandaria* **46**, 1–13.

Hall, F.R. (1985) The opportunities for improved pesticide application. *Proceedings of the National Peach Council Annual Convention* **44**, 69–73.

Haney, P.B. and Morse, J.G. (1988) Chemical and physical trunk barriers to exclude adult Fuller rose beetles (Coleoptera: Curculionidae) from skirt pruned citrus trees. *Applied Agricultural Research* **3**, 65–70.

Haney, P.B., Morse, J.G., Luck, R.F., Griffiths, H., Grafton-Cardwell, E.E. and O'Connell, N.V. (1992) *Reducing Insecticide Use and Energy Costs in Citrus Pest Management*. University of California IPM Publication No. 15.

Haydu, J.J. (1981) *Pesticide Use in US Citrus Production 1977*. United States Department of Agriculture, Economic Statistics Service Publication. Washington DC.

Haynes, K.F. and Baker, T.C. (1988) Potential for evolution of resistance to pheromones. Worldwide and local variation in chemical communication system of pink bollworm moth, *Pectinophora gossypiella*. *Journal of Chemical Ecology* **14**, 1547–1560.

Headley, J.C. and Hoy, M.A. (1987) Benefit/cost analysis of an integrated mite management program for almonds. *Journal of Economic Entomology* **80**, 555–559.

Hollingsworth, C.S. and Coli, W.M. (1992) The integrated pest management education and certification project, 1992 update. *University of Massachusetts Fruit Notes* **57**(4), 1–2.

Hollingsworth, C.S., Coli, W.M. and Van Zee, V. (1992) Massachusetts grower attitudes toward a certification program for integrated pest management. *University of Massachusetts Fruit Notes* **57**(4), 7–11.

Honiball, F., Giliomee, J.H. and Randall, J.H. (1979) Mechanical control of red scale, *Aonidiella aurantii* (Mask.) on harvested oranges. *Citrus Subtropical Fruit Journal* **549**, 17–18.

Hoy, M.A., Barnett, W.W., Reil, W.O., Castro, D., Cahn, D., Hendricks, L.C., Coviello, R. and Bentley, W.J. (1982) Large-scale releases of pesticide-resistant spider mite predators. *California Agriculture* **36**(1–2), 8–10.

Hoy, M.A., Barnett, W.W., Hendricks, L.C., Castro, D., Cahn, D. and Bentley, W.J. (1984) Managing spider mites in almonds with pesticide-resistant predators. *California Agriculture* **38**(7–8), 18–20.

Hoyt, S.C. (1969) Integrated chemical control of insects and biological control of mites on apple in Washington. *Journal of Economic Entomology* **62**, 74–86.

Huffaker, C.B. (1985) Biological control in integrated pest management: An entomological perspective. In Hoy, M.A. and Herzog, D.C. (Eds) *Biological Control in Agricultural IPM Systems*. Academic Press, New York, pp. 67–88.

Hunter, C.D. (1994) *Suppliers of Beneficial Organisms in North America*. California Environmental Protection Agency, Department of Pesticide Regulation, PM 94-03. Sacramento, CA.

Johnson, D.T. (1988) Peach scouting program in Arkansas. *Proceedings of the Arkansas State Horticultural Society* **109**, 144–149.

Jones, V.P. (1990) Sampling and dispersion of the twospotted spider mite (Acari: Tetranychidae) and the western orchard predatory mite (Acari: Phytoseiidae) on tart cherry. *Journal of Economic Entomology* **83**, 1376–1380.

Jones, V.P. and Davis, D.W. (1989) Evaluation of traps for apple maggot (Diptera: Tephritidae) populations associated with cherry and hawthorn in Utah. *Environmental Entomology* **18**, 521–525.

Kido, H., Kasimatis, A.N. and Jensen, G.L. (1982) Grape phylloxera. In Flaherty, D.L., Jensen, F.L., Kasimatis, A.N., Kido, H. and Moller, W.J. (Eds) *Grape Pest Management*. University of California Division of Agriculture and Natural Resources, Publication No. 4105. pp. 170–175.

Klonsky, K., Zalom, F.G. and Barnett, W.W. (1990) California's almond IPM program. *California Agriculture* **44**(5), 21–24.

Knight, A.L. and Hull, L.A. (1988) Area-wide population dynamics of *Platynota idaeusalis* in southcentral Pennsylvania pome and stone fruits. *Environmental Entomology* **17**, 1000–1008.

Kovach, J. and Tette, J.P. (1988) A survey of the use of IPM by New York apple producers. *Agriculture, Ecosystems and Environment* **20**, 101–108.

Larsen, K.J. and Whalon, M.E. (1987) Crepuscular movement of *Paraphlensiuus irroratus* between groundcover and cherry trees. *Environmental Entomology* **18**, 1103–1106.

Madsen, H.F., Vakenti, J.M. and Peters, F.E. (1976) Codling moth: Suppression by male removal with sex pheromone traps in an isolated apple orchard. *Journal of Economic Entomology* **69**, 597–599.

Meagher, R.I. and Meyer, J.R. (1990) Influence of ground cover and herbicide treatments on *Tetranychus urticae* populations in peach orchards. *Experimental and Applied Acarology* **9**, 149–158.

Messing, R.H. and AliNiazee, M.T. (1989) Introduction and establishment of *Trioxys pallidus* in Oregon, USA, for control of filbert aphid *Myzocallis coryli* (Homoptera; Aphididae). *Entomophaga* **34**, 153–163.

Michelbacher, A.E. and Bacon, O.G. (1952) Walnut insect control in northern California. *Journal of Economic Entomology* **45**, 1020–1027.

Moericke, V., Prokopy, R.J,, Berlocher, S. and Bush, G.L. (1975) Visual stimuli eliciting attraction of *Rhagoletis pomonella* (Diptera: Tephritidae) flies to trees. *Entomologica Experimentalis et Applicata* **18**, 497–507.

Moffitt, H.R. and Westigard, P.H. (1984) Suppression of the codling moth (Lepidoptera: Totricidae) population on pears in southern Oregon through mating disruption with sex pheromone. *Journal of Economic Entomology* **77**, 1513–1519.

Morse, J.G. and Klonsky, K. (1994) Pesticide use on California citrus: A baseline to measure progress in adoption of IPM. *California Grower* **18**(4): 16–26.

National Research Council (1989) *Alternative Agriculture*. National Academy Press, Washington, DC.

Oatman, E.R. (1965) Predaceous mite control of twospotted spider mite on strawberry. *California Agriculture* **19**(2), 6–7.

Oatman, E.R. (1970) Integration of *Phytoseiulus persimilis* with the native predators for control of the twospotted spider mite on rhubarb. *Journal of Economic Entomology* **63**, 1177–1180.

Oatman, E.R., Gilstrap, F.E. and Voth, V. (1976) Effect of different release rates of *Phytoseiulus persimilis* (Acarina: Phytoseiidae) on the twospotted spider mite on strawberry in southern California. *Entomophaga* **21**, 269–273.

Paschall, M.J., Hollingsworth, C.S., Coli, W.M. and Cohen, N.L. (1992) Attitudes and perceptions of New England consumers toward a certification program for integrated pest management. *University of Massachusetts Fruit Notes* **57**(4), 3–6.

Pence, R.A. and Grieshop, J.I. (1990) Leaf removal in wine grapes: A case study in extending research to the field. *California Agriculture* **45**(6), 28–30.

Phelan, P.L. and Baker, T.C. (1987) An attracticide for control of *Amyelois transitella* (Lepidoptera: Pyralidae) in almonds. *Journal of Economic Entomology* **80**, 779–783.

Pimentel, D., McLaughlin, L., Zepp, A., Lakitan, B., Kraus, T., Kleinman, P., Vancini, F., Roach, W.J., Graap, E., Keeton, W.S. and Selig, G. (1991) Environmental and economic impacts of reducing US agricultural pesticide use. In Pimentel, D. (Ed.) *Handbook of Pest Management in Agriculture*. Vol. I. CRC Press, Boca Raton, FL, pp. 679–718.

Prokopy, R.J. (1994) Integration in orchard pest and habitat management: A review. *Agriculture, Ecosystems and Environment* **50**, 1–10.

Prokopy, R.J. and Owens, E.D. (1976) Visual generalist with visual specialist phytophagous insects: Host selection behavior and application to management. *Entomologica Experimentalis et Applicata* **24**, 409–420.

Prokopy, R.J., Hislop, R.G., Adams, R.G., Hauschild, K.J., Owens, E.O., Ackes, C.A. and Ross, A.W. (1978) Towards integrated management of apple insects and mites. *Proceedings of the Annual Meeting of the Massachusetts Fruit Growers Association* **84**, 38–43.

Prokopy, R.J., Christie, M.M., Gamble, J., Heckscher, D. and Mason, J. (1991) Nonpesticidal control of summer codling moths through habitat management. *Massachusetts Fruit Notes* **56**(1), 16–17.

Rajotte, E.G., Kazmierczak, R.F., Norton, G.W., Lambur, M.T. and Allen, W.A. (1987) *The National Evaluation of Extension's Integrated Pest Management (IPM) Programs*. Virginia Cooperative Extension Service Publication No. 491–010.

Rice, R.E. and Kirsch, P.A. (1990) Mating disruption of oriental fruit moth in the United States. In Ridgway, R.L., Silverstein, R.M. and Inscoe, M.N. (Eds) *Behavior-Modifying Chemicals for Insect Management*. Marcel Dekker, New York, pp. 193–211.

Rice, R.E., Sadler, L.L., Hoffman, M.L. and Jones, R.A. (1976) Egg traps monitor navel orangeworm. *California Agriculture* 31(3), 21–22.

Rissler, J. and Mellon, M. (1993) *Perils Amidst the Promise. Ecological Risks of Transgenic Crops in a Global Market*. Union of Concerned Scientists, Cambridge, MA.

Roltsch, W.J., Zalom, F.G., Barry, J.W., Kirfman, G.W. and Edstrom, J.P. (1995) ULV aerial applications of *Bacillus thuringiensis* var. *kurstaki* for control of peach twig borer in almond trees. *Applied Engineering Agriculture* 11(1), 25–30.

Rossi, D., Dhillon, P.S. and Hoffman, L. (1983) Apple pest management: A cost analysis of alternative practices. *Journal of the Northeastern Agricultural Economic Council* 12(2), 77–81.

Schlinger, E.I., Hagen, K.S. and van den Bosch, R. (1960) Parasite of walnut aphid. *California Agriculture* 14(10), 3–4.

Seiber, J.N. (1988) California's initiative in support of improved application efficacy. In *Improving On-Target Placement of Pesticides*. Agricultural Research Institute, Bethesda, MD, pp. 59–67.

Snow, J.W. (1990) Peachtree borer and lesser peachtree borer control in the United States. In Ridgway, R.L., Silverstein, R.M. and Inscoe, M.N. (Eds) *Behavior-Modifying Chemicals for Insect Management*. Marcel Dekker, New York, pp. 241–253.

Sorenson, A.A. (1994) *Proceedings of the National Integrated Pest Management Forum*. American Farmland Trust Center for Agriculture in the Environment, De Kalb, IL.

Steinhaus, E.A. (1954) The effects of disease on insect populations. *Hilgardia* 23, 197–261.

Tamaki, G. (1972) The biology of *Geocoris bullatus* inhabiting orchard floors and its impact on *Myzus persicae* on peaches. *Environmental Entomology* 1, 559–565.

Tamaki, G. (1975) Weeds in orchards as important alternate sources of green peach aphid in late spring. *Environmental Entomology* 4, 958–960.

Tedders, W.L., Payne, J.A. and Inman, J. (1984) A migration of *Tetranychus urticae* from clover into pecan trees. *Journal of the Georgia Entomological Society* 19, 498–502.

Tette, J.P. and Jacobson, B.J. (1992) Biologically intensive pest management in the tree fruit system. In Zalom, F.G. and Fry, W.E. (Eds) *Food Crop Pests, and the Environment: The Need and Potential for Biologically Intensive Integrated Pest Management*. American Phytopathological Society Press, St. Paul, MN, pp. 83–105.

University of California (1985) *Integrated Pest Management for Almonds*. University of California Division of Agriculture and Natural Resources, Publ. 3308.

USDA (1994) *Integrated Pest Management Practices on 1991 Fruits and Nuts*. United States Department of Agriculture Economic Research Service RTC Updates, August 1994, No. 2.

Van den Baan, H.E. and Croft, B.A. (1990) Factors influencing insecticide resistance in *Psylla pyricola* and susceptibility in the predator *Dereocoris brevis*. *Environmental Entomology* 19, 1223–1228.

van den Bosch, R., Frazer, B.D., Davis, C.S., Messenger, P.S. and Hom, R. (1970) *Trioxys pallidus*: An effective walnut aphid parasite from Iran. *California Agriculture* 24(11), 8–10.

van den Bosch, R., Hom, R., Matteson, P., Frazer, B.D., Messenger, P.S. and Davis, C.S. (1979) Biological control of the walnut aphid in California: Impact of the parasite *Trioxys pallidus*. *Hilgardia* 47, 1–13.

Weakley, C.V., Kirsch, P.A. and Rice, R.E. (1987) Control of oriental fruit moth by mating disruption. *California Agriculture* 41(5), 7–8.

Westigard, P.H. and Moffitt, H.R. (1984) Natural control of the pear psylla (Homoptera: Psyllidae): Impact of mating disruption with the sex pheromone for control of the codling moth (Lepidoptera: Tortricidae). *Journal of Economic Entomology* 77, 1520–1523.

Whalon, M.E. and Croft, B.A. (1984) Apple IPM implementation in North America. *Annual Review of Entomology* 29, 435–470.

Wilson, L.T., Hoy, M.A., Zalom, F.G. and Smilanick, J.M. (1984) The within-tree distribution and clumping pattern of mites in almond orchards: Comments on predator–prey interactions.

Hilgardia **52**(7), 1–13.

Zalom, F.G. and Fry, W.E. (1992) *Food, Crop Pests, and the Environment: The Need and Potential for Biologically Intensive Integrated Pest Management.* American Phytopathological Society Press, St. Paul, MN.

Zalom, F.G., Hoy, M.A., Wilson, L.T. and Barnett, W.W. (1984) Presence–absence sequential sampling for web-spinning mites in almonds. *Hilgardia* **52**(7), 14–24.

Zalom, F.G., Kennett, C.E., O'Connell, N.V., Flaherty, D.L., Morse, J.G. and Wilson, L.T. (1985) Distribution of *Panonychus citri* (McGregor) and *Euseius tularensis* Congdon on central California orange trees with implications for binomial sampling. *Agricultural Environment and Ecosystems* **14**, 119–129.

CHAPTER 16

IPM Practices for Reducing Fungicide Use in Fruit Crops

Maurizio G. Paoletti

Padova University, Padova, Italy

INTRODUCTION

Fungi affect plants and their fruits before and after harvest. In evolutionary terms, fungi were living long before the vascular plants became widespread in terrestrial environments, and they are generally described as eterotrophic, since they live at the expense of other organisms. If most species are considered parasitic to plants, others are symbiotic, living especially in the root systems of plants like the VAM (Vescicular–arbuscular mycorrhiza). In fact, most fungi can live on the surface of vegetation without consistent damage to the plant tissue (Fokkema, 1991). Cultivated fruit crops have acquired a number of pathogenic fungi, but in general wild varieties, or wild relatives of cultivates trees, are less affected. An example is the crab apple, *Malus floribunda*, which is rarely or never affected by the apple scab, *Venturia inaequalis* (Cke.) Wint., and only slightly affected by the powdery mildew *Podosphaera leuchotricha* (Well. & Ev.) Sal. (AA, 1990), which are still key pathogens in most apple orchards worldwide. Accessions resistant to fungi are sometimes resistant to some insect pests as well, as in the case of apples (Alston and Briggs, 1970, 1977; Goonewardene and Pouish, 1988; Crosby et al., 1992). In general, cultivated grapes are more affected by the key pathogen *Plasmopara viticola* and other fungi than their wild ancestors such as *Vitis riparia* or *Vitis labrusca* (Alleweldt and Possingham, 1988). The wild apricot varieties endemic to the hilly woodlands near Beijing, China, are also almost unaffected by fungi or other pathogens (data from personal observations and discussions with Chinese plant pathologists at BAU, September, 1995).

The domestication process of fruit plants, with the introduction of more intensive monospecific associations, breeding for larger, more succulent, sweeter and better coloured fruits, and the process of refrigeration in order to extend their storage have, in general, made plants less resistant or tolerant to insect pests and plant pathogens compared with their wild relatives (Russell, 1978; Smith, 1989; Farquhar, 1992). Nevertheless, the control of parasitic fungi has become an agronomic problem for agriculture. By-products such as wood ash and amurca (the liquid waste that remains once olives have been processed) have been used extensively in the Mediterranean area by ancient Roman,

Techniques for Reducing Pesticide Use. Edited by D. Pimentel
© 1997 John Wiley & Sons Ltd.

Greek and Spanish farmers for disease control. Amurca, for instance, was used to control the esca disease of grape (*Stereum hirsutum*?). Ashes were used as dust applied to plants and also to fruit in storage to prevent food deterioration (Thurston, 1992). Sulfur, copper sulphate, the so-called 'Bordeaux mixture', lime and lime sulphate have been used since the last century on many plants affected by fungi, and are still in orchards, even though some growers have recently substituted chemical fungicides. Chemical fungicides, including the copper derivatives, have been used extensively, especially since World War II, and they are effective against fungal diseases, but they also create problems for non-target organisms such as insect pollinators, parasitoids, predators, and an array of soil microbes and invertebrates. Intensive fungicide use has also induced resistance in plant pathogens (Chaboussou, 1980; Brown, 1978, pp. 442–443). Earthworms have almost disappeared from vineyards containing high copper and zinc residues in the soils as the result of intensive fungicidal control (Paoletti and Bentocello, 1985; Paoletti et al., 1988, 1995b). The same has happened in apple orchards (Paoletti et al., 1995a). Most modern fungicides are harmful to both humans and domestic animals; the carcinogenetic risk from fungicides is the highest of all the pesticides and accounts for 70% of human health problems associated with pesticide exposure (NAP, 1987; Culliney et al., 1993).

Van den Bosch (entomologist at the University of California, Berkeley, CA) has stated that the heavy use of pesticides (including fungicides) to control pathogens and pests has always been linked to 'academic connections' and is still present in most 'schools' of agriculture, worldwide:

> The corruptive and coercive influence of the pesticide mafia is widespread in the land-grant universities, where much of the nation's pest-control research is conducted and from which most of the pest-control recommendations emanate. In the agricultural experiment stations and the Agricultural Extension Service, deans, directors, department chairmen, division heads, or whatever titles they go by, too often knuckle under to the political pressures directly or indirectly generated by the agri-chemical industry and its allies. (Van Den Bosch, 1978, p. 61)

This strong statement suggests the need to consider integrated strategies, including socio-economic ones, in order to change our methods of controlling fungi and also other pathogens and insect pests. It also underlines the need to preserve the economy of farmers and the environment, and protect consumers from harmful residues in food. However, the major trends in plant breeding have been in the other direction explained by Russell (1978, p. 48): 'New cheaper and more effective fungicides will undoubtedly play an important part in reducing damage by diseases so that the breeder can concentrate more on increasing the productivity of crop species.' As is underlined here, mainstream breeding has evolved together with the development of fungicides, and in general with fertilizers and pesticides being applied to experimental field plots at high rates. Most of the current, highly productive cultivated plants have gone through this bottleneck, and much work is needed in order to modify this dominant trend. Breeding must be considered as a tool to obtain plants resistant to insect pests and plant pathogens, as well as yielding high-quality fruits (AA, 1989; Francis, 1990).

AMOUNTS OF FUNGICIDES USED AND TRENDS

Fruit and vegetable crops are routinely treated with large amounts of fungicides. In the United States, apple orchards receive an average of $18 \, \mathrm{kg \, ha^{-1}}$ and grapes up to

$29\,kg\,ha^{-1}$, (McEwen and Stephenson, 1979). Traditional vineyards in northern Italy receive between 8 and 32 kg of copper (as copper sulfate) $ha^{-1}\ year^{-1}$ (Paoletti et al., 1991). When more modern fungicides are included, the input is even higher (see Appendix A).

The majority of available fungicides are applied to fruit crops. In the United States, where 38 000 tons of fungicides are used annually (Pimentel et al., 1991), 90% is applied to apples, peaches, citrus, grape and other fruit crops (Pimentel and Levitan, 1986; Epson, 1989).

Reducing the use of pesticides including fungicides, by 50% is the objective of places such as Sweden, Denmark, The Netherlands and Ontario, Canada (Pettersson, 1993; Surgeoner and Roberts, 1993). In the United States, a 25% reduction seems easily feasible for fruit crops (Pimentel et al., 1991) without any major investment or additional costs, by simply increasing monitoring and improving the spraying equipment (e.g. nozzle design). The reduction of pesticides and fungicides has also been suggested by the European Commission assessment on the European situation (Jordan, 1993), but without any clear target objective. In Sweden for instance, the use of fungicides increased from 1973 to 1988 from 5.8 to 629.6 tons. In an opposite trend, the use of other pesticides (herbicides and insecticides) decreased from 1850.3 to 349.6 tons during the same period of time (Surgeoner and Roberts, 1993). Comparing trends in fungicide use, the ecological effects of pesticides and the doses used seems to be the best way to assess the environmental benefits of any pesticide policy. During the past 20 years, many IPM projects to decrease the levels of pesticides use, and ban those with the highest toxicity to non-target organisms have been tried. However, the degree of toxicity varies considerably in different groups of animal organisms. For instance, copper- and zinc-based fungicides are considered to be safe for many beneficial insects and vertebrates, including poultry. Cooper-based fungicides are recommended in many organic farming protocols and IPM headlines (Paoletti et al., 1988, 1991), but they have a consistent toxic effect when residues accumulate in grape and apple orchard soils, destroying most earthworms, detritovore organisms such as isopods, and some polyphagous predators like the carabids and other microfauna (Paoletti et al., 1995a,b, Paoletti and Sommaggio, 1996).

Another tendency in the use of new pesticides is to decrease the amount of active ingredients per hectare to a level which will have the same fungicidal effect as the previous generation of products. However, this weight reduction must result in reduced pollution. Comparative assessments need to be developed to monitor the improvements in non-target organisms and ecosystem biodiversity. For example, if Bordeaux mixture is replaced by benomyl, the quantities applied are about one-half but the degree of toxicity, for numerous non-target beneficial organisms (Table 16.1) (Southerton, 1989), including earthworms, is increased (Theiling and Croft, 1989; Edwards and Bohlen, 1992). Accurate bioindicator protocols need to be developed in order to evaluate the continual use of fungicides (and of other pesticides as well) on a large range of non-target organisms (Paoletti, 1993; Paoletti and Bressan, 1996).

Fungicide use in different crops and post-harvest

Table 16.2 (from Pimentel et al., 1991) summarizes the current use of fungicides in the United States and its possible reduction (following the personal evaluation of all authors), based on the assumption that modern plants were not replaced by cultivars tolerant to

Table 16.1   Laboratory screening of 27 single-active-ingredient foliar fungicides against two species of beneficial insect. Fungicides were applied topically via a Potter spray tower at maximum field dose rates at dilutions approved for use for UK cereal crops (from Southerton, 1989)

| Active ingredient | Mortality after 48 h (%) | | | |
| | Hoverfly larvae | | Leaf beetle larvae | |
	Control	Fungicide	Control	Fungicide
Benodanil	0.0	4.0	0.0	4.0
Benomyl	0.0	10.0	3.0	10.0
Captafol	0.0	0.0	7.0	19.8
Carbendazim	6.0	8.0	4.0	4.5
Chlorthalonil	4.0	4.0	2.0	4.5
Ditalimfos	12.0	26.0	2.0	13.0
Dithiocarbamids	12.0	26.0	0.0	1.0
Ethirimol	0.0	2.0	5.0	2.0
Fenpropidin	0.0	4.0	0.0	9.0
Fenpropimorpf	4.0	4.0	1.5	3.5
Flutriafol	4.0	12.0	5.5	8.8
Iprodione	0.0	4.0	6.0	6.0
Mancozeb	0.0	4.0	6.0	5.0
Maneb	8.0	6.0	4.0	13.0
Nuarimol	0.0	0.0	4.0	8.0
Polyram	12.0	6.0	4.0	7.0
Prochloraz	0.0	6.0	0.0	7.0
Propiconazole	0.0	4.0	4.0	13.5
Propineb	12.0	12.0	2.0	2.0
Pyrazophos	4.0	72.0	1.0	100.0
Sulphur	4.0	4.0	6.0	3.5
Thiophanate-methyl	6.0	6.0	5.0	5.0
Triadimefon	8.0	10.0	3.0	4.5
Triadimenol	10.0	6.0	3.5	13.5
Tridemorph	6.0	4.0	5.0	8.0
Triforine	0.0	4.0	9.0	3.5
Zineb	2.0	0.0	5.0	4.04
Toxic standard	0.0	96.0	1.0	100.0

fungi, and that more accurate monitoring and spray-nozzle equipment were being used. Under these conditions they estimated a possible reduction by 25%. However, it has already been demonstrated in experimental fields and farms that some varieties of apples which are resistant to apple scab and powdery mildew (such as 'Prima', 'Priscilla', 'Enterprise', 'Golden Rush', etc.) may allow a reduction of fungicides by 80% (Crosby et al., 1992; Janick, 1993) (Table 16.3). Similar reductions are possible in grape production. The process of developing new resistant plants requires time as well as financial invest-ment. At the same time, appropriate policies are needed to promote these new options to farmers and to the market. Consumers also need to be educated about these new varieties. Table 16.4 summarizes the estimated use of pesticides in different countries. Appendix A shows treatment data from different countries from conventional, integrated and biologi-cal farming systems. Note that integrated and biological farming systems use different products and amounts.

Table 16.2 Fruit and nut crop losses in the USA from plant pathogens with current fungicide use, and estimated costs if fungicides were reduced and several alternatives were substituted (from Pimentel et al., 1991)

Crop	Crop area (ha × 10³)	Total fungicide use (kg × 10⁶)		Fungicide treatment		Total costs ($ × 10⁶)	Current crop pestloss (%)	Added alternative control ($ ha⁻¹)	Total added alternative control cost ($ × 10⁶)
		Current	Reduced	Area treated (ha)	Cost ($ ha⁻¹)				
Apples	198	3.5	2.8	90	170	30.3	8	0	0
Cherries	54	0.5	0.4	85	70	3.2	24	0	0
Peaches	94	2.2	1.2	90	205	17.3	21	0	0
Pears	34	0.14	0.11	86	60	1.8	17	0	0
Plums	57	3	2.4	35	60	1.2	10	0	0
Grapes	308	9	7	95	50	14.6	27	0	0
Oranges	276	1.5	1	84	60	13.9	16	0	0
Grapefruit	98	0.5	0.4	94	60	5.5	2	0	0
Lemons	32	0.3	0.2	48	60	0.92	29	0	0
Other fruit	100	3	2.4	26	50	1.3	20	0	0
Pecans	155	1	0.8	46	50	3.6	21	0	0
Other fruits	170	4	2.6	46	50	3.9	12	0	0
Total		28.64	21.3			97.52			0

Table 16.3   Pesticide use in one 2 ha scab-resistant experimental (recently planted) apple orchard in north eastern Italy for 2 years (varieties: TSR, Florina, Coop 8, Coop 16, Coop 17, Golden Rush) (see Plate 16.1)

Variety	Application number	1994 total (kg ha^{-1})	Application number	1995 total (kg ha^{-1})
Sulphur*	2	3	3	6.5
CUSO$_4$**	2	3.2	3	8
Mineral oil***	0	–	1	1.5
Bacillus thuringiensis[t]	1	1	2	1.5
Piretrum[tt]	2	2	3	3

*   Against powdery mildew.
**   Against apple scab and other fungi.
***   Against scales and aphid eggs.
[tt]   Against aphids.
[t]   Against the defoliator *Iphantria cunea*.
   In both years, codling moth was not controlled, so damaged apples were about 18% in 1994 and about 38% in 1995.

Apples

The largest amounts of fungicide applied in United States and other countries are on fruit crops, especially apple orchards. From 8 to 35 applications of fungicides are necessary for apples which are susceptible to fungi (Tables 16.3 and 16.5). The key pathogenic fungi are apple scab and powdery mildew. From 8 to over 180 kg of fungicides are applied ha^{-1} year^{-1}. Improving the equipment, especially the design of spray nozzles, may reduce this figure by 50% (Van der Sheer, 1984). Appropriate scouting and monitoring, where applicable, may also consistently reduce the applications.

   Plant architecture, and the distances between plants, can also influence the presence of fungi. The introduction of new varieties of plants which are resistant to fungi is considered to be essential (Williams and Kuc, 1969; Crosby et al., 1992; Janick, 1993). These varieties would reduce the use of fungicides to only a few applications annually, and a consistent reduction in the use of fungicides could be achieved (author's personal observations in Italy, 1995) (Table 16.3). Adopting the new resistant varieties of apples in apple orchards, in addition to an appropriate spray technology, would reduce fungicide by about 70–80%. However in order to obtain these results, investments are needed to change the cultivars currently planted in fruit orchards.

   Post-harvest fungicides have also been used extensively on apples to decrease infection during their long refrigeration periods (AA, 1990).

Pears

Pear varieties have been developed which are resistant to fungal pathogens (Bell et al., 1976). Industrial pear orchards are heavily treated with pesticides. For example, in the Pianura Padana, Italy, up to 53 different applications of fungicides and other pesticides can be applied per year (author's personal observations, 1991). Fungicide applications may range from 15 to 90 kg ha^{-1} (Appendix A). Market forces are encouraging 'high cosmetic standards' and heavy pesticide use. These restricted standards are sometimes adopted in organic farming also!

Plate 16.1   Apples which are resistant or susceptible to parasitic fungi and which have been cultivated under low fungicide input (only sulphur and copper sulphate) (see Table 16.3)

(a)   TSR; scab-resistant.

(b)   Coop 17; scab-resistant.

(c)   Querina – Florina; scab-resistant.

(d)   Querina – Florina branches.

(e) TSR branches.

(d)–(f) Branches of Querina – Florina and TSR with a severe attack of powdery mildew on 40 – 60% of branches (July – August 1995).

(g) Comparison of leaves of Golden, which is susceptible to scab, and Querina – Florina, which is scab resistant.

(h) Golden; fungi-susceptible, affected by scab.

Table 16.4 World pesticide use*, in 1984

	Quantity (% of world use)	Total in 1984 (US$ million)	Breakdown (%)			
			Herbicide	Insecticide	Fungicide	Other
North America	39	6350	67	21	6	6
USA		5500	67	21	6	6
Canada		600	81	9	6	4
Mexico and Central America		250	25	55	15	5
South America	7	1175	53	26	20	1
Brazil		675	53	26	20	1
Other		500	54	25	20	1
Western Europe	21	3355	44	18	29	7
France		1050	46	16	31	7
British Isles		610	58	8	28	6
Federal Republic of Germany		500	54	14	27	5
Italy		490	33	27	33	7
Spain		300	26	38	20	16
The Netherlands		165	46	15	32	7
Scandinavia		180	45	17	30	7
Other		60	40	30	20	10
Eastern Europe	7	1190	53	30	20	5
Russia		545	58	22	15	5
Hungary		240	43	· 26	29	2
German Democratic Republic		180	53	19	22	6
Eurasia etc.		225	56	18	21	5
Asia	20	3335	28	44	24	4
Japan		1825	30	38	30	2
Balance of Soviet Union		225	57	21	14	7
India		500	6	79	9	6
Indonesia		160	11	71	11	7
South Korea		225	30	37	30	3
China		400	30	36	28	6
Other	6	895	52	25	17	6
Total	100	16300	52	26	17	5

*Includes agricultural, home, garden and industrial uses.
Source: Results of discussions with several producers (from Helsel, 1992).

Peaches and apricots

Peach orchards need some fungicides to control a few fungal pathogens such as *Taphrina deformans*, mildew, *Coryneum* and a few others (Appendix A). Biological and integrated peach orchards can reduce the quantities of fungicides applied (Table 16.6).

Grapes

Grape vineyards receive the largest amounts of fungicides, and wine grapes are highly susceptible to pathogenic fungi (the most common in Europe are *Plasmopara viticola*, which became a serious pest in 1878, and *Uncinula necator*, *Oidium tuckeri* and *Botrytis cinerea*, which became pests in 1845). The amounts of fungicides used vary according to the region in which the grapes are cultivated (Table 16.2, Appendix A). This is due to the

Table 16.5   Pesticides applied in a conventional apple orchard in 1987, near Lagundo, Bolzano, Italy (from Paoletti et al., 1995a)

Date	Active ingredients	Quantity $(\text{kg ha}^{-1}\,\text{year}^{-1})$
7 April	Mineral oil (I+A) + Ziram	33.3 + 2
16 April	Closentezine (A) + wettable sulphur + Ziram (F)	0.3 + 2.7 + 1.3
23 April	Ziram (F) + Triadimefon (F)	1.3 + 0.7
4 May	Fenarimol (F) + Zineb (F)	0.7 + 1.3
11 May	Zineb (F) + wettable sulphur (F)	2 + 2
18 May	Zineb (F) + wettable sulphur (F)	2 + 2
26 May	Zineb (F) + wettable sulphur (F)	2 + 2
31 May	Wettable sulphur + Captan + Zineb (F)	2 + 2 + 4.7
1 June	Paraquat (E) + Diquat (E) + Metoxitriazin (E)	2 + 2 + 2
11 June	Zineb (F) + Omethoate (I+A) + >wettable sulphur	2 + 2
23 June	Zineb (F) + wettable sulphur (F)	1.3 + 0.7
13 June	Ziram (F) + Triadimefon (F)	1.3 + 0.4 + 0.3
11 August	Ziram (F) + Propiconazol (F) + Poliglicole (B)	0.7 + 1.3 + 0.3
26 August	Ziram (F) + Poliglicole (B) + Genarimol (F)	1.3 + 0.3 + 0.7
4 September	Captan + Poliglicole (B)	1.3 + 0.3
Total		84.5

F, fungicidal activity; E, herbicidal activity; I, insecticidal activity; B, wetting agent; A, acaricidal activity.

fact that grapes are grown in areas where moisture is high. The persistence of moisture on the leaves, together with high temperatures, is related to the degree of fungal infection. Reduced use of animal manure has been reported to be another possible cause of high infection rates in Western countries (Howard, 1949). Reducing fungicide inputs as well as those of other pesticides is the main objective of integrated sustainable farming (AA, 1989; Boller et al., 1990; Haui, 1990). However, in viticulture, high rates of fungicides are generally used. The biodiversity of *Vitis* is over 70 species, which come from North America, South Asia and Asia Minor–South Europe (Alleweldt and Possingham, 1988; Smith, 1995). New interspecific hybrids, with resistance to fungal pathogens have been condemned (at least for wine production) owing to their undesirable flavours. Continuous breeding has resulted in new grape cultivars which are resistant to most important fungus diseases in middle Europe, and a wine quality which is indistinguishable from the *vinifera* wines. These acceptable new varieties are free from the undesirable flavours of the American *Vitis* species. At least five of these resistant cultivars have already been tested in Germany (Olmo, 1986; Alleweldt and Possingham, 1988).

Cherry

Resistance to brown rot (*Monilia fruticosa*), the key pathogen in the United States, has been evaluated and some possibilities for improvement have been discussed (Brown and Wilcox, 1989).

Table 16.6 Pesticides input in six peach orchards in one intensive fructicultural area in Emilia Romagna, Italy, during 1991. Quantities are expressed in kg/ha^{-1} of active ingredient. In biological Farms (B1, B2) and integrated farms (I2), pheromone dispensers are present in order to control the presence of *Grapholita (Cydia) molesta* and *Anarsia lineatella* through sexual confusion. (from Paoletti et al., 1992)

Treatments	Toxicity class	Biological farms		Integrated farms		Conventional farms	
		B1	B2	I1	I2	C1	C2
Fungicides							
Penconazol	III	—	—	0.08	—	—	—
Triforine	IV	—	—	—	—	0.6	—
Dithiocarbamates, Thiram	III	—	—	—	—	—	6.2
Dithiocarbamates, Ziram	III	—	—	9.0	8.8	8.8	2.2
Thiophanates, Thiophanate-methyl	III	—	—	0.8	—	—	—
Copper hydroxide	III	3.1	3.1 4	—	—	—	—
Copper sulphate	IV	—	2.0	—	—	—	—
Sulphate + melassa	IV	—	1.2	—	—	—	—
Sulphur as wettable dust	IV	4.0	1.0	1.2	6.0	—	2.0
Propoli	–	0.8	1.2	—	—	—	—
Insecticides							
Acephate	III	—	—	0.4	1.0	1.0	—
Azinphos-methyl	I	—	—	0.4	—	0.5	1.2
Methamidophos	I	—	—	—	—	—	0.5
White oil	III	—	—	—	—	—	12.0
Calcium polysulphide	III	—	9.4	—	—	—	—
Other substances and integrators							
Brown algae		1.0	1.1	—	—	—	—
Bentonite (hydrated aluminium silicate)		—	3.0	—	—	—	—
Mixture of nutritious elements							
Sequestrene (g per plant)		—	—	—	25	25	—
Sodium silicate		—	4.0	—	—	—	—

TOXICITY OF FUNGICIDES AND ENVIRONMENTAL INPUTS

Vertebrates

In general, the toxicity of fungicides is not high (Table 16.7) when compared with that of insecticides and herbicides (Pimentel, 1971; Brown, 1988). However, mutagenic and carcinogenic risks to humans are higher for fungicides than for insecticides or herbicides (NAP, 1987).

Insects

Tables 16.1, 16.7 and 16.8 summarize data on fungicide toxicity to insects compared with that of insecticides and herbicides (from Southerton, 1989; Theiling and Croft, 1989; AA, 1990). In general, fungicides seem to have a lower degree of toxicity to insects than

Table 16.7 Average toxocity ratings and variances of pesticide classes to all arthropod natural enemies, predators and parasitoids from the SELCTV database (from Theiling and Croft, 1989)

Pesticide class	Predators			Parasitoides			All		
	Mean*	Variance	No.**	Mean*	Variance	No.**	Mean*	Variance	No.**
Insecticide	3.611	1.78	(7326)	3.74	1.74	(2989)	3.65	1.77	(10315)
Fungicide	2.59	1.51	(781)	2.58	1.45	(357)	2.59	1.49	(1138)
Acaricide	2.76	1.81	(747)	2.83	1.50	(144)	2.77	1.76	(176)
Herbicide	2.83	1.73	(92)	3.10	1.77	(84)	2.95	1.76	(176)
All pesticides	3.43	1.92	(8968)	3.57	1.86	(3583)	3.47	—	(12551)

* Average toxocity rating and its associated variance were calculated for predators, parasitoids and all natural enemies using the TOX:RATING field in the SELCTV database.

** Number of database records used in the calculation.

Table 16.8   Relative toxicity of fungicides to various beneficial arthropods*

	Stethorus punctum		Aphidoletes	T. pyri	A. fallacis
	Adult	Larvae			
Bayleton (triadimefon)	No data	No data	No data	+	+
Benlate (benomyl)	+	+	+	+	+ +[2]
Captan	No data	No data	+	+	+
Dodine	No data	No data	+	+[1]	+[1]
Dikar (mancozeb + dinocap)	+	+	+	+	+[1]
Funginex (triforine)	No data	No data	No data	+	+
Glyodin	No data	No data	No data	+	+ +[1]
Karathane (dinocap)	+	+	No data	+	+
Manzate (mancozeb)	No data	No data	No data	+	+
Nustar (flusilazol)	No data	No data	No data	+	+
Polyram (metiram)	No data	No data	+	+[1]	+[1]
Ronilan (vinclozolin)	No data	No data	No data	No data	+
Rovral (iprodione)	No data	No data	No data	No data	+
Rubigan (fenarimol)	No data	No data	No data	+	+
Sulphur	No data	No data	No data	+[1]	+[1]
Thiram	No data	No data	No data	No data	+[1]

*   Amblyscius fallacis and Typhlodromus pyri (predatory mites), Stethorus punctum (ladybird beetle mite predator) and Aphidoletes aphidimyza (Cecidomyiid aphid predator) (from USDA, 1990).
+,   Low impact on population (less than 30% mortality after 48 h).
+ +, Moderate impact on population (between 30 and 70% mortality after 48 h).
[1]   This information is derived from 24-h slide-dip tests conducted at NYSAES. (New York State Agricultural Experiment Station Ithaca N.Y.)
[2]   Benlate supresses egg-laying almost completely and is harmful to immature A. fallacis.

insecticides. However, the majority of tests reviewed in Tables 16.1, 16.7 and 16.8 were run on organisms mainly living above the soil. In the soil the impact of fungicides is not well known because soil is not a targeted environment to which fungicides are applied. Fungicides may develop high levels of toxicity for many soil-dwelling, non-target organisms.

Earthworms

Fungicides are, in general, highly toxic to earthworms when they are present as copper and zinc residues, e.g. copper sulphate or carbamates containing zinc. Soil fumigants, nematicides and fungicides such as D–D mixture (dichloropropane:dichloropropene), metham-sodium and methyl bromide are also highly toxic to earthworms. In general, the majority of fumigants and contact nematicides are toxic to earthworms as well (Edwards and Bohlen, 1992), along with carbamate fungicides such as benomyl and carbendazim (Stringer and Wright, 1976; Brown, 1978). In England, it has been reported that about 1.8 kg ha^{-1} year^{-1} of benomyl could destroy all the Lumbricus terrestris and most of Allolobophora present in apple orchards. The degree of toxicity of the more traditional copper sulphates is controversial. At the laboratory level, tests especially developed for Eisenia foetida (a lubricid species living only in composting manure) suggest that higher doses of copper sulphates (over 1000 p.p.m.) are lethal (Malecki et al., 1982). However, this

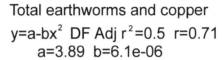

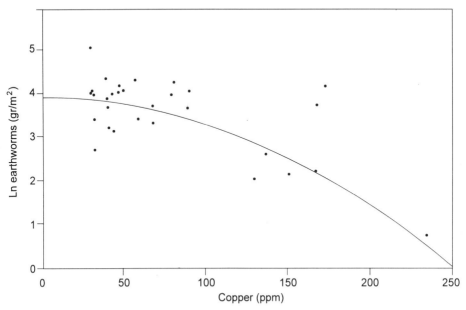

Figure 16.1   Total earthworm living biomass and copper concentration in soil in 32 orchards near Forli', Emilia Romagna, Italy (from Paoletti and Sommaggio, 1996). Each point represents the mean of five samples both of earthworms and of copper concentration in the soil. The copper is related to the copper sulphate used as a fungicide

species is completely absent in the field and is not representative of the earthworm fauna in the rural landscape. In addition, some researchers have demonstrated that copper toxicity for earthworms decreases when the level of organic matter in the soil is high (Jaggy and Streit, 1982). Direct experience has demonstrated that if the total concentration of copper present in the soil is between 100 and 150 p.p.m. it will be sufficient to decrease the earthworm population in most cases (Figures 16.1 and 16.2). Numerous orchards worldwide have higher concentrations of copper in the soil.

MONITORING FUNGICIDE EFFECTS WITH NON-TARGET ORGANISMS

Wood ash, sulphur, lime, lime sulphate and copper sulphate have been applied for many years to plants attacked by fungi, including fruit crops such as grape, apple, pear, peach, apricot and citrus fruits (Debach, 1974). In general, they have been considered not to be very toxic to humans, vertebrates and non-target beneficial insects such as predators and parasitoids (Pimentel, 1971; Brown, 1978) (Tables 16.1 and 16.7). However if we consider the agroecosystem as a whole rather than only the organisms present at canopy level, some effects have been reported, especially in soil organisms. Detritivores such as earthworms are severely affected by products containing copper. Soils containing over

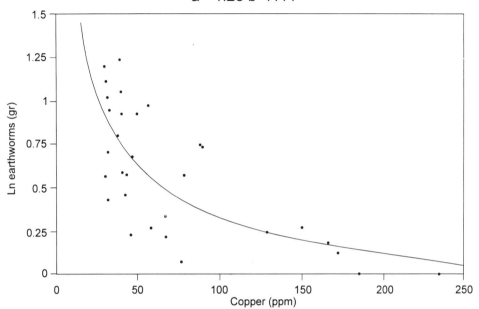

Figure 16.2 The same relationship of before, for only the endogeic earthworms and copper. Observe that for endogenic species the loss is more severe than for all earthworms

100–150 p.p.m. of copper, a condition surpassed in numerous intensive and traditional grape areas in Europe, such as France, Spain, Italy, Switzerland and Germany, are responsible for severe damage to earthworms, since only few species can survive in such conditions. An example is given by *Aporrectodea* (= *Allolobophora*) *rosea* (Figure 16.3) or some epigeic species such as *Lumbricus castaneus* and *L. rubellus*. Most endogenic species disappear (Paoletti et al., 1988) and a number of large burrowing species drastically decrease, as is the case with *Lumbricus terrestris*. The soil tends to accumulate undecomposed organic material at the surface and become less permeable. Sometimes the level of the biomass of oribatids, collembola and dipteran larvae increases as well. Other polyphagous predators can be affected, as in the case of the Carabid beetles, which may decrease in both numbers and species; some species seem to be particularly sensitive to copper (Figure 16.4). The zinc contained in more recent fungicides, such as carbamates, is also very harmful to non-target soil organisms, but zinc concentrations in orchard soils are never as high as those with copper (Figure 16.5).

Few data exist on the impacts of sulphur, lime and lime sulphate compared with those of copper and other currently used active ingredients. However, changes in soil pH as a result of the presence of sulphur can influence soil detritivores such as earthworms and other invertebrates (Paoletti and Bressan, 1996).

Using small invertebrates as a tool to evaluate the input and the environmental damage of pesticides appears to be a good strategy. Figure 16.6 shows the results of a

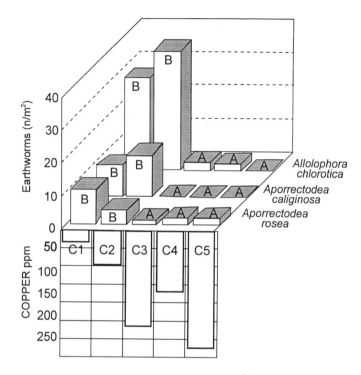

Figure 16.3   Earthworms loss in a lowland agroecosystem (Pegolotte di Cona, Venezia, Italy) under different copper inputs in five different fields which had or had not contained vineyards. The copper concentration in the soils (C1–C5) is inversely related to earthworm numbers, note that the endogenic *A. rosea* is also present in the very contaminated C5 plot (from Paoletti et al., 1995b)

2-year evaluation of six different peach orchards to record taxa diversity. The biological and integrated farms have a higher number of species than conventional orchards. A consistent reduction of species affects both integrated and conventional farms (Paoletti and Sommaggio, 1996).

ALTERNATIVES TO FUNGICIDES

From the available literature and our personal experience, we have compiled a list of possible ways to reduce fungicide levels in fruit orchards.

Cultivation and equipment

Tillage and cultivation between plant rows can reduce the presence on the soil surface of some pathogens that could affect the pathogen reservoir for fruit plants. Tillage itself increases soil disturbance and can affect and damage earthworms, whose major role is the incorporation of organic matter into the soil. Disking appears to be more damaging to earthworms than the use of any other type of rotary mechanism. Superficial tillage is less severe to earthworms than deep ploughing (Lee, 1985).

    The introduction of new designs for spray nozzles, on application equipment has

DISTRIBUTION OF MAIN SPECIES OF GROUND BEETLES

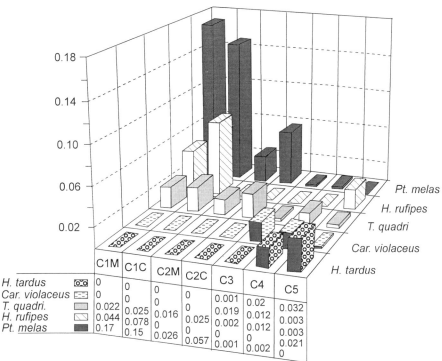

	C1M	C1C	C2M	C2C	C3	C4	C5
H. tardus	0	0	0	0	0.001	0.02	0.032
Car. violaceus	0	0	0	0	0.019	0.012	0.003
T. quadri.	0.022	0.025	0.016	0.025	0.002	0.012	0.003
H. rufipes	0.044	0.078	0	0	0	0	0.021
Pt. melas	0.17	0.15	0.026	0.057	0.001	0.002	0

Figure 16.4 Carabid species and copper in the same situation as in Figure 16.3. C1C and C2C mean field center; C1M and C2M is field margin

demonstrated that the amount of fungicides applied for apple scab control could be reduced by 50% (Van der Scheer, 1984).

Resistance

Resistance to diseases is a normal condition, and good sources of resistance are readily available (Harlan, 1977). The main aim of breeding for resistance is to seek varieties which are less susceptible to disease or suffer less damage. Resistance to pathogens, disease immunity and disease tolerance are fundamental areas to be investigated (Russell, 1978). A good example of resistance is still present, after many years, in some apple varieties which are resistant to apple scab (Russell, 1978; Crosby et al., 1992). In the case of grapes, some native cultivars are very resistant to fungi and crossing such grapes with the European ones seems promising except for some possibly undesirable flavour changes. Recently produced grape accessions appear to be free from this flavour problem (Alleweldt and Possingham, 1988). However, it is not easy to convince traditional producers, linked to particular varieties of grape for wine production, to abruptly change their stocks or the accessions without strong pressure from the consumers in association with innova-

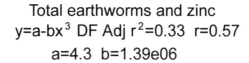

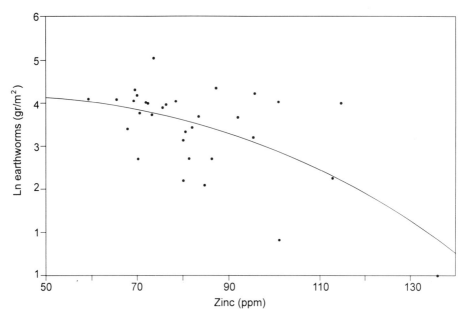

Figure 16.5   Zinc and earthworm concentrations in soils in the same orchards as Figure 16.1

tive policies and massive investments. New resistant varieties are not indefinitely resistant to the target pathogens or insect pests. For example, the European grapes grafted onto American stock at the beginning of the century are still resistant to the key insect pest (*Phyloxera vastatrix*). However, some apple-scab-resistant varieties ('Baujade', 'Coop 28', 'Florina', 'Liberty', 'Priscilla' and 'X4171') seem to have lost their resistance against at least one German fungal type (Parisi et al., 1993; Schmidt et al., 1995). Fungi overcoming resistance in hosts appear to be a common trend, but the resistance factors involved in apples are not well understood. If the concentration of flavan-3-ols in scab-resistant varieties is high (Treutter and Feucht, 1990) the cuticular membrane of the leaves is not a key factor in preventing infection as had been expected (Valsangiacomo and Gessler, 1988). More intensive work on resistance factors is required in order to have accessions which can be multiplied and used.

Mulching

Introducing permanent or temporary cover for annual or perennial weeds may increase the level of organic matter and nitrogen in the soil. Potential benefits include an increase of biodiversity into the system with an increased presence of beneficial insects such as predators and parasitoids (Favretto et al., 1988, 1992). The soil ecosystem will have an

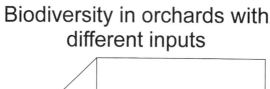

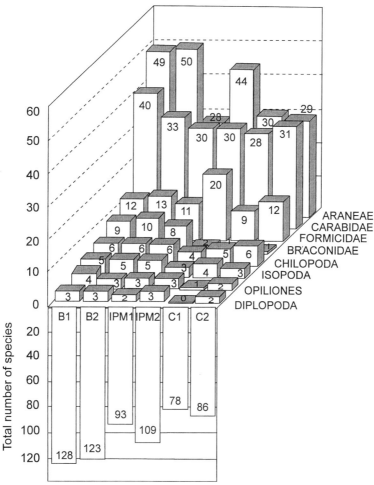

Figure 16.6 Number of arthropod species and input strategies (see Table 16.6) in three groups of peach orchards in Emilia Romagna, Italy. B1 and B2 are biological farms, IPM1 and IPM2 are integrated farms and C1 and C2 are conventional, high-input farms. Note that there is a decreased number of invertebrate species (in 2 years) in integrated and conventional farms compared with biological farms (from Paoletti and Sommaggio, 1996)

increased biomass of detritivores such as earthworms. Earthworms incorporate surface litter and organic residues into the soil, and this may decrease the infectivity of dormant pathogens (such as apple scab in apple orchards). An adverse effect of mulching can be the competition for water and nutrients, especially when fruit is initially planted. Higher moisture in the understory of orchards, which is associated with permanent mulching, may promote pathogenic fungi. Some pests can be encouraged by mulching. For instance, slugs and earwigs can increase and may damage fruit in apricot and peach orchards.

Fertilizer application

In India, grapes are less susceptible to fungal diseases because of the use of large amounts of animal manure and the drier condition of the soil in the east (Howard, 1949). A heavy use of chemical fertilizers may increase pathogenic fungi and insect pests. A decrease in the level of nitrogen may reduce the pathogenicity of fungi and the damage to cultivated plants (Smith, 1989).

Adjusting crop density

Increasing plant density promotes the risk of plant pathogenic and insect pest infections. However, the current tendency is to increase the number of plants per hectare. From the traditional 300–1300 ha^{-1}, new intensive orchards have 4000–5000 plants ha^{-1} (apples, pears, etc.). Cultivars that are highly susceptible to powdery mildew should not be grown with mildew-resistant cultivars (such as 'Liberty' in apples orchards). In fact, even cultivars resistant to diseases can become heavily infected with secondary mildew if they are planted close to cultivars providing abundant inoculum (AA, 1990).

Altering plant and crop architecture

Different densities of branches and leaves in relation to air space affects moisture levels and therefore pathogen inoculation. Improving the air circulation promotes plant health.

Increasing biological soil activity

Earthworms have been shown to be responsible for the incorporation of litter into the soil, and because of this activity they can disseminate some beneficial fungi, such as the VAM (Rabatin and Stinner, 1988). They can also decrease the surface reservoir of some plant pathogens such as the apple scab, or reduce the virulence of others (Raw, 1962; Brown, 1978). Experiences with earthworm augmentation in orchards have been successful (Wu Wenliang, Agricultural University of Beijing, personal communication, 1994), (Figure 16.7). The diversity and abundance of invertebrates can also promote biological control (Paoletti et al., 1994) (Fig 16.7).

Biological control using mychorrizae

Many different microorganisms and mychorrizae have been reported in literature as possible ameliorators of soil conditions and plant sanitation, including the basidiomicete *Tricoderma*.

Wood ash

The extensive traditional use of wood ash as a fungicide spread on plants would suggest

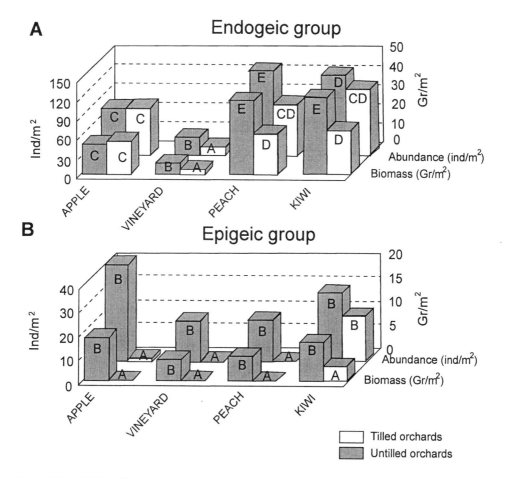

Figure 16.7 (A) Distribution of endogenic earthworms in 64 orchards (in Emilia Romagna, Italy) with different chemical input: apple orchards with high input; vineyards with medium chemical but high copper input; kiwi orchards with low chemical input. (B) Distribution of epigeic earthworms in orchards with chemical input. Note that tillage can also affect earthworm populations, especially among epigeic earthworms

that research into its effectiveness would be a worthwhile objective for governments and development agencies (Thurston, 1992).

Antagonistic effects against pathogenic fungi

Different microorganisms, including fungi, have been suggested as possible supressors or antagonists to plant pathogenic fungi (Hoitink and Fahy, 1986). Organic manure has also been considered as another possible ingredient to reduce pathogenic fungi (Howard, 1949). Weltzien (1988) has demonstrated the effect of watery extracts of composts in reducing pathogenic fungi in both laboratory and field trials, particularly on the grape downy mildew (*Plasmopara viticola*), powdery mildew (*Uncinula nectator–Oidium tuckeri*) and leaf blight (*Pseudopeziza tracheiphila*). In the field of antagonistic activity among different microorganisms, more extensive research is needed before the large-scale use of such alternatives enter common practice (Weltzien, 1988).

Genetically engineered organisms

Plants modified by genetic engineering seems promising in the world of agriculture. No notable innovations to agriculture have been released yet, except a long-lasting type of tomato and the engineered growth hormone to increase milk production in cows. Plants resistant to herbicides and plants engineered with BT endotoxine to resist pest attack are among future developments to be released in the market. Improving resistance to fungi in order to decrease the level of fungicide use and improve the environment are a priority (Paoletti and Pimentel, 1995). However, to date, insufficient research has been carried out to produce effective results against fungi. Some institutions are working on grape resistance (G. Alleweldt, personal communication, October 1992).

CONCLUSIONS

Reducing pesticide use on crops to improve the environment is a target for farmers, consumers and the majority of dealers. It is not an easy step to achieve without additional research and extension efforts. In general, agriculturalists do not feel that fungicides, insecticides or herbicides are really an environmental problem or the cause of risks to farmers and consumers. To achieve a reduction in fungicides and other pesticides in fruit crops will require a great deal of research.

(1) Breeders will have to focus on selecting and promoting plants which are less prone to pests and pathogens, and producing better flavoured and more nutrient-rich fruits instead of concentrating on some 'cosmetic' or aesthetic standards.

(2) Genetic engineering must work more intensively to produce varieties that require fewer pesticides than are now used (Paoletti and Pimentel, 1995).

(3) Agriculturalists have to understand that today, in both developed and undeveloped countries, reducing pesticides is a way to improve the agroecosystem as well as the profitability of farming activities.

(4) Researchers, especially ecotoxicologists, have to monitor not only the target arthropods in orchards, but also the large array of organisms living in the soil.

(5) Policy makers have to be more active in promoting policies and subsidies to help farmers who adopt low-input strategies, including low levels of chemical pesticides. They need to ban chemical products that have severe environmental effects, and discourage plant varieties that are highly susceptible to plant pathogens and insect pests.

(6) In selecting varieties, farmers should give priority to resistant genotypes.

(7) Consumers, and consumers' associations, have to be more aware that quality standards of fruit, including those that are simply cosmetic, may influence the amount of pesticide used, and in turn the biodiversity in agroecosystems.

(8) In some cases, biological (organic), biodynamic and other low-input agricultural systems tend to have similar cosmetic standards for the market to those of conven-

tional products. Substituting chemical inputs with other inputs which include large amounts of copper and sulphur is not, by itself the best way to decrease environmental pollution.

(9) The sustainability of farms must be further considered and biodiversity encouraged. This is the only way to avoid pesticide contamination and risks for the future.

ACKNOWLEDGEMENTS

I am indebted to many people who have offered information regarding fungicides, in particular M. Borgo, J. Kinzle, E. Jorg, Li Ji, M. Morten and Dr Malavolta. I extent very special thanks to D. Pimentel for his information and thoughtful comments on the manuscript, and to R. D'Orazio for help with the tables and improving the manuscript.

REFERENCES

AA (1989) *Alternative Agriculture*. National Research Council, National Academy Press, Washington, DC, 448 pp.

AA (1990) *Management Guide for Low-Input Sustainable Apple Production*. In: USDA–LISA Apple Production Project, 83 pp.

Alleweldt, G. and Possingham, J.V. (1988) Progress in grapevine breeding. *Theoretical and Applied Genetics* **75**, 669–673.

Alston, F.H. and Briggs, J.B. (1970) Inheritance of hypersensitivity to rosy apple aphid *Dysaphis plantaginea* in apple. *Canadian Journal of Genetic Cytology* **12**, 257–258.

Alston, F.H. and Briggs, J.B. (1977) Resistance genes in apple and biotypes of *Dysaphys devecta*. *Annals of Applied Biology* **87**, 75–81.

Bell, R.L. and Janick, J., Zimmerman, R.G. and van der Zwet, T. (1976) Relationship between fire blight resistance and fruit quality in pear. *HortScience* **11**(5), 500–502.

Boller, E.F., Baster, P. and Koblet, W. (1990) Integrated production in viticulture of eastern Switzerland: Concepts and organization (the 'Wadenswil Model'). *Schweizerische Landwirtschaftliche Forschung* **29**(4), 287–291.

Brown, A.W.A. (1978) *Ecology of Pesticides*. Wiley, New York. 525 pp.

Brown, K.S. and Wilcox, W.F. (1989) Evaluation of cherry caenotypes for resistance to fruit infection by *Monilia fruticosa*. *HortScience* **24**, 1013–1015.

Chaboussou, F. (1980) *Les Plantes Malades des Pesticides*. Debard, Paris, 265 pp.

Crosby, J.A., Janick, J., Pecknold, P.C., Korban, S.S., O'Connor, P.A., Ries, S.M., Goffreda, J. and Voordeckers, A. (1992) Breeding apples for scab resistance. *Fruit Varieties Journal* **46**(3), 145–166.

Culliney, T.W., Pimentel, D. and Pimentel, M.H. (1993) Pesticides and natural toxicants in foods. In Pimentel, D. and Lehman, H. (Eds) *The Pesticide Question: Environment, Economics and Ethics*. Chapman & Hall, New York, pp. 126–150.

DeBach, P. (1974) *Biological Control by Natural Enemies*. Cambridge University Press, Cambridge, 323 pp.

Edwards, C.A. and Bohlen, P.J. (1992) The effects of toxic chemicals on earthworms. *Reviews of Environmental Contamination and Toxicology* **125**, 23–99.

Farquhar, L. (1992) Diversity and disease. Has biotechnology got it all wrong? *Agricultural Biotechnology Network Bulletin* **January**, 4–5.

Favretto, M.R., Pavletti, M.G., Lorenzi, G.G. and E., Dioli (1988) Loscambio di invertebrati tra un relitio di bosco planiziale ed. agro ecosi stemi contigui. *l'Extropodofauna del Bosco di Lison*. Nova thalassha, pp. 329–358.

Favretto, M.R., Paoletti, M.G., Caporali, F., Nannipieri, P., Onnis, A. and Tomei, P.E. (1992) Invertebrates and nutrients in a Mediterranean vineyard mulched with subterranean clover (*Trifolium subterraneum* L.). *Biology and Fertility of Soils*, Springer-Verlag, **14** 151–158.

Fokkema, N.J. (1991) The phyllosphere as an ecologically neglected milieu: A plant pathologist's point of view. In Andrews, J.B. and Hirano, S.S. *National Ecology of Leaves*, New York, Springer-Verlag, pp. 499.

Francis, C.A. (1990) Breeding hybrids and varieties for sustainable systems. In Francis, C.A., Butler, F.C. and King, L.D. (Eds) *Sustainable Agriculture in Temperate Zones*. Wiley, pp. 24–27.

Goonewardene, H.F. and Povish, W.R. (1988) Arthropod resistance in plant introduction accessions of *Malus* spp. to some arthropod pests of economic importance. *Fruit Varieties Journal* **42**(3), 88–91.

Harlan, J.R. (1977) Sources of genetic defence. *Annals of the New York Academy of Sciences* **287**, 345.

Haui, F. (1990) Farming systems research at Ipsach, Switzerland. The 'Third Way' Project. *Schweizerische Landwirtschaftliche Forschung* **29**, 257–271.

Helsel, Z.R. (1992) Energy and alternatives for fertilizer and pesticide use. In Fluck, R.C. (Ed.) *Energy in Farm Production*. Elsevier, Amsterdam, pp.

Hoitink, H.A.J. and Fahy, P.C. (1986) Basis for the control of soil-borne plant pathogens with composts. *Annual Review of Phytopathology* **24**, 93–114.

Howard, A. (1949) *An Agricultural Testament*. Oxford University Press, Oxford, 253 pp.

Jaggy, A. and Streit, B. (1982) Toxic effects of soluble copper on *Octolasium cyaneum* Sav. *Revue Suisse de Zoologie* **4**, 881–889.

Janick, J. (1993) Disease-resistant apples. Tasty new cultivars require fewer sprays. *Fine Gardening* **May/June**, 55–57.

Jepson, P.C. (1989) *Pesticides and Non-target Invertebrates*. Intercept, 240 pp. London

Jordan, V.W.L. (1993) *Agriculture. Scientific Basis for Codes of Good Agricultural Practice*. European Communities Commission Directorate General XIII, Luxembourg, 163 pp.

Lee, K.E. (1985) *Earthworms. Their Ecology and Relationships with Soils and Land Use*. Academic Press, New York, 411 pp.

Malecki, M.R., Neuhauser, E.F. and Goehr, R.C. (1982) The effects of heavy metals on the growth and production of *Eisenia foetida*. *Pedobiologia* **24**, 129–137.

McEwen, F.L. and Stephenson, G.R. (1979) *The Use and Significance of Pesticides in the Environment*. J. Wiley, New York, 538pp.

NAP (1987) *Regulating Pesticides in Food. The Delaney Paradox*. National Academy Press, Washington, DC, 272 pp.

Olmo, H.F. (1986) The potential role of (*Vinifera* × *Rotundifolia*) hybrids in grape variety improvement. *Experientia* **42**(8), 921–926.

Paoletti, M.G. (1993) Bioindicators of sustainability and biodiversity in agroecosystems. In Paoletti, M.G., Napier, T., Ferro, O., Stinner, B.R. and Stinner, D. (Eds) *Socio-economic and Policy Issues for Sustainable Farming Systems*. Coop Erativa Amicizia, Padova, pp. 21–36.

Paoletti, M.G. and Bertoncello Brotto, G. (1985) Side-effects of fungicide residues (Cu, Zn) on soil invertebrates in a vineyard and meadow agrosystem in N-E Italy. *Les Colloques de l'INRA* **31**, 233–254.

Paoletti, M.G. and Bressan, M. (1996) Soil invertebrates as bioindicators of human disturbance. *Critical Reviews in Plant Sciences* 1–42. Vol. N. **15**(1): 21–102.

Paoletti, M.G. and Pimentel, D. (1995) The environmental and economic costs of herbicide resistance and host - plant resistance to plant pathogens and insects. *Technological Forecasting and Social Change* **50**, 9–23.

Paoletti, M.G. and Sommaggio, D. (1996) Biodiversity indicators for sustainability. Assessment of rural landscapes. In Van Straalen, N.M. and Krivolutskii, D.A. (Eds) *New Approaches to the Development of Bioindicator Systems for Soil Pollution*. Publ. 16: 123–140 Nato Ast Series. Kluwer.

Paoletti, M.G., Iovane, E. and Cortese, M. (1988) Pedofauna bioindicators as heavy metals in five agroecosystems in north-east Italy. *Review d' Ecologie et de Biologie du Sol* **25**(1), 33–58.

Paoletti, M.G., Favretto, M.R., Stinner, B.R., Purrington, F.F. and Bater, J.E. (1991) Invertebrates as bioindicators of soil use. *Agriculture, Ecosystems and Environment* **34**, 341–362.

Paoletti, M.G., Favretto, M.R., Bressa, M., Marchiorato, A. and Babetto, M. (1992) Biodiversità in pescheti forlivesi. In Paoletti, M.G. et al. (Eds) *Biodiversità agli Agroecosistemi*. Osservatorio Agroambientale, Forli, pp. 33–80.

Paoletti, M.G., Schweigl, U. and Favretto, M.R. (1995a) Soil microinvertebrates, heavy metals and organochlorines in low and high input apple orchards and coppiced woodland. *Pedobiologia* **39**, 20–33.

Paoletti, M.G., Sommaggio, D., Petruzzelli, G., Pezzarossa, B. and Barbafieri, M. (1995b) Soil invertebrates as monitoring tools for agricultural sustainability. *Bull. Entomol. Pologne* **64**(1–2),

1–10.

Paoletti, M.G., Favretto, M.R., Marchiorato, A., Bressan, M. and Babetto, M. (1993) Biodiversità in peacheti forliveai. In Paoletti et al. *Biodiversità negli Agroecosistemi. Osservatorio Agroembiantale*, Centrale Ortofruiticola, Forli, pp. 20–56.

Polskie Piamo, *J. entomologiczne*, **104**, 113–122.

Parisi, L., Lespinasse, Y., Guillaumes, and Kruger, J. (1993) A new race of *Venturia inaequalis* Virulent to apples with resistance due to the Vf gene. *Phytopathology* **83**, 533–537.

Petterson, O. (1993) Swedish pesticide policy in a changing environment. In Pimentel, D. and Lehman, H. (Eds) *The Pesticide Question. Environment, Economics and Ethics*. Chapman & Hall, New York, pp. 182–205.

Pimentel, D. (1971) *Ecological Effects of Pesticides on Non-Target Species*. Executive Office of the President, US Government Printing Office, Washington, DC, 220 pp.

Pimentel, D. and Levitan, L. (1986) Pesticides: Amounts applied and amounts reaching pests. *BioScience* **36**, 86–91.

Pimentel, D., McLaughlin, L., Zepp, A., Lakitan, B., Kraus, T., Kleinman, P., Vancini, F., Roach, W.J., Graap, E., Keeton, W.S. and Seling, G. (1991) *Environmental and Economic Impacts of Reducing US Agricultural Pesticide Use*. CRC Press, Boca Raton, FL, pp. 679–718.

Rabatin, S.L. and Stinner, B.R. (1988) Indirect effects of interactions between VAM fungi and soil - inhabiting invertebrates on plant processes. *Agriculture, Ecosystems and Environment* **24**, 135–146.

Raw, F. (1962) Studies of earthworm populations *In orchards I*. Leaf burial in apple orchards. *Ann. Appl. Biol* **50**, 389–404.

Russell, G.E. (1978) *Plant Breeding for Pest and Disease Resistance*. Butterworth, London, 485 pp.

Schmidt, H., Kruger, J., Dunemann, F. and Parisi, L. (1995) Anomalous scab data from Ahrensburg. *European Apple* **3**. 9.

Shen Der Hong (1996) Microbial diversity and its application in agriculture in China. *Agriculture, Ecosystems and Environment* submitted for publication.

Smith, B.D. (1995) The emergence of agriculture, *Scientific American Library*, Washington DC. 231pp.

Smith, C.M. (1989) *Plant Resistance to Insects. A Fundamental Approach*. Wiley, New York, 286 pp.

Southerton, N.W. (1989) Farming practices to reduce the exposure of non-target invertebrates to pesticides. In Jepson, P. (Ed.) *Pesticides and Non-Target Invertebrates*. Intercept, London, UK, pp. 195–212.

Stringer, A. and Wright, M.A. (1976) The toxicity of benomyl and some related 2-substituted benzimidazoles to earthworm *Lumbricus terrestris*. *Pesticide Science* **7**, 459–464.

Surgeoner, G.A. and Roberts, W. (1993) Reducing pesticide use by 50% in the Province of Ontario: Challenges and progress. In Pimentel, D. and Lehman, H. (Eds) *The Pesticide Question. Environment, Economics and Ethics*. Chapman & Hall, New York, pp. 206–222.

Theiling, K.M. and Croft, B.A. (1989) Toxicity, Selectivity and Sublethal Effects of Pesticides on Arthropod Natural Enemies: A Data base Summary. In Jepson, P. (Ed.) Intercept, London. pp. 213–232.

Thurston, H.D. (1992) *Sustainable Practices for Plant Disease Management in Traditional Farming Systems*. Westview Press, Boulder, CO, 279 pp.

Treutter, D. and Feucht, W. (1990) The patter of flavan-3-ols in relation to scab resistance of apple cultivars. *Journal of Horticultural Science* **65**(5), 511–517.

USDA, 1990 Cornell University, Geneva, N.Y.

Valsangiacomo, C. and Gessler, C. (1988) Role of cuticular membrane in ontogenic and Vf-resistance of apple leaves against *Venturia inaequalis*. *Phytopathology* **87**(8), 1066–1069.

Van Den Bosch, R. (1978) *The Pesticide Conspiracy*. Doubleday, Garden City, 228 pp. New Jersey.

Van Den Bosch, R., Messenger, P.S. and Gutierrez, A.P. (1982) *An Introduction to Biological Control*. Plenum Press, New York, 247 pp.

Van Der Scheer, H.A. (1984) *Testing of Crop Protection Chemicals in Fruit Growing*. Annual Report of the Research Station for Fruit Growing, Wilhelminadorp, The Netherlands, 70 pp.

Weltzien, H.C. (1989) Some effects of composted organic materials on plant health. *Agriculture, Ecosystems and Environment* **27**, 439–446.

Williams, E.B. and Kuc, J. (1969) Resistance in *Malus* to *Venturia inaequalis*. *Annual Review of Pathology* **7**, 223–224.

APPENDIX A

List of some fungicides which are currently used in orchards with different input-rates. Although the list is not exhaustive, the major chemicals are included. The annual usage rates for a given fungicide with respect to the number of treatments is given. The product of the quantity (g ha^{-1}) and the mean number of treatments gives the total amount of fungicide used, for each product, per ha per year.

Fruit	Country[1]	Farm[2]	Mean number of treatments	Quantity[3] (g ha^{-1})	Active ingredient	Trademark	Line[7]
Apples	C	CF	–	135000	Copper sulphate	–	–
	C	CF	–	30000	Sulphur	–	–
	ER	IF	0.7	928	Captan[4]	–	a
	ER	IF	0.2	323	Carbaryl	–	a
	ER	IF	0.1	69	Chlorothalonil[4]	–	a
	ER	IF	1.6	4894	Copper salts	–	a
	ER	IF	0.1	94	Diclofluanide	–	a
	ER	IF	0.0	5	Dinocap	–	a
	ER	IF	1.3	1553	Dithianon	–	a
	ER	IF	0.3	299	DNOC	–	a
	ER	IF	2.7	2772	Dodina	–	a
	ER	IF	0.5	19	Hexaconazole	–	a
	ER	IF	0.1	12	Fenarimol	–	a
	ER	IF	8.8	18 294	Mancozeb[4]	–	a
	ER	IF	1.4	3299	Metiram[4]	–	a
	ER	IF	0.3	23	Miclobutanyl	–	a
	ER	IF	1.4	94	Penconazole	–	a
	ER	IF	0.0	45	Propineb	–	a
	ER	IF	1.1	1911	Sulphur	–	a
	ER	IF	0.00	5	Tiofanate-methyl	–	a
	ER	IF	0.20	54	Triforine	–	a
	ER	IF	0.0	14	Zineb[4]	–	a
	ER	IF	0.4	1156	Ziram	–	a
	ER	IF	0.1	192	Captan[4]	–	b
	ER	IF	0.1	105	Carbaryl	–	b
	ER	IF	2.1	4249	Copper salts	–	b
	ER	IF	0.0	18	Diclofluanide	–	b

ER	IF	1.0	1000	Dithianon	—	b
ER	IF	2.3	2099	Dodina	—	b
ER	IF	0.1	11	Fenarimol	—	b
ER	IF	0.6	24	Hexaconazole	—	b
ER	IF	6.1	11 688	Mancozeb[4]	—	b
ER	IF	3.6	7714	Metiram[4]	—	b
ER	IF	0.1	6	Miclobutanyl	—	b
ER	IF	0.6	38	Penconazole	—	b
ER	IF	1.3	2094	Sulphur	—	b
ER	IF	0.2	143	Triforine	—	b
ER	IF	0.0	92	Zineb[4]	—	b
ER	IF	0.1	250	Ziram	—	b
G1	CF	1.3	2000	Dichlofluanid	Euparen	—
G1	CF	1.5	500	Dithianon	Delan	—
G1	CF	3.5	525	Flusiazol	Benocap	—
G1	CF	1.5	2200	Mancozeb[4]	Dithane Ultra	—
G1	CF	1.0	3000	Sulphur	Cosan 80	—
G1	CF	0.7	1000	Thiophanate-methyl	Carcobin FL	—
G1	CF	2.0	3000	Triadimenol	Bayfidan	—
G1	IF	1.3	1500	Dichlofluanid	Euparen	—
G1	IF	1.5	500	Dithianon	Delan	—
G1	IF	3.0	125	Flusiazol	Benocap	—
G1	IF	1.5	2000	Mancozeb[4]	Dithane Ultra	—
G1	IF	1.0	3000	Sulphur	Cosan 80	—
G1	IF	0.7	70	Thiophanate-methyl	Cercobin FL	—
G1	IF	2.0	500	Triadimenol	Bayfidan	—
G2	BF	1–2	3000	Equisetum arvense (extract)	—	—
TAA	BF	4–7	800–2500	Ca polysulphur	Mycosin	—
TAA	BF	–	500–1000	Clay + equisetum extract + yeast	—	—
TAA	BF	3–8	50–500	Copper	Bordeaux-mixture	—
TAA	BF	2–5	20–100	Copper oxychloride	—	—
TAA	BF	2–5	300	Herb extract[5]	—	—
TAA	BF	–	500	NAB	—	—
TAA	BF	1–2	100	Propolis	—	—
TAA	BF	15.0	100–400	Sulphur	—	—
TAA	BF	1–3	1000–2500	Sulphur + molasses	Sulfar	—
TAA	BF	3–4	900	ZAB	ZAB	—

Appendix A (cont.)

Fruit	Country[1]	Farm[2]	Mean number of treatments	Quantity[3] (g ha^{-1})	Active ingredient	Trademark	Line[7]
Apricots	ER	IF	0.3	97	Benomyl[4]	–	a
	ER	IF	0.1	127	Carbaryl	–	a
	ER	IF	0.2	49	Carbedazin	–	a
	ER	IF	0.0	54	Dodina	–	a
	ER	IF	1.0	2848	Copper salts	–	a
	ER	IF	2.8	5222	Sulphur	–	a
	ER	IF	0.4	384	Thiram	–	a
	ER	IF	0.2	172	Triforine	–	a
	ER	IF	0.0	14 551	Ba polysulphur	–	b
	ER	IF	0.1	5541	Ca polysulphur	–	b
	ER	IF	0.3	181	Carbedazin	–	b
	ER	IF	0.0	5	Dinocap	–	b
	ER	IF	0.0	62	Dodina	–	b
	ER	IF	1.0	2875	Copper salts	–	b
	ER	IF	3.4	9308	Sulphur	–	b
	ER	IF	1.6	3147	Thiram	–	b
	ER	IF	0.9	332	Triforine	–	b
	ER	IF	0.5	2310	Ziram	–	b
Cherries	G1	CF	2.0	1500	Bifertanol	Baycor	–
	G1	CF	1.0	5000	Cupric oxychloride	Grun Kupfer	–
	G1	CF	1.0	1500	Propineb	Antracol	–
	G1	CF	1.0	1500	Triforine	Saprol Neu	–
	G1	CF	1.0	1500	Vinclozolin	Ronilan	–
	G1	IF	1.0	1500	Bifertanol	Baycor	–
	G1	IF	1.0	5000	Cupric oxychloride	Grun Kupfer	–
	G1	IF	1.0	500	Dithianon	Delan	–
	G1	IF	2.5	1500	Triforine	Saprol Neu	–
	G2	BF	3.0	10 000	Clay	Ulmasud	–
	G2	BF	3.0	10 000	–	Neudovital	–
	G2	BF	3.0	10 000	–	Mycosan	–

Grapes	C	CF	—	22 500	Copper sulphate	—	—
	C	CF	—	7500	Mancozeb[4]	—	—
	C	CF	—	30 000	Sulphur	—	—
	C	CF	—	7500–22 500	Thiophanate-methyl	—	—
	ER	IF	0.2	31	Menalaxyl	—	—
	ER	IF	0.0	4	Benomyl[4]	—	—
	ER	IF	0.0	15	Chlorothalonil[4]	—	—
	ER	IF	0.3	68	Dinocap	—	—
	ER	IF	0.2	4	Hexaconazole	—	a
	ER	IF	0.1	2	Fenarimol	—	a
	ER	IF	6.0	407	Folpet[4]	—	a
	ER	IF	0.0	30	Iprodione	—	a
	ER	IF	3.9	4735	Mancozeb[4]	—	a
	ER	IF	0.9	1648	Metiram[4]	—	a
	ER	IF	0.6	22	Miclobutanyl	—	a
	ER	IF	0.3	9	Penconazole	—	a
	ER	IF	0.1	68	Procymidone	—	a
	ER	IF	0.2	341	Propineb	—	a
	ER	IF	0.0	1	Pyrazophos	—	a
	ER	IF	4.1	6879	Copper salts	—	a
	ER	IF	6.3	11 009	Sulphur	—	a
	ER	IF	0.3	175	Vinclozolin	—	a
	ER	IF	0.3	144	Zineb[4]	—	a
	ER	IF	0.2	33	Benalaxyl	—	b
	ER	IF	0.0	337	Bordeaux mixture	—	b
	ER	IF	0.0	2	Carbendazin	—	b
	ER	IF	0.0	46	Chlorothalonil	—	b
	ER	IF	0.0	16	Dinocap	—	b
	ER	IF	0.0	3	Hexaconazole	—	b
	ER	IF	0.0	260	Folpet[4]	—	b
	ER	IF	3.0	4058	Mancozeb[4]	—	b
	ER	IF	1.5	2310	Maneb[4]	—	b
	ER	IF	1.5	2310	Metiram[4]	—	b
	ER	IF	0.1	6	Miclobutanyl	—	b
	ER	IF	0.2	5	Penconazole	—	b
	ER	IF	0.0	40	Potassium sulphate	—	b
	ER	IF	0.0	224	Procymidone	—	b

Appendix A (cont.)

Fruit	Country[1]	Farm[2]	Mean number of treatments	Quantity[3] (g ha^{-1})	Active ingredient	Trademark	Line[7]
Grapes	ER	IF	0.0	27	Propineb	–	b
	ER	IF	4.6	7824	Copper salts	–	b
	ER	IF	6.9	15150	Sulphur	–	b
	ER	IF	0.0	4	Triforine	–	b
	ER	IF	0.1	41	Zineb[4]	–	b
	V1	CF	3–4	2500	Benalaxil	Galben	a
	V1	CF	3	10000	Bordeaux mixture	–	a
	V1	CF	9	3000	Cymoxanil + mancozeb[4] + folpet[4]	Curzate/Curit Zeb	a
	V*	CF	3	10000	Copper sulphate	Solfato Cu	a
	V*	CF	9	3000	Dimetomorf	Forum MZ	a
	V*	CF	3	10000	Copper idroxides	Kocide	a
	V*	CF	11–12	2000	Mancozeb[4]	Crittox MZ 80	a
	V*	CF	2–3	2500	Metalaxil	Eucritt	a
	V*	CF	11–12	2000	Metiram[4]	Polycram DF	a
	V*	CF	2–3	2500	Oxadixil	Sandofan	a
	V*	CF	3–4	2500	Phosethyl Al[4]	R6 Triplo	a
	V*	CF	11–12	2000	Propineb	Antracol	a
	V*	IF	3	2500	Benalaxil	Galben	a
	V*	IF	6–7	10000	Bordeaux mixture	–	a
	V*	IF	3	3000	Cymoxanil + mancozeb[4] + folpet[4]	Curzate/Curit Zeb	a
	V*	IF	2–3	10000	Copper idroxides	Kocide	a
	V*	IF	2–3	3000	Dimetomorf	Forum MZ	a
	V*	IF	2–3	2000	Mancozeb[4]	Crittox MZ	a
	V*	IF	2–3	2500	Metalaxil	Eucritt	a
	V*	IF	2–3	2000	Metiram[4]	Polycram DF	a
	V*	IF	2–3	2500	Oxadixil	Sandofan	a
	V*	IF	3	2500	Phosethyl Al[4]	R6 Triplo	a
	V*	IF	2–3	2000	Propineb	Antracol	a
	V*	BF	12–13	10000	Bordeaux mixture	–	a

				Active ingredient	Product	
V*	IF	2	2000	Mancozeb[4]	Crittox	b
V*	IF	2	2000	Metiram[F44]	Polyram	b
V*	IF	1–2	1500–2000	Procimidone	Sialex	b
V*	IF	1–2	1500–2000	Vinclozolin	Ronilan	b
V*	CF	5–6	1000	Dinocap	Karatane	c
V*	CF	2	300–600	Fenarimol	Rubigan	c
V*	CF	2	300–600	Flusilazol	Nustar	c
V*	CF	2	300–600	Miclobutanyl	Systhane	c
V*	CF	2	300–600	Penconazole	Topas	c
V*	CF	12–13	1000–2000	Sulphur	—	c
V*	CF	12–13	1000–2000	Sulphur 1[6]	—	c
V*	CF	12–13	1000–2000	Sulphur 2[6]	—	c
V*	CF	12–13	1000–2000	Sulphur 3[6]	—	c
V*	IF	2	1000	Dinocap	Karatane	c
V*	IF	2	300–600	Fenarimol	Rubigan	c
V*	IF	2	300–600	Flusilazol	Nustar	c
V*	IF	2	300–600	Miclobutanyl	Systhane	c
V*	IF	2	300–600	Penconazole	Topas	c
V*	BF	2	300–600	Miclobutanyl	Systhane	c
V*	BF	2	300–600	Penconazole	Topas	c
V*	BF	12–13	1000–2000	Sulphur	—	c
V*	BF	12–13	1000–2000	Sulphur 1[6]	—	c
V*	BF	12–13	1000–2000	Sulphur 2[6]	—	c
V*	BF	12–13	1000–2000	Sulphur 3[6]	—	c
V*	CF	3.0	700	Hexaconazole	—	a
V**	CF	1.0	—	Mancozeb[4]	Dithane	a
V**	CF	3.0	4000	Mancozeb[4]	Dithane	a
V**	CF	3.0	4000	Phosetyl-Al	R6 Triplo Blu	a
V**	CF	11.0	2000–3000	Sulphur	—	a
V**	CF	3.0	450	Copper	Rame Plant	b
V**	CF	3.0	500	Cymoxanil	Curzate MZ	b
V**	CF	4.0	3000	Folpet[4]	—	b
V**	CF	4.0	2000	Mancozeb[4]	Asar 80	b
V**	CF	4.0	3000	Metalaxyl	Ridomil	b
V**	CF	4.0	500	Miclobutanyl	Thiocor	b
V**	CF	2.0	650	Bordeaux mixture	BRS	b
V**	CF	3.0	1000	Sulphur	Primosol	b

Appendix A (cont.)

Fruit	Country[1]	Farm[2]	Mean number of treatments	Quantity[3] (g ha[-1])	Active ingredient	Trademark	Line[7]
Grapes	V**	CF	3.0	1000	Sulphur 2[6]	Kumulus	b
	V**	CF	1.0	4000	Copper-oxychloride	–	c
	V**	CF	4.0	2500	Cymoxanil + mancozeb[4]	Curzate M	c
	V**	CF	6.0	3000	Cymoxanil + copper	Curzate R	c
	V**	CF	3.0	75	Flusilazol	–	c
	V**	CF	2.0	1500	Iprodione	–	c
	V**	CF	8.0	2000	Sulphur 80	–	c
	V**	IF	4.0	5000	Copper oxychloride	Cuprocaffaro Pb	a*
	V**	IF	4.0	3000	Cymoxanil + copper oxychloride	Curzate R	a*
	V**	IF	3.0	4000	Cymoxanil + mancozeb[4] + phosethyl Al[4]	R6 Triplo Blu	a*
	V**	IF	1.0	300	Fenarimol	Rubigan	a*
	V**	IF	3.0	700	Hexaconazole	Anvil	a*
	V**	IF	3.0	2000	Mancozeb[4]	Dithane DG	a*
	V**	IF	7.0	2000–3000	Sulphur 80	Microthiol WDG	a*
	V**	IF	2.0	1000	Sulphur 2[6]	Kumulus DF	a*
	V**	IF	1.0	1500	Vinclozolin	Ronilan	a*
	V**	IF	4.0	5000	Copper oxychloride	Cuprocaffaro Pb	b*
	V**	IF	4.0	3000	Cymoxanil + copper oxychloride	Curzate R	b*
	V**	IF	3.0	4000	Cymoxanil + mancozeb[4] + phosethyl Al[4]	R6 Triplo Blu	b*
	V**	IF	3.0	700	Hexaconazole	Anvil	b*
	V**	IF	2.0	2000	Mancozeb[4]	Dithane DG	b*
	V**	IF	–	2000–3000	Sulphur 80	Microthiol WDG	b*
	V**	IF	2.0	1000	Sulphur 2[6]	Kumulus DF	b*
	V**	IF	3.0	3500	Copper oxychloride	–	b*
	V**	IF	2.0	3000	Cymoxanil + copper	Curzate R	c*

V**	IF	2.0	4000	Cymoxanil + mancozeb[4] + phosethyl Al[4]		c*
V**	IF	1.0	40	Dinocap	Sialite	c*
V**	IF	1.0	1500	Hexaconazole	Anvil	c*
V**	IF	2.0	2000	Mancozeb[4]	Micene HZ	c*
V**	IF	4.0	2000	Sulphur 80	Microthiol WDG	c*
V**	IF	1.0	1500	Sulphur 2[6]	Kumulus DF	c*
V**	IF	3.0	3500	Copper oxychloride	–	e*
V**	IF	2.0	3000	Cymoxanil + copper	Curzate R	e*
V**	IF	–	4000	Cymoxanil + mancozeb[4] + phosethyl Al[4]	R6 Triplo Blu	e*
V**	IF	3.0	700	Hexaconazole	Anvil	e*
V**	IF	2.0	2000	Mancozeb[4]80	Micene HZ	e*
V**	IF	4.0	2000	Sulphur 80	Tiosol	e*
V**	IF	2.0	1500	Sulphur 2[6]	Kumulus DF	e*
V**	IF	1.0	–	Copper	–	e*
V**	IF	1.0	–	Mancozeb[4]	–	f*
V**	IF	5.0	3000	Copper	Idrorame Fl	f*
V**	IF	1.0	2000	Cymoxanil + copper	Curzate R	g*
V**	IF	2.0	2000	Cymoxanil + mancozeb[4]	Curzate M	g*
V**	IF	3.0	500	Dinocap	Karathane	g*
V**	IF	2.0	2500	Metalaxil + mancozeb[4]	Ridomil MZ	g*
V**	IF	7.0	2000	Sulphur 80	–	g*
V**	IF	1.0	1500	Vinclozolin	Ronilan Fl	g*
V**	IF	2.0	4000	Copper oxychloride	KupperKalk	g*
V**	IF	3.0	2500	Cymoxanil + copper oxychloride	Curzate R	h*
V**	IF	2.0	4000	Cymoxanil + mancozeb[4] + phosethyl Al[4]	R6 Triplo Blu	h*
V**	IF	2.0	700	Hexaconazole	Anvil	h*
V**	IF	1.0	2000	Mancozeb[4]	Micene MZ	h*
V**	IF	5.0	2000	Sulphur B 80	Tiosol	h*

Appendix A (cont.)

Fruit	Country[1]	Farm[2]	Mean number of treatments	Quantity[3] (g ha^{-1})	Active ingredient	Trademark	Line[7]
Grapes	V**	IF	1.0	1300	Sulphur S 80	Tiovit	h*
	V**	IF	1.0	4000	Copper oxychloride	KupperKalk	i*
	V**	IF	2.0	2500	Cymoxanil+copper oxychloride	Curzate R	i*
	V**	IF	2.0	4000	Copper oxychloride	Cuprocaffaro Pb	i*
	V**	IF	2.0	700	Hexaconazole	Anvil	i*
	V**	IF	3.0	2000	Mancozeb[4]	Micene MZ	i*
	V**	IF	5.0	2000	Sulphur B 80	Top 90	i*
	V**	IF	1.0	1300	Sulphur S 80	Tiovit	i*
	V**	IF	4.0	–	Copper oxychloride	KupperKalk	l*
	V**	IF	1.0	–	Cymoxanil+copper oxychloride	Curzate R	l*
	V**	IF	3.0	–	Dithiocarbamates	–	l*
	V**	IF	2.0	–	Bordeaux mixture	–	l*
	V**	IF	9.0	–	Sulphur 80	–	l*
	V**	BF	11.0	4000–9000	Copper	Bordeaux mixture	m*
	V**	BF	1.0	18000	Sulphur powder	–	m*
	V**	BF	11.0	1000	Wettable sulphur	–	m*
	V**	CF	4.0	5000	Copper oxychloride	Cuprocaffaro Pb	n*
	V**	CF	6.0	3500	Copper DMM	Acrobat R	n*
	V**	CF	4.0	700	Hexaconazole	Anvil	n*
	V**	CF	2.0	2000	Mancozeb[4] 80	Dithane DG	n*
	V**	CF	6.0	2000–3000	Sulphur 80	Microthiol WDG	n*
	V**	CF	5.0	5000	Copper oxychloride	Cuprocaffaro Pb	o*
	V**	CF	2.0	3000	Cymoxanil+copper oxychloride	Curzate R	o*
	V**	CF	3.0	4000	Cymoxanil+mancozeb[4]+phosethyl Al[4]	R6 Triplo Blu	o*
	V**	CF	4.0	700	Hexaconazole	Anvil	o*
	V**	CF	2.0	2000	Mancozeb[4] 80	Dithane DG	o*
	V**	CF	8.0	2000–3000	Sulphur 80	Microthiol WDG	o*
	FVG	–	4.0	–	Copper oxychloride	–	a

		Compound					
FVG	–	Cymoxanil + folpet[4]	1.0	–	–	a
FVG	–	Fenarimol	2.0	–	–	a
FVG	–	Hexaconazole	2.0	–	–	a
FVG	–	Mancozeb[4]	3.0	–	–	a
FVG	–	Metiram[4]	3.0	–	–	a
FVG	–	Propineb	3.0	–	–	a
FVG	–	Sulphur B	8.0	–	–	a
TAA	BF	Clay	4–6	800–1000	–	–
TAA	BF	Copper oxychloride	2–5	100–500	–	–
TAA	BF	Bordeaux mixture	5–10	200–1000	–	–
TAA	BF	Sulphur	10–15	150–400	–	–
TAA	BF	ZAB	1–4	900	–	–
C	CF	Sulphur	–	30 000	–	–
ER	IF	Sulphur	0.1	5445	–	a
ER	IF	Ba polysulphur	0.0	993	–	a
ER	IF	Ca polysulphur	0.5	787	–	a
ER	IF	Carbaryl	0.3	127	–	a
ER	IF	Carbendazim	0.0	12	–	a
ER	IF	Dithianon	0.0	36	–	a
ER	IF	Dnoc	0.4	464	–	a
ER	IF	Dodina	0.1	3	–	a
ER	IF	Fenarimol	0.1	4	–	a
ER	IF	Hexaconazole	0.0	8	–	a
ER	IF	Mancozeb[4]	0.0	3	–	a
ER	IF	Maneb	0.1	5	–	a
ER	IF	Miclobutanyl	0.1	44	–	a
ER	IF	Procymidone	0.0	7	–	a
ER	IF	Pyrazophos	2.3	5411	–	a
ER	IF	Sulphur	1.1	1511	–	a
ER	IF	Thiram	0.7	244	–	a
ER	IF	Triforine	3.2	11 445	–	b
ER	IF	Ziram	0.0	17 445	–	b
ER	IF	Ba polysulphur	0.0	845	–	b
ER	IF	Ca polysulphur	0.0	18	–	b
ER	IF	Carbendazim	0.1	700	–	b
ER	IF	Copper salts				

Peaches

Appendix A (cont.)

Fruit	Country[1]	Farm[2]	Mean number of treatments	Quantity[3] (g ha^{-1})	Active ingredient	Trademark	Line[7]
Peaches	ER	IF	0.0	18	Dithianon	–	b
	ER	IF	0.5	583	Dodina	–	b
	ER	IF	0.1	3	Hexaconazole	–	b
	ER	IF	0.2	11	Penconazole	–	b
	ER	IF	0.0	2	Procymidone	–	b
	ER	IF	2.2	5341	Sulphur	–	b
	ER	IF	0.1	212	Thiram	–	b
	ER	IF	0.0	6	Tiophanate-methyl	–	b
	ER	IF	2.4	9603	Ziram	–	b
	G1	CF	0.5	1500	Dichlofluanid	Euparen	–
	G1	Cf	2.0	1000	Dithianon	Delan	–
	G1	IF	0.5	1500	Dichlofluanid	Euparen	–
	G1	IF	2.0	1000	Dithianon	Delan	–
	G2	BF	6.0	500	Copper	–	–
Plums	ER	IF	0.2	11 862	Ca polysulphur	–	a
	ER	IF	0.3	342	Carbaryl	–	a
	ER	IF	1.1	3303	Copper salts	–	a
	ER	IF	1.1	1564	Diclofluanide	–	a
	ER	IF	0.3	322	Dithianon	–	a
	ER	IF	0.4	1700	DNOC	–	a
	ER	IF	0.5	275	Dodina	–	a
	ER	IF	0.1	14	Fenarimol	–	a
	ER	IF	0.0	18	Folpet[4]	–	a
	ER	IF	0.2	12	Hexaconazole	–	a
	ER	IF	1.7	4076	Mancozeb[4]	–	a
	ER	IF	0.9	1904	Metiram[4]	–	a
	ER	IF	0.1	7	Miclobutanyl	–	a
	ER	IF	0.3	13	Menconazole	–	a
	ER	IF	1.8	1429	Procymidone	–	a
	ER	IF	0.0	26	Propineb	–	a
	ER	IF	0.1	56	Sulphur	–	a
	ER	IF	0.2	54	Tetradifon	–	a

ER	IF	5.7	1 042 200	Thiram	—		a
ER	IF	0.0	3	Tiophanate-methyl	—		a
ER	IF	0.0	730	Ba polysulphur	—		b
ER	IF	2.1	3260	Sali di Cu vari	—		b
ER	IF	0.0	5	Diclofluanide	—		b
ER	IF	0.1	150	Dithianon	—		b
ER	IF	0.0	94	DNOC	—		b
ER	IF	0.4	293	Dodina	—		b
ER	IF	0.1	7	Fenarimol	—		b
ER	IF	0.1	5	Hexaconazole	—		b
ER	IF	0.0	7	Iprodione	—		b
ER	IF	1.3	2306	Mancozeb[4]	—		b
ER	IF	1.5	3472	Metiram[4]	—		b
ER	IF	0.0	3	Miclobutanyl	—		b
ER	IF	0.1	7	Penconazole	—		b
ER	IF	1.0	662	Procymidone	—		b
ER	IF	0.2	203	Sulphur	—		b
ER	IF	0.1	5	Tetradifon	—		b
ER	IF	4.9	8562	Thiram	—		b
ER	IF	0.0	1	Triforine	—		b
ER	IF	0.0	37	Zineb[4]	—		b
ER	CF	4.8	10 091	Ziram	—		—
G*	IF	>4	500	Dithianon	—	Delan	—
G*	IF	4.0	500	Dithanon	—	Delan	a
ER	IF	0.1	9918	Ca polysulphur	—		a
ER	IF	1.9	4839	Sali di Cu vari	—		a
ER	IF	0.4	233	Carbendazim	—		a
ER	IF	0.1	9918	Ca polysulphur	—		a
ER	IF	0.0	12	Procymidone	—		a
ER	IF	0.2	56 200	Sulphur	—		a
ER	IF	0.8	1330	Thiram	—		a
ER	IF	1.1	416	Triforine	—		b
ER	IF	0.0	76	Ba polysulphur	—		b
ER	IF	0.2	4066	Ca polysulphur	—		b
ER	IF	0.2	98	Carbendazim	—		b
ER	IF	0.0	8	Iprodione	—		b
ER	IF	0.0	11 699	Procymidone	—		b

Appendix A (cont.)

Fruit	Country[1]	Farm[2]	Mean number of treatments	Quantity[3] (g ha^{-1})	Active ingredient	Trademark	Line[7]
Plums	ER	IF	1.8	4782	Sali di Cu vari	–	b
	ER	IF	0.0	57	Sulphur	–	b
	ER	IF	0.1	14	Thiram	–	b
	ER	IF	0.1	124	Tiophanate-methyl	–	b
	ER	IF	0.7	291	Triforine	–	b
	ER	IF	0.2	599	Ziram	–	b
	G1	CF	1.0	5000	Bifertanol	Baycor	–
	G1	CF	>0.2	1500	Cupric oxychloride	Grunkupfer	–
	G1	CF	0.5	2000	Mancozeb[4]	Dithane Ultra	–
	G1	CF	0.5	2000	Metiram[4]	Polyram WG	–
	G1	IF	1.0	5000	Bifertanol	Baycor	–
	G1	IF	>0.2	1500	Cupric oxychloride	Grunkupfer	–
	G1	IF	0.5	2000	Mancozeb[4]	Dithane Ultra	–
	G1	IF	0.5	2000	Metiram[4]	Polyram WG	–

[1] Country codes: C, China; ER, Emilia Romagna (Italy); G1, Stuttgart (Germany); G2, Mainz (Germany); FVG, Friuli Venezia Giulia (Italy); TAA, Trentino Alto Adige (Italy); V, Veneto (Italy). Information from: National Plant Protection Network, China; Assessorato all'Agricoltura e Alimentazione, Regione Emilia Romagna (ER), Italy; Rheinlandpflaz-Landensanstalt fur Pflanzenbau und Pfaluzenschutz (G*), Germany; CO-DI-TV (V*), Treviso, Italy; Istituto Sperimentale per la Viticoltura (FVG, V**), Conegliano Veneto, TV, Italy; Regione Trentino Alto Adige (TAA), Italy.–, no data available.

[2] Target farm keys: CF, conventional farm; IF, integrated farm; BF, biological farm.

[3] Quantity: The annual input rate for fungicides is given on the first reference to each product. This does not indicate the season of use.

[4] Active ingredient designated by the EPA as oncogenic (NAP, 1987). The quantitative oncogenic potency factor (Q) for each ingredient is as follows: benomyl $Q = 2.065 \times 10^{-3}$; captafol, $Q = 2.50 \times 10^{-2}$; captan, $Q = 2.30 \times 10^{-3}$; chlorothalonil, $Q = 2.4 \times 10^{-2}$; folpet, $Q = 2.4 \times 10^{-2}$; mancozeb, $Q = 1.76 \times 10^{-2}$; maneb, $Q = 1.76 \times 10^{-2}$; metiram, $Q = 1.76 \times 10^{-2}$; phosetyl A1 $Q = 4.3 \times 10^{-3}$; zineb, $Q = 1.76 \times 10^{-2}$.

[5] Allium spp, Equisetum spp, Quercus spp, Rafanus spp, Tanacetum spp, Urtica spp and fern spp.

[6] Sulphur = wettable sulphur, Sulphur 2 = colloidal sulphur; Sulphur 3 = micronized sulphur.

[7] a, b, c, d, e, f, g, h, i, l, m, n, o, different spraying protocols in the same or different farms in the same area having different pathogen damages.

CHAPTER 17

Reducing Insecticide, Fungicide and Herbicide Use on Vegetables and Reducing Herbicide Use on Fruit Crops

David Pimentel, Jason Friedman and David Kahn

Cornell University, Ithaca, NY, USA

INTRODUCTION

Several studies have suggested that it is technologically feasible to reduce pesticide use in the United States by 35–50% without reducing crop yields (OTA, 1979; NAS, 1989; Pimentel et al., 1991). US farmers use an estimated 400 million kg of pesticides annually, at an approximate cost of $5.3 billion (Osteen, 1993; USBC, 1994). About 3.5 kg of pesticide are applied per hectare of treated land, at a cost of about $47 per hectare. The dollar returns for direct benefits to farmers have been estimated to be about $4 for every $1 invested in the use of pesticides (Pimentel et al., 1978, 1993). However, these cost figures do not reflect the indirect costs of pesticide chemical use, such as human pesticide poisonings, reduction of fish and wildlife populations, livestock and honeybee losses, destruction of susceptible crops and natural vegetation, destruction of natural enemies, evolved pesticide resistance in pests, and creation of secondary pest problems (Pimentel et al., 1993). These environmental costs were estimated to be more than $8 billion annually (see Chapter 4).

The objective of this chapter is to estimate the potential agricultural and environmental benefits and costs of reducing insecticide, fungicide and herbicide use on vegetables, and reducing herbicide use on fruit crops. The goal will be two-fold: to evaluate current crop losses to insects, plant pathogens and weeds, and to estimate the agricultural benefits and costs of reducing pesticide use by substituting currently available biological, cultural and environmental pest-control technologies for some of the current chemical-control practices.

EXTENT OF PESTICIDE USE

Of the estimated 400 million kg of pesticides used in US agriculture, 69% are herbicides, 19% are insecticides and 12% are fungicides (Pimentel et al., 1991). Approximately 62%

Techniques for Reducing Pesticide Use. Edited by D. Pimentel
© 1997 John Wiley & Sons Ltd.

of the agricultural land that is planted receives some treatment, while the remaining 38% of crops receives no pesticides.

The application of pesticides for pest control is not evenly distributed among crops. Overall, 93% of the hectarage of row crops such as corn, soybeans and cotton is treated with some type of pesticide (Pimentel et al., 1991). In contrast, less than 10% of the forage crop hectarage is treated.

Herbicides are currently used on approximately 90 million ha in the US—more than half of the nation's cropland. Field corn alone accounts for 53% of agricultural herbicide use; almost three-quarters of the herbicides used are applied to corn and soybeans (Pimentel et al., 1991).

TECHNIQUES TO REDUCE PESTICIDE USE

Some of the increased crop losses associated with recent changes in agricultural technologies suggest that some alternative strategies exist that might be utilized to reduce pesticide use (Pimentel et al., 1991). Two important practices that apply to all agricultural crops include the widespread use of monitoring pest and natural enemy populations, and the use of improved application equipment. Currently, a significant number of pesticide treatments are applied unnecessarily and at improper times owing to a lack of effective monitoring programs. In addition, pesticides are unnecessarily lost during application. For example, less than 0.1% of the pesticide that is applied reaches the target pests (Pimentel, 1995).

REDUCING INSECTICIDE USE ON VEGETABLES

A significant amount of insecticide (4.2 million kg) is applied to vegetable crops each year (Table 17.1). The largest amount is used on potatoes, the crop which also has the largest hectarage. Various known practices that might be employed to reduce insecticide use on certain vegetable crops are discussed below.

Lettuce

The principal insect pests of lettuce are aphids and cabbage loopers. Loopers can be controlled with *Bacillus thuringiensis* (Bt), scouting and a nuclear polyhedrosis virus which became the first baculovirus patented by the US Government Patent and Trademark Office, on 27 March 1990 (Kimberly and Puttler, 1991). For control of aphids, the alternative is scouting. By employing a combination of these techniques, it might be possible to reduce insecticide use by about 25% at an estimated cost of $10 ha^{-1} (Table 17.1).

Cole

The primary pests of cole crops include cabbage maggots, cabbage loopers, cabbage butterflies and diamondback moths (Kirby and Slosser, 1984). By incorporating granular insecticide into the potting soil of seedlings, it has been demonstrated that the quantity of insecticide used for control of early season pests could be reduced by more than 50% (Straub, 1988). However, about 70% of insecticides are used for control of caterpillar pests. Scouting is an important means of increasing the effectiveness of sprays and eliminating needless treatments (Kirby and Slosser, 1984).

Both the cabbage looper and the cabbage butterfly are highly susceptible to virus diseases, which could be effectively used against these pests (Falcon, 1976; Jaques, 1988). To date, these viruses have not been approved for use on food crops, but there is no evidence of risk to public health or to the environment (Summers and Kawanishi, 1978; Pimentel et al., 1984). Bt can be used against all three caterpillar species (Jaques, 1988). Thus it might be possible to reduce insecticide use in cole crops by an estimated 50% (Table 17.1). The added cost of these alternatives was estimated to be $10 ha^{-1} (Table 17.1).

Carrots

Losses to the carrot fly, carrot beetle and carrot weevil are estimated to be about 7% (Table 17.1). By implementing a sound scouting programme, it might be possible to reduce insecticide use in carrot production by about one-half, at an added estimated cost of $5 ha^{-1} (Table 17.1).

Potatoes

The principal insect pests of potatoes are Colorado potato beetles, aphids, potato flea beetles and potato leafhoppers. The dominant pest is the potato beetle, and alternative control methods include rotations, early-maturing varieties, short-season potatoes, mulches, plastic-lined trench barriers (Boiteau et al., 1994) and scouting (Shields et al., 1984; Wright, 1984; Wright et al., 1986; CR, 1987; Zehnder and Evanylo, 1989; Radcliffe et al., 1991). The use of short-season potatoes and scouting may reduce insecticide use by 33% (Shields et al., 1984). However, E.B. Radcliffe (personal communication, 1989) reports that insecticide use on potatoes could be reduced by 75% with effective scouting. Also, Bt has been found to be effective in controlling the potato beetle (Cantwell and Cantello, 1984; Jaques and Laing, 1989; Gelernter, 1990). Additionally, plastic-lined trenches reduce the number of potato beetles by as much as 45% (Boiteau et al., 1994). Using the fungus *Beauberia bassiana* for Colorado beetle control demonstrated the potential for an 80% reduction in insecticide use (Roberts et al., 1981); however, a recent study reported that the fungus was ineffective (Jaques and Laing, 1989). By employing pest control combinations such as rotations, short-season potatoes, scouting and Bt, it might be possible to reduce insecticide use by 30% at an estimated cost of $10 ha^{-1} (Table 17.1).

Tomatoes

Tomatoes are a high-value crop (about $7500 ha^{-1}). To protect this valuable crop, about $64 ha^{-1} in insecticide is applied and 95% of the hectarage is treated (Table 17.1). The primary insect pests are the tomato fruitworm and the tomato hornworm; however, potato beetles and aphids are also occasional pests (Farrar et al., 1986; Zehnder and Linduska, 1987; Walgenbach and Estes, 1992). Insecticide applications to control these pests can be reduced by an estimated 20% through scouting and by another 60–80% by substituting Bt and other natural enemies (Krishnaiah et al., 1981; Antle and Park, 1986;

Table 17.1 Vegetable crop losses from insects with current insecticide use and estimated costs if insecticides were reduced and several alternatives were substituted

Crop	Area (ha – 10³)[a]	Total insecticide use (kg – 10⁶) Current[b]	Total insecticide use (kg – 10⁶) Reduced[c]	Insecticide treatment Hectares treated (%)[d]	Insecticide treatment Cost ($ ha⁻¹)[e]	Insecticide treatment Total cost ($ – 10⁶)	Current crop pest loss (%)[c]	Added alternative cost ($ ha⁻¹)[c]	Total added alternative control cost ($ – 10⁶)[c]
Lettuce	90	0.35	0.26	97	68	5.9	7	10	0.70
Cole	111	0.40	0.20	62	30	2.1	13	10	0.70
Carrots	39	0.02	0.01	37	10	0.1	7	5	0.08
Potatoes	570	1.60	1.12	88	46	23.1		10	5.40
Tomatoes	145	0.20	0.15	95	26	3.6		0	0.00
Sweetcorn	206	0.27	0.05	84	70	12.1	19	10	2.00
Onions	54	0.75	0.50	79	18	0.8	4	5	0.27
Cucumbers	42	0.02	0.01	34	12	0.2	21	5	0.10
Beans	132	0.11	0.07	72	9	0.9	12	5	0.33
Cantaloupe	50	0.08	0.05	78	40	1.6	8	0	0.00
Peas	135	0.02	0.01	49	5	0.3	4	5	0.61
Peppers	25	0.09	0.06	85	80	1.7	7	5	0.09
Sweet potatoes	31	0.26	0.02	100	0	1.3	16	5	0.22
Watermelons	72	0.06	0.04	53	14	0.5	4	5	0.30
Other vegetables	100	0.01	0.006	40	30	1.2	13	5	0.20
Total		4.24	2.556			54.40			11.00

[a] USDA, 1992.
[b] Converted from USDA, 1993.
[c] Pimentel et al., 1991.
[d] USDA, 1993.
[e] Calculated.

Farrar et al., 1986; Hoffman et al., 1986; Horn, 1988; Jimenez et al., 1988). By employing these techniques it might be possible to reduce insecticide use on tomatoes by 25% (Table 17.1). Although additional labour is needed for scouting, total control costs remain the same because of the savings from reduced insecticide applications (Antle and Park, 1986; Jimenez et al., 1988).

Sweetcorn

The primary insect pest in sweet corn production is the corn earworm (McLeod, 1986). Several alternative techniques exist for controlling this pest. These include a highly effective nuclear polyhedrosis virus (Oatman et al., 1970), Bt, pheromone traps (Drapek et al., 1992), mineral oil treatment of ears (Barber, 1942; Johns, 1966), early-maturing varieties (Huffaker, 1980) and rotations. More reasonable cosmetic standards could also greatly reduce the need for high levels of pesticide use (Straub and Heath, 1983). Using a rotation sequence of sweet corn and soybeans in Georgia, with effective management practices, Tew et al. (1982) reported that, at a minimum management level, pesticide costs decreased 17-fold and net profits increased significantly compared with conventional methods. By employing a combination of several of these alternative technologies, it might be possible to reduce insecticide use in sweet corn by more than three-quarters (Table 17.1). Because the results of Tew et al. (1982) were for a particular rotation system, we estimated that the added costs for the alternatives would be $10 ha^{-1} (Table 17.1).

Onions

The primary insect pests of onions are the onion maggot and onion thrips (Ritcey and McEwen, 1984; Edelson et al., 1986). Losses of onions where insecticide treatments were made averaged about 4% (Stemeroff and George, 1983) as opposed to losses of 39% for untreated onions (Tolman et al., 1986). Alternatives for control of these pests include rotations, scouting and sanitation (Comin, 1946; Cadoux, 1984). Recently, D. Haynes (personal communication, 1988) reported that onions could be produced without insecticides by raising cattle adjacent to the onion field and mulching the onions with straw. A parasitic wasp species uses the maggots in the cattle manure as an alternative host, and the straw protects a predaceous beetle that preys on the onion maggot. Onion losses to maggots and other insects in the alternative system were only 2–3% compared with an average of 4% in insecticide-treated plots (Table 17.1). It has also been reported that preventing injuries to the onion bulbs during the growing season will help reduce maggot attack (Cadoux, 1984). By employing several of these alternatives in combination, it might be possible to reduce insecticide use by about one-third at an added estimated cost of $5 ha^{-1} (Table 17.1).

Beans

About 72% of bean hectarage is treated with insecticides for two primary insect pests, the Mexican bean beetle and the pea moth (Table 17.1). Insecticide use might be reduced by one-third through the use of scouting, planting short-season varieties, resistant cultivars, the application of Bt, and by employing a vacuum apparatus to remove pests (Krishnaiah et al., 1981; Karel and Rweyemamu, 1985; Karel and Schoonhoven, 1986; Mahrt et al.,

1987; Stockwin, 1988; Street, 1989) (Table 17.1). Insecticide use might also be reduced by using a vacuum-operated delivery system (VONDS), which has been developed to allow bean plants to grow greater leaf area with increased root, stem, leaf and pod dry weights (Brown et al., 1992). Scouting and the vacuum techniques were estimated to increase control costs by $5 ha^{-1} (Table 17.1).

Cucumbers and watermelons

The primary pests of cucumbers and watermelons are cucumber beetles and pickleworms (Douce and Suber, 1985). The most practical substitutes for insecticides are scouting, rotations, pheromone traps and reflective mulches (Schalk et al., 1979; Elsey et al., 1991; Nordblom et al., 1994). Using these alternatives, it might be possible to reduce insecticide use by one-third to one-half, at an estimated cost of $5 ha^{-1} (Table 17.1).

Peas

The major insect pests of peas are pea aphids and pea moths (Metcalf and Metcalf, 1993). Scouting can help reduce insecticide treatments to about one per season for the aphid (Maiteki and Lamb, 1985). Pea moths can be controlled by deep ploughing, early threshing and pheromone traps (Metcalf and Metcalf, 1993; Witgall et al., 1993). Through scouting and the improved targeting of insecticides, insecticide use can be reduced by 50% (E.B. Radcliffe, personal communication, 1989; Cranshaw and Radcliffe, 1984). By employing a combination of the various alternatives, it might be possible to reduce insecticide use by about one-half at an added estimated cost of $5 ha^{-1} (Table 17.1).

Sweet potatoes

The sweet potato weevil is reported to be the most serious pest of this crop. Rotation, sanitation, pheromone traps and scouting are suitable alternative control methods (Metcalf and Metcalf, 1993; Smits et al., 1994). Many other root-feeding larval species, such as wireworms, corn rootworms, flea beetles and white grubs, also cause significant losses. The most effective method to prevent damage from these pests is the use of cultivars with multiple insect resistance (Schalk and Jones, 1985; Jones et al., 1987). The reduction in insecticide cost by the use of these cultivars would amount to about $138 ha^{-1}. Using these methods, it might be possible to reduce insecticide use by about 90% at an estimated $5 ha^{-1} (Table 17.1).

REDUCING FUNGICIDE USE ON VEGETABLES

An estimated 7.1 million kg of fungicide is used on vegetable crops, with most being applied to potatoes and tomatoes (Table 17.2). These two crops combined also have more hectarage than most of the other crops. Various known practices that might be employed to reduce fungicide use on some vegetable crops are discussed below.

Cole

About 43% of cole crop hectarage receives fungicide treatments (Table 17.2). Cole crop diseases can be reduced by purchasing disease-resistant seeds, using proper crop rotations, improving sanitation and using appropriate fertilizer—especially lime (Roberts and Boothroyd, 1972). By employing a combination of several of these technologies, it may be possible to reduce fungicide use on cole crops by about two-thirds, at an estimated cost of $5 ha^{-1} (Table 17.2).

Potatoes

About 80% of potato hectarage is treated with fungicide (Table 17.2). Without fungicide treatments, losses from diseases ranged from 5 to 25%, while losses with fungicide treatments were reported to be about 20% (Teng and Bissonnette, 1985; Tolman et al., 1986; Love and Tauer, 1987). Shields et al. (1984) reported that the planting of short-season potatoes in Wisconsin reduced the number of fungicide applications by one-third. Correct storage, handling and planting of seed tubers, and proper management of soil moisture and fertility minimize losses to most diseases (UC, 1983). Forecasting and scouting might also be employed to reduce fungicide use by 15–25% (Royle and Shaw, 1988; Tette and Koplinka-Loehr, 1989). Fungicides should be applied before infection, and at the same time, the crop should be monitored for the appearance of disease symptoms. By employing a combination of these controls, it might be possible to reduce fungicide use on potatoes by about one-third, at an estimated cost of $5 ha^{-1} (Table 17.2).

Tomatoes

A forecasting system employed with tomatoes in Pennsylvania indicated that fungicide use could be reduced by 55% while maintaining excellent pathogen control (Madden et al., 1978). Thus, forecasting and scouting methods may allow a 55% reduction in fungicide use in tomato production. This could provide savings of about $65 ha^{-1} in use of fungicides; however, we estimated that the added alternative control would cost an estimated $10 ha^{-1} (Table 17.2).

Sweetcorn

Only about 1% of sweetcorn is treated with fungicides because yield losses to plant pathogens are relatively low (Table 17.2). Thus, it was assumed that an effective treat-when-necessary programme could reduce fungicide use by 50%, at an estimated cost of $6 ha^{-1} (Table 17.2.).

Onions

Diseases are a major limitation in onion production. Losses of onions with fungicide treatments average 21% (USDA, 1965), whereas losses without fungicide treatment average 24% (Tolman et al., 1986). Alternative practices which are available to reduce the use of fungicides include improved sanitation, rotations and scouting (Ellerbrock and Lorbeer, 1977; Shoemaker and Lorbeer, 1977). With these methods, fungicide use might be reduced by one-third at an estimated cost of about $5 ha^{-1} (Table 17.2).

Table 17.2 Vegetable crop losses from plant pathogens with current fungicide use and estimated costs if fungicides were reduced and several alternatives were substituted

Crop	Area (ha × 10³)[a]	Total fungicide use (kg − 10⁶)		Fungicide treatment			Current crop pest loss (%)[c]	Added alternative cost ($ ha⁻¹)[c]	Total added alternative control cost ($ − 10⁶)[c]
		Current[b]	Reduced[c]	Hectares treated (%)[d]	Cost ($ ha⁻¹)[e]	Total cost ($ − 10⁶)			
Lettuce	90	0.32	0.26	76	59	4.00	12	5	0.42
Cole	111	0.20	0.06	43	20	0.95	9	5	0.24
Carrots	39	0.31	0.20	79	31	0.96	8	5	0.07
Potatoes	570	2.30	1.60	80	57	26.00	20		2.70
Tomatoes	145	2.00	1.30	86	37	4.61	21	10	1.66
Sweetcorn	206	0.10	0.05	1	9	0.01	8	6	0.014
Onions	54	0.35	0.20	83	74	3.32	21	5	0.19
Cucumbers	42	0.08	0.05	32	21	0.28	15	5	0.06
Beans	132	0.21	0.15	55	8	0.58	20	5	0.41
Cantaloupe	50	0.60	0.40	73	8	0.29	21	5	0.19
Peas	135	0.06	002	20	1	0.03	23	5	0.39
Peppers	25	0.23	0.15	66	52	1.09	14	5	0.06
Sweet potatoes	31	0.01	0.006	1	20	0.01	18	5	0.002
Watermelons	72	0.35	0.20	71	50	2.56	14	5	0.26
Other vegetables	100	0.001	0.001	10	20	0.20	10	0	0.00
Total		7.121	5.097			44.89			6.766

[a] USDA, 1992.
[b] Converted from USDA, 1993.
[c] Pimentel et al., 1991.
[d] USDA, 1993.
[e] Calculated.

Beans

About 55% of the bean hectarage is treated with fungicides (Table 17.2). Bean rust and white mould are the major diseases of beans. Through improved forecasting, scouting and biocontrol, and the use of resistant and mixed varieties (Baker et al., 1985; Schwartz et al., 1987; Mukishi and Trutman, 1988; Stavely, 1988), it is estimated that fungicide use might be reduced 30%, with an estimated cost of \$5 ha^{-1} (Table 17.2).

Cucumbers

About 32% of cucumber hectarage is treated with fungicides (Table 17.2). Thompson and Jenkins (1985) reported that improved forecasting and scouting may reduce fungicide use by 50%. Other alternative techniques for reducing diseases in cucumbers include using resistant cucumber varieties and rotations (Lloyd and McCollum, 1940; Sitterly, 1969; Thompson and Jenkins, 1985; Sumner and Phatak, 1987). In addition, the use of photo-degradable plastic was found to be significantly more effective than fungicides or other control technologies for control of several diseases (Lewis and Papavizas, 1980). By employing combinations of these alternatives, it might be possible to reduce fungicide use in cucumber production by about one-half, at an estimated cost of \$5 ha^{-1} (Table 17.2).

Sweet potatoes

Growers of sweet potatoes normally do not apply pesticides for the control of fungal and viral diseases. However, nematocides are routinely applied for nematodes. An effective method of control for diseases and nematodes is the use of resistant cultivars (Jones et al., 1985, 1989). The use of these cultivars should reduce or eliminate dependency on pesticides.

REDUCING HERBICIDE USE ON VEGETABLE AND FRUIT CROPS

A variety of possible techniques are available to commercial farmers for the reduction of herbicide use in vegetable and fruit crops. While some non-chemical practices are more expensive than herbicide weed control, some non-chemical controls lead to better yields and improved environmental benefits (Pimentel et al., 1991; Bridges, 1992). Certain practices, such as scouting or improved application techniques, can save the farmer money because less herbicide is necessary to achieve the same level of weed control (Pimentel et al., 1991). Various known practices that might be employed to reduce herbicide use on vegetable and fruit crops are discussed below.

Vegetable crops

The amount of herbicide applied to vegetable crops totals about 2.2 million kg per year (Table 17.3).

Cole

Weeds are a serious problem in cole crop production, and about 84% of the crop is

Table 17.3 Vegetable crop losses from weeds with current herbicide use and estimated costs if herbicides were reduced and several alternatives were substituted

Crop	Area (ha − 10³)[a]	Total herbicide use (kg − 10⁶)		Herbicide treatment			Current crop pest loss (%)[c]	Added alternative cost ($ ha⁻¹)[c]	Total added alternative control cost ($ − 10⁶)[c]
		Current[b]	Reduced[c]	Hectares treated (%)[d]	Cost ($ ha⁻¹)[e]	Total cost ($ − 10⁶)			
Lettuce	90	0.07	0.04[f]	68	10	0.61	8	10	0.90
Cole	111	0.20	0.10[e]	68	10	0.61	12	15	1.40
Carrots	39	0.04	0.03[e]	67	76	1.99	12	10	0.39
Potatoes	570	1.05	0.74[e]	83	9	4.26	7	0	0.00
Tomatoes	145	0.05	0.03[f]	75	10	1.09	10	0	0.00
Sweetcorn	206	014	0.07[e]	75	7	1.08	11	15	2.32
Onions	54	0.36	0.25[e]	86	24	1.11	9	0	0.00
Cucumbers	42	0.02	0.01[e]	74	20	0.62	12	0	0.00
Beans	132	0.22	0.11[e]	95	8	1.00	10	5	0.66
Cantaloupe	50	0.04	0.02[e]	44	4	0.09	10	10	0.50
Peas	135	0.11	0.06[e]	91	6	0.74	13	10	1.23
Peppers	25	0.02	0.01[e]	65	6	0.10	11	10	0.16
Sweet potatoes	31	0.03	0.02[g]	81	20	0.50	9	10	0.25
Watermelons	72	0.025	0.01[e]	37	20	0.50	9	10	0.27
Other vegetables	100	0.04		50	20	1.00	9	0	0.00
Total		2.175	1.4			17.49			8.08

[a] USDA, 1992.
[b] Converted from USDA, 1993.
[c] USDA, 1993.
[d] Calculated.
[e] Pimentel et al., 1991.
[f] Curtis et al., 1991
[g] Vos, 1992.

treated with herbicides (Table 17.3). Herbicide use may be reduced by using band applications (Hicks and Rehm, 1986) as well as mechanical cultivation and rope-wick application technology (Dale, 1979). Additional methods to reduce weed problems include planting early-maturing varieties and using large transplants which give the cole plants a competitive advantage over the weeds (Agamalian, 1984). It is projected that herbicide use might be reduced by one-half, with an estimated cost of $15 ha^{-1} (Table 17.3).

Potatoes

Losses of potato crops are estimated to be 7% with herbicides and other controls. Effective control of weeds can be accomplished without herbicides if mechanical tillage systems are employed, especially in drier climates (William et al., 1995). Most potato crops in the US are harvested in the Midwestern region, where the average rainfall is lower than in other parts of the country. In wet climates, pesticides are often more effective than tillage systems (William et al., 1995), but there also is the potential for the biological control of weeds. Strains of pathogenic fungi can significantly decrease the amount of herbicide needed in wet areas (Pimentel et al., 1991). However, crops in which pathogenic fungi have been introduced for weed control cannot be treated with fungicides; if pathogenic fungi of the potato are also a problem, biological control may not be feasible (Pimentel et al., 1991). We estimate that the use of a variety of alternative weed controls can reduce herbicide use by about one-third with no added cost (Table 17.3).

Tomatoes

Despite all controls, weeds reduce potential tomato yields by about 10% (Table 17.3). Mulches are often effective for weed control in tomato crop fields. For example, black plastic mulch provides effective weed control while at the same time producing a cleaner, higher-quality tomato (Pimentel et al., 1991). At a cost of $600 ha^{-1}, however, the black plastic technique is generally too costly for large-scale use (Pimentel et al., 1991).

Another option is mechanical cultivation, which costs about $15 ha^{-1} (Pimentel et al., 1991). A combination of cultivation and herbicides are a viable option. This approach reduced both herbicide use and the costs of production, as well as potential weed resistance problems.

In addition, Pimentel et al. (1991) suggest that wiper-application technologies can reduce the amount of herbicides applied by increasing the amount of herbicides that reach the target weeds. Combining more effective application techniques with a scouting system and spot-treatment of weedy areas could significantly decrease the amount of herbicides applied. These techniques could reduce herbicide use by about 40% in tomatoes, with all added costs for tillage and labour being offset by the savings on herbicide expenditure (Table 17.3).

Sweetcorn

Tillage systems in corn production are more expensive than production in no-till systems; however, these costs may be offset by the need for added herbicides and other pesticides in no-till systems (William et al., 1995). Overall, comparisons between mechanical systems and no-till systems show similar yields and similar costs associated with weed control (Pimentel et al., 1991). A certain type of no-till system called 'ridge-till' can provide benefits similar to no-till without the added costs of high pesticide treatments.

Better application techniques, such as band-spraying, can help reduce the amount of herbicides applied by almost 50% at no added cost to the grower (Pimentel et al., 1991). In addition, experiments have shown that sweet corn grows more successfully when rotated with sorghum as compared with continuously planted corn. Rotational systems also lead to lower weed infestations, as does planting peas in rotation with sweet corn (Pimentel et al., 1991). A reduction of about one-half in herbicide use might be possible in sweet corn, with the added cost of about $10–15 ha^{-1} (Table 17.3).

Onions

Weeds are a major problem in onion production because of competition within the row; hand-weeding appears to be the only alternative (Boldt et al., 1981). Because of the high cost of hand-weeding, it was assumed that there was no alternative weed control technology for large-scale onion production (Table 17.3).

Beans

Bean losses to weeds are estimated to be 10% despite herbicides and all other weed controls. Mechanical cultivation and better planting techniques, such as timed planting, can significantly decrease the amount of herbicides needed in bean production. In addition, better application procedures, including rope-wick applicators and band-spraying applications, can drastically reduce the amount of herbicides applied (Pimentel and Levitan, 1986). About 95% of bean crops are treated with herbicides. However, we estimate that through the use of various alternative techniques, a 50% reduction in herbicide use is possible, with an added cost for the mechanical cultivation of only $5 ha^{-1} (Table 17.3).

Cucumbers

Cucumber losses to weeds are about 12% with herbicides and other weed-control techniques. Black plastic mulch has been shown to be effective in controlling weeds in cucumber fields without the use of post-harvest herbicides (Pimentel et al., 1991). While this is expensive ($600 ha^{-1}), the costs are more than offset by the profits garnered from increased yields and higher market values for mulch-treated cucumbers (Pimentel et al., 1991).

A number of other weed-control alternative practices have been shown to be effective in controlling weeds in cucumbers. These include crop rotations, canopy planting and the development of weed-resistant cultivars (Pimentel et al., 1991). Using these methods, herbicide use in cucumbers might be reduced by 50% with no added cost to the grower (Table 17.3).

Sweet potatoes

Cultivation can improve weed control in herbicide-treated sweet potatoes (Glaze et al., 1981). We estimated that one-third of the herbicide used in sweet potatoes could be substituted by mechanical cultivation at an estimated cost of $10 ha^{-1} (Table 17.3).

Fruit crops

The amount of herbicide applied to fruit crops totals about 9.6 million kg per year (Table 17.4).

Apples

Apple losses to weeds are a relatively low 4% (Table 17.4). The commercial apple industry depends upon herbicides as its main mechanism for weed control (Hardman et al., 1987). However, a variety of possible weed control alternatives are available in apple orchards. Mechanical cultivation and other tillage systems are some of the most useful alternatives. While mechanical weed control usually carries higher production costs owing to increased labour requirements, this burden may be offset by combining it with a mulching system such as straw, which has been shown to increase yields and the commercial market value of the apples (Pimentel et al., 1991). Therefore, a reduction of one-half in herbicide use is estimated to add no cost to apple production (Table 17.4).

Peaches and pears

Potential losses in peach and pear production are estimated to be 7 and 4%, respectively (Table 17.4). A reduction of one-half in herbicide use is possible in both of these crops, at an estimated added cost of $5 ha^{-1} as a result of the mechanical cultivation necessary to achieve both the reduction and effective weed control (Table 17.4).

Plums

Plum production losses to weeds are estimated to be 6% with current herbicides and other weed controls. Based on the evidence, Pimentel et al. (1991) propose that herbicide use on plums might be reduced by 50%, with a saving of $10 ha^{-1} to the grower. However, since the major alternative control listed is extra cultivation, it is doubtful that the savings in herbicide expenditure will offset the added labour cost incurred by multiple cultivations. We consider a more conservative estimate to be a cost of $5 ha^{-1} for effective weed control (Table 17.4).

Citrus fruits

Citrus losses to weeds range from 4–7% using current weed control practices (Table 17.4). Mechanical cultivation in citrus fruits is effective in controlling weeds while maintaining high yields (Pimentel et al., 1991). By combining cultivation practices with better herbicide application techniques, we estimate that herbicide use might be reduced by one-half with no added cost (Table 17.4).

Table 17.4 Fruit and nut crop losses from weeds with current herbicide use and estimated costs if herbicides were reduced and several alternatives were substituted

Crop	Area (ha − 10³)[a]	Total herbicide use (kg − 10⁶)		Herbicide treatment			Current crop pest loss (%)[e]	Added alternative cost ($ ha⁻¹)[f]	Total added alternative control cost ($ − 10⁶)[f]
		Current[b]	Reduced[c]	Hectares treated (%)[d]	Cost ($ ha⁻¹)[e]	Total cost ($ − 10⁶)			
Apples	190	0.70	0.35[e]	42	30	2.4	4	0	0
Cherries	50	0.12	0.06[e]	47	35	0.8	8	5	0.171
Peaches	92	0.25	0.125[e]	49	80	3.6	7	5	0.211
Pears	27	0.13	0.065[e]	44	30	0.4	4	5	0.03
Plums	55	0.01	0.005[e]	70	100	3.8	6	−5	−0.085
Grapes	301	0.25	0.125[f]	64	8	1.5	15	5	0.77
Oranges	272	7.15	3.575[f]	94	42	10.7	5	0	0.00
Grapefruit	95	1.50	0.75[e]	93	42	3.7	7	0	0.00
Lemons	30	0.02	0.01[e]	71	42	0.9	4	0	0.00
Other fruit	100	0.10	0.05[e]	20	30	0.6	5	5	0.00
Pecans	155	0.10	0.05[e]	39	30	1.8	11	5	0.00
Other nuts	120	0.10	0.05[e]	31	100	5.3	5	−10	−0.53
Total		9.61	5.150			35.5			0.567

[a] USDA, 1992.
[b] Converted from USDA, 1994.
[c] Calculated.
[d] USDA, 1994.
[e] Pimentel et al., 1991.
[f] Curtis et al., 1991

Other fruit crops

The use of herbicides and the potential to reduce herbicide use through alternative weed control practices are listed in Table 17.4

REFERENCES

Agamalian, H. (1984) Selective weed control in cole crops. *Proceedings of the California Weed Conference* **36**, 118–120.

Antle, J.M. and Park, S.K. (1986) The economics of IPM in processing tomatoes. *California Agriculture* **40**(3/4), 31–32.

Baker, C.J., Stavely, J.R. and Mock, N. (1985) Biocontrol of bean rust by *Bacillus subtilis* under field conditions. *Plant Disease* **69**, 770–772.

Barber, G.W. (1942) Mineral oil treatment of sweet corn for earworm control. US Department of Agriculture Circular No. 657, Washington, DC, 16 pp.

Boiteau, G., Pelletier, Y., Misener, G.C. and Bernard, G. (1994) Development and evaluation of a plastic trench barrier for protection of potato from walking adult Colorado potato beetles (Coleoptera: Chrysomelidae). *Journal of Economic Entomology* **87**(5), 1325.

Boldt, P., Putnam, A. and Binning, L. (1981) Economic analysis of nitrogen use on onions grown in Minnesota, Michigan, and Wisconsin. *Proceedings of the North Central Weed Control Conference* **36**, 57–58 (abstract).

Bridges, D.C. (Ed.) (1992) *Crop Losses Due to Weeds in the United States 1992.* Weed Science Society of America, Champaign, IL.

Brown, C.S., Cox, W.M., Dreschel, T.W. and Chetirkin, P.V. (1992) The vacuum-operated nutrient delivery system: Hydroponics for microgravity. *HortScience* **27**(11), 1183.

Cadoux, M. (1984) Chlorpyrifo dissipation in muck soil and maggot resistance in onions: Implications for management of the onion maggot. MPS Project Report, Department of Vegetable Crops, Cornell University, Ithaca, NY, 16 pp.

Cantwell, G.E. and Cantello, W.W. (1984) Control of the Colorado potato beetle with *Bacillus thuringiensis* variety *thuringiensis. American Potato Journal* **61**, 451–459.

Comin, D. (1946) *Onion Production.* Orange Dudd, New York.

CR (1987) *Cornell Recommendations for Commercial Vegetable Production.* NY State College of Agriculture and Life Science, Cornell University, Ithaca, NY, 99 pp.

Cranshaw, W.S. and Radcliffe, E.B. (1984) Insect contaminants of Minnesota processed peas. Technical Bulletin AD-T-2211, University of Missouri Agricultural Experiment Station.

Curtis, J., Mott, L. and Kuhnle, T. (1991) *Harvest of Hope.* Natural Resources Defense Council, Washington, DC.

Dale, J. (1979) A non-mechanical system of herbicide application with a rope wick. *PANS* **25**, 431–436.

Douce, G.K. and Suber, E.F. (1985) *Summary of Losses from Insect Damage and Costs of Control in Georgia, 1985.* Special Publication No. 40, Georgia Agricultural Experiment Station, College of Agriculture, University of Georgia, December 1988.

Drapek, R.J., Croft, B.A. and Fisher, G.C. (1992) Relationship of corn earworm (Lepidoptera: Noctuidae) pheromone catch and silking to infestation levels in Oregon sweet corn. *Journal of Economic Entomology* **85**(1), 240.

Edelson, J.V., Cartwright, B. and Royer, T.A. (1986) Distribution and impact of *Thrips tabaci* on onion. *Journal of Economic Entomology* **79**, 502–505.

Ellerbrock, L.A. and Lorbeer, J.W. (1977) Sources of primary inoculum of *Botrytis squamosa. Phytopathology* **67**, 363–372.

Elsey, K.D., Klun, J.A. and Schwarz, M. (1991) Forecasting pickleworm (Lepidoptera: Pyralidae) larval infestations using sex pheromone traps. *Journal of Economic Entomology* **84**(6), 1837.

Falcon, L.A. (1976) Problems associated with the use of arthropod virus pest control. *Annual Review of Entomology* **21**, 305–324.

Farrar, C.A., Perring, T.M. and Toscano, N.C. (1986) A midge predator of potato aphids on tomatoes. *California Agriculture* **40**(11), 9–10.

Gelernter, W.D. (1990) Targeting insecticide-resistant markets: New Developments in microbial-based products. ACS Symposium Series, American Chemical Society, Vol. 421, p. 105.

Glaze, N.C., Harman, S.A. and Phatak, S.C. (1981) Enhancement of herbicidal weed control in sweet potatoes (*Ipomoea batatas*) with cultivation. *Weed Science* 29, 275–281.

Hardman, J.M., Rogers, R.E.L. and MacLellan, C.R. (1987) Pesticide use and levels of insect and scab injury on fruit in Nova Scotia apple orchards. *Journal of Economic Entomology* 80(4), 979–984.

Hicks, D.R. and Rehm, G.W. (1986) Corn production costs. *Crops Soils Magazine* 38, 17–19.

Hoffman, M.P., Wilson, L.T., Zalom, F.G. and McDonough, L. (1986) Lures and traps for monitoring tomato fruitworm. *California Agriculture* 4(9/10), 17–18.

Horn, D.J. (1988) *Ecological Approach to Pest Management*. Guilford Press, New York.

Huffaker, C.B. (1980) *New Technology of Pest Control*. Wiley, New York.

Jaques, R.P. (1988) Field tests on control of the imported cabbageworm (Lepidoptera: Pieridae) and the cabbage looper (Lepidoptera: Noctuidae) by mixtures of microbial and chemical insecticides. *Canadian Entomology* 120, 575–580.

Jaques, R.P. and Laing, D.R. (1989) Effectiveness of microbial and chemical insecticides in control of the Colorado potato beetle (Coleoptera; Chrysomelidae) on potatoes and tomatoes. *Canadian Entomology* 121, 1123–1131.

Jimenez, M.J., Toscano, N.C., Flaherty, D.L., Ilic, P., Zalom, F.G. and Kido, K. (1988) Controlling tomato pinworm by mating disruption. *California Agriculture* 42(6), 10–12.

Johns, G.F. (Ed.) (1966) *On the Way to Plant Protection*. Rodale, Emmaus, PA, 355 pp.

Jones, A., Dukes, P.D., Schalk, J.M., Hamilton, M.G., Mullen, M.A., Baumgardner, R.A., Paterson, D.R. and Boswell, T.E. (1985) 'Regal' sweet potato. *HortScience* 20, 781–782.

Jones, A., Schalk, J.M. and Dukes, P.D. (1987) Control of soil insect injury by resistance in sweet potato. *Journal of the American Society for Horticultural Science* 112, 195–197.

Jones, A., Dukes, P.D., Schalk, J.M. and Hamilton, M.G. (1989) 'Excel' sweet potato. *HortScience* 24, 171–172.

Karel, A.K. and Rweyemamu, C.L. (1985) Resistance to foliar beetle, *Ootheca bennigseni* (Coleoptera: Chrysomelidae) in common beans. *Environmental Entomology* 14, 662–664.

Karel, A.K. and Schoonhoven, A.V. (1986) Use of chemical and microbial insecticides against pests of common beans. *Journal of Economic Entomology* 79, 1692–1696.

Kimberly, I.D. and Puttler, B. (1991) A new broad-host-spectrum nuclear polyhedrosis virus isolated from a celery looper, *Anagrapa falcifera* (Kirby), Lepidoptera: Noctuidae. *Environmental Entomology* 20(5), 1480.

Kirby, R.D. and Slosser, J.E. (1984) Composite economic threshold for three lepidopterous pests of cabbage. *Journal of Economic Entomology* 77, 725–733.

Krishnaiah, K., Mohan, N.J., and Prasad, V.G. (1981) Efficacy of *Bacillus thuringiensis* Ber. for the control of lepidopterous pests of vegetable crops. *Entomon* 6, 87–93.

Lewis, J.A. and Papavizas, G.C. (1980) Integrated control of *Rhizoctonia* fruit rot of cucumber. *Phytopathology* 70, 85–89.

Lloyd, J.W. and McCollum, J.P. (1940) Fertilizing onion sets, sweet corn, cabbage and cucumbers in a four-year rotation. *Bulletin of the University of Illinois Agricultural Experiment Station* 464, 217–235.

Love, J. and Tauer, L.W. (1987) Crop biotechnology research: The case of viruses. *Agricultural Economics Research* 87-15. Cornell University, Ithaca, NY.

Madden, L., Pennypacker, S.P. and MacNab, A.A. (1978) FAST, a forecast system for *Alternaria solani* on tomato. *Phytopathology* 68, 1354–1358.

Mahrt, G.G., Stoltz, R.L., Blickenstaff, C.C. and Holtzer, T.O. (1987) Comparisons between blacklight and pheromone traps for monitoring the western bean cutworm (Lepidoptera: Noctuidae) in south central Idaho. *Journal of Economic Entomology* 80, 242–247.

Maiteki, G.A. and Lamb, R.J. (1985) Spray timing and economic threshold for the pea aphid, *Acyrthosiphon pisum* (Homoptera; Aphididae), on field peas in Manitoba. *Journal of Economic Entomology* 78, 1449–1454.

McLeod, D. (1986) Economic effectiveness: An alternate approach to insecticide evaluation in sweet corn. *Journal of Agricultural Entomology* 3, 272–279.

Metcalf, R.L. and Metcalf, R.A. (1993) *Destructive and Useful Insects*. 5th edn. McGraw-Hill, New

York, 1987 pp.

Mukishi, P. and Trutman, P. (1988) Can diseases be effectively controlled in traditional varietal mixtures using resistant varieties? *Annual Report of the Bean Improvement Cooperative* **31**, 104–105.

NAS (1989) *Alternative Agriculture.* National Academy of Sciences, Washington, DC.

Nordblom, T.L., Pannell, D.J., Christiansen, S., Nersoyan, N. and Bahhady, F. (1994) From weed to wealth? Prospects for medic pastures in the Mediterranean farming system of north-west Syria. *Agricultural Economics* **11**(1), 29.

Oatman, E.R., Hall, I.M., Arakawa, K.Y., Plantner, G.R., Bascom, L.A. and Beagle, L.L. (1970) The corn earworm on sweet corn in southern California with a nuclear polyhedrosis virus and *Bacillus thuringiensis. Journal of Economic Entomology* **63**, 415–421.

Osteen, C. (1993) Pesticide use trends and issues in the United States. In Pimentel, D. and Lehman, H. (Eds) *The Pesticide Question: Environment, Economics and Ethics.* Chapman & Hall, New York, pp. 309–336.

OTA (1979) *Pest Management Strategies.* Vol. II. Working Papers, Office of Technology Assessment, Washington, DC, 169 pp.

Pimentel, D. (1995) Amounts of pesticides reaching target pests: Environmental impacts and ethics. *Journal of Agricultural and Environmental Ethics* **8**, 17–29.

Pimentel, D. and Levitan, L. (1986) Pesticides: Amount applied and amount reaching pests. *BioScience* **36**(1), 86–91.

Pimentel, D., Krummel, J., Gallahan, D., Hough, J., Merrill, A., Schreiner, I., Vittum, P., Koziol, F., Back, E., Yen, D. and Fiance, S. (1978) Benefits and costs of pesticide use in US food production. *BioScience* **28**, 772, 778–784.

Pimentel, D., Glenister, C., Fast, S. and Gallahan, D. (1984) Environmental risks of biological pest controls. *Oikos* **42**, 283–290.

Pimentel, D., McLaughlin, L., Zepp, A., Lakitan, B., Kraus, T., Kleinman, P., Vancini, F., Roach, W.J., Graap, E., Keeton, W.S. and Selig, G. (1991) Environmental and economic impacts of reducing US agricultural pesticide use. In Pimentel, D. (Ed.) *Handbook of Pest Management in Agriculture.* Vol. I. 2nd edn. CRC Press, Boca Raton, FL, pp. 679–718.

Pimentel, D., Acquay, H., Biltonen, M., Rice, P., Silva, M., Nelson, J., Lipner, V., Giordano, S., Horowitz, A. and D'Amore, M. (1993) Assessment of environmental and economic impacts of pesticide use. In Pimentel, D. and Lehman, H. (Eds) *The Pesticide Question: Environment, Economics and Ethics.* Chapman & Hall, New York, pp. 47–84.

Radcliffe, E.B., Flanders, K.L., Ragsdale, D.W. and Noetzel, D.M. (1991) Potato insects: Pest management systems for potato insects. In Pimentel, D. (ed.) *Handbook of Pest Management in Agriculture.* 2nd edn. CRC Press, Boca Raton, FL, pp. 587–622.

Ritcey, G. and McEwen, F.L. (1984) Control of the onion maggot with furrow treatments. *Journal of Economic Entomology* **77**, 1580–1584.

Roberts, D.A. and Boothroyd, C.W. (1972) *Fundamentals of Plant Pathology.* W.H. Freeman, San Francisco, CA, 402 pp.

Roberts, D.W., Lebrun, R.A. and Semel, M. (1981) Control of the Colorado potato beetle with fungi. In Lashcomb, J.H. and Casagrande, R. (Eds) *Advances in Potato Pest Management.* Hutchinson Ross, Stroudsburg, PA, pp. 119–137.

Royle, D.J. and Shaw, M.W. (1988) The costs and benefits of disease forecasting in farming practice. In *Control of Plant Diseases: Costs and Benefits.* Clifford, B.C. and Lester, E. (Eds) Blackwell Scientific, Palo Alto, CA, pp. 231–246.

Schalk, J.M. and Jones, A. (1985) Major insect pests. In *Sweet Potato Products: A Natural Resource for the Tropics.* Boww Kamp, J.C. (Ed.) CRC Press, Boca Raton, FL, pp. 59–78.

Schalk, J.M., Creighton, C.S., Fery, R.L., Sitterly, W.R., Davis, B.W., McFadden, T.L. and Day, A. (1979) Reflective film mulches influence insect control and yield in vegetables. *Journal of the American Society for Horticultural Science* **104**(6), 759–762.

Schwartz, H.F., Casciano, D.H., Asenga, J.A. and Wood, D.R. (1987) Field measurement of white mold effects upon dry beans with genetic resistance or upright plant architecture. *Crop Science* **27**, 699–702.

Shields, E.J., Hygnstrom, J.R., Curwen, D., Stevenson, W.R., Wyman, J.A. and Binning, L.K. (1984) Pest management for potatoes in Wisconsin: A pilot program. *American Potato Journal* **61**,

508–516.

Shoemaker, P.B. and Lorbeer, J.W. (1977) Timing initial fungicide application to control botrytis leaf blight epidemics on onion. *Phytopathology* **67**, 412–413.

Sitterly, W.R. (1969) Effect of rotation on cucumber gummy stem blight. *Plant Disease Reporter* **53**, 417–449.

Smits, N.E.J.M., Magenya, O. and Parker, B.L. (1994) Biology and pheromone studies with the sweet potato weevils: *Cylas puncticollis* (Bohe.) and *C. brunneus* (Fabr.). *Acta Horticulturae* **380**, 299.

Stavely, J.R. (1988) Bean rust resistance in the United States in 1987. *Annual Report of the Bean Improvement Cooperative* **31**, 130–131.

Stemeroff, M. and George, J.A. (1983) The benefits and costs of controlling destructive insects on onions, apples, and potatoes in Canada 1960–1980: Summary. *Bulletin of the Entomological Society of Canada* **15**, 91–97.

Stockwin, W. (1988) Sweeping away pests with BugVac. *American Vegetable Grower* **36**(11), 34–38.

Straub, R.W. (1988) Suppression of cabbage root maggot (Diptera: Anthomyiidae) damage to cruciferous transplants by incorporation of granula insecticide into potting soil. *Journal of Economic Entomology* **81**, 578–581.

Straub, R.W. and Heath, J.C. (1983) Patterns of pesticide use on New York State produced sweet corn. *New York Food Life Science Bulletin* **102**, 1–6.

Street, R.S. (1989) The big sucker. *Agrichemical Age* **33**(3), 38–39.

Summers, M. and Kawanishi, C.Y. (1978) *Viral Pesticides: Present Knowledge and Potential Effects on Public and Environmental Health.* Symposium Proceedings Health Effects Research Laboratory, Office of Health and Ecological Effects, USEPA, Research Triangle Park, NC, 311 pp.

Sumner, D.R. and Phatak, S.C. (1987) Control of foliar diseases of cucumber with resistant cultivars and fungicides. *Applied Agricultural Research* **2**, 324–329.

Teng, P.S. and Bissonnette, H.L. (1985) Potato yield losses due to early blight in Minnesota fields, 1981 and 1982. *American Potato Journal* **62**, 619–628.

Tette, J.P. and Koplinka-Loehr, C. (1989) *New York State Integrated Pest Management Program: 1988 Annual Report.* IPM House, New York State Agricultural Experiment Station, Geneva, NY, 66 pp.

Tew, B.V., Wetzstein, M.E., Epperson, J.E. and Robertson, J.D. (1982) Economics of selected integrated pest management production systems in Georgia. Research Report 395, University of Georgia College Agricultural Experiment Station, Athens, GA, 12 pp.

Thompson, D.C. Jenkins, S.F. (1985) Influence of cultivar resistance, initial disease, environment, and fungicide concentration and timing on anthracnose development and yield loss in pickling cucumbers. *Phytopathology* **75**, 1422–1427.

Tolman, J.H., McLeod, D.G.R. and Harris, C.R. (1986) Yield losses in potatoes, onions, and rutabagas in southwestern Ontario, Canada: The case for pest control. *Crop Protection* **5**, 227–237.

UC (1983) *Integrated Pest Management for Rice.* Division of Agricultural Science Publication No. 3280, University of California, Berkeley, CA, 94 pp.

USBC (1994) *Statistical Abstract of the United States 1993.* US Bureau of the Census, US Government Printing Office, Washington, DC.

USDA (1965) *Losses in Agriculture.* Agricultural Handbook No. 291, Agricultural Research Service, US Government Printing Office, Washington, DC.

USDA (1992) *Agricultural Statistics 1992.* US Government Printing Office, Washington, DC.

USDA (1993) *Agricultural Chemical Usage. Vegetables 1992. Summary.* Economic Research Service, Washington, DC.

USDA (1994) *Agricultural Chemical Usage. Fruits 1993. Summary.* Economic Research Service, Washington, DC.

Vos, J. (1992) A case history: Hundred years of potato production in Europe with special reference to The Netherlands. *American Potato Journal* **69**(11), 731–751.

Walgenbach, J.F. and Estes, E.A. (1992) Economics of insecticide use on staked tomatoes in western North Carolina. *Journal of Economic Entomology* **85**(3), 888.

William, R.D., Burrill, L.C., Ball, D. and Miller, T.L. (1995) *1995 Pacific Northwest Weed Control Handbook.* Oregon State University Extension Service.

Witzgall, P., Bengtsson, M., Unelius, C.R. and Lofqvist, J. (1993) Attraction of pea moth *Cydia nigricana* F. (Lepidoptera: Tortricidae) to female sex hormone (E,E)-8,10-dodecadien-1-y1 acetate is inhibited by geometric isomers EZ, ZE, and ZZ. *Journal of Chemical Ecology* **19**(9), 1917.

Wright, R.J. (1984) Evaluation of crop rotation for control of Colorado potato beetles in commercial potato fields on Long Island. *Journal of Economic Entomology* **77**, 1254–1259.

Wright, R.J., Kain, D.P., Loria, R., Sieczka, J.B. and Mayer, D.D. (1986) Final report of the 1985 Long Island potato integrated pest management pilot program. Cornell L.I. Horticultural Research Laboratory, Riverhead, NY, 22 pp.

Zehnder, G.W. and Evanylo, G.K. (1989) Influence of extent and timing of Colorado potato beetle Coleoptera: Chrysomelidae) defoliation on potato tuber production in Eastern Virginia. *Journal of Economic Entomology* **82**, 948–953.

Zehnder, G.W. and Linduska, J.J. (1987) Influence of conservation tillage practices on population of Colorado potato beetle (Coleoptera: Chrysomelidae) in rotated and non-rotated tomato fields. *Environmental Entomology* **16**, 135–139.

CHAPTER 18

IPM Techniques for Greenhouse Crops

Jennifer A. Grant

Cornell University, Ithaca, NY, USA

INTRODUCTION

In the greenhouse, integrated pest management (IPM) includes many practices such as scouting of crops, emphasis on least toxic pesticides, optimal timing of applications, mechanical pest exclusion and the use of biological controls. IPM strives to obtain the best management of pests while minimizing environmental impacts, and has been adopted to varying degrees throughout the world in countries with greenhouse industries. Differences in crops grown, major pests, level of technology, climate, market standards and greenhouse structures determine which practices are feasible in each region and individual growing operation.

Closed growing systems facilitate IPM by allowing for containment of natural enemies, exclusion of pests and modification of environmental conditions. However, challenges are presented by successive and continual cropping, and the year-round climate moderation that protects and harbour pests as well as the crop. In addition, greenhouse ornamentals are bought for their aesthetic value, and therefore must meet extremely high quality standards.

Pest monitoring, pesticide management and a diversity of non-chemical control tactics, including cultural, mechanical and biological controls, are used when implementing IPM. Monitoring is the foundation on which all pest management decisions are made, and is the first step for any operation embarking on an IPM programme. Secondly, pesticides must be used minimally and to their optimal effect before the benefits of non-chemical management options can be fully realized. These components are discussed below.

PEST MONITORING

'Monitoring' or 'scouting' is a systematic method of looking for pests (insects, diseases and weeds), and is the backbone of any pest management programme. Information on the pests present, and their stage of development and abundance is gathered by a 'scout', utilizing insect traps, plant inspections and indicator plants. Greenhouse crops should be monitored a minimum of once a week, employing the techniques described below.

Traps are a passive method of pest monitoring that detect the presence of insects and

Techniques for Reducing Pesticide Use. Edited by D. Pimentel
© 1997 John Wiley & Sons Ltd.

help determine populations levels and locations. In greenhouses, sticky cards are commonly used to trap adult whiteflies, thrips, fungus gnats, aphids and shore flies. The scout identifies and counts the insects, revealing the presence of new insect infestations, population trends and treatment efficacy. Yellow is the preferred colour for trapping a broad spectrum of these pests, although other colours have been found to be more attractive to specific target pests (e.g. blue for monitoring thrips) (Brødsgaârd, 1989; Gillespie and Vernon, 1990; Mateus and Mexia, 1995). Generally, 3 inch × 5 inch cards are the most efficient traps, but these or larger cards may be subdivided for equally reliable results (Heinz and Parrella, 1991; Heinz et al., 1992). For most insects, the cards should be placed a few inches above the crop canopy, at a minimum spacing of 1 card per 1000 square feet of greenhouse area (Gillespie and Vernon, 1990; Topliff et al., 1992; Ferrentino et al., 1993). Other types of traps are appropriate in some situations, e.g. potato disks for monitoring fungus gnat larvae (Gill et al., 1994; Sanderson and Ferrentino, 1994).

Plant inspections are an important aspect of monitoring, and give information on plant health, diseases, crawling and sessile insects, and pest damage. While walking through the greenhouse, the scout takes a macro-view of the crop to look for damaged or unhealthy plants. Suspect plants are then examined on the upper and lower leaf surfaces, stems and roots for specific signs of disease, malnutrition or insect infestation. In addition, the scout randomly inspects plants throughout the greenhouse area. Recommendations for the number of plants in this random check vary by crop and time into the cropping cycle (Topliff et al., 1992a; Ferrentino et al., 1993; Sanderson and Ferrentino, 1994; Hausbeck, 1995; Shipp, 1995). Plant inspections are instrumental in detecting a variety of pest and plant problems in their early stages, before they have spread through the crop.

Indicator plants complement insect traps and plant inspections to give a full picture of pest establishment and development. The term 'indicator plant' commonly refers to plants that are highly susceptible to particular pests. For example, tospoviruses (i.e. tomato spotted wilt virus and impatiens necrotic spot virus) can be detected early in crops such as impatiens or tomatoes by interspersing petunias or faba bean plants within the crop (Allen and Matteoni, 1991; Pundt et al., 1991; Daughtrey, 1994). These indicator plants show signs of viral infection much quicker and more conclusively than the crop.

The term indicator plant is also used synonymously with 'sentinel plant' to refer to plants that are known to be infested (or infected) with a particular pest. They are used to track pest development and evaluate control on selected plants in individual greenhouses. When a scout discovers an infested plant, the pest is monitored to determine the average development time under the specific environmental conditions found in each greenhouse (Sanderson and Ferrentino, 1991). For example, whiteflies are predominately sessile insects whose development can be tracked on individual leaves or plants. This information is critical for the timing of control actions and evaluating post-treatment efficacy.

PESTICIDE MANAGEMENT

The next step in greenhouse IPM is pesticide management, which includes the use of thresholds, discrimination in the selection of control products, optimal timing of applications, and limited area or 'spot' treatments. These practices are a dramatic change for growers following a traditional pest management programme with a calendar spray schedule. For these operations, implementation of monitoring and pesticide management usually results in significant initial reductions in pesticide use (Neuhauser et al., 1991; McCoy et al., 1993, 1994).

Growers decide when pest intervention is necessary by using scouting reports that track trends in insect and disease populations. Pest management experience, aesthetic standards and customer demands usually dictate when such action is taken (Mumford, 1992). However, action thresholds that predict when pest populations will reach damaging or aesthetically unacceptable levels have been determined for some greenhouse pest situations (Helgesen and Tauber, 1974; Sanderson and Ferrentino, 1993; Shipp, 1995). Scouting information also helps locate the sources of pest problems, allowing containment and limited-area treatments.

Pesticide management includes selecting chemicals that will cause the most damage to pests and the least damage to plants, humans, beneficial organisms and the environment (Lindquist, 1993). Materials that best meet these criteria are sometimes referred to as 'least toxic' or 'soft' pesticides, such as horticultural oil, soap, microbials and insect growth regulators. Optimal selection depends on specific greenhouse conditions, such as the crop(s) being grown, pest species, stage and abundance, the time in the growing cycle, the equipment available, other management practices and the greenhouse structure.

As discussed, many factors determine when pest intervention is warranted. However, that decision may be overridden or postponed because of the time in the pest's life cycle or the crop production cycle. Controls should be aimed at the pest's most vulnerable life stage (Sanderson and Ferrentino, 1991). For example, whiteflies are usually most susceptible to insecticide sprays in the young nymphal stages. If scouting records reveal that eggs are the dominate stage in the greenhouse, the pest manager should wait for the eggs to hatch into nymphs before spraying. The waiting time can be calculated if whitefly indicator plants have been monitored. If other stages predominate, a different pest control measure should be chosen. Time in the plant production cycle can similarly influence management strategies.

DIVERSIFIED CONTROL STRATEGIES

Diversification of pest control strategies is the principal method of reducing reliance on pesticides. The following methods are currently available for use in greenhouses.

Cultural

Cultural pest controls are among the IPM practices most commonly adopted throughout the world (Gullino, 1992). For example, good sanitation is standard greenhouse procedure, and an important aspect of IPM (Topliff et al., 1992b). In the case of diseases, it is necessary that known inoculum be removed and destroyed, containers and equipment be properly sterilized, and workers use care when moving from potentially infected to non-infected plants. Insects and mites are commonly harboured in crop plants and weeds, and in soil in and under benches. These materials should be inspected and removed from the greenhouse if infested.

Resistant varieties are used as cultural pest controls in all crops and production systems. They are widely available for vegetable crops in greenhouses, but less so for ornamentals (Gullimo, 1992). Manipulation of soil fertility, air flow, temperature and humidity, plant spacing and irrigation practices also affect pest establishment and dissemination. For example, increasing air circulation using forced hot air, reducing relative humidity, minimizing free standing water, and mulches are known to significantly reduce *Botrytis* inoculum (Hausbeck, 1993). In many cases, cultural practices will not eliminate pest problems but they will work in cooperation with other pest control practices.

Biological

Biological control in the greenhouse includes the use of insect predators and parasites, insect- and disease-attacking fungi, and other beneficial microbial organisms. In many cases, development and implementation of biological controls were instigated by an increase in pesticide resistance and the failure of chemical pesticides to control important pests (van Lenteren, 1992; Bishop, 1995). The majority of commercial biological control successes have been in crops with one, or only a few, primary pests, where concurrent management of pesticide use is easiest to coordinate. In an ideal IPM programme, chemical pesticides are unnecessary and the biological control agents act in concert with cultural and mechanical controls to suppress, reduce and eliminate pest populations.

Over 15 species of natural enemies are available worldwide for use in greenhouse environments (van Lenteren, 1992), and are used in a majority of the protected vegetable crops in many countries (van Lenteren and Woets, 1988; Enkegaard, 1993). However, use on ornamentals is often more difficult for reasons such as the low tolerance for both pest and beneficial insects (especially in export markets), a larger number of pests, and the common influx of outside plant materials into greenhouses (van Lenteren and Woets, 1988; Fransen, 1992; Enkegaard, 1993; Osborne et al., 1995). These challenges are being overcome, as evidenced by the fact that biological control is used on 15% of the greenhouse ornamentals grown in Denmark (Enkegaard, 1993). Worldwide, protected crops utilizing natural enemies for pest control rose from approximately 400 ha in 1970 to 14 000 in 1991 (van Lenteren, 1992).

Microbial biological control agents include pest-attacking fungi, bacteria and nematodes. In European greenhouses, the fungus *Verticillium lecanii* is a standard control for thrips, whiteflies and aphids (Heyler et al., 1992; Enkegaard, 1993; Brownbridge et al., 1994). Other fungi that have been introduced as insect pathogens include *Metarhizium anisopilae*, *Paecilomyces fumosoroseus* and *Beauveria bassiana* (Enkegaard, 1993; Brownbridge et al., 1994; Price, 1995). Fungi are also used to control disease pathogens; examples of these mycopesticides include *Trichoderma (Gliocladium) virens*, marketed for control of root-rotting fungi of ornamentals (Jacobsen and Backman, 1993), and *Streptomyces griseoviridis* for management of seed-, root- and stem-rotting organisms (Horst, 1993). The ability to manipulate greenhouse environments facilitates the use of beneficial fungi that usually require high relative humidities.

Entomopathic nematodes, *Steinernema* spp., and the bacteria *Bacillus thuringiensis* are used for control of fungus gnat larvae in protected crops (Enkegaard, 1993; Lindquist, 1995; Price, 1995). A new microbial product, Spod-X1®, is derived from a nuclear polyhedrosis virus of the beet armyworm (Price, 1995). It is effective against the beet armyworm only, but has the advantage that infected larvae will inoculate their neighbours. Naturally occurring microbes are also used in IPM programmes. For example, bark mulches and composted potting mixes are used for disease suppression in greenhouses and nurseries (Hoitink et al., 1991; Hoitink and Boehm, 1993).

Mechanical

Mechanical controls include hand-removal of weeds, rouging and isolation of infested and infested leaves and plants, vacuuming and physical exclusion of pests. Barrier screening of vents, fans and open walkways has become an important tool for excluding thrips, whiteflies, aphids and leaf miners from immigrating into greenhouses (Bethke et al., 1994; Robb, 1995). However, the finely meshed screening materials severely restrict air flow, which must be considered when constructing new houses or retro-fitting older structures.

DOCUMENTATION

Accurate records of pests, cultural management and pest control activities are essential for the success of an IPM programme (Topliff et al., 1992a; Ferrentino et al., 1993). Pest records should reflect species, development stage, host plants and locations of infestations within each greenhouse. Management information, including fertilization, irrigation, environmental conditions, plant movement, pesticide applications (rate, formulations, etc.) and natural enemy releases, should also be documented. Examination of both sets of records determines when, what and where pest management actions are warranted. This analysis is also critical after taking a pest-control action to evaluate the success of pest management strategies. At the end of a cropping cycle, the information can be reviewed in order to plan and prioritize scouting and management activities for future crops.

COST

IPM programmes usually save money in expenditure on pesticides and labour saved in application time. However, the increased labour expense of scouting is significant. For example, vegetable scouting time in the UK was estimated as 12 h week^{-1} ha^{-1} (Jacobsen, 1995), and poinsettia scouting is even more intensive and has been estimated at 18–27 h week^{-1} ha^{-1} (10–15 min per 1000 square feet) for an experienced scout (Ferrentino et al., 1993). The availability of time and trained labour is often a major factor in limiting how many sticky cards, random plant inspections and indicator plants are included in a greenhouse monitoring plan.

Additional expenses are accrued in IPM programmes because alternative control products are often more expensive than their pesticide counterparts. For example, whiteflies were controlled in a poinsettia crop with regular releases of a parasite and predator, but the cost was five times greater than the comparable insecticide treatments (Heinz and Parrella, 1994). On the other hand, van Lenteren (1992) reviewed several situations where biological control was less expensive than chemical regimes. Many of the costs associated with biological control should subside as the technologies become more popular and production systems are established and refined.

MODEL INTEGRATED PROGRAMS

The goal of IPM is to maintain or improve quality while reducing pesticide use. Success was documented in New York State's IPM programme for poinsettias, where IPM growers had 3–5 times fewer whitefly-infested plants than non-IPM growers (Sanderson and Ferrentino, 1993, 1994). At the same time, IPM participants decreased pesticide use by 40–75% (Freeman and Gilrein, 1993; Grant and Ferrentino, 1995). An impatiens IPM programme in Maryland reduced insecticide use by 10–50%, and losses to viral infections were held to 1% as compared with 3% the previous year (Gill and Dutky, 1994). In Ontario, Canada, pesticide use was reduced by up to 80% in chrysanthemums and 30–35% in poinsettias (Broadbent, 1994a; Murphy and Broadbent, 1994). Other successes have been reported in botanical garden greenhouses (Kole and Hennekam, 1990), chrysanthemums (Hesselein et al., 1993) and multi-cropped commercial greenhouse operations (Graeb, 1991; Broadbent, 1994b).

IPM practices have been incorporated into greenhouse production systems through the world. Growers have adopted new technologies and developed new approaches to pest management. However, further adoption will clearly be necessary to compensate for the loss of traditional control materials and to meet the pest challenge of the 21st century.

REFERENCES

Allen, W.A. and Matteoni, J.A. (1991) Petunia as an indicator plant for use by growers to monitor for thrips carrying the tomato spotted wilt virus in greenhouses. *Plant Disease* **75**, 78–82.

Bethke, J.A., Redak, R.A. and Paine, TD. (1994) Screens deny specific pests entry to greenhouses. *California Agriculture* **48**(3), 37–40.

Bishop, A.L. (1995) Putting pest management to work. In *Proceedings for the 11th Conference on Insects and Disease Management on Ornamentals*. Society of American Florists, Alexandria, VA, pp. 51–57.

Broadbent, A.M. (1994a) Integrated pest management of thrips in greenhouse floriculture. In McAvoy, R.J. (Ed.) *Proceedings of the 1994 New England Greenhouse Conference*. Sturbridge, MA, pp. 175–177.

Broadbent, A.B. (1994b) The integrated pest management of western flower thrips. In *Proceedings of the 10th Conference on Insects and Disease Management on Ornamentals*. Society of American Florists, Alexandria, VA, pp. 68–76.

Brødsgaard, H.F. (1989) Coloured sticky traps for *Frankliniella occidentalis* (Pergande) (Thysanoptera, Thripidae) in glasshouses. *Journal of Applied Entomology* **107**, 136–140.

Brownbridge, M., McLean, D.L., Parker, B.L. and Skinner, M. (1994) Use of fungal pathogens for insect control in greenhouses. In *Proceedings of the 10th Conference on Insects and Disease Management on Ornamentals*. Society of American Florists, Alexandria, VA, pp. 7–20.

Daughtrey, M. (1994) Suggestions for management of tomato spotted wilt tospovirus (TSWV) and impatiens necrotic spot tospovirus (INSV) in the greenhouse. In McAvoy, R.J. (Ed.) *Proceedings of the 1994 New England Greenhouse Conference*. Sturbridge, MA, pp. 120–121.

Enkegaard, A. (1993) Biological/integrated control of pests in Danish glasshouse ornamentals. *GrønViden* 79.

Ferrentino, G.W., Grant, J.A., Heinmiller, M., Sanderson, J.P. and Daughtrey, M. (1993) *IPM for Poinsettias in New York: A Scouting and Pest Management Guide*. New York State Integrated Pest Management Program, Publication No. 403.

Fransen, J.J. (1992) Development of integrated crop protection in glasshouse ornamentals. *Pesticide Science* **36**(4), 329–333.

Freeman, R. and Gilrein, D. (1993) Integrated pest management strategies for commercial poinsettia production on Long Island. In *Ornamentals: Reports Pertinent to the Integrated Pest Management Effort at Cornell University 1992*, pp. 10–14.

Gill, S. and Dutky, E. (1994) Total plant management for greenhouse crops. In McAvoy, R.J. (Ed.) *Proceedings of the 1994 New England Greenhouse Conference*. Sturbridge, MA, pp. 19–22.

Gill, S., Dutky, E., Davidson, J. and Raupp, M. (1994) *Managing Fungus-Eating Gnats, Humpbacked Flies and Shore Flies in the Greenhouse*. University of Maryland Fact Sheet No. 633.

Gillespie, D.R. and Vernon, R.S. (1990) *Journal of Economic Entomology* **83**(3), 971–975.

Graeb, J. (1991) Solve your pest problems. *GrowerTalks* August, 21–39.

Grant, J.A. and Ferrentino, G.W. (1995) Integrated pest management implementation in New York greenhouses. In Parker, B.L., Skinner, M. and Lewis, T. (Eds) *Thrips Biology and Management.* Plenum, New York, p. 417.

Gullino, L. (1992) Integrated control of diseases in closed systems in the sub-tropics. *Pesticide Science* **36**(4), 335–340.

Hausbeck, M.K. (1993) Biological control and environmental manipulation for control of *Botrytis.* In *Proceedings of the 9th Conference on Insects and Disease Management on Ornamentals.* Society of American Florists, Alexandria, VA, pp. 73–79.

Hausbeck, M. (1995) Managing powdery mildew on poinsettias. In *Proceedings of the 11th Conference on Insects and Disease Management on Ornamentals.* Society of American Florists, Alexandria, VA, pp. 7–13.

Heinz, K.M. and Parrella, M.P. (1991) A shortcut with sticky traps. *GrowerTalks,* August, 40–45.

Heinz, K.M. and Parrella, M.P. (1994) Biological control of *Bemesia argentifolii* (Homoptera: Aleyrodidae) infesting *Euphorbia pulcherrima*: Evaluations of releases of *Encarsi luteola* (Hymenoptera: Aphelinidae) and *Delphastus pusillus* (Coleoptera: Coccinellidae). *Environmental Entomology* **23**(5), 1346–1353.

Heinz, K.M., Parrella, M.P. and Newman, J.P. (1992) Time-efficient use of yellow sticky traps in monitoring insect populations. *Journal of Economic Entomology* **85**(6), 2263–2269.

Helgesen, R.G. and Tauber, M.J. Biological control of greenhouse whitefly, *Trialeurodes vaporariorum* (Aleyrodidae: Homoptera), on short-term crops by manipulating biotic and abiotic factors. *Canadian Entomology* **106**, 1175–1188.

Hesselein, C., Robb, K., Newman, J., Evans, R. and Parrella, M. (1993) Demonstration/integrated pest management program for potted chrysanthemums in California. In *Proceedings of the 9th Conference on Insects and Disease Management on Ornamentals.* Society of American Florists, Alexandria, VA, pp. 15–25.

Heyler, N., Gill, G., Bywater, A. and Chambers, R. (1992) Elevated humidities for control of chrysanthemum pests with *Verticillium lecanii. Pesticide Science* **36**(4), 373–378.

Hoitink, H.A. and Boehm, M.J. (1993) Potting mixes naturally suppressive to soilborne disease of floricultural crops. In *Proceedings of the 9th Conferences on Insects and Disease Management on Ornamentals.* Society of American Florists, Alexandria, VA, pp. 80–90.

Hoitink, H.A., Inbar, Y. and Boehm, M.J. (1991) Status of compost-amended potting mixes naturally suppressive to soilborne diseases of floricultural crops. *Plant Disease* **75**(9), 869–873.

Horst, R.K. (1993) Current trends in ornamentals. *Plant Health Guide.* Meister, pp. 138–139.

Jacobsen, B.J. and Backman, P.A. (1993) Biological and cultural plant disease controls: Alternatives and supplements to chemicals in IPM systems. *Plant Disease* **77**, 311–315.

Jacobsen, R.J. (1995) Resources to implement biological control in greenhouses. In *Thrips Biology and Management,* Parker, B.L., Skinner, M. and Lewis, T. (Eds) Plenum, New York, pp. 211–219.

Kole, M. and Hennekam, M. (1990) Update: Six years of successful biological control in interior plantscapes in The Netherlands. *IPM Practitioner* **12**(1), 1–4.

Lindquist, R.K. (1993) Integrated insect mite and disease management programs on greenhouse crops: Pesticides and application methods. In *Proceedings of the 9th Conference on Insects and Disease Management on Ornamentals.* Society of American Florists, Alexandria, VA, pp. 38–44.

Lindquist, R.K. (1995) Control of fungus gnats and shore flies. In *Proceedings of the 11th Conference on Insects and Disease Management on Ornamentals.* Society of American Florists, Alexandria, VA, pp. 69–74.

Mateus, C. and Mexia, A. (1995) Western flower thrips response to color. In Parker, B.L., Skinner, M. and Lewis, T. (Eds) *Thrips Biology and Management.* Plenum, New York, pp. 567–570.

McCoy, R., Grant, J. and Ferrentino, G. (1993) 1992 Poinsettia IPM report for Erie County. In *Ornamentals: Reports Pertinent to the Integrated Pest Management Effort at Cornell University 1992,* pp. 15–19.

McCoy, R., Ferrentino, G., Sanderson, J. and Grant, J. (1994) 1993 Poinsettia IPM report for Erie County. In *Ornamentals: Reports Pertinent to the Integrated Pest Management Effort at Cornell University 1993,* pp. 44–49.

Mumford, J.D. (1992) Economics of integrated pest control in protected crops. *Pesticide Science* **36**(4), 379–383.

Murphy, G. and Broadbent, B. (1994) Implementation of IPM in Ontario, Canada. Paper presented to the Northeast Greenhouse IPM Workshop, Ithaca, NY, July.

Neuhauser, W., Grant, J. and Ferrentino, G. (1991) Implementation of an IPM poinsettia program in central New York greenhouses. In *Ornamentals: Reports from the 1990 IPM Research, Development and Implementation Projects in Ornamentals*, pp. 30–35.

Osborne, L.S., Pena, J., Petitt, F.L. and Fan, Y.Q. (1995) Biological control of mites. In *Proceedings of the 11th Conference on Insects and Disease Management on Ornamentals*. Society of American Florists, Alexandria, VA, pp. 41–49.

Price, J.F. (1995) New biological and chemical insecticides for management of arthropod pests of floriculture. In *Proceedings of the 11th Conference on Insects and Disease Management on Ornamentals*. Society of American Florists, Alexandria, VA, pp. 101–104.

Pundt, L.S., Sanderson, J.P. and Daughtrey, M.L. (1991) The use of Petunia as a rapid indicator for tomato spotted wilt virus in NYS greenhouses. In *Proceedings of the 53rd Annual New York State Pest Management Conference*, pp. 7–13.

Robb, K.L. (1995) Controlling thrips and the use of exclusion devices. In *Proceedings of the 11th Conference on Insects and Disease Management on Ornamentals*. Society of American Florists, Alexandria, VA, pp. 113–120.

Sanderson, J.P. and Ferrentino, G.W. (1991) Whiteflies 101: A primer on biology and control. *GrowerTalks* August, 48–61.

Sanderson, J.P. and Ferrentino, G.W. (1993) Basic IPM: Whiteflies on poinsettias. In *Proceedings of the 9th Conference on Insects and Disease Management on Ornamentals*. Society of American Florists, Alexandria, VA, pp. 1–14.

Sanderson, J.P. and Ferrentino, G.W. (1994) Biological insect pest management on poinsettias. In *Ornamentals: Reports Pertinent to the Integrated Pest Management Effort at Cornell University 1993*. pp. 52–64.

Shipp, J.L. (1995) Monitoring of western flower thrips on glasshouse and vegetable crops. In Parker, B.L., Skinner, M. and Lewis, T. (Eds) *Thrips Biology and Management*. Plenum, New York, pp. 547–555.

Topliff, L.A., Schnelle, M.A., Pinkston, K.N., Cuperus, G.W. and von Broembsen, S. (1992a) *Scouting and Monitoring for Pests in Commercial Greenhouses*. Oklahoma State University Extension Facts No. 6711.

Topliff, L.A., Schnelle, M.A., von Broembsen, S., Cuperus, G.W. and Pinkston, K.N. (1992b) *Integrated Pest Management for Commercial Greenhouses: A General Overview of IPM Principles and Practices*. Oklahoma State University Extension Facts No. 6710.

van Lenteren, J.C. (1992) Biological control in protected crops: Where do we go? *Pesticide Science* **36**(4), 321–327.

van Lenteren, J.C. and Woets, J. (1988) Biological and integrated pest control in greenhouses. *Annual Review of Entomology* **33**, 239–269.

CHAPTER 19

Environmental Aspects of 'Cosmetic Standards' of Foods and Pesticides*

David Pimentel and David Kahn

Cornell University, Ithaca, NY, USA

INTRODUCTION

The American supermarkets feature nearly perfect fruits and vegetables. Gone are the apples with an occasional blemish, a slightly russetted orange, or fresh spinach with a leaf miner. Less apparent but present in fresh and processed fruits and vegetables are a few small insects and mites. This increase in the 'cosmetic standards' of fruits and vegetables has resulted from the development of new pesticide technologies and the efforts of the Food and Drug Administration (FDA) and US Department of Agriculture (USDA) to limit the levels of insects and mites in fruits and vegetables, plus new standards established by wholesalers, processors and retailers. Consumer preferences have probably influenced these changes.

The Food and Drug Administration sets defect action levels (DAL) for insects and mites allowed in fruits and vegetables, and in products made from these foods. During the past 40 years, as the FDA and USDA have been lowering these tolerance levels, more pesticides have been used to ensure that crop produce meets the more stringent defect levels (FDA, 1972a,b, 1974, 1989a; Zalom and Jones, 1994). In addition, wholesalers, processors and retailers have been increasing their cosmetic standards for various reasons, including perceived consumer demand. The results have been higher economic costs for pest control, widespread environmental and human health problems caused by pesticides, and higher contamination levels of insecticides and miticides in fruits and vegetables (Steinman, 1990). Clearly, the economic, public health and environmental values behind these changes need to be re-examined.

In this study, the following factors are examined: the legal tolerance levels of non-harmful insects and mites allowed in foods; the health and nutritional aspects of consuming these insects and mites; related trends in pesticide use and crop loss; fossil energy costs of producing pesticides. In addition, the environmental and health hazards associated

---

* This chapter is a revision of a paper entitled "The Relationship between 'Cosmetic Standards' for Foods and Pesticide Use", by D. Pimentel, C. Kirby and A. Shroff, published in *The Pesticide Question: Environment, Economics, and Ethics*. Routledge, Chapman & Hall, New York, 1993.

with increased pesticide use are compared with the benefits of having fewer insects and mites in foods.

GOVERNMENTAL REGULATION OF INSECTS AND MITES FOUND IN FOOD

Because American consumers appear to be strongly disposed to purchase produce which is not damaged by pests or does not show the presence of insects and mites in or on their produce, the FDA established *defect action levels* (DAL) to keep insects and mites in foods to a minimum (FDA, 1974). In addition to the visual prejudice against insects, there is the well-placed concern that some insects, such as house flies and cockroaches, may transmit disease organisms.

The dominant consideration for establishing the DALs (FDA, 1972b) was to reduce insect and mite infestations to a reasonable level, based on the existing state of insect and mite control technology (provided that the insects and mites are not easily seen). DALs serve the dual purpose of ensuring the uniform quality of the final product, while accepting the inevitability of at least some level of unavoidable defects and arthropods (Zalom and Jones, 1994). As detailed in *Food Purity Perspectives* (Anonymous, 1974), FDA standards for small insects and mites in fruits, vegetables and products made from them were established because the presence of insects and mites indicated that the crop had sufficient insect and mite control, was improperly washed or was unsatisfactorily inspected, and/or contained small insects and mites harmful to human health.

FDA and USDA inspectors check food lots during processing and before transport to market. If any lot is found to have an insect infestation above the DAL, the lot is seized and destroyed. During 1950, for example, one of the peak years for quantities of food seized, only about 0.2% of the total crop of spinach and broccoli was seized (FDA, 1944–66). At that time, neither food processors nor the FDA issued reports as to the actual level of insects and mites found in or on fresh or processed fruits and vegetables. The defect action levels for insects and mites present in broccoli, spinach and other crops for 1949 and 1950 were listed only in restricted FDA administrative guidelines. The established DALs were not published by FDA until 1972, but have been available to the public since then (FDA, 1972a, 1989a).

Even under the DAL regulatory guidelines, a few insects and mites remain in or on produce. For instance, the DAL for apple butter is an 'average of 5 whole insects or equivalents ... per 100 grams' ... 'not counting mites, aphids, thrips, or scale insects' (FDA, 1989a).

The DAL for canned sweet corn is similar. It states that if '2 or more 3 mm or longer larvae, cast skins, larval or cast skin fragments of corn ear worm or corn borer and the aggregate length of such larvae, cast skins, larval or cast skin fragments exceeds 12 mm in 24 pounds' (11 kg) then the DAL is exceeded (FDA, 1989a). For shelled peanuts, the DALs are an average of 5% insect-infested, while an 'average of 30 insect fragments per 100 grams' is permitted in peanut butter (FDA, 1989a).

Tomatoes are commonly infested with insects, especially by fruit flies (*Drosophila*). DALs for processed tomatoes are established only for *Drosophila* eggs and maggots (Zalom and Jones, 1994). Industry marketing research suggests that consumers prefer foods without insect damage or contaminants, but it is impossible to produce, harvest and process crops that are totally free of such natural defects (Zalom and Jones, 1994).

For processed tomato paste and other sauces the DALs are an 'average of 30 or more fly eggs per 100 grams; or 15 or more fly eggs and 1 or more maggots per 100 grams; or 2 or more maggots per 100 grams' (FDA, 1989a). It is difficult to detect all of the insects or damage present in a sample of tomato paste or juice, and the paste and juice surveys used for processing tomatoes do not remove all insect fragments (Zalom and Jones, 1994). Likewise, for processed spinach, for which the DAL is an 'average of 50 or more aphids and/or thrips and/or mites per 100 grams' or 'leaf miners of any size average 8 or more per 100 grams or leaf miners 3 mm or longer average 4 or more per 100 grams' (FDA, 1989a).

Indeed, thrips, aphids and mites, all minute in size, are practically impossible to eliminate from most vegetables as well as fruits. Consider raspberries and blackberries, which consist of clusters of many individual fruits from which it is impossible entirely to exclude these tiny organisms. Recognizing that it is impossible to spray and/or clean these berries intensely without destroying the fruit, the DAL for such berries permits an 'average of 4 or more larvae [insect] per 500 grams' (not counting mites, aphids, thrips or scale insects) (FDA, 1989a).

The DALs for other fruits and vegetable are similar to those listed above but are tailored to reflect the pests of particular crops. It is obvious that although the number of insects and their parts are severely limited, some will remain, generally, unseen, and will be eaten.

CHANGES IN THE DALs AND PESTICIDE USE

The DALs have become more rigorous over time, according to a statement by FDA administrators published in the *Federal Register* (1973). The reduced DALs for broccoli and spinach, which were especially well documented, illustrate this. Between 1938 and 1973 the DALs for aphids, thrips and/or mites in broccoli were 80 per 100 grams (R. Angelotti, Associate Director, Compliance, Bureau of Foods, FDA, personal communication, 19 January 1976). Then in 1974 the level was reduced to 60 aphids/thrips/mites per 100 grams, which continues in effect (FDA, 1974, 1989a; USDA, 1983).

During the 1930s the FDA's 'confidential figure' for spinach was 110 aphids allowed per 100 grams (FDA, 1972b). This guideline was based on 'information on market sample findings'. Successful aphid control was achieved in 'fresh spinach by immersion in a dilute pyrethrum [insecticide] solution to loosen the insects from the leaves, followed by a detergent wash' (FDA, 1972b). In the early 1940s, pressure for stricter standards from FDA's District Laboratories resulted in a reduction in the DALs to a level of 60 aphids allowed per 100 grams of spinach (FDA, 1972b). This DAL remained in effect until 1974, when it was further reduced to 50 aphids per 100 grams (FDA, 1974), or to less than half the 1930s guidelines for aphids in spinach.

Over time the DALs for leaf miners in spinach have also been reduced. During the 1930s, based on what was termed 'a guide to repulsiveness', 40 leaf miners were allowed per 100 grams of spinach (USDA, 1969; FDA, 1972b). The FDA reported that 'numerous seizures of leaf miners in spinach were effected as early as 1938 based on findings which "appear sufficiently repulsive to warrant consideration of regulatory action"' (FDA, 1972b; and the same was true for the USDA [Memo of R. Angellotti, FDA, 1972]). After the severe 'leaf miner outbreak in California in 1956', a lower level of 9 leaf miners per 100 grams of spinach was adopted (FDA, 1972a,b). The DAL remained at this level until 1974, when it was further reduced to 8 leaf miners per 100 grams (FDA, 1974), a level five times lower than that allowed in the 1930s. The 8 leaf miners allowed per 100 grams-level continues today for spinach (USDA, 1983; FDA, 1989a).

In an effort to meet the FDA and USDA DAL regulations and their increasing stringency, farmers have used increasing amounts of pesticides on their crops and also instituted other pest control measures. The FDA (1972b) reported that the 1956 pest outbreak in the spinach crop 'stimulated research by the University of California (Davis) Department of Entomology to develop more effective field programs to control leaf miner damage in spinach. Control programs have apparently been effective since we [FDA] have had little or no regulatory action on this problem in recent years [through 1972].' Altered FDA DALs appear to have influenced insect control procedures and the amount of insecticide used in spinach production, and probably on other crops as well. W.H. Lange, Jr. (University of California [Davis], personal communication, 1976) reported that the reduction of the leaf miner problem since 1956 was made possible because of several interrelated programmes: a 3-fold increase in the use of insecticides on spinach, from 1–2 to 3–6 treatments per season; a new bait-spray programme for control of adult leaf miners; planting in spring and late fall, instead of in fall when leaf miners are most severe; growing fewer crops that act as alternate hosts for leaf miners in the spinach production area.

Is it realistic to aim for ever more stringent DALs, until no pests are allowed? The possibility of reducing the presence of apple maggots in apple sauce to zero has been discussed in New York State, but if this were accomplished, the amount of insecticide used in apple orchards would increase substantially and would undermine the current integrated pest management (IPM) programme now operating in New York State (Tette and Koplinka-Loehr, 1989). IPM systems must be dynamic rather than static in order to deal with the development of pesticide resistance in pests (Glass, 1992). This currently used IPM program is generally successful in controlling major apple pests, while keeping pesticide applications to a relatively low but effective level. Only with greatly augmented insecticide use could insect-free apple sauce be produced.

If a *zero* tolerance for insects and mites in fresh and processed foods were established, many foods such as raspberries and strawberries would be totally eliminated from the market because it is impossible to produce these products without any insects or mites present. Furthermore, as mentioned above, the absolute elimination of insects and mites from other fruits and vegetables would require enormous amounts of insecticides and miticides. This would result in a 'very real danger' of exposing the public and the environment to hazardous levels of pesticides (FDA, 1974; Zalom and Jones, 1994). Pesticides applied for the control of insects and mites often disrupt other biological control systems as well (Tette and Jacobsen, 1992). For example, IPM practitioners have discovered that some conventional chemicals used to control leafminers and leafrollers—both major insect pests of apples—are ineffective because natural enemies have been destroyed and both leafminers and leafrollers have become resistant to pesticides (Cowles et al., 1995). Even with high levels of pesticides, it is probably impossible to reach the goal of no insects or mites in fruits and vegetables and, as discussed here, this may be not only an unattainable goal but also an unwarranted one.

COSMETIC APPEARANCE

In addition to restricting the numbers of pests found in and on produce, the minor surface blemishes found on fruits and vegetables and caused by pests are a part of the cosmetic standards. The growing emphasis given to the cosmetic appearance of fruits and vegetables is alleged to reflect consumer preference (van den Bosch et al., 1975). Since 1945, food processors, wholesalers and retailers, following the lead of the FDA and USDA, have placed increasing importance on improving the cosmetic appearance of fruits and vegetables. As mentioned, the achievement of producing almost 'perfect' produce has been possible because of the increased availability and use of insecticides and miticides. As a result apples, oranges, tomatoes, cabbage and other fruits and vegetables found in US supermarkets today have little or no insect damage on their surface.

Clearly, the cosmetic appearance of produce is one of the primary factors used by consumers in assessing the overall quality of the produce they buy. Certainly visually perfect produce is appealing. Unfortunately consumers are not provided with more substantive measures of quality such as nutritional values or pesticide-residue levels. In considering produce to purchase, the consumer is left to make selections based solely on surface cosmetic appearance and, of course, the price of the produce (EPA, 1990).

In general, the public has not been aware of the connection between cosmetic appearance and increased pesticide use. However, some recent evidence suggests that when consumers are made aware of the trade-offs, they will purchase produce that is not cosmetically perfect because it has less or no pesticide residue (Lynch, 1991).

Evidence suggests that distributors and wholesalers desire to propagate the idea that consumers will not tolerate any cosmetic damage on their produce. With fresh produce, contracts between growers and buyers (i.e. distributors and wholesalers) permit buyers to make subjective evaluations of produce based on cosmetic appearance. This enables buyers to reject produce when supply is excessive. Growers agree to such contracts because of the market power of buyers. That is, many small growers face monopolistic buyers who are dominant in the market place (EPA, 1990). Given such contractual agreements, growers feel assured of sales, and to achieve this have a strong incentive to produce cosmetically perfect produce and thus resort to heavy pesticide use (EPA, 1990).

Marketing order arrangements currently present in the produce industry also play a role in increasing pesticide use by growers. Although the original intent of marketing orders was to improve price stability and grower profitability, marketing orders have had the unintentional result of raising the cosmetic standards of produce. The establishment of voluntary grading standards by federal marketing orders has resulted in grading of produce (e.g. USDA Extra Fancy) that, over time, has evolved into mandatory industry requirements. For example, although the federal grade standard for Extra Fancy Red Delicious apples requires only two-thirds red colour, industry buyers typically insist that Extra Fancy Delicious be 95–100% red (GPO, 1992). Distributors and wholesalers supplying retail supermarket chains will only purchase the highest grade of produce (i.e. cosmetically perfect produce) from growers, especially during times of abundant supply. In this way marketing orders raise the cosmetic standards of produce and exclude cosmetically less perfect, but nutritious, produce from entering the fresh market (Curtis et al., 1991; GPO, 1992).

Moreover, retail supermarket chains, claiming to satisfy consumer preference for perfect cosmetic appearance, demand fresh produce with a maximum specified level of cosmetic damage. To meet the demands of the supermarket chains, distributors and wholesalers require growers to meet an even higher cosmetic standard. Growers, in turn, must set even higher standards to have a margin of safety. The result is that growers must apply more pesticide to achieve these marketplace demands (EPA, 1990).

The presence of surface blemishes on fruits and vegetables generally does not affect their nutritional content, storage life or even their flavour (van den Bosch et al., 1975; Curtis et al., 1991). For example, citrus rust mites cause 'russetting' or 'bronzing' on Florida oranges. Unless the mite population is extremely high (Allen, 1979; McCoy et al., 1988), the juice quality in the endocarp, determined by the content of sugars and other nutrients, is virtually unaffected by the russetting (Tisserat and Galetta, 1992). About 80% of the citrus acreage in Florida (Krummel and Hough, 1980) is sprayed for rust mites, usually about three times during the season, at a cost to the grower of about $200 ha^{-1}. The rust mites cause little or no reduction in the yield of oranges unless they become highly abundant (Lye et al., 1990). Russetting of the fruits due to mite damage is well correlated with the moisture content of the fruits (Kamau et al., 1992).

Marketing conditions continued to favour larger fruit in 1989, especially first-grade fruit (Hare et al., 1992). Thus, although effective citrus mite suppression resulted in an increase in total fruit yield, most of that increase occurred in the less valuable fruit size classes (Hare et al., 1992).

One has to question why Florida oranges are excessively treated for rust mites when 95% of these oranges are ground-up for juice. Apparently the reason orange growers continue to treat their oranges for rust mites is because juice processors require a russet-free external appearance (Ziegler and Wolfe, 1975; NAS, 1980). Cultivars that have developed russetting in the past have been intentionally eliminated by growers and breeders (Tisserat and Galletta, (1992). Further, it appears that processors use the presence of rust mite injury to downgrade the price of oranges purchased from growers when the supply of oranges is abundant. In this way the processors use the presence of rust-mite injury to their economic advantage. This seems to be a common strategy among processors who process other fruits and vegetables (EPA, 1990).

In contrast to oranges, fresh grapefruit with russetting is classified as 'golden' and sells for a higher price in the market than unblemished grapefruit that is classified as 'bright' (Krummel and Hough, 1980). For example, they reported that in the Chicago and Boston markets, 'golden' grapefruit sold for more than $2 per box higher than 'bright' grapefruit. The reason for the price differential is that russetted fruit are reported to be sweeter than 'bright' fruits. Some evidence supports this idea—mite russetting is reported to allow some moisture to escape from the grapefruit (Krummel and Hough, 1980; McCoy et al., 1988). When this occurs the sugars and solids in the russetted grapefruit and oranges are concentrated. This suggests that education of the public is possible for new, sound cosmetic standards.

Another example of how stringent cosmetic standards have influenced pesticide use concerns the control of citrus thrips on California oranges that 'scar' the skin of the fruit. Scarred fruit receives a lower grade from wholesalers/distributors and therefore sells at a lower price in the marketplace (Flaherty et al., 1973; Tanigoshi et al., 1985). As with mite blemishes, thrip blemishes do not affect the nutritional or eating quality of oranges as measured by percentage moisture and ratio of soluble solids to acid (van den Bosch et al., 1975). Nonetheless, citrus thrips are considered one of the most serious pests of oranges because of the scarring problem, and as a result large quantities of insecticides are applied for thrips control. Currently, control of thrips and other pests in California orange groves is estimated to average about $600 ha^{-1} year^{-1} (Teague et al., 1988).

Similarly, on tomatoes grown for processing, about two-thirds of all insecticide applied is to control the tomato fruitworm (EPA, 1990), which is 'essentially a cosmetic pest' because it damages only the tomato skin (van den Bosch et al., 1975; Walgenbach et al., 1989). Most processors allow no more than 0.5–2% fruitworm damage to the surface of tomatoes by weight, while many accept only perfect fruit (van den Bosch et al., 1975; Metcalf, 1986; Feenstra, 1988; Zalom and Jones, 1994). However, 90% of processed tomatoes are peeled and then used for paste, sauce, catsup, juice and puree, products with no skins (van den Bosch et al., 1975). Also, a study of insecticide use on the tomato fruitworm has found that many treatments and insecticides applied in tomato fields failed to decrease net profit losses due to fruitworm injury significantly compared with un-treated tomato fields (Walgenbach and Estes, 1992).

To date, consumers, processors and regulators have not clearly understood that the nutritional quality of surface-scarred or blemished fruit and vegetables is not inferior to the perfect fruit or vegetable (van den Bosch et al., 1975), except under conditions of excessive pest damage when nutritional quality may be affected (Gorham, 1981; McCoy et al., 1988). They also seem to be unconcerned about the hazards of ingesting pesticide residues and/or that there is a direct correlation between perfect produce and pesticide residues. However, there is evidence that this perception is changing. In a recent FMI survey, 80% of consumers ranked pesticide and herbicide residues as a major health hazard associated with food (Swanson and Lewis, 1993). However, concern over the possible presence of pesticides in fresh fruits and vegetables, Americans are reluctant to buy produce that is not aesthetically appealing (Russell, 1991). Although a significant majority of US consumers identify themselves as environmentally conscious it is probable that fewer are substantially motivated by environmental concerns (including lower pesticide residues) at the point of purchase (Jolly and Norris, 1991). If consumers could overcome their aesthetic desire for perfect produce with a full awareness of the real dangers of pesticide residues, then perhaps more people would accept slightly blemished produce, which is less likely to contain insecticide and miticide residues, than perfect produce. There is some evidence that suggests that changes are in progress. Recently several state farm bills and the federal farm bill defined what food can be certified as 'organically' grown (Gates, 1990).

CONSUMING INSECTS—HEALTH EFFECTS

According to the FDA (1974), the DALs for insects and mites in produce were established to prevent any 'hazard to health'. This goal appears to have been met, because in recent reports no mention is made of health hazards related to the presence of insects and mites in foods (USDA, 1983; FDA, 1989a,c). The only exception to this would be for insects such as house flies and cockroaches, which could invade foods stored prior to processing (Gorham, 1989, 1992; Kopanic et al., 1994). Indeed all herbivorous insects and mites that are found in and on harvested fruits and vegetables are harmless to humans (Phelps et al., 1975; Taylor, 1975; Pimentel et al., 1977; Defoliart, 1989; Gorham, 1992). Further evidence of their safety is demonstrated in countries throughout the world where insects are a part of the normal diet and contribute important nutrients to peoples' daily nutrition (Boden-heimer, 1951; Gorham, 1976; Dufour, 1987; Posey, 1987; Brickey and Gorham, 1989).

In many places, pest insects of crops also are important foods for humans. Defoliart (1989) suggests that harvesting insect pests for food could be a part of pest management programmes and thereby reduce the need for insecticides. Although some insects, such as cockroaches that invade produce during processing, may present a health hazard (Gorham, 1989, 1992), it is the harmless herbivores (pests of crops) that are the target of increased insecticide use designed to produce perfect fruits and vegetables. If the choice is between perfect produce with increased insecticide residues or less than perfect produce and the presence of a few insects and mites, then indeed it would be safer to tolerate a few insects and mites (Pimentel et al., 1977; Gorham, 1992).

The EPA regulates pesticide use. The USDA and FDA regulate the levels of insects and mites in fresh and processed foods, respectively (USDA, 1983; FDA, 1989a). Because the FDA regulations identify the types of insect/mite contaminants it allows in foods, it should be able to develop specific regulations for house flies and cockroaches, while enabling harmless insect and mite residues to be regulated less stringently (Gorham, 1992). There appears to be a lack of consideration about the trade-offs of pesticide use and insect and mite levels in foods both within and between the federal agencies.

Defoliart (1975), Taylor (1975), Finke et al. (1987, 1989), Nakagaki et al. (1987), Gorham (1989) and Van-Wright (1991) have assembled data on the nutritional values of several insects, which compare favourably with those of shrimp, lobster and crawfish. The latter are also arthropods, but are often considered to be food delicacies. The digestible protein content of the insect ranges from 40 to 65% (Defoliart, 1975; Taylor, 1975; Kok et al., 1988), that of shrimp, lobsters and crawfish from 75 to 84% (USDA, 1986), and that of trimmed beef, lamb, pork, chicken and fish from 30 to 75% (USDA, 1986). Insect protein could be a substitute for almost all the vertebrate protein on which many of us now depend (Vane-Wright, 1991). Besides containing protein, some insects have been found to be 'leaner' than trimmed beef, while others offer abundant fat calories (Vane-Wright, 1991). Insects also contain calcium, iron, other minerals and some vitamins, notably riboflavin (Vane-Wright, 1991). Using calorific values, Vane-Wright calculated that about 8 kg of *Orthoptera* has the same calorific value as 87 'chilli dogs', 49 slices of pizza, or 43 Big Macs (Van-Wright, 1991). Some insects such as termites could also serve as a food supplement. Termites are rich in lysine, an essential amino acid in which cereals are deficient (Benhura and Chitsaku, 1992).

Given the conclusion that most insects found on produce are probably not any more of a health hazard than beef or chicken, consumers must decide whether they are willing to tolerate the presence of a few insects rather than have perfect produce which has required the use of high levels of pesticides.

PESTICIDE USAGE

An estimated 434 million kg of pesticides are used in the United States annually (Figure 19.1). These pesticides consist of 69% herbicides, 19% insecticides and 12% fungicides (Pimentel et al., 1991). Of this, agriculture uses about 320 million kg of pesticides, with about 3 kg applied per hectare on 100 million ha of farm land (Pimentel and Levitan, 1986). The remaining pesticides are used by the public, industries and government.

The application of pesticides for pest control is not evenly distributed among crops. For example, 94% of all row crop hectarage, such as corn, cotton and soybeans, is treated

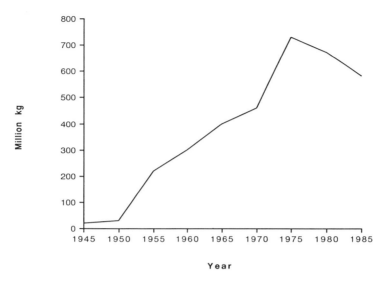

Figure 19.1   The amount of synthetic pesticides—insecticides, herbicides, and fungicides—produced in the United States. About 90% is sold in the United States. The decline in the total amount produced since 1975 is in large part due to the 10– to 100–fold increased toxicity and effectiveness of the newer pesticides (based on Pimentel et al. 1993)

with some type of pesticide (Osteen, 1993). In contrast, less than 10% of forage crop hectarage is treated (Osteen, 1993).

About 62 million kg of insecticides are applied to 5% of the US agricultural land (Pimentel and Levitan, 1986). Vegetable crops that have from 85 to 100% of their acreage treated include potatoes, tomatoes, sweetcorn, onions and sweet potatoes (Pimentel et al., 1991). The heaviest insecticide-treated fruit crops include apples, cherries, peaches, pears and grapefruit.

Fungicides are primarily applied to fruit and vegetable crops. Vegetable crops having 85–97% of their acreages treated with fungicide include lettuce, potatoes, tomatoes and onions (Pimentel et al., 1991). Fruit Crops having 85% or more of their acreages treated include apples, cherries, peaches, pears, grapes, oranges and grapefruit (Pimentel et al., 1991).

Various experts estimate that from 60 to 80% of the pesticide applied to oranges and 40–60% of the pesticide applied to tomatoes to be processed are used only to improve cosmetic standards (CALPIRG, 1991; GPO, 1992). Overall, we estimate that from 10 to 20% of insecticides and fungicides used on fruits and vegetables are applied to comply with the strict cosmetic standards now in force.

PESTICIDES AND CROP LOSSES

Synthetic pesticide use in the United States is about 33 times higher than it was in 1945 (Figure 19.1). The amounts of insecticides, herbicides and fungicides used have changed with time not only because of changes in agricultural practices, but also because cosmetic standards have been raised. Herbicides work so effectively that some relegate mechanical weed control to secondary importance (Forcella and Burnside, 1992, in Forcella et al.,

1993). However, corn production, that uses about 40% of all herbicides, still utilizes mechanical weed control on more than 90% of the corn acreage (Pimentel et al., 1993). Concurrently, the toxicity of new pesticides to pests and their biological effectiveness have increased at least 10-fold. For example, in 1945 DDT was applied at a rate of about $2 \, kg \, ha^{-1}$; at present, similar effective insect control is achieved with pyrethroids and aldicarb applied at $0.1 \, kg \, ha^{-1}$ and $0.05 \, kg \, ha^{-1}$, respectively.

In spite of the current use of pesticides plus some non-chemical controls, an estimated 37% of all crop production is lost annually to all pests (13% to insects and mites, 12% to plant pathogens and 12% to weeds) (Pimentel et al., 1991). The share of crop yields lost to insects has nearly doubled (from 7% to 13%) during the last 40 years despite a more than 10-fold increase in the amount and toxicity of synthetic insecticides used (Arrington, 1956; USBC, 1970, 1990).

This dramatic rise in crop losses despite increased insecticide use can be partially explained by some of the major changes that have taken place in agricultural practices. These include: the reduction in crop rotations; the increase in monocultures and reduced crop diversity (Pimentel, 1961; Pimentel et al., 1977; Tonhasca and Byrne, 1994); a reduction in tillage, with more crop residues left on the land; the planting of some crop varieties that are highly susceptible to insect and mite pests; the increased use of aircraft for pesticide application; the reduction in field sanitation (Liquido, 1993); pesticides causing some destruction of natural enemies, thereby creating the need for additional pesticide treatments (van den Bosch and Messenger, 1973; Timti, 1991; Kerns and Gaylor, 1993); the increase in the number of pests becoming resistant to pesticides (Georghiou, 1986; Kilman, 1991); the culture of crops in climatic regions where insects are serious pests; the lowering of FDA tolerances for insects and mites in foods, plus the enforcement of more stringent cosmetic standards by fruit and vegetable processors and retailers (Pimentel et al., 1977); the use of pesticides that have altered the physiology of crop plants, making them more susceptible to insect attack (Oka and Pimentel, 1976).

The best example of how changes in agricultural practices have led to greater crop losses, despite increased pesticide use, is illustrated with the culture of field corn. In 1945, most field corn was grown in rotation after soybeans, wheat, oats and other non-corn host crops (USDA, 1954). During the early 1940s, little or no insecticide was applied to corn, and losses to insects averaged only 3.3% (USDA, 1954). Corn rootworm larvae was not a widespread problem in the corn belt at this time, because all corn was grown in rotation with other crops (Steffey et al., 1992). Since then, a major portion of the corn has been cultured without rotation, insecticide use on corn has grown more than 1000-fold, and losses due to insects have increased to 12% or nearly 4 times the 1940 levels (Schwartz and Klassen, 1981). With no rotation, the corn rootworm population continued to increase on the stands of corn. Many farmers are forced to culture corn as a monoculture because of a 'base acreage' requirement in the price support program, and thus they are forced to use pesticides (Jennings, 1991). Crop rotation has also been ignored by some growers because of a few cases of limited land availability or because of marketing constraints (Noling and Becker, 1994).

HEALTH HAZARDS

Increased public health and environmental hazards have been associated with the increased use of insecticides and other pesticides on fruits and vegetables (NAS, 1987, 1989).

About 67 000 humans are poisoned by pesticides annually (Litovitz et al., 1990). Yearly about 27 fatalities (Blondell, 1987; Litovitz et al., 1990) and almost 10 000 cases of cancer are associated with pesticide use (Pimentel et al., 1991). This sharply contrasts to no known case of human poisoning or death from ingesting insects or mites in or on foods.

The US Environmental Protection Agency (EPA) sets food tolerances for pesticide residues under the Federal Insecticide, Fungicide, and Rodenticide Act to ensure that any residues remaining on produce will be below what are considered to be safe levels (Zalom and Jones, 1994). Annual studies are conducted by FDA to determine the kinds and amounts of pesticide residues in typical daily diets (FDA, 1990; Prokopy, 1992). They report that about 35% of the foods eaten contain detectable pesticide residues (FDA, 1990); about 1–3% of these foods contain pesticide residues above the legal tolerance level (Hundley et al., 1988; FDA, 1990).

However, a major concern about the 'acceptable tolerance levels' remains because significant gaps exist in the data concerning tumour production in animals for a majority of the pesticides that are currently registered and used in agriculture (GAO, 1986). Also, the Congressional Office of Technology Assessment found that the FDA primary laboratory method can only detect about half the pesticides registered for use on food (OTA, 1988, in Curtis et al., 1991). Thus, the absolute safety of the currently accepted levels of pesticide residues that occur in our foods has not been proven.

Under existing regulations foods are seized for exceeding FDA regulatory tolerances for pesticide residues as well as for exceeding the DALs for insects and mites. From 45 to 60% of this food is sold and consumed before it can be recalled (GAO, 1986; Mott and Snyder, 1987). Note that the OTA (1988) reports that of 743 pesticides and their breakdown products that can be found in foods, the best analytical chemical methods are capable of detecting less than one-third of them. The NAS (1987) reports that because nine oncogonic compounds are found in 90% of all fungicides sold, fungicides constitute 60% of the oncogonic risk among all the pesticides used on food (NAS, 1987, in Wilson et al., 1991). The five major methods used by the FDA only detect 290 pesticides and their breakdown products, or 40% of all those that can be found in foods.

ECONOMIC COSTS AND RETURNS

Each year US farmers use an estimated 320 million kg of pesticides for all crops at an approximate cost of $4.1 billion (Pimentel et al., 1991). This investment in pesticides saves farmers less than $16 billion in crops, or about 10% of their total crop yield (Pimentel et al., 1978). This saving, however, does not take into account the indirect or environmental and social costs associated with pesticide use, which may have a total annual cost of nearly $4 billion (Pimentel et al., 1991).

The direct benefits of pesticides are about $4 per dollar invested in pesticides, as indicated above. However, a much higher return could be realized through the implementation of non-chemical alternatives for pest control. For example, crop rotations, biological control and breeding for host-plant resistance would return on average about $30 per dollar invested in pest control (Pimentel, 1986). Alternative techniques such as banding herbicide applications, ridge-till, crop rotations, strip intercropping, narrow row production and corn rootworm baits can potentially reduce herbicide use by 50% and insecticide use by 80% or more (Curtis et al., 1991).

However, US federal and state policies hinder the adoption of alternative farming systems. Many farmers receive a large portion of their income from farm subsidies dispersed by the federal government. The rules by which these payments are distributed prevent reductions in pesticide use by penalizing farmers who employ crop rotations (Curtis et al., 1991). In addition, current pesticide regulations hinder the rapid registration of biological control materials that could substitute for chemical pesticides (Curtis et al., 1991). Several recent studies suggest that it is technologically feasible to reduce pesticide use in the United States by 35–50% without reducing yields (PSAC, 1965; OTA, 1979; NAS, 1989; Palladino, 1989; Pimentel et al., 1991). This type of pest control policy was begun in 1985 in Denmark, where an action plan was developed to reduce the use of pesticides by 50% before 1997 (B. Mogensen, National Environmental Research Institute, Denmark, personal communication, 1989). Subsequently, in 1987, Sweden approved a programme to reduce pesticide use by 50% within 5 years (NBA, 1988). Sweden achieved their goal of a 50% reduction in 1992, and is now working on another 50% reduction in pesticide use. The Netherlands has also developed a programme to reduce pesticide use by 50% within the next 10 years (A. Pronk, Wageningen University, The Netherlands, personal communication, 1990). Similarly, in 1987, the Province of Ontario, Canada, developed a programme to reduce pesticide use by 50% during the following 15-year period (G. Surgeoner, University of Guelph, Canada, personal communication, 1990). Indonesia successfully banned the use of 57 pesticides on rice in 1986 and rice yields were not affected (Steba and James, 1990, in Curtis et al., 1991). In fact, pesticide use in rice has been reduced by 65% with a 12% increase in rice yield (I.N. Oka, Bogar Food Research Institute, Bogar, Indonesia, personal communication, 1995).

The Australian Apple and Pear Grower's Association recently made the reduction of pesticide use its number one research priority, and has committed apple growers to the goal of reducing pesticide use by 50% by 1996, and by 75% by the year 2000 (Anonymous, 1991, in Penrose et al., 1994). Both the New Zealand Kiwifruit and Apple and Pear Marketing Board have announced their intention to significantly reduce pesticide residues on produce (Manktelow, personal communication, in Penrose et al., 1994).

An assessment of the impact of a programme to reduce pesticide use in the United States by 50% suggested that it would cause no reduction in crop yields, and for some crops it would increase yields (Pimentel et al., 1991). One example is that of citrus fruits. Although high citrus mite populations may reduce yields slightly, an increase has been found in the size of the remaining fruit (Hare et al., 1992). This observation is novel only because the effects of the citrus mites on fruit size were never previously considered (Hare et al., 1992). Similarly, the economically beneficial aspects of an increase in fruit size are novel only because previous investigators neglected to ascertain the impact of the citrus mites in any way which was at all relevant to the economic returns of commercial citrus growers (Hare et al., 1992). As a result, growers with well-maintained groves are now encouraged to tolerate citrus mite populations in excess of the old threshold of two adult females per leaf, thereby minimizing acaricide treatments and preserving populations of predaceous mites and other natural enemies which are valuable in controlling other more important citrus insect pests (Pehrson, et al., 1991, in Hare et al., 1992).

The estimated coincident increase in food costs to the consumer as a result of a 50% reduction in pesticide use in the United States would be only 0.6% (Pimentel et al., 1993). This marketplace cost increase does not take into account the positive environmental and social benefits that would accrue if pesticide use were reduced. If the environmental and social benefits are considered, the 0.6% increase in consumer costs would be more than offset by the environmental and social benefits associated with reduced pesticide use.

ETHICAL AND MORAL ISSUES

Recently, an FDA (1989b) survey found that 97% of the public preferred food without pesticides. Other reports also indicate that the public is becoming truly concerned about pesticide use and residues in its foods (Lecos, 1984; Steele and Chandler, 1990; Chou, 1991). In another survey, from 50 to 66% of the people polled indicated that they would be willing to pay higher prices for pesticide-free food (Ott, 1990; Anonymous, 1991). Few doubt the desire of the public for foods untreated by pesticides or treated with minimum amounts of pesticides, but the unanswered question is: Will they purchase foods that have a few blemishes? One survey showed that three out of every four people would buy fruits and vegetables with some blemishes on the outer surface if they were grown with fewer pesticides than perfect-looking produce (Chou, 1991). However, another study showed that consumers are more willing to accept pesticide residues in produce if the produce is free of surface blemishes and insects (Goldman and Clancy, 1991). This dilemma illustrates the different values held by the individual people who make up the population.

However, another related message from the public concerning pesticides is clearer than it has been in recent decades. The public has less confidence in government and less confidence that food is safe because of current levels of pesticides. Sachs et al. (1987) published the results of a recent survey showing that from 1965 to 1985 the public became increasingly concerned about the safety of the food they purchased. In 1965, about 98% of those surveyed were confident that pesticide regulations were sound and were being effectively implemented (Sachs et al., 1987). In 1985, however, less than 46% felt that their food was safe and that pesticide regulations were being effectively implemented.

Several major incidents associated with pesticides may have contributed to the public's changing views. For example, the 'watermelon poisonings' that took place in California in 1985, when farmers illegally treated watermelons with aldicarb and over 1000 people were poisoned, clearly shook public confidence in government regulations (Taylor, 1986). Equally important was the more recent 'Alar incident' (Hathaway, 1988, 1989). Although Alar is a plant growth regulator, it is regulated by EPA as a pesticide. Alar was used not only to keep fruit on the tree until harvest time, but also for cosmetic purposes to enhance the redness of apples. Many years prior to its final removal from use, questions had been raised concerning its safety by New York State Health officials and the EPA, which had surveyed several scientists (D. Pimentel, unpublished data, 1987). The results of the EPA survey were never published. However, alarm over the use of Alar continued to grow, particularly because apple juice and apple sauce are consumed in large quantities by infants and young children. Public alarm escalated, but the EPA did not act. It was not until 1990 that Uniroyal Company, the producer of Alar, decided to withdraw Alar from the market. With hindsight, government action to restrict Alar should have been taken when enough data had accumulated to suggest that Alar was suspected to be a carcinogen. Because of the delay, farmers lost millions of dollars once the danger of Alar was aired in the press, and the public boycotted apples and apple products whether they were treated with Alar or not.

Without question the inept handling of Alar further eroded public confidence in the government's ability to regulate pesticide use. Furthermore, the public now appears to be of the opinion that chemical industries determine pesticide policies in the United States. One example is EPA's proposal in October 1992 to continue the use of the cancer-causing herbicide amitrole and terminate its review of the chemical (EPA, 1992, in McCarthy, 1992). Also, the Delaney Clause, designed to prevent the addition of cancer-causing substances to processed foods, is threatened by powerful food and agrichemical businesses who have vowed to see its demise in Congress (Curtis, 1994). Surely the EPA, FDA, USDA and other agencies of the federal government and their state counterparts have the obligation to represent and understand public concerns about pesticide use. Although national pesticide policy derived from the Federal Insecticide, Fungicide, and Rodenticide Act (FIFRA) and regulations of the Environmental Protection Agency, inadequacies in current legislation and regulation are becoming increasingly apparent (Higley et al., 1992). Yet in fairness, it must be pointed out that individuals appear to have differing values concerning blemish-free produce versus pesticide-free produce. These opinions send conflicting messages to government regulators.

All concerned, including farmers and chemical companies, should be heard and their viewpoints should receive attention. However, the general impression given by government agencies is that their primary concern is fear of lawsuits from chemical companies because of their regulatory decisions rather than public safety and health concerns.

The following example illustrates the problem. Insects, such as apple maggots, are a concern of processors because they fear consumer lawsuits or negative publicity. This is the view of processors but not that of the FDA (1989a). To achieve zero tolerance of apple maggots in apple sauce requires the use of enormous amounts of insecticides. This would cause higher insecticide-residue contamination of fruit sold in the market and of the environment, and also high economic pest control costs. In addition, substantially increasing use on apples would undermine various IPM programs that have been established in apple-growing states such as New York. In regulating pesticides, government agencies should carefully consider the views and concerns of chemical companies, processors, farmers and consumers. Equally vital is the consideration of all the environmental and public health aspects that are adversely affected by heavy pesticide use.

One example of questionable US policy in public health is the incident involving the fungicide procymidone. In 1990, the United States seized imports of European wine containing residues of procymidone because there was no legal tolerance established for the fungicide. European exporters and US importers claimed that the ban would result in up to $300 million in lost market sales. In April, 1991, the EPA issued an internal tolerance for procymidone, which allowed for the distribution of the imported wine (GAO, 1991).

Local governments may soon begin drafting pesticide legislation also. In June 1991, the US Supreme Court ruled that FIFRA does not prohibit state or local governments from enacting pesticide legislation that is more restrictive than that of FIFRA itself (Higley et al., 1992). As a result, public concern over pesticide safety is increasingly being expressed in legislation. For example, California's 'Big Green' Proposition 128 (which was defeated) would have cancelled registration of all carcinogenic pesticides even if no viable alternatives were available (Higley et al., 1992). Even Congress, in an attempt to alleviate public concerns over pesticide residues in foods, passed the Pesticide Monitoring Improvements Act (Gorham, 1992).

Hopefully state and federal governments will work together to develop educational programs that keep the public better informed about the relationships that exist between blemish-free fruits and vegetables and heavy pesticide use. Another consideration is the high economic costs associated with heavy pesticide use. For example, both farmers and consumers understand the fact that the overall quality of russetted oranges is not lower than the quality of 'perfect' oranges. Further, of particular interest to farmers is the fact that russetted orange yields are not reduced and pesticide treatment costs could be reduced by \$200 ha^{-1}. With this situation both the farmer and consumer would benefit from less pesticide and lower production costs.

Government policies need to be carefully monitored to avoid the possibility of inadvertently encouraging pesticide use. For example, past price support policies have encouraged the use of pesticide in cotton, corn and numerous other crops because various high-pesticide-use technologies were inadvertently legislated for (NAS, 1989).

CONCLUSIONS

From the 1930s to 1976, the FDA and USDA regulations gradually reduced the defect action levels (DLAs) for insects and mites found in foods, even though there was no proven health hazard associated with the presence of small herbivorous arthropods in foods. Since then, both the FDA and USDA have maintained the established DLAs. This is encouraging. However, food processors, wholesalers and retailers seem to be placing even greater emphasis on blemish-free, perfect produce. Not only has this pressure caused substantial crop losses because large portions of food crops are now being classified as unsuitable for commercial sale, but it has also been contributing to heavy pesticide usage by farmers who feel compelled to spray to reduce the incidence of insects and mites in foods to meet these 'cosmetic appearance' standards (Lichtenberg and Zilberman, 1986).

The estimated 10–20% additional insecticide and/or miticide used on fruits and vegetables to meet the new cosmetic appearance standards (van den Bosch et al., 1975; Pimentel et al., 1977) has caused substantial increases in pesticide use. This has resulted in a greater portion of the foods being contaminated with pesticide residues. Concurrently, the number of human pesticide poisonings and illnesses has increased and there has also been further contamination of the environment. In addition, more fossil energy was used for spraying and producing pesticides, and food costs for the consumer increased.

Further investigation of the widespread impact that cosmetic appearance standards are having on pesticide usage is needed in order to understand the possible trade-offs between insect and mite damage of food and quantities of pesticide residues in food. Although the presently enforced DALs are stringent, overall the FDA and USDA appear to be realistic in their regulations. However, for some crops the DALs could safely be raised. In addition, food processors and wholesalers need to reassess their market policies concerning 'perfect' produce, especially as related to how the produce is to be used, e.g. whole tomatoes skinned and made into sauce. Retailers must also be realistic in assessing the trade-offs between perfect produce, pesticide contamination and the relative prices of foods. Such changes certainly involve consumers, who need to understand the pesticide consequences of buying only perfect produce.

In this chapter, many of the factors related to maintaining stringent cosmetic standards for produce were analyzed. All sectors of society are involved in the relationships that developed in response to government regulations. To find safe and equitable solutions will require knowledge and compromise. We hope that this analysis will be helpful to all concerned as they endeavor to make wise, safe and fair choices for the betterment of agriculture, the environment, public health and society as a whole.

ACKNOWLEDGEMENTS

We thank the following people for reading an earlier draft of this manuscript, for their many helpful suggestions, and, in some cases, for providing additional information: H. Lehman and G.A. Surgeoner, University of Guelph; O. Pettersson, The Swedish University of Agricultural Sciences; J.R. Gorman, US Food and Drug Administration; E.R. Figueroa, Cornell University.

REFERENCES

Allen, J.C. (1979) The effect of citrus rust mite damage on citrus tree growth. *Journal of Economic Entomology* **72**, 195–201.

Anonymous, (1974) Amounts and kinds of filth in foods and the parallel methods for assessing filth and insanitation. *Food Purity Perspectives* **3**, 19–20.

Anonymous (1991) Appearance of produce versus pesticide use. *Chemecology* **20**(4), 11.

Arrington, L.G. (1956) *World Survey of Pest Control Products*. Government Printing Office, Washington, DC.

Benhura, M.A.N. and Chitsaku, I.C. (1992) Seasonal variation in the food consumption patterns of the people of Mutambara District of Zimbabwe. *Central Africa Journal of Medicine* **38**, 8–13.

Blondell, J. (1987) *Accidental Pesticide Related Deaths in the United States, 1980 to 1985*. Report of US Environmental Protection Agency, Office of Pesticides and Toxic Substances, Health Effects Division, Washington, DC.

Bodenheimer, F.S. (1951) *Insects as Human Food*. Junk, The Hague.

Brickey, P.M. and Gorham, J.R. (1989) Preliminary comments on federal regulations pertaining to insects as food. *Food Insects Newsletter* **2**, 1, 7.

CALPIRG (1991) *Who Chooses Your Food?* CALPIRG, Los Angeles, CA.

Chou, M. (1991) Two years after the Alar crisis. *Cereal Foods World* **36**, 526–527.

Cowles, E.A., Yunowitz, H., Charles, J.F., Gill, S.S. (1995) Comparison of toxin overlay and solid phase binding assays to identify diverse CryIA(c) toxin-binding proteins in *Heliothis virescens* midgut. *Applied and Environmental Microbiology* **61**(7), 2738–2744.

Curtis, J. (1994) Getting cancer-causing pesticides out of food! *Journal of Pesticide Reform* **14**, 13.

Curtis, J., Mott, L. and Kuhnie, T. (1991) *Harvest of Hope. The Potential for Alternative Agriculture to Reduce Pesticide Use*. National Resources Defense Council, New York, pp. 3–97.

Defoliart, G.R. (1975) Insects as a source of protein. *Bulletin of the Entomological Society of America* **21**, 161–163.

Defoliart, G.R. (1989) The human use of insects as food and as animal feed. *Bulletin of the Entomological Society of America* **35**, 22–35.

Dufour, D.L. (1987) Insects as food: A case study from the northwest Amazon. *American Anthropologist* **89**, 383–397.

EPA (1990) *An Overview of Food Cosmetic Standards and Agricultural Pesticide Use*. Office of Policy, Planning and Evaluation, US Environmental Protection Agency, Washington, DC.

FDA (1944–66) *Notices of Judgement under the Federal Food, Drug, and Cosmetic Act*. Issued monthly by the US Department of Health Education and Welfare, Washington, DC.

FDA (1972a) *Current Levels for Natural or Unavoidable Defects in Food for Human Use that Present No Health Hazard*. Office of the Assistant Commissioner for Public Affairs, Food and Drug Administration, Rockville, MD, 31 October.

FDA (1972b) *Revision of Defect Action Levels for Spinach*. In-House Memorandum, US Department of Health Education and Welfare, Washington, DC, 14 December.

FDA (1974) *Current Levels for Natural or Unavoidable Defects in Food for Human Use that Present No Health Hazard*. (5th revision). Department of Health Education and Welfare, US Public Health Service, Rockville, MD.

FDA (1989a) *Defect Action Levels.* Food and Drug Administration, US Department of Health, Education, and Welfare, Washington, DC.

FDA (1989b) Food and Drug Administration Pesticide Program Residues in Foods—1988. *Journal of the Association of Official Analytical Chemists* 72, 133A–152A.

FDA (1989c) *The Food Defect Action Levels.* Food and Drug Administration, Washington, DC.

FDA (1990) Food and Drug Administration Pesticide Program Residues in Foods—1989. *Journal of the Association of Official Analytical Chemists* 73, 127A–146A.

Federal Register (1973) Human foods: Current good manufacturing practice (sanitation) in manufacture, processing, packaging, or holding. *Federal Register* 38(3 (Part 1)), 854–855.

Feenstra, G.A. (1988) *Who Chooses Your Food? A Study of the Effects of Cosmetic Standards on the Quality of Produce.* CALPIRG, Los Angeles, CA.

Finke, M.D., Defoliart, G.R. and Benevenga, N.J. (1987) Use of a four-parameter logistic model to evaluate the protein quality of mixtures of Mormon cricket meal and corn gluten meal in rats. *Journal of Nutrition* 117, 1740–1750.

Finke, M.D., Defoliart, G.R. and Bennevenga, N.J. (1989) Use of a four-parameter logistic model to evaluate the quality of the protein from three insect species when fed to rats. *Journal of Nutrition* 119, 864–871.

Flaherty, D.L., Pehrson, J.E. and Kennett, C.E. (1973) Citrus pest management studies in Tolare County. *California Agriculture* 27, 3–7.

Forcella, F., Eradat-Oskoui, K. and Wagner, S.W. (1993) Application of weed seedbank ecology to low-input crop management. *Ecological Applications* 3, 74–83.

GAO (1986) *Pesticides: EPA's Formidable Task to Assess and Regulate their Risks.* US General Accounting Office, Washington, DC.

GAO (1991) *International Food Safety. Comparison of US and Codex Pesticide Standards*: Report to Congressional Requestora, US General Accounting Office, Washington, DC.

Gates, J.P. (1990) *Organic Certification.* Special Reference Briefs of the National Agriculture Library, US, Beltsville, MD. The Library, January 1990, Issue 90-04, 9 pp.

Georghiou, G.P. (1986) *The Magnitude of the Resistance Problem. Pesticide Resistance, Strategies and Tactics for Management.* National Academy of Sciences, Washington, DC.

Glass, E.H. (1992) Constraints to the implementation and adoption of IPM. In Zalom, F.G. and Fry, W.E. (Eds) *Food, Crop Pests, and the Environment: The Need and Potential for Biologically Intensive Integrated Pest Management.* American Phytopathological Society, St. Paul, MN, pp. 167–174.

Goldman, B.J. and Clancy, K.L. (1991) A survey of organic produce purchases and related attitudes of food cooperative shoppers. *American Journal of Alternative Agriculture* 6, 89–92.

Gorham, J.R. (1976) Insects as food. *Bulletin of the Society for Vector Ecology* 3, 11–16.

Gorham, J.R. (1981) Principles of food analysis for filth-decomposition and foreign matter. In Gorham, J.R. (Ed.) *Principles of Food Analyses.* Food and Drug Administration Technical Bulletin 1, US Department of Health and Human Services, Washington, DC, Publication No. (FDA) 80-2128, pp. 63–124.

Gorham, J.R. (1989) Foodborne filth and human disease. *Journal of Food Protection* 52, 674–677.

Gorham, J.R. (1992) Filth and extraneous matter in food. In Hui, Y.H. (Ed.) *Encyclopedia of Food Science and Technology.* Wiley, New York, pp. 847–868.

GPO (1992) *Cosmetic Standards and Pesticide Use on Fruits and Vegetables.* US Government Printing Office, Washington, DC.

Hare, J.D., Pehrson, J.E., Clements, T., Menge, J.A. and Coggins, Jr., C.W. (1992) Effect of citrus red mite *Acari tetranychidae* and cultural practices on total yield fruit size and crop value of navel orange years 3 and 4. *Journal of Economic Entomology* 85, 486–495.

Hathaway, J.S. (1988) *Agriculture and Public Health: Why We Aren't Protected from Pesticides in Food.* New England Fruit Meeting, Proceedings of the Annual Meeting of the Massachusetts Fruit Growers Association. North Amherst MA.

Hathaway, J.S. (1989) An environmentalist's perspective on the magnitude of the health risk from pesticide residues in foods. *Food, Drug Cosmetic Law Journal* 44, 659–670.

Higley, L.G., Zeiss, M.R., Wintersteen, W.K. and Pedigo, L.P. (1992) National pesticide policy: A call for action. *American Entomology* 38, 139–146.

Hundley, H.K., Cairns, T., Luke, M.A. and Masumoto, H.T. (1988) Pesticide residue findings by the

Luke method in domestic and imported foods and animal feeds for fiscal years 1982–1986. *Journal of the Association of Official Analytical Chemists* **71**, 875–877.

Jennings, A.L. (1991) Some economic and social aspects of pesticide use. *ACS Symposium Series, American Chemical Society* **446**, 31–37.

Jolly, D.A. and Norris, K. (1991) Marketing prospects for organic and pesticide-free produce. *American Journal of Alternative Agriculture* **6**, 174–179.

Kamau, A.W., Mueke, J.M. and Khaemba, B.M. (1992) Resistance of tomato varieties to the tomato russet mite *Aculops-lycopersici massee acarina eriophyidae. Insect Science Applications* **13**, 351–356.

Kerns, D.L. and Gaylor, M.J. (1993) Induction of cotton aphid outbreaks by insecticides in cotton. *Crop Protection* **12**, 387–393.

Kilman, S. (1991) Winter's melons, leafy vegetables face threat from pesticide-resistant insect. *Wall St. Journal*, East edn., A2.

Kok, R., Lomaliza, K. and Shivhare, U.S. (1988) The design and performance of an insect farm chemical reactor for human food production. *Canadian Agricultural Engineering* **30**, 307–318.

Kopanic, R.J. Jr., Sheldon, B.W. and Wright, C.G. (1994) Cockroaches as vectors of *Salmonella*: Laboratory and field trials. *Journal of Food Protection* **57**, 125–132.

Krummel, J. and Hough, J. (1980) Pesticides and controversies: Benefits versus costs. In Pimentel, D. and Perkins, J.H. (Eds) *Pest Control: Cultural and Environmental Aspects.* Westview Press, Boulder, CO.

Lecos, C. (1984) Pesticides and food public worry No. 1. *FDA Consumer* **18**, 12–15.

Lichtenberg, E. and Zilberman, D. (1986) Problems with pesticide regulation: Health and environment versus food and fiber. In Phipps, T.T. and Crosson, P.R. (Eds) *Agriculture and the Environment: An Overview.* Resources for the Future, Washington, DC, pp. 123–145.

Liquido, N.J. (1993) Reduction of oriental fruit fly. (*Diptera: Tephritidae*) populations in papaya orchards by field sanitation. *Journal of Agricultural Entomology* **10**, 163–170.

Litovitz, T.L., Schmitz, B.F. and Bailey, K.M. (1990) 1989 Annual report of the American Association of Poison Control Centers National Data Collection System. *American Journal of Emergency Medicine* **8**, 394–442.

Lye, B.H., McCoy, C.W. and Fojtik, J. (1990) Effect of copper on the residual efficacy of acaricides and population dynamics of citrus mite, Acari, Eriophyidae. *Florida Entomology* **73**, 230–237.

Lynch, L. (1991) Consumers choose lower pesticide use over picture-perfect produce. *Food Review* January–March, 9–11.

McCarthy, S. (1992) More regulation and more cancer risks from amitrole and 2,4-D: Wouldn't supporting alternatives make more sense? *Journal of Pesticide Reform* **12** 27.

McCoy, C.W., Albrigo, L.G. and Allen, J.C. (1988) The biology of citrus rust mites and its effects on fruit quality. *The Citrus Industry* September, 44–54.

Metcalf, R. (1986) The ecology of insecticides and the chemical control of insects. In Kogan, M. (Ed.) *Ecological Theory and Integrated Control of Insects.* Wiley, New York, pp. 251–297.

Mott, L. and Snyder, K. (1987) *Pesticide Alert: A Guide to Pesticides in Fruit and Vegetables.* Sierra Club Books, San Francisco, CA.

Nakagaki, B.J., Sunde, M.L. and Defoliart, G.R. (1987) Protein quality of the house cricket, *Acheta domesticus* when fed to broiler chicks. *Poultry Science* **66**, 1367–1371.

NAS (1980) *Regulating Pesticides.* National Academy of Sciences, Washington, DC.

NAS (1987) *Regulating Pesticides in Food.* National Academy of Sciences, Washington, DC.

NAS (1989) *Alternative Agriculture.* National Academy of Sciences, Washington, DC.

NBA (1988) *Action Programme to Reduce the Risks to Health and the Environment in the Use of Pesticides in Agriculture.* The National Board of Agriculture, Stockholm.

Noling, J.W. and Becker, J.O. (1994) The challenge of research and extension to define and implement alternatives to methyl bromide. *Journal of Nematology* **26**, 573–586.

Oka, I.N. and Pimentel, D. (1976) Herbicide (2,4-D) increases insect and pathogen pests on corn. *Science* **193**, 239–240.

Osteen, C. (1993) Pesticide use trends and issues in the United States. In Pimentel, D. and Lehman, H. (Eds.) *The Pesticide Question: Environment, Economics and Ethics.* Routledge, Chapman & Hall, New York, pp. 307–336.

OTA (1979) *Pest Management Strategies.* Office of Technology Assessment, Congress of the United

States, Washington, DC, 2 vols.

OTA (1988) *Pesticide Residues in Food: Technologies for Detection*. Office of Technology Assessment, US Congress, Washington, DC.

Ott, S.L. (1990) Supermarket shopper's pesticide concerns and willingness to purchase certified pesticide residue-free fresh produce. *Agribusiness* **6**(6), 593–602.

Palladino, P.S.A. (1989) Entomology and ecology: The ecology of entomology. The 'insecticide crisis' and the entomological research in the United States in the 1960s and 1970s: Political, institutional, and conceptual dimensions. Ph.D. Thesis, University of Minnesota, St. Paul, MN.

Penrose, L.J., Thwaite, W.G., and Bower, C.C. (1994) Rating index as a basis for decision making on pesticide use reduction and for acredation of fruit produced under integrated pest management. *Crop Protection* **13**(2), 146–152.

Phelps, R.J., Struthers, J.K. and Moyo, S.J.L. (1975) Investigations into the nutritive value of *Macrotermes falciger* (Isoptera: Termitidae). *Zoology of Africa* **10**, 123–132.

Pimentel, D. (1961) Species diversity and insect population outbreaks. *Annals of the Entomology Society of America* **54**, 76–86.

Pimentel, D. (1986) Agroecology and economics. In Kogen (Ed.) *Ecological Theory and Integrated Pest Management Practice*. Wiley, New York.

Pimentel, D. and Levitan, L. (1986) Pesticides: Amounts applied and amounts reaching pests. *BioScience* **36**, 86–91.

Pimentel, D., Terhune, E., Dritschilo, W., Gallahan, D., Kinner, N., Nafus, D., Peterson, R., Zareh, N., Misiti, J. and Haber-Schaim, O. (1977) Pesticides, insects in foods, and cosmetic standards. *BioScience* **27**, 178–185.

Pimentel, D. Krummel, J., Gallahan, D., Hough, J., Merrill, A., Schreiner, I., Vittum. P., Koziol, F., Back, E., Yen, D. and Fiance, S. (1978) Benefits and costs of pesticide use in US food production. *BioScience* **28**, 778–784.

Pimentel, D., McLaughlin, L., Zepp, A., Lakitan, B., Kraus, T., Kleinman, P., Fancini, F., Roach, W.J., Graap, E., Keeton, W.S. and Selig, G. (1991) Environmental and economic impacts of reducing US agricultural pesticide use. In Pimentel, D. (Ed.) *Handbook of Pest Management in Agriculture*. CRC Press, Boca Raton, FL, pp. 679–718.

Pimentel, D., McLaughlin, L., Zepp, A., Lakitan, B., Kraus, T., Kleinman, P., Vancini, F., Roach, W.J., Graap, E., Keeton, W.S. and Selig, G. (1993) Environmental and economic impacts of reducing US agricultural pesticide use. In Pimentel, D. and Lehman, H. (Eds) *The Pesticide Question: Environment, Economics, and Ethics*. Routledge, Chapman & Hall, New York, pp. 223–280.

Posey, D.A. (1987) Ethoentomological survey of Brazilian Indians. *Entomological Genetics* **12**, 191–202.

Prokopy, R.J. (1992) Future trends in orchard pest control. *New England Fruit Meeting, Proceedings of the Annual Meeting of the Massachusetts Fruit Growers Association* **98**, 102–109.

PSAC (1965) *Restoring the Quality of our Environment*. President's Science Advisory Committee, The White House, Washington, DC.

Russell, S. (1991) Organic foods: Consumer viewpoint. *Food Australia* **43**(1), 14–15.

Sachs, C., Blair, D. and Richter, C. (1987) Consumer pesticide concerns: A 1965 and 1984 comparison. *Journal of Consumer Affairs* **21**, 96–107.

Schwartz, P.H. and Klassen, W. (1981) Estimate of losses caused by insects and mites to agricultural crops. In Pimentel, D. (Ed.) *Handbook of Pest Management in Agriculture*. CRC Press, Boca Raton, FL, pp. 15–77.

Steele, J.H. and Chandler, P.S. (1994) Pesticides and food safety: perception vs. reality. *Agri-Practice* **15**(3), 14–17.

Steffey, K.L., Gray, M.E. and Kuhlman, D.E. (1992) Extent of corn rootworm *Coleoptera Chrysomelidae* larval damage in corn after soybeans: Search for the expression of the prolonged diapause trait in Illinois. *Journal of Economic Entomology* **85**, 268–275.

Steinman, D. (1990) *Diet for a Poisoned Planet*. Harmony Books, New York.

Swanson, R.B. and Lewis, C.E. (1993) Alaskan direct-market consumers: Perception of organic produce. *Home Economics Research Journal* **22**, 138–155.

Tanigoshi, L.K., Fargerlund, J. and Nishio-Wong, J.Y. (1985) Biological control of citrus thrips, *Scirtothrips citri* (Thysanoptera: Thripidae), in southern California citrus groves. *Environmental*

Entomology **14**, 733–741.

Taylor, R.B. (1986) State sues three farmers over pesticide use on watermelons. *Los Angeles Times* I (CC)-3-4.

Taylor, R.L. (1975) *Butterflies in my Stomach.* Woodbridge Press, Santa Barbara, CA.

Teague, P.W., Smith, G.S., Swietlik, D. and French, J.V. (1988) Survey of citrus producers in the Rio Grande Valley: Results and analysis. *Journal of the Rio Grande Valley Horticultural Society* **41**, 97–109.

Tette, J.P. and Jacobsen, B.J. (1992) Biologically intensive pest management in the tree fruit system. In Zalom, F.G. and Fry, W.E. (Eds) *Food, Crop Pests, and the Environment: The Need and Potential for Biologically Intensive Integrated Pest Management.* American Phytopathological Society, St. Paul, MN, pp. 83–105.

Tette, J.P. and Koplinka-Loehr, C. (1989) *New York State Integrated Pest Management Program: 1988 Annual Report.* IPM House, New York State Agricultural Experiment Station, Geneva, NY.

Timti, I.N. (1991) Control of *Coelaenomenodera minuta uhlmann* with *Crematogaster* spp. *Tropical Pest Management* **37**, 403–408.

Tisserat, B. and Galletta, P.D. (1992) Adventitious juice vesicles produced from the exocarp in the *Citrinae aurantioidae. Hortscience* **27**, 843–846.

Tonhasca, A., Jr. and Byrne, D.N. (1994) The effects of crop diversification of herbivorous insects: A meta-analysis approach. *Ecological Entomology* **19**, 239–244.

USBC (1970) *Statistical Abstracts of the United States.* Bureau of the Census, US Department of Commerce, Washington, DC.

USBC (1990) *Statistical Abstracts of the United States.* US Bureau of the Census, US Department of Commerce, Washington, DC.

USDA (1954) *Losses in Agriculture.* US Department of Agriculture, Agricultural Research Service, Washington, DC.

USDA (1969) *Consumer Marketing Service.* US Department of Agriculture, Fruit and Vegetable Division, Processed Products Standardization and Inspection Branch, Washington, DC.

USDA (1983) *Inspection Procedures for Foreign Material.* US Department of Agriculture, Agricultural Marketing Service, Fruit and Vegetable Division, Processed Products Branch, Washington, DC.

USDA (1986) *Nutritive Value of Foods.* US Department of Agriculture, Human Nutrition Information Service, Home and Garden Bulletin No. 72, Washington, DC.

van den Bosch, R. and Messenger, P.S. (1973) *Biological Control.* Intext Educational, New York.

van den Bosch, R., Brown, M., McGowan, C., Miller, A., Moran, M., Petzer, D. and Swartz, J. (1975) *Investigation of the Effects of Food Standards on Pesticide Use.* US Environmental Protection Agency, Draft Report, Washington, DC.

Vane-Wright, R.I. (1991) Why not eat insects? *Bulletin of Entomological Research* **81**, 1–4.

Walgenbach, J.F. and Estes, E.A. (1992) Economics of insecticide use on staked tomatoes in western North Carolina. *Journal of Economic Entomology* **85**, 888–894.

Walgenbach, J.F., Shoemaker, P.B. and Sorensen, K.A. (1989) Timing pesticide applications for control of *Heliothis zea* (Boddie) (Lepidoptera Noctuidae) *Alternaaria solani* (Ell. and G. Martin) Sor., and *Phytophthora infestans* (Mont.) de Bary on tomatoes in western North Carolina USA. *Journal of Agricultural Entomology* **6**, 159–168.

Wilson, C.L., Wisniewski, M.E., Biles, C.L., McLaughlin, R., Chalutz, E. and Droby, S. (1991) *Crop Protection* **10**, 172–177.

Zalom, F.G. and Jones, A. (1994) Insect fragments in processed tomatoes. *Journal of Economic Entomology* **87**, 181–186.

Ziegler, L.W. and Wolfe, H.S. (1975) *Citrus Growing in Florida.* The University Presses of Florida, Gainesville, FL.

Index